Edited by Jürgen Dinges and Clemens Lamberth

Bioactive Heterocyclic Compound Classes

Related Titles

Lamberth, C., Dinges, J. (eds.)

Bioactive Heterocyclic Compound Classes

Agrochemicals

2012
Hardcover
ISBN: 978-3-527-33396-7

Anderson, R., Groundwater, P., Todd, A., Worsley, A.

Antibacterial Agents

Chemistry, Mode of Action, Mechanisms of Resistance and Clinical Applications

Hardcover
ISBN: 978-0-470-97244-1

Majumdar, K. C., Chattopadhyay, S. K. (eds.)

Heterocycles in Natural Product Synthesis

2011
Hardcover
ISBN: 978-3-527-32706-5

Alvarez-Builla, J., Vaquero, J. J., Barluenga, J. (eds.)

Modern Heterocyclic Chemistry

2011
Hardcover
ISBN: 978-3-527-33201-4

Brahmachari, G.

Handbook of Pharmaceutical Natural Products

2010
Hardcover
ISBN: 978-3-527-32148-3

Dunn, P., Wells, A., Williams, M. T. (eds.)

Green Chemistry in the Pharmaceutical Industry

2010
Hardcover
ISBN: 978-3-527-32418-7

Royer, J. (ed.)

Asymmetric Synthesis of Nitrogen Heterocycles

2009
Hardcover
ISBN: 978-3-527-32036-3

Fattorusso, E., Taglialatela-Scafati, O. (eds.)

Modern Alkaloids

Structure, Isolation, Synthesis and Biology

2008
Hardcover
ISBN: 978-3-527-31521-5

Edited by Jürgen Dinges and Clemens Lamberth

Bioactive Heterocyclic Compound Classes

Pharmaceuticals

WILEY-VCH

WILEY-VCH Verlag GmbH & Co. KGaA

The Editors

Dr. Jürgen Dinges
Abbott Laboratories
Global Pharmaceutical R&D
200, Abbott Park Road
Abbott Park, IL 60064-6217
USA

Dr. Clemens Lamberth
Syngenta Crop Protection AG
Research Chemistry
Schaffhauserstr. 101
4332 Stein
Schweiz

Library of Congress Card No.: applied for

British Library Cataloguing-in-Publication Data
A catalogue record for this book is available from the British Library.

Bibliographic information published by the Deutsche Nationalbibliothek
The Deutsche Nationalbibliothek lists this publication in the Deutsche Nationalbibliografie; detailed bibliographic data are available on the Internet at <http://dnb.d-nb.de>.

Bioactive Heterocyclic Compound Classes (Pharmaceuticals and Agrochemicals, 2 Volume Set) ISBN: 978-3-527-32993-9

Print ISBN: 978-3-527-33395-0
ePDF ISBN: 978-3-527-66448-1
ePub ISBN: 978-3-527-66447-4
mobi ISBN: 978-3-527-66446-7
oBook ISBN: 978-3-527-66445-0

Cover Design Adam-Design, Weinheim
Typesetting Laserwords Private Limited, Chennai, India
Printing and Binding Markono Print Media Pte Ltd, Singapore

Printed in Singapore
Printed on acid-free paper

Contents

Preface

Approximately 70% of all the 2400 pharmaceuticals listed in the online version of "Pharmaceutical Substances" (A. Kleemann *et al.*, Thieme) bear at least one heterocyclic ring; the latest edition of the "Pesticide Manual" (C. D. S. Tomlin, BCPC) contains a similar percentage of heterocyclic agrochemicals among its about 900 entries. This vast number of known pharmaceuticals and agrochemicals makes the field of commercialized active ingredients an unmanageable jungle. Only specialists are able to understand the connectivities of these active ingredients, many of which are heterocycles.

Therefore, we decided to put this book together, which tries to show the relationship of those heterocyclic active ingredients, which belong together, forming a biologically active heterocylic chemistry class. According to our definition, such a heterocyclic family is built from at least three compounds that fulfill the following conditions: (i) same heterocyclic scaffold, (ii) same mode of action, and (iii) similar substitution pattern.

Although the strength of this concept is that for the first time the members of the most important heterocyclic active ingredient families, their historical background, chemical syntheses, biochemical modes of action, and biological activities are discussed in detail, there are also some limitations. For instance, there are some heterocyclic families of drugs or crop protection agents, such as the analgesic COX-2 inhibitors celecoxib, valdecoxib, and rofecoxib or the dicarboxamide fungicides vinclozolin, iprodione, and procymidone, which are closely related by structure and possess the same mode of action, but bear different heterocyclic scaffolds and therefore could not be considered.

We would like to thank the authors of the 40 chapters of this book, all of them experts in their field, for spending their scarce time summarizing their area of interest. They all agreed to write the chapters according to the same outline: (i) introduction, (ii) history, (iii) synthesis, (iv) mode of action, and (v) structure–activity relationship. Only the agrochemical chapters possess an additional section "biological activity," mainly describing the target spectrum of the active ingredients. This book would definitely not exist without your engagement!

Furthermore, we also would like to thank Anne Brennführer and Stefanie Volk of Wiley-VCH, who from the beginning guided us very efficiently through all different phases of this exciting project.

The introductory chapter about "The significance of heterocycles for pharmaceuticals and agrochemicals" tries to explain the different roles of heterocyclic scaffolds in active ingredients, e.g. as framework of biologically active substances, as prodrugs, as tool for fine-tuning the physicochemical properties, as isosteric replacements of functional groups, alicyclic rings or other heterocyclic rings. As this is demonstrated at the hand of many prominent and characteristical examples of pharmaceuticals as well as of agrochemicals, also pointing out the many similarities, but also some differences between the two big classes of active ingredients, we decided to put this chapter in front of both volumes.

Although currently living in two different continents, both of us enjoyed exactly the same excellent education, a Ph.D. in organic chemistry from the Technical University at Darmstadt, Germany, and a subsequent postdoctoral fellowship at the chemistry department of the University of California at Berkeley. We are very grateful to our teachers, mentors, and research advisors at both universities, who built the foundation for our successful work in the research departments of the agrochemical and pharmaceutical industry.

Finally, we are deeply indebted to our wives Annette and Petra, who continuously supported us, as always, and tolerated that we spent many hours of our spare time, which should have belonged to our families, working on this book. You really made this possible!

Clemens Lamberth
Switzerland
Jürgen Dinges
USA

List of Contributors

Irini Akritopoulou-Zanze
Abbott Laboratories
Global Pharmaceutical R&D
R4CP, AP10-1
100 Abbott Park Road
Abbott Park
IL 60064-6099
USA

Milan Bruncko
Abbott Laboratories
Global Pharmaceutical R&D
Department R4N6, AP10
100 Abbott Park Road
IL 60064-6101
USA

Sanjay R. Chemburkar
Abbott Laboratories
GPO
Department 045B NCR13-2
1401 Sheridan Road
North Chicago
IL 60064
USA

Kevin P. Cusack
Abbott Laboratories
Global Pharmaceutical R&D
Immunology
381 Plantation Street
Worcester
MA 01605
USA

Jürgen Dinges
Abbott Laboratories
Global Pharmaceutical R&D
Department R4CP
200 Abbott Park Road
IL 60064-6217
USA

Alastair Donald
Harlow
United Kingdom

Maurizio Franzini
Exelixis Pharmaceuticals
210 East Grand Avenue
South San Francisco
CA 94080
USA

Dawn George
Abbott Laboratories
Global Pharmaceutical R&D
381 Plantation Street
Worcester
MA 01605
USA

Michael Z. Hoemann
Abbott Laboratories
381 Plantation Street
Worcester
MA 01605
USA

Ingo Janser
Eastern Michigan University
Department of Chemistry
Ypsilanti
MI 48197
USA

Romy Janser
Ypsilanti
MI 48197
USA

Vikram G. Kalthod
Abbott Laboratories
GPO
Department 045B NCR13-2
1401 Sheridan Road
North Chicago
IL 60064
USA

Clemens Lamberth
Syngenta Crop Protection AG
Research Chemistry
Schaffhauserstrasse 101
4332 Stein
Switzerland

James C. Lanter
Johnson and Johnson R&D
Drug Discovery
Welsh and McKean Roads
Spring House
PA 19477
USA

Edward C. Lawson
Johnson and Johnson R&D
Drug Discovery
Welsh and McKean Roads
Spring House
PA 19477
USA

Craig W. Lindsley
Vanderbilt University Medical Center
Department of Chemistry and Pharmacology
2213 Garland Avenue
Nashville
TN 37232-6600
USA

Robert Mah
Global Discovery Chemistry - Oncology
Novartis Institutes for BioMedical Research
Novartis Pharma AG
Basel
Switzerladn

Derek W. Nelson
Abbott
Global Pharmaceutical Research and Development
Abbott Park
IL 60064
USA

Stephanie M. Ng
Exelixis
210 East Grand Avenue
South San Francisco
CA 94080
USA

Rajarathnam E. Reddy
Abbott Laboratories
GPO
Department 045B NCR13-2
1401 Sheridan Road
North Chicago
IL 60064
USA

Steven Richards
Exelixis
210 East Grand Avenue
South San Francisco
CA 94080
USA

Michael L. Schulte
Vanderbilt University Medical Center
Department of Chemistry
2213 Garland Avenue
Nashville
TN 37232-6600
USA

Brian C. Shook
Johnson and Johnson R&D
Drug Discovery
Welsh and McKean Roads
Spring House
PA 19477
USA

Steve Swann
Fragment Based Drug Discovery
Translational Sciences and Technology
Eli Lilly
San Diego, CA
USA

Stacy Van Epps
Abbott Laboratories
Global Pharmaceutical R&D
381 Plantation Street
Worcester
MA 01605
USA

Ying Wang
Abbott
Department R4CP
BLDG. AP10
100 Abbott Park Road
North Chicago
IL 60064
USA

Cristiana A. Zaharia
644 Topaz Street
Redwood City
CA 94061
USA

and

Exelixis Inc.
210 East Grand Avenue
South San Francisco
CA 94083
USA

Hongyu Zhao
Abbott Laboratories
R4CP, 100 Abbott Park Road
Abbott Park
IL 60064
USA

Introduction

Bioactive Heterocyclic Compound Classes: Pharmaceuticals,
First Edition. Edited by Jürgen Dinges and Clemens Lamberth.

1
The Significance of Heterocycles for Pharmaceuticals and Agrochemicals*

Clemens Lamberth and Jürgen Dinges

1.1
Introduction

Heterocycles, their preparation, transformation, and properties, are undoubtedly a cornerstone of organic chemistry. Several books not only on heterocyclic chemistry [1–6] but also on some special aspects, such as heterocyclic name reactions [7], heterocyclic palladium-catalyzed reactions [8], heterocyclic carbene complexes [9], and fluorinated heterocycles [10], have been published recently.

Approximately more than 70% of all pharmaceuticals and agrochemicals bear at least one heterocyclic ring. In addition, some of the biggest commercial products to date, such as the blockbuster blood cholesterol reducer atorvastatin (Lipitor®, **1**) [11] for the treatment of dyslipidemia and the prevention of cardiovascular diseases and the broad-spectrum fungicide azoxystrobin (Amistar®, **2**) [12], currently applied against diseases of more than 100 different crops in more than 100 different countries, belong to this huge heterocyclic group of active ingredients (Figure 1.1).

There are two major reasons for the tremendous value of heterocycles for the lead optimization of pharmaceuticals and agrochemicals. The heterocyclic scaffold of a drug often has a positive impact on its synthetic accessibility and its physicochemical properties, driving these values of lipophilicity and solubility toward the optimal balanced range regarding uptake and bioavailability. Furthermore, heterocycles seem to be perfect bioisosteres of other iso- or heterocyclic rings as well as of several different functional groups, in most cases, delivering through their similarity in structural shape and electronic distribution equal or even better biological efficacy [13].

* Identically published in both volumes of "Bioactive Heterocyclic Compound Classes", as different roles of heterocycles in pharmaceuticals and agrochemicals are explained in this introductory chapter.

Bioactive Heterocyclic Compound Classes: Pharmaceuticals,
First Edition. Edited by Jürgen Dinges and Clemens Lamberth.

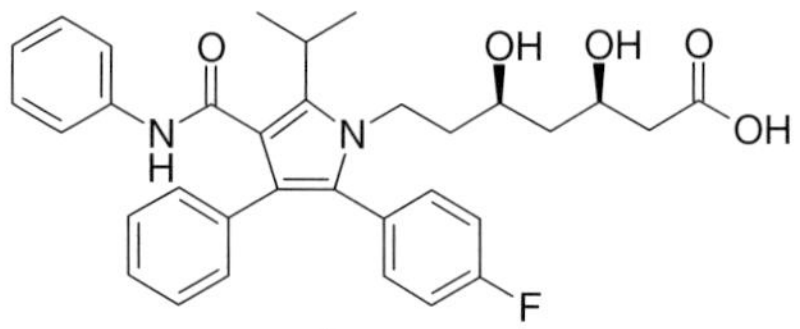

1

Atorvastatin
antilipemic

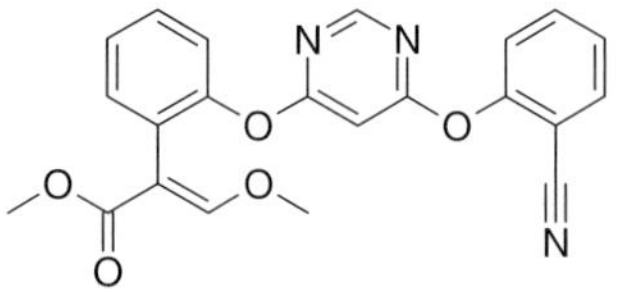

2

Azoxystrobin
fungicidal

Figure 1.1 Atorvastatin (**1**) and azoxystrobin (**2**), two of the currently most successful pharmaceuticals and agrochemicals.

1.2
Heterocycles as Framework of Biologically Active Compounds

Several heterocycles possess excellent biological activity almost without bearing any substituents, which means that their heterocyclic core is definitely part of the pharmacophore. Examples of such scarcely substituted and highly active heterocycles are the two bipyridyl derivatives such as amrinone (**3**) [14], which is used in the treatment of congestive heart failure, and paraquat (**4**) [15], which is applied as a total herbicide (Figure 1.2).

Another important role of the heterocyclic core of several pharmaceuticals and agrochemicals is that of an easily accessible scaffold, which carries the substituents that are responsible for the biological activity in the right orientation. There are several highly active per-substituted heterocycles, as demonstrated by the pyrazole derivatives propyphenazone (**5**) [16] and fipronil (**6**) [17], which are widely applied as efficient analgesic and insecticide, respectively, and synthetically available in only few steps (Figure 1.3).

Even simple aliphatic heterocycles display astonishing biological activities. The *gem*-diethyl-substituted barbituric acid derivative barbital (**7**) has been widely applied as a sleeping aid [18]. The pentamethylated piperidine pempidine (**8**) is used as a ganglionic blocker [19]. The trithiane thiocyclam (**9**), in comparison to the marine natural product nereistoxin enlarged by one additional ring sulfur atom, has been

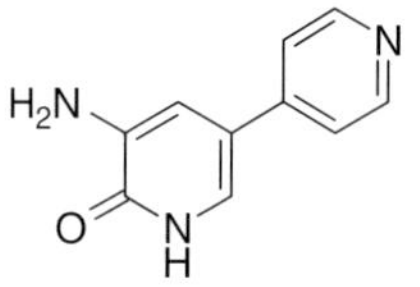

3

Amrinone
cardiotonic

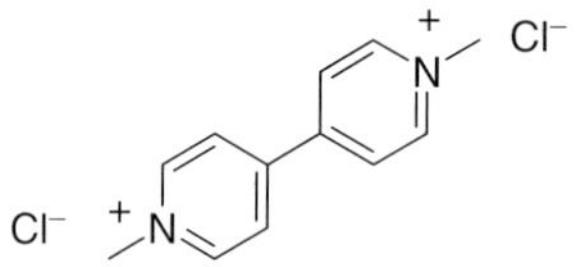

4

Paraquat
herbicidal

Figure 1.2 The highly active bipyridyl derivatives amrinone (**3**) and paraquat (**4**), each carrying only two small substituents.

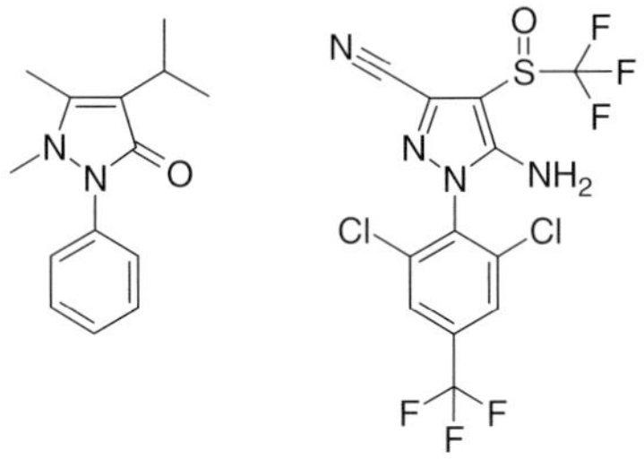

5 Propyphenazone analgesic — 6 Fipronil insecticidal

Figure 1.3 The persubstituted pyrazole derivatives propyphenazone (**5**) and fipronil (**6**).

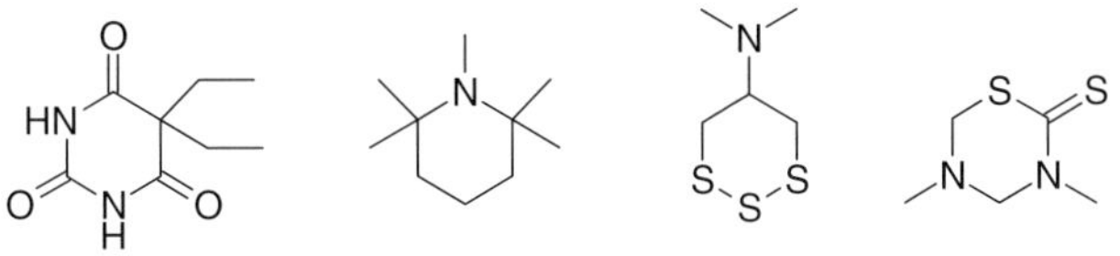

7 Barbital sedative — 8 Pempidine antihypertensive — 9 Thiocyclam insecticidal — 10 Dazomet nematicidal

Figure 1.4 The saturated bioactive heterocycles barbital (**7**), pempidine (**8**), thiocyclam (**9**), and dazomet (**10**) [18–21].

developed as a broad-spectrum insecticide [20]. The cyclic dithiocarbamate dazomet (**10**) is a soil fumigant, which readily decomposes, yielding methyl isothiocyanate as principal toxicant against nematodes (Figure 1.4) [21].

Not only monocyclic heterocycles but also annelated bicyclic ring systems are applied as pharmaceuticals and crop protection agents, regardless of whether the biheterocyclic core consists of aliphatic, aliphatic and aromatic, or purely aromatic rings. The tetrahydroimidazothiazole levamisole (**11**) has been used as anthelmintic and immunomodulator [22]. The dopamine agonist talipexole (**12**) combines a five- and seven-membered ring and has been proposed as an antiparkinsonian agent [23]. The triazolopyrimidine sulfonanilide flumetsulam (**13**) is used for the control of broadleaf weeds in corn and soybean (Figure 1.5) [24].

Finally, there are also several examples of active ingredients, which bear two or more heterocycles in completely different positions of the molecule. For instance, the nonsteroidal anti-inflammatory drug meloxicam (**14**) consists of an amide with a benzothiazine-dione acid moiety and a thiazole amine component [25]. In addition, the agrochemical fungicide ethaboxam (**15**) contains an amide functionality, combining a thiazole carboxylic acid with a thiophene-containing amine (Figure 1.6) [26].

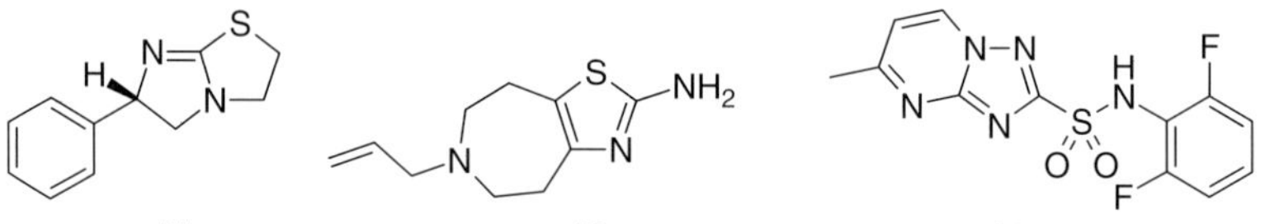

11 Levamisole anthelmintic

12 Talipexole antiparkinsonian

13 Flumetsulam herbicidal

Figure 1.5 The highly active annelated bicyclic heterocycles levamisole (**11**), talipexole (**12**), and flumetsulam (**13**) [22–24].

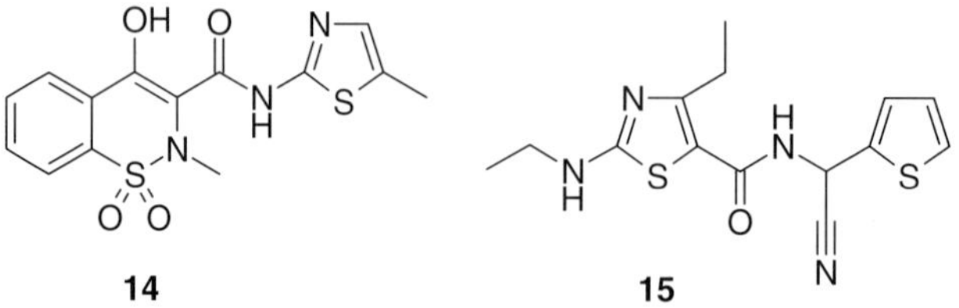

14 Meloxicam anti-inflammatory

15 Ethaboxam fungicidal

Figure 1.6 Meloxicam (**14**) and ethaboxam (**15**), two active ingredients carrying heterocycles in different parts of the molecule [25, 26].

1.3 Fine-Tuning the Physicochemical Properties with Heterocycles

The fact that in most cases aromatic heterocycles are more polar than their isocyclic analogs is often used for the lead optimization of pharmaceuticals and agrochemicals. For example, the replacement of the 4-trifluoromethylphenyl moiety of the herbicidal lead structure **16** by a 5-CF_3-pyrid-2-yl group resulting in the postemergence herbicide fluazifop-butyl (**17**) did not lead to any considerable enhancement of the herbicidal activity but significantly improved the ability of the target grass weeds to translocate into the plant tissue because of an optimum partition coefficient [27]. Furthermore, the replacement of the furane scaffold of the antiulcer histamine H_2-receptor antagonist ranitidine (**18**) by a thiazole resulted in nizatidine (**19**), which possesses not only a considerably lower log *P* value than ranitidine but also a much higher human oral bioavailability (Figure 1.7) [28].

1.4 Heterocycles as Prodrugs

The efficacy of several heterocyclic active ingredients is based on the fact that the heterocycle is acting as a prodrug, itself being not efficacious against the target enzyme or organism but delivering the intrinsically active compound by

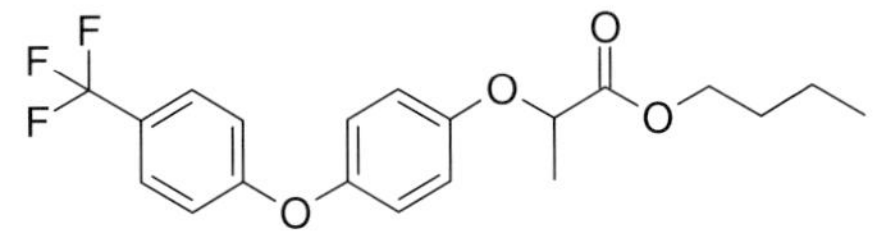

16

Herbicidal

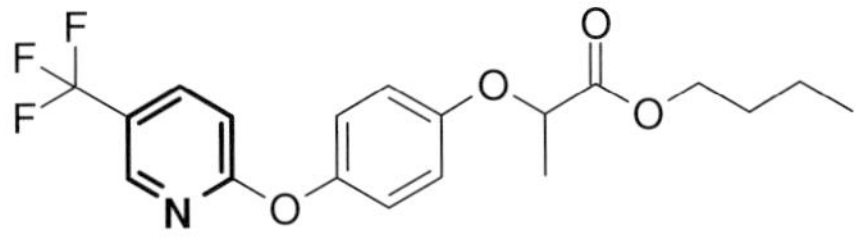

17

Fluazifop-butyl
herbicidal

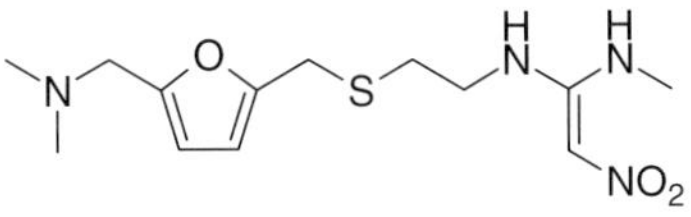

18

Ranitidine
antiulcer

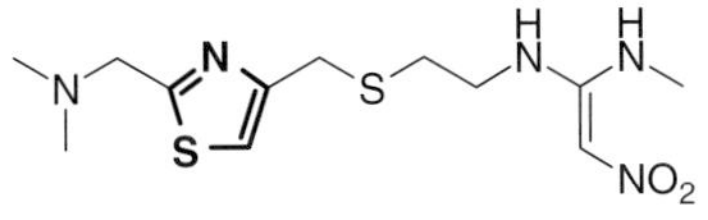

19

Nizatidine
antiulcer

Figure 1.7 Fluazifop-butyl (**17**) and nizatidine (**19**) possess optimum physicochemical properties to transport their high intrinsic activity to the target [27, 28].

UV light, heat, moisture, or a metabolic transformation. Leflunomide (**20**), for example, is a prodrug against transplant rejection, which ring-opens quantitatively in the cellular system to the hydroxypropenamide (**21**), which is responsible for the immunosuppressive efficacy [29]. In addition, the isoxazole ring of the herbicide isoxaflutole (**22**) is metabolically converted in plants and soil to the 2-cyano-1,3-diketone (**23**), which is a potent inhibitor of *p*-hydroxyphenylpyruvate dioxygenase (HPPD), one of the most important molecular targets for herbicides [30]. The fungicidal activity of the benzothiadiazine derivative **24** originates from its ability to be converted by sulfur extrusion in aqueous solutions and in plants into the benzimidazole fungicide carbendazim (**25**) [31]. The *in vivo* isomerization of fluthiacet-methyl (**26**) by *glutathione-S-transferase* leads to the urazole derivative **27**, which is entirely responsible for the strong herbicidal activity (Figure 1.8) [32].

1.5
Heterocycles as Peptidomimetics

Several different heterocyclic rings have a proven record as perfect isosteric replacement of the amide function in peptides [33]. The highly active HIV-1 protease inhibitors saquinavir (**29**) [34] and (**30**) [35] are close analogs of telinavir (**28**) [36], in which part of its urea function have been replaced by either a decahydroisoquinoline or a tetrazole (Figure 1.9).

Also, other five-membered heterocycles have been applied as amide isosteres in HIV-1 protease inhibitors for the treatment of AIDS. Examples are the imidazole derivative **32** [37] and the pyrrolinone (**34**) [38], in which the heterocyclic ring replaces the amide function of the corresponding di- or tripeptides **31** and **33** (Figure 1.10). All four HIV-1 protease inhibitors, the peptidic drugs, as well as

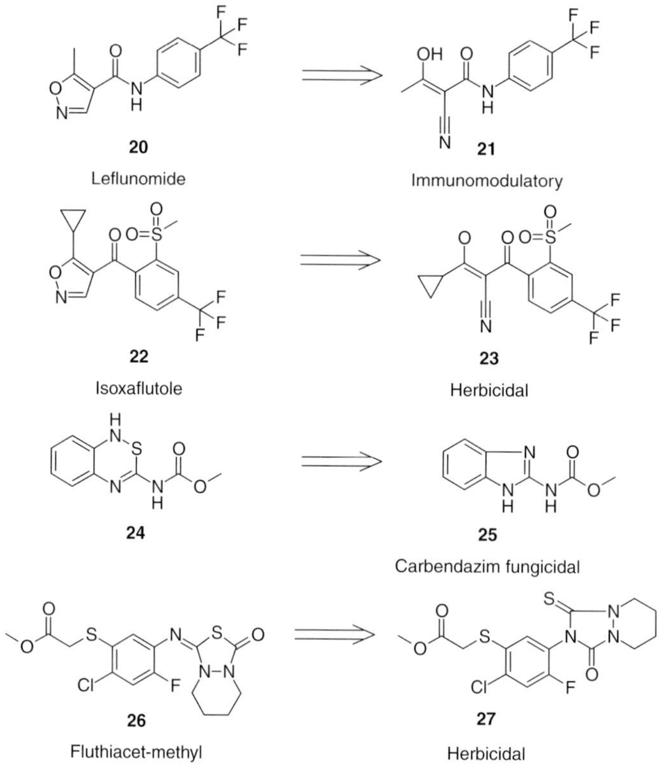

Figure 1.8 The heterocyclic prodrugs leflunomide (**20**), isoxaflutole (**22**), and fluthiacet-methyl (**26**) and (**24**).

their heterocyclic isosteres are active in the nanomolar range. The pyrrolidinone peptidomimetic **36** is 100 times more potent than the open-chain thrombin inhibitor NAPAP (**35**) [39]. The pyridine-based peptidomimetic **38** is a potent analog of PLG (**37**) (Pro-Leu-Gly-NH_2), an endogenous tripeptide found in the central nervous system, which is known to exert its pharmacological effects through the modulation of dopamine D2 receptors [40].

Further heterocycles, which have been successfully applied as amide isosteres, are pyrroles [41], thiazolidines [42], isoxazolines [43], imidazolines [44], oxazoles [45], triazoles [46], oxadiazoles [47], and benzimidazoles [48].

1.6
Heterocycles as Isosteric Replacement of Functional Groups

Heterocycles are also capable of mimicking other functional groups, besides the above-mentioned amide group. The most prominent examples are 5-substituted 1*H*-tetrazole as carboxylic acid replacements [49]. One of the success stories of

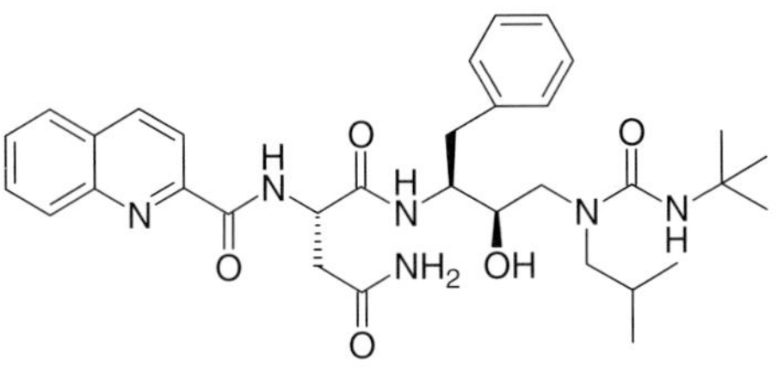

28

Telinavir
anti-HIV

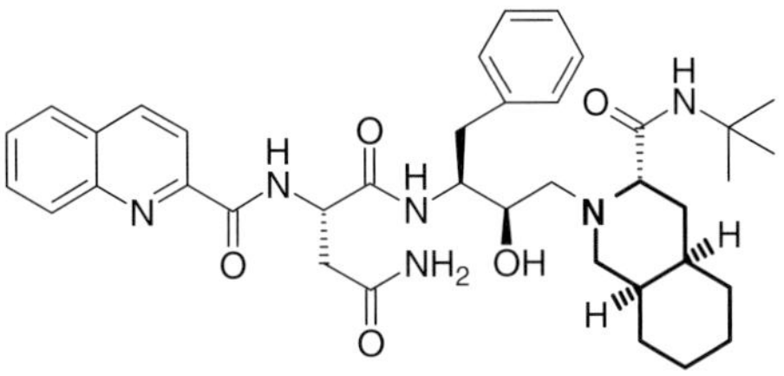

29

Saquinavir
anti-HIV

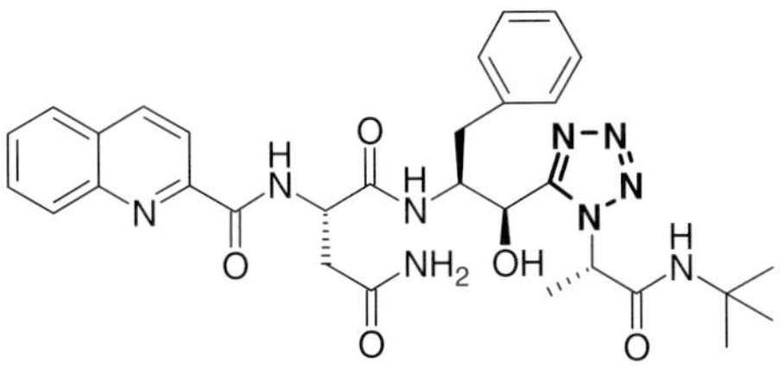

30

Anti-HIV

Figure 1.9 Telinavir (**28**) and its peptidomimetics saquinavir (**29**) and (**30**).

the tetrazole-carboxylate isosterism is the angiotension II receptor antagonist losartan (**40**). This drug for the treatment of hypertension and its carboxylic acid lead structure **39** possess similar acidity (pK_a of **39**: 4.5, losartan: 5.0) but differ significantly in lipophilicity (log P of **39**: 1.2, losartan: 4.5). The higher lipophilicity of losartan results in considerably improved oral bioavailability [49]. Also, the two gamma-aminobutyric acid (GABA) agonists isoguvacine (**41**) and gaboxadol (THIP, **42**) possess similar pharmacological properties due to comparable acidity (p$K_a \approx 4$) (Figure 1.11) [50].

Moreover, triazoles [51], hydroxythiadiazoles [13a], hydroxychromones [52], oxadiazolones [53], and thiazolidinediones [54] have been reported as heterocyclic carboxylic acid bioisosteres.

If tetrazole is an excellent carboxylic acid replacement, then alkylated tetrazoles should be able to mimick esters. This is demonstrated by azimsulfuron (**44**), which shows longer persistence in rice paddy fields than its ethyl ester analog pyrazosulfuron-ethyl (**43**) [55]. Also, oxazoles [56] and oxadiazoles [57] have been successfully applied as bioisosteres of esters (Figure 1.12).

In search for more potent and selective dopamine D2 agonists for the treatment of psychiatric and neurological diseases such as schizophrenia and Parkinson's disease, the indole moiety in **46** turned out to be an excellent bioisosteric replacement of the metabolically labile phenol function of the lead structure **45** [58].

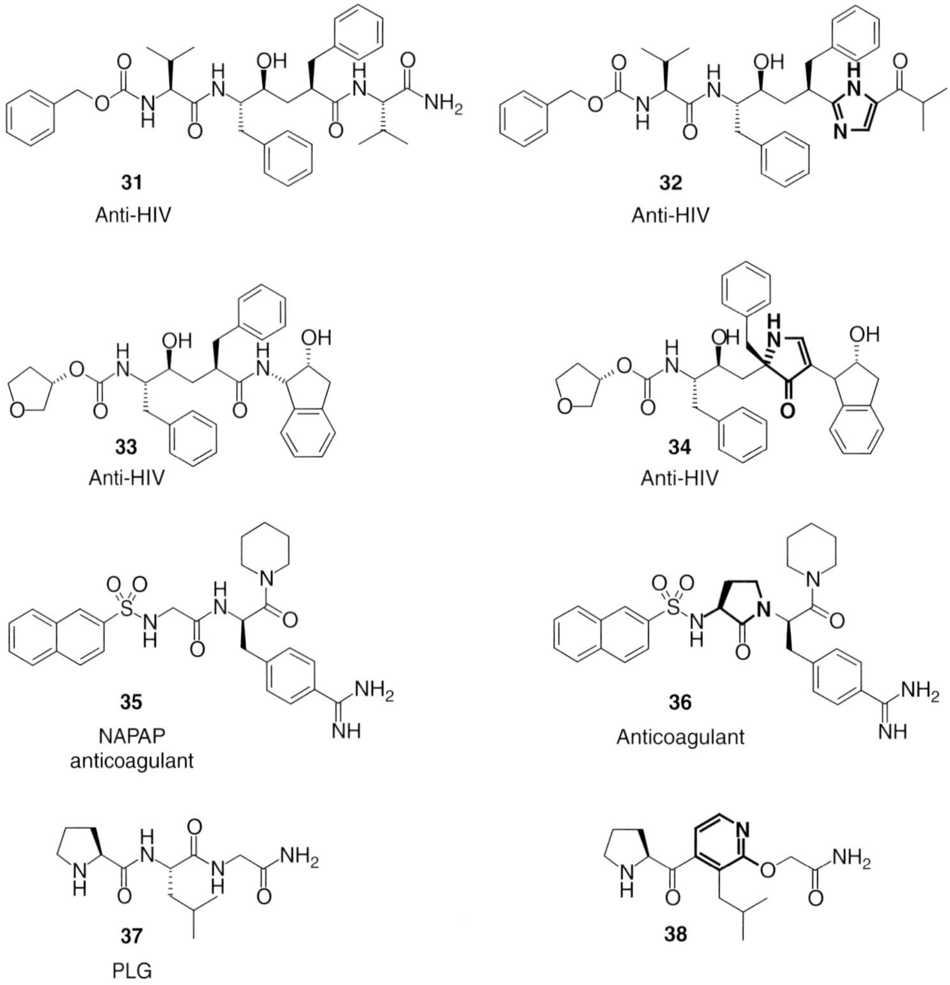

Figure 1.10 The heterocyclic peptidomimetics **32**, **34**, **36**, and **38**.

A widely used trick in lead optimization makes use of the fact that a carbon atom bearing a cyano function is often isosteric with an azomethine, often the ring nitrogen of an aromatic heterocycle. The potassium channel openers BMS182264 (**47**) and pinacidil (**48**), only differing by the replacement of a cyanophenyl ring by pyridine are both highly potent aortic smooth muscle relaxants [59].

The replacement of the highly basic benzamidine group in the thrombin inhibitor NAPAP (**35**) by a moderately basic 1-aminoisoquinoline moiety provides **49**, which displays potent enzyme inhibition and significant improvements in membrane transport and oral bioavailability [60].

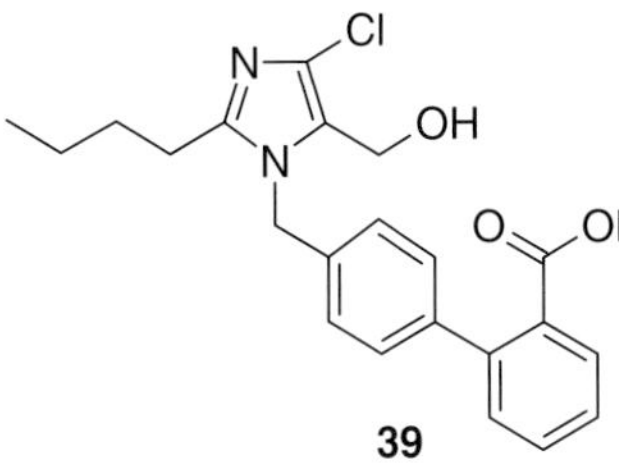

39

Antihypertensive

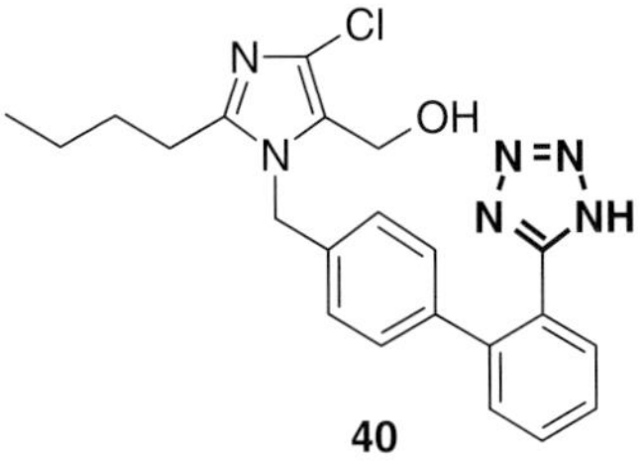

40

losartan
antihypertensive

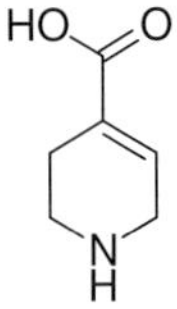

41

Isoguvacine

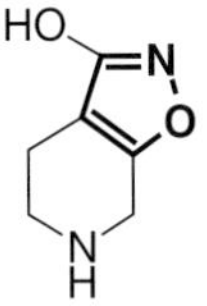

42

Gaboxadol (THIP)
sedative

Figure 1.11 The tetrazole derivative losartan (**40**) and the hydroxyisoxazole derivative gaboxadol (**42**) as carboxylic acid bioisosteres.

1.7
Heterocycles as Isosteric Replacement of Alicyclic Rings

A phenyl ring in biologically active compounds can often be replaced by a thiophene without any loss of activity because the sulfur atom is equivalent to an ethylenic group with respect to size, mass, and capacity to provide an aromatic lone pair [61]. For instance, a phenyl ring of the biologically active compound piroxicam (**50**) can be exchanged by thiophene, leading to tenoxicam (**51**) with similar anti-inflammatory activity (Figure 1.13) [62]. The thiophene derivative sufentanil (**53**) is at least five times more potent than its phenyl-analog fentanyl (**52**) [63]. The replacement of the *o,o′*-dialkylated phenyl ring of the chloroacetamide herbicide metolachlor (**54**) by a 2,4-dimethylthiophene results in dimethenamid (**55**) with comparable biological activity [64]. Also, in the area of *acetolactate-synthase*-inhibiting sulfonylurea herbicides, the ester-substituted phenyl ring could be successfully replaced by thiophene, leading from metsulfuron-methyl (**56**) to thifensulfuron-methyl (**57**) [65].

In addition, other heterocycles are able to mimic the phenyl ring of biologically active compounds. The substitution of one of the benzene rings of promazine's phenothiazine scaffold by pyridine led to prothipendyl (**59**) with improved neuroleptic activity and reduced undesired sedative and extrapyramidal effects (Figure 1.14)

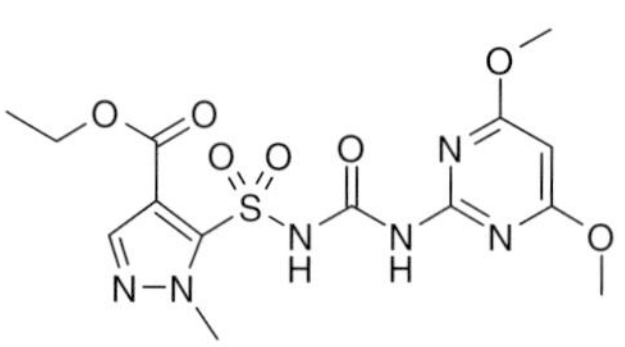

43

Pyrazosulfuron-ethyl
herbicidal

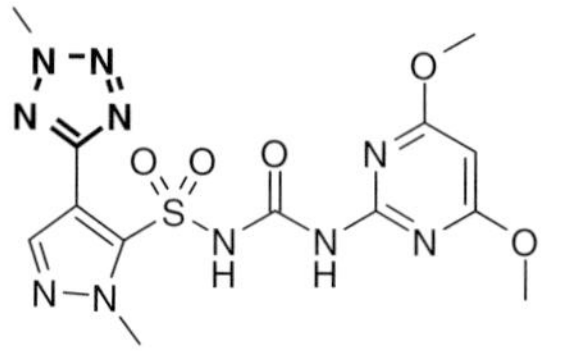

44

Azimsulfuron
herbicidal

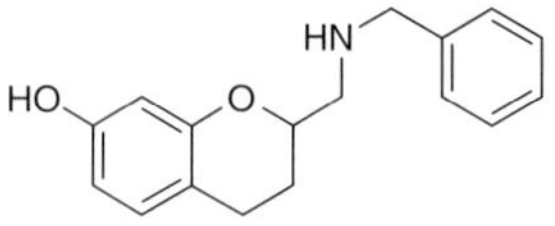

45

Dopaminergic

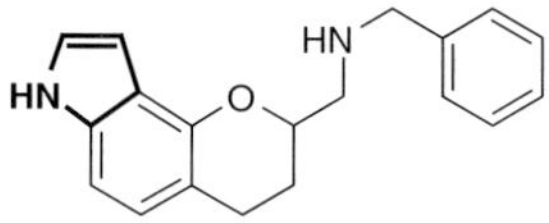

46

Dopaminergic

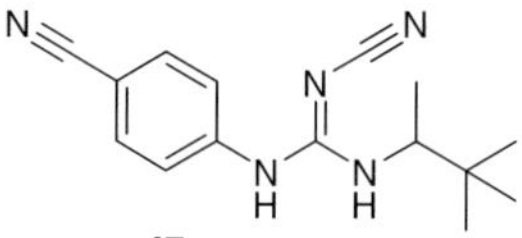

47

BMS182264
vasodilatory

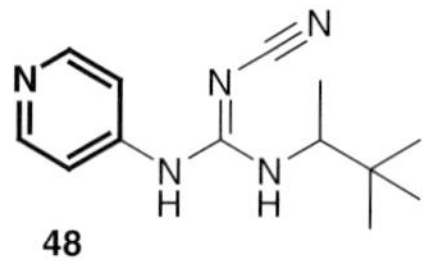

48

Pinacidil
vasodilatory

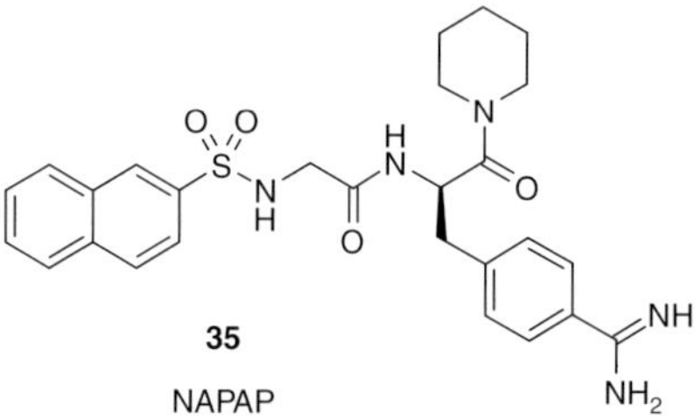

35

NAPAP
anticoagulant

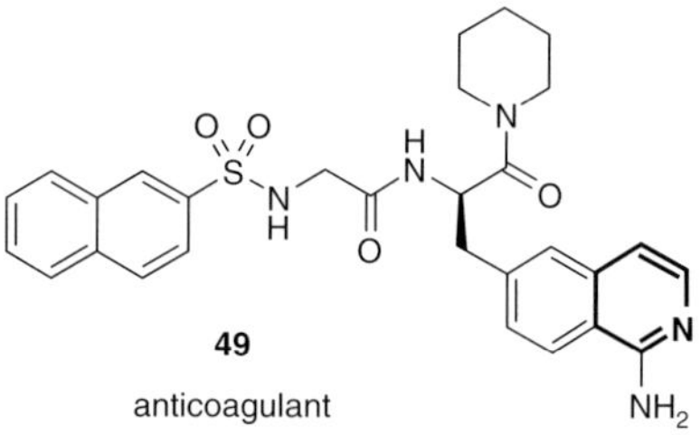

49

anticoagulant

Figure 1.12 Ring nitrogen atoms of heterocycles **44**, **46**, **48**, and **49** are able to mimic functional groups such as ester, phenol, nitrile, and amidine, respectively.

[66]. Both compounds are structurally related to the antidepressants maprotiline (**60**) and imipramine (**61**), the latter also a heterocyclic isostere of the tetracarbocyclic maprotiline (**60**) [67]. Interestingly, molecular geometry is determining the direction of pharmacological activity of these four psychotropic drugs [13b]. A dihedral angle between both planes of the two annelated phenyl rings higher than 50°, as is the case for the dibenzobicyclo[2.2.2]octane **60** and the dibenzazepine **61**, results in the preponderance of antidepressive activity [68]. If the same angle is only around 25°, as in the phenothiazines **58** and **59**, then neuroleptic efficacy prevails.

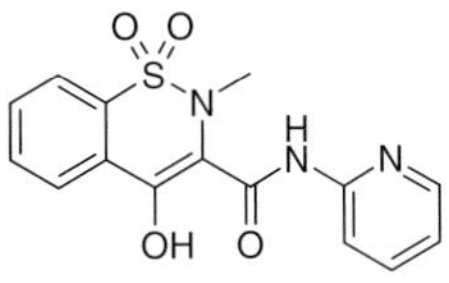

50
Piroxicam
anti-inflammatory

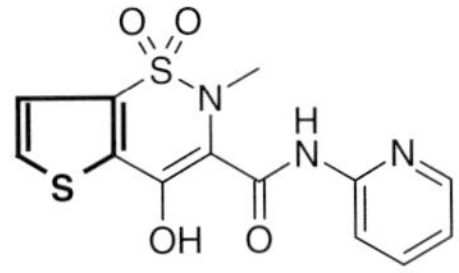

51
Tenoxicam
anti-inflammatory

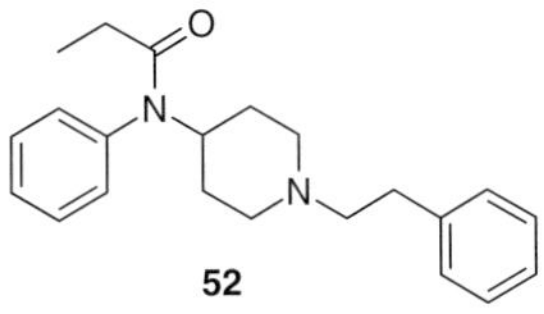

52
Fentanyl
analgetic

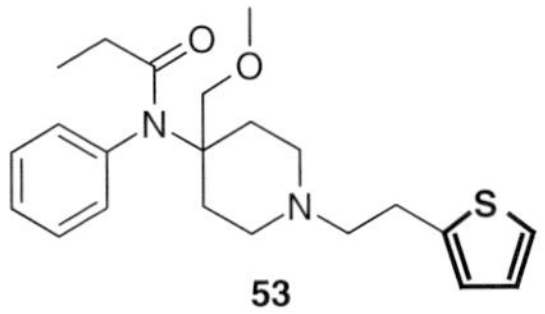

53
Sufentanil
analgetic

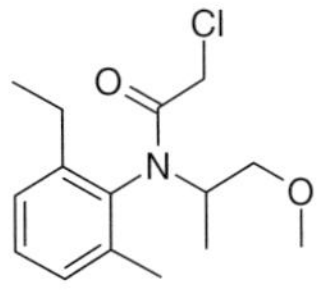

54
Metolachlor
herbicidal

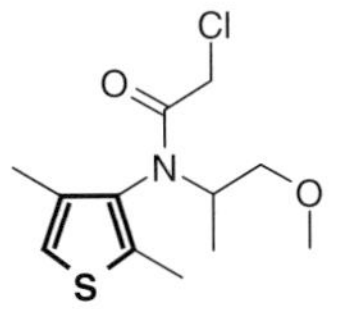

55
Dimethenamid
herbicidal

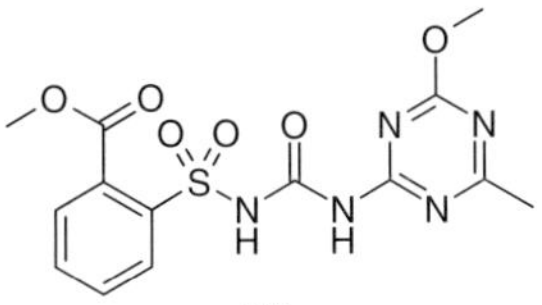

56
Metsulfuron-methyl
herbicidal

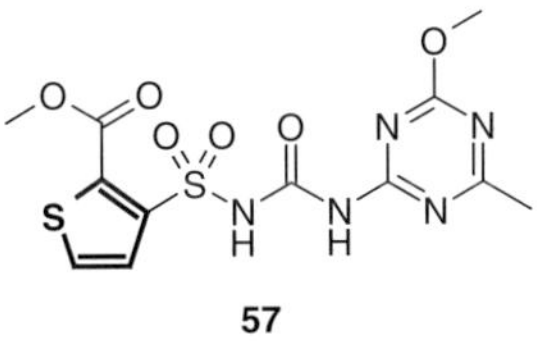

57
Thifensulfuron-methyl
herbicidal

Figure 1.13 The thiophene derivatives **51**, **53**, **55**, and **57** as highly active heterocyclic isosteres of the corresponding phenyl analogs **50**, **52**, **54**, and **56**.

1.8
Heterocycles as Isosteric Replacement of other Heterocyclic Rings

It seems that all kinds of heterocyclic rings, aromatic and nonaromatic ones, can be replaced by other heterocycles, resulting in similar biological activity. The fact that aromatic heterocycles with a similar boiling point are often suitable bioisosteres is an interesting observation [13b]. For instance, a pyridazine (b.p. 208 °C) can be replaced successfully by an aromatic heterocycle with one additional ring nitrogen (1,2,4-triazine, b.p. 200 °C) but not by a ring with one nitrogen atom less (pyridine, b.p. 115–116 °C) or another diazine, in which one of

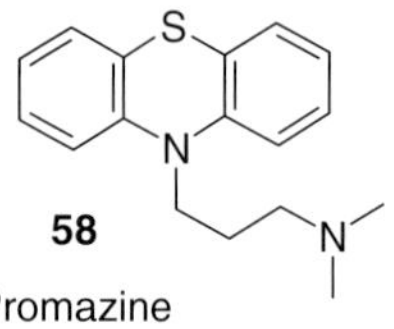

58

Promazine
neuroleptic

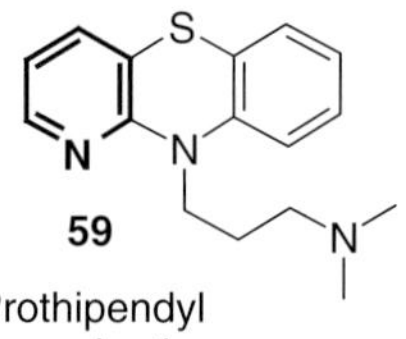

59

Prothipendyl
neuroleptic

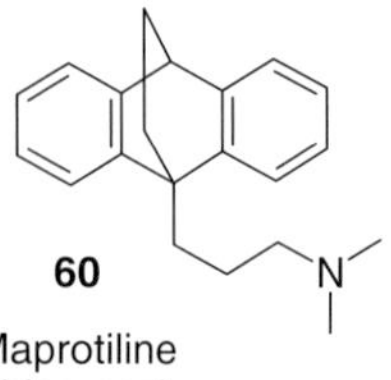

60

Maprotiline
antidepressive

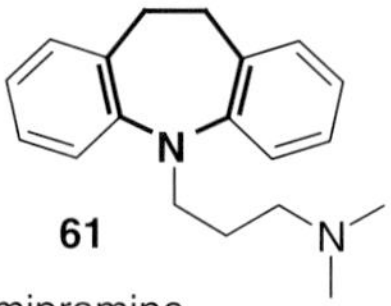

61

Imipramine
antidepressive

Figure 1.14 Prothipendyl (**59**) and imipramine (**61**), two psychotropic heterocyclic isosteres of promazine (**58**) and maprotiline (**60**).

the ring nitrogen is moved to another place (pyrimidine, b.p. 123–124 °C, or pyrazine, b.p. 115–118 °C) [13b].

One ring nitrogen and one carbon atom changing places turns a dihydropyrazolo[4,3-d]pyrimidine scaffold into a dihydroimidazo[5,1-f][1,2,4]triazine framework, thereby producing vardenafil (**63**) from sildenafil (**62**) [69]. The C-nucleosides oxazofurin (**64**) and selenazofurin (**65**) both inhibit the NAD-dependent inosine monophosphate dehydrogenase and show antiviral activities because of the impact of this inhibition of DNA synthesis (Figure 1.15). In contrast to oxazofurin, selenazofurin is also highly active against certain types of leukemia because it is readily metabolized to analogs of NAD, which may be attributed to the higher basicity of selenazole moiety [70]. The muscarinic agonist pilocarpine (**66**) is widely employed as topical miotic for lowering the elevated intraocular pressure associated with glaucoma, but the duration of this effect lasts only about 3 h, which is mainly due to the hydrolytic instability of the lactone ring. Replacement of one of the chiral carbon atoms in pilocarpine's dihydrofuranone ring by nitrogen results in the cyclic carbamate **67**, which is equipotent with pilocarpine and less susceptible to hydrolysis [71]. A ring contraction, which has been successfully applied in pharmaceutical lead optimization, is the replacement of the heptacyclic dihydrobenzodiazepine scaffold of the anticonvulsant α-amino-3-hydroxy-5-methyl-4-isoxazole-propionic acid (AMPA) receptor antagonist GYKI-53655 (**68**) by a dihydrophthalazine ring system, as in SYM-2207 (**69**) [72].

The bleaching herbicide fluridone (**70**) as well as its tetrahydropyrimidinone analog **71**, a cyclic urea, are very active against monocotyledonous and dicotyledonous weeds [73]. In animal health, the triaminated triazine cyromazine (**72**)

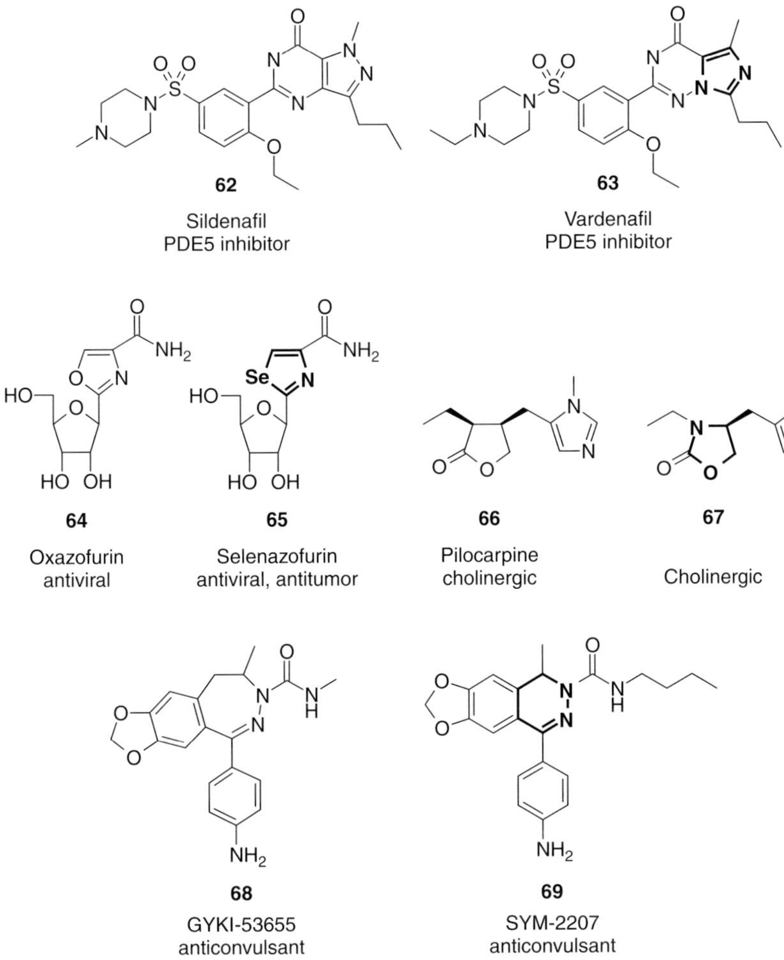

Figure 1.15 The biologically active compounds **63**, **65**, **67**, and **69**, bearing a slightly modified heterocycle compared to their analogs **62**, **64**, **66**, and **68**.

and its pyrimidine analog dicyclanil (**73**) are both very efficient against blowfly strike on sheep and screwworm infestation of cattle [74]. Both compounds are insect growth regulators, inhibiting the biosynthesis of chitin. Both aromatic and aliphatic heterocycles of imidacloprid (**74**) are replaced in the second-generation neonicotinoid thiamethoxam (**75**) by an isosteric ring with a different ring size [75]. The sulfonium salt **77**, an ionized thiane mimicking successfully the N-protonated fenpropidin (**76**), which is the active form of this fungicidal sterol biosynthesis inhibitor, shows activity against different wheat phytopathogens (Figure 1.16) [76].

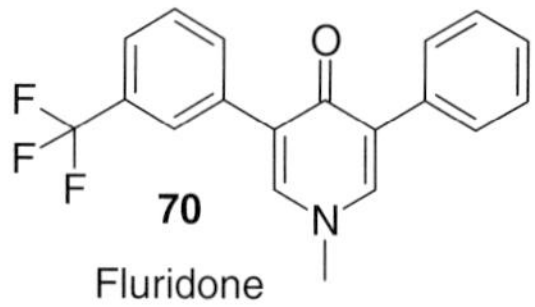

70

Fluridone
herbicidal

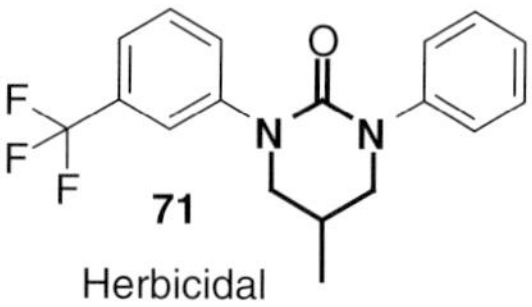

71

Herbicidal

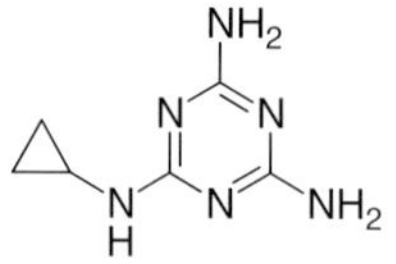

72

Cyromazine
insect growth regulator

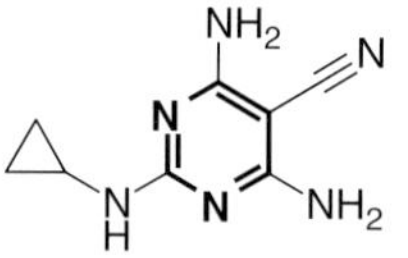

73

Dicyclanil
insect growth regulator

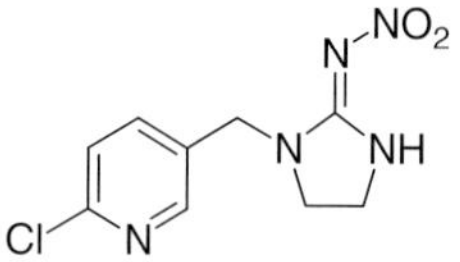

74

Imidacloprid
insecticidal

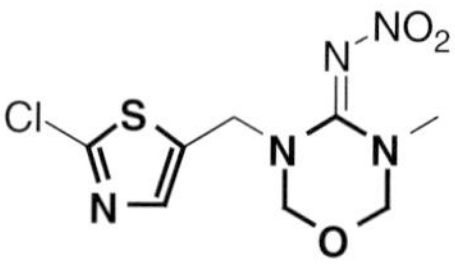

75

Thiamethoxam
insecticidal

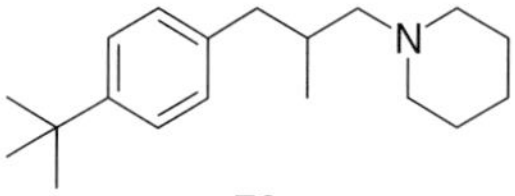

76

Fenpropidin
fungicidal

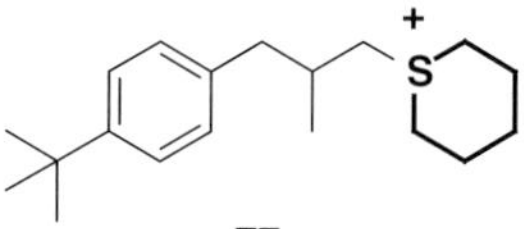

77

Fungicidal

Figure 1.16 The biologically active compounds **71**, **73**, **75**, and **77**, bearing a slightly modified heterocycle compared to their analogs **70**, **72**, **74**, and **76**.

References

1. Alvarez-Builla, J., Vaquero, J.J., and Barluenga, J. (eds) (2011) *Modern Heterocyclic Chemistry*, Wiley-VCH Verlag GmbH, Weinheim.
2. Joule, J.A. and Mills, K. (2010) *Heterocyclic Chemistry*, Wiley-Blackwell, Oxford.
3. Quin, L.D. and Tyrell, J. (2010) *Fundamentals of Heterocyclic Chemistry*, Wiley-Blackwell, Oxford.
4. Katritzky, A.R., Ramsden, C.A., Joule, J.A., and Zhdankin, V.V. (2010) *Handbook of Heterocyclic Chemistry*, Elsevier, Amsterdam.
5. Bansal, R.K. (2008) *Heterocyclic Chemistry*, Anshan, Tunbridge Wells.
6. Eicher, T. and Hauptmann, S. (2003) *The Chemistry of Heterocycles*, Wiley-VCH Verlag GmbH, Weinheim.
7. Li, J.J. (ed.) (2005) *Name Reactions in Heterocyclic Chemistry*, John Wiley & Sons, Inc., Hoboken.
8. Li, J.J. and Gribble, G.W. (eds) (2007) *Palladium in Heterocyclic Chemistry – A*

Guide for the Synthetic Chemist, Elsevier, Oxford.
9. Kühl, O. (2010) *Functionalised N-Heterocyclic Carbene Complexes*, John Wiley & Sons, Ltd, Chichester.
10. Petrov, V.A. (ed.) (2009) *Fluorinated Heterocyclic Compounds*, John Wiley & Sons, Inc., Hoboken.
11. Roth, B.D. (2002) *Prog. Med. Chem.*, **40**, 1–22.
12. Bartlett, D.W., Clough, J.M., Godwin, J.R., Hall, A.A., Hamer, M., and Parr-Dobrzanski, B. (2002) *Pest Manag. Sci.*, **58**, 649–662.
13. For reviews on bioisoterism in pharmaceutical and agrochemical lead optimisation see: (a) Meanwell, N.A. (2011) *J. Med. Chem.*, **54**, 2529–2591; (b) Ciapetti, P. and Giethlen, B. (2008) in *The Practice of Medicinal Chemistry* (ed. C.G. Wermuth), Elsevier, Amsterdam, pp. 290–342; (c) Chen, X. and Wang, W. (2003) *Ann. Rep. Med. Chem.*, **38**, 333–346; (d) Olesen, P.H. (2001) *Curr. Opin. Drug Discov. Dev.*, **4**, 471–478; (e) Patani, G.A. and LaVoie, E.J. (1996) *Chem. Rev.*, **96**, 3147–3176; (f) Koyanagi, T. and Haga, T. (1995) in *Synthesis and Chemistry of Agrochemicals IV*, ACS Symposium Series, Vol. 584 (eds D.R. Baker, J.G. Fenyes, and G.S. Basarab), American Chemical Society, Washington, DC, pp. 15–24.
14. Ward, A., Brogden, R.N., Heel, R.C., Speight, T.M., and Avery, G.S. (1983) *Drugs*, **26**, 468–502.
15. Calderbank, A. and Farrington, J.A. (1984) *Drug Chem. Toxicol.*, **10**, 89–106.
16. Brogden, R.N. (1986) *Drugs*, **32**, 60–70.
17. Lamberth, C. (2007) *Heterocycles*, **71**, 1467–1502.
18. Ho, I.K. and Hoskins, B. (1986) *Mech. Drug Action*, **1**, 177–201.
19. Boura, A.L.A. and Green, A.F. (1984) *Discov. Pharmacol.*, **2**, 73–105.
20. Jacobsen, N. and Pedersen, L.-E.K. (1983) *Pestic. Sci.*, **14**, 90–97.
21. (a) Guile, C.T. (1965) *Plant Pathol.*, **14**, 179–187; (b) Munnecke, D.E. and Martin, J.P. (1964) *Phytopathology*, **54**, 941–945.
22. Janssen, P.A.J. (1976) *Prog. Drug Res.*, **20**, 347–383.
23. Plosker, G.L. and Benfield, P. (1997) *CNS Drugs*, **7**, 410–416.
24. Kleschick, W.A., Gerwick, B.C., Carson, C.M., Monte, W.T., and Snider, S.W. (1992) *J. Agric. Food Chem.*, **40**, 1083–1085.
25. Gates, B.J., Nguyen, T.T., Setter, S.M., and Davies, N.M. (2005) *Exp. Opin. Pharmacother.*, **6**, 2117–2140.
26. Kim, D.-S., Chun, S.-J., Jeon, J.-J., Lee, S.-W., and Joe, G.-H. (2004) *Pest Manag. Sci.*, **60**, 1007–1012.
27. Haga, T., Fujikawa, K., Sakashita, N., and Nishiyama, R. (1987) *Nippon Noyaku Gakkaishi*, **12**, 311–325.
28. Veber, D.F., Johnson, S.R., Cheng, H.-Y., Smith, B.R., Ward, K.W., and Kopple, K.D. (2002) *J. Med. Chem.*, **45**, 2615–2623.
29. Prakash, A. and Jarvis, B. (1999) *Drugs*, **58**, 1137–1164.
30. (a) Pallett, K.E., Cramp, S.M., Little, J.P., Veerasekaran, P., Crudace, A.J., and Slater, A.E. (2001) *Pest Manag. Sci.*, **57**, 133–142; (b) Garcia, I., Job, D., and Matringe, M. (2000) *Biochemistry*, **39**, 7501–7507.
31. Lamberth, C. (2004) *J. Sulf. Chem.*, **25**, 39–62.
32. (a) Shimizu, T., Hashimoto, N., Nakayama, I., Nakao, T., Mizutani, H., Unai, T., Yamaguchi, M., and Abe, H. (1995) *Plant Cell Physiol.*, **36**, 625–632; (b) Sato, Y., Hoshi, T., Iida, T., Ogino, C., Nicolaus, B., Wakabayashi, K., and Boeger, P. (1994) *Z. Naturforsch. C*, **49**, 49–56.
33. For reviews on heterocyclic peptidomimetics see: (a) Abell, A.D. (2002) *Lett. Peptide Sci.*, **8**, 267–272; (b) Venkatesan, N. and Kim, B.H. (2002) *Curr. Med. Chem.*, **9**, 2243–2270; (c) Bohacek, R.S. and Skakespeare, W.C. (2000) in *The Amide Linkage: Selected Structural Aspects in Chemistry, Biochemistry and Material Science* (eds A. Greenberg, C.M. Breneman, and J.F. Liebman), John Wiley & Sons, Ltd, Chichester, pp. 377–407; (d) Giannis, A. and Kolter, T. (1993) *Angew. Chem.*, **105**, 1303–1326; (1993) *Angew. Chem., Int. Ed. Engl.*, **32**, 1244–1267.
34. La Porte, C.J.L. (2009) *Exp. Opin. Drug Metabol. Toxicol.*, **5**, 1313–1322.

35. May, B.C.H. and Abell, A.D. (2002) *J. Chem. Soc., Perkin Trans. 1*, 172–178.
36. Getman, D.P., DeCrescenzo, G.A., Heintz, R.M., Reed, K.L., Talley, J.J., Bryant, M.L., Clare, M., Houseman, K.A., Marr, J.J., Mueller, R.A., Vazquez, M.L., Shieh, H.-S., Stallings, W.C., and Stegeman, R.A. (1993) *J. Med. Chem.*, **36**, 288–291.
37. (a) Thompson, S.K., Murthy, K.H.M., Zhao, B., Winborne, E., Green, D.W., Fisher, S.M., DesJarlais, R.L., Tomaszek, T.A., Meek, T.D., Gleason, J.G., and Abdel-Meguid, S.S. (1994) *J. Med. Chem.*, **37**, 3100–3107; (b) Thompson, S.K., Eppley, A.M., Frazee, J.S., Darcy, M.G., Lum, R.T., Tomaszek, T.A., Ivanoff, L.A., Morris, J.F., Sternberg, E.J., Lambert, D.M., Fernandez, A.V., Petteway, S.R., Meek, T.D., Metcalf, B.W., and Gleason, J.G. (1994) *Bioorg. Med. Chem. Lett.*, **4**, 2441–2446.
38. (a) Smith, A.B., Hirschmann, R., Pasternak, A., Yao, W., Sprengeler, P.A., Holloway, M.K., Kuo, L.C., Chen, Z., Darke, P.L., and Schleif, W.A. (1997) *J. Med. Chem.*, **40**, 2440–2444; (b) Gosh, A.K., Thompson, W.J., McKee, S.P., Duong, T.T., Lyle, T.A., Chen, J.C., Darke, P.L., Zugay, J.A., Emini, E.A., Schleif, W.A., Huff, J.R., and Anderson, P.S. (1993) *J. Med. Chem.*, **36**, 292–294.
39. Mack, H., Pfeiffer, T., Hornberger, W., Bohm, H.-J., and Hoeffken, H.W. (1995) *J. Enzyme Inhib.*, **9**, 73–86.
40. (a) Saitton, S., Del Tredici, A.L., Saxin, M., Stenström, T., Kihlberg, J., and Luthman, K. (2008) *Org. Biomol. Chem.*, **6**, 1647–1654; (b) Saitton, S., Del Tredici, A.L., Mohell, N., Vollinga, R.C., Boström, D., Kihlberg, J., and Luthman, K. (2004) *J. Med. Chem.*, **47**, 6595–6602.
41. (a) Bolton, D., Boyfield, I., Coldwell, M.C., Hadley, M.S., Johns, A., Johnson, C.N., Markwell, R.E., Nash, D.J., Riley, G.J., Scott, E.E., Smith, S.A., Stemp, G., Wadsworth, H.J., and Watts, E.A. (1997) *Bioorg. Med. Chem. Lett.*, **7**, 485–488; (b) Van Wijngaarden, I., Kruse, C.G., van Hes, R., van der Heyden, J.A.M., and Tulp, M.T.M. (1987) *J. Med. Chem.*, **30**, 2099–2104.
42. (a) Steinmetzer, T., Zhu, B.Y., and Konishi, Y. (1999) *J. Med. Chem.*, **42**, 3109–3115; (b) Mimoto, T., Kato, R., Takaku, H., Nojima, S., Terashima, K., Misawa, S., Fukazawa, T., Ueno, T., Sato, H., Shintani, M., Kiso, Y., and Hayashi, H. (1999) *J. Med. Chem.*, **42**, 1789–1802.
43. Kim, B.H., Chung, Y.J., Keum, G., Kim, J., and Kim, K. (1992) *Tetrahedron Lett.*, **33**, 6811–6814.
44. Jones, R.C.F. and Ward, G.J. (1988) *Tetrahedron Lett.*, **29**, 3853–3856.
45. Einsiedel, J., Thomas, C., Hübner, H., and Gmeiner, P. (2000) *Bioorg. Med. Chem. Lett.*, **10**, 2041–2044.
46. Liu, Y., Li, Y., Myles, D.C., Claypool, M., Carreras, C.W., and Shaw, S.J. (2010) *Bioorg. Med. Chem.*, **18**, 7651–7658.
47. (a) Legeay, J.C., Vanden Eynde, J.J., and Bazureau, J.P. (2007) *Tetrahedron Lett.*, **48**, 1063–1068; (b) Feng, D.D., Biftu, T., Candelore, M.R., Cascieri, M.A., Colwell, L.F., Deng, L., Feeney, W.P., Forrest, M.J., Hom, G.J., MacIntyre, D.E., Miller, R.R., Stearns, R.A., Strader, C.D., Tota, L., Wyvratt, M.J., Fisher, M.H., and Weber, A.E. (2000) *Bioorg. Med. Chem. Lett.*, **10**, 1427–1429; (c) Andersen, K.E., Jorgensen, A.S., and Braestrup, C. (1994) *Eur. J. Med. Chem.*, **29**, 393–399.
48. (a) Zhang, D., Wang, Z., Xu, W., Sun, F., Tang, L., and Wang, J. (2009) *Eur. J. Med. Chem.*, **44**, 2202–2210; (b) Chen, J.J., Zhang, Y., Hammond, S., Dewdney, N., Ho, T., Lin, X., Browner, M.F., and Castelhano, A.L. (1996) *Bioorg. Med. Chem. Lett.*, **6**, 1601–1606.
49. For reviews on tetrazoles as carboxylic acid bioisosteres see: (a) Zych, A.J. and Herr, R.J. (2007) *PharmaChem*, **6**, 21–24; (b) Herr, R.J. (2002) *Bioorg. Med. Chem.*, **10**, 3379–3393.
50. (a) Frolund, B., Ebert, B., Kristiansen, U., Liljefors, T., and Krogsgaard-Larsen, P. (2002) *Curr. Top. Med. Chem.*, **2**, 817–832; (b) Krogsgaard-Larsen, P., Hjeds, H., Falch, E., Jorgensen, F.S., and Nielsen, L. (1988) in *Advances in Drug Research* (ed. B. Testa), Academic Press, London, pp. 381–456.

51. Drysdale, M.J., Pritchard, M.C., and Horwell, D.C. (1992) *J. Med. Chem.*, **35**, 2573–2581.

52. Atkinson, J.G., Girard, Y., Rokach, J., Rooney, C.S., McFarlane, C.S., Rackham, A., and Share, N.N. (1979) *J. Med. Chem.*, **22**, 99–106.

53. (a) Gezginci, M.H., Martin, A.R., and Franzblau, S.G. (2001) *J. Med. Chem.*, **44**, 1560–1563; (b) Kohara, Y., Kubo, K., Imamiya, E., Wada, T., Inada, Y., and Naka, T. (1996) *J. Med. Chem.*, **39**, 5228–5235; (c) Bock, M.G., DiPardo, R.M., Mellin, E.C., Newton, R.C., Veber, D.F., Freedman, S.B., Smith, A.J., Patel, S., Kemp, J.A., Marshall, G.R., Fletcher, A.E., Chapman, K.L., Anderson, P.S., and Freidinger, R.M. (1994) *J. Med. Chem.*, **37**, 722–724.

54. (a) Henke, B.R. (2004) *J. Med. Chem.*, **47**, 4118–4127; (b) Hulin, B., McCarthy, P.A., and Gibbs, E.M. (1996) *Curr. Pharm. Des.*, **2**, 85–102.

55. Marquez, T., Joshi, M.M., Fader, T.P., and Massasso, W. (1995) *Brighton Crop Prot. Conf. Weeds*, **1**, 65–72.

56. Kim, K.S., Kimball, S.D., Misra, R.N., Rawlins, D.B., Hunt, J.T., Xiao, H.-Y., Lu, S., Qian, L., Han, W.-C., Shan, W., Mitt, T., Cai, Z.-W., Poss, M.A., Zhu, H., Sack, J.S., Tokarski, J.S., Chang, C.Y., Pavletich, N., Kamath, A., Humphreys, W.G., Marathe, P., Bursuker, I., Kellar, K.A., Roongta, U., Batorsky, R., Mulheron, J.G., Bol, D., Fairchild, C.R., Lee, F.Y., and Webster, K.R. (2002) *J. Med. Chem.*, **45**, 3905–3927.

57. (a) Petukhov, P.A., Zhang, M., Johnson, K.J., Tella, S.R., and Kozikowski, A.P. (2001) *Bioorg. Med. Chem. Lett.*, **11**, 2079–2083; (b) Sauerberg, P., Kindtler, J.W., Nielsen, L., Sheardown, M.J., and Honore, T. (1991) *J. Med. Chem.*, **34**, 687–692; (c) Orlek, B.S., Blaney, F.E., Brown, F., Clark, M.S.G., Hadley, M.S., Hatcher, J., Riley, G.J., Rosenberg, H.E., Wadsworth, H.J., and Wyman, P. (1991) *J. Med. Chem.*, **34**, 2726–2735; (d) Street, L.J., Baker, R., Book, T., Kneen, C.O., MacLeod, A.M., Merchant, K.J., Showell, G.A., Saunders, J., Herbert, R.H., Freedman, S.B., and Harley, E.A. (1990) *J. Med. Chem.*, **33**, 2690–2697; (e) Saunders, J., Cassidy, M., Freedman, S.B., Harley, E.A., Iversen, L.L., Kneen, C., MacLeod, A.M., Merchant, K.J., Snow, R.J., and Baker, R. (1990) *J. Med. Chem.*, **33**, 1128–1138.

58. Mewshaw, R.E., Zhao, R., Shi, X., Marquis, K., Brennan, J.A., Mazandarani, H., Coupet, J., and Andree, T.H. (2002) *Bioorg. Med. Chem. Lett.*, **12**, 271–274.

59. (a) Butera, J.A., Antane, M.M., Antane, S.A., Argentieri, T.M., Freeden, C., Graceffa, R.F., Hirth, B.H., Jenkins, D., Lennox, J.R., Matelan, E., Norton, N.W., Quagliato, D., Sheldon, J.H., Spinelli, W., Warga, D., Wojdan, A., and Woods, M. (2000) *J. Med. Chem.*, **43**, 1187–1202; (b) Atwal, K.S., Moreland, S., McCullough, J.R., O'Reilly, B.C., Ahmed, S.Z., and Normandin, D.E. (1992) *Bioorg. Med. Chem. Lett.*, **2**, 83–86.

60. Rewinkel, J.B.M., Lucas, H., van Galen, P.J.M., Noach, A.B.J., van Dinther, T.G., Rood, A.M.M., Jenneboer, A.J.S.M., and van Boeckel, C.A.A. (1999) *Bioorg. Med. Chem. Lett.*, **9**, 685–690.

61. (a) Press, J.B. (1991) *Chem. Heterocycl. Comp.*, **44**, 397–502; (b) Drehsen, G. and Engel, J. (1983) *Sulfur Rep.*, **3**, 171–214.

62. Bird, H.A. and Naden, M.A. (1989) *J. Drug Dev.*, **2**, 119–135.

63. Monk, J.P., Beresford, R., and Ward, A. (1988) *Drugs*, **36**, 286–313.

64. Couderchet, M., Bocion, P.F., Chollet, R., Seckinger, K., and Böger, P. (1997) *Pestic. Sci.*, **50**, 221–227.

65. Cuomo, J., Gee, S.K., and Hartzell, S.L. (1991) in *Synthesis and Chemistry of Agrochemicals II*, ACS Symposium Series, Vol. **443** (eds D.R. Baker, J.G. Fenyes, and W.B. Moberg), American Chemical Society, Washington, DC, pp. 62–73.

66. Ellenbroek, B., Prinssen, E., and Cools, A. (1992) *Neurosci. Res. Commun.*, **11**, 155–161.

67. Gurguis, G.N., Blakeley, J.E., Antai-Otong, D., Vo, S.P., Orsulak, P.J., Petty, F., and Rush, A.J. (1999) *J. Psychiatr. Res.*, **33**, 309–322.

68. Wilhelm, M. (1975) *Pharm. J.*, **214**, 414–416.

69. Stamford, A.W. (2002) *Ann. Rep. Med. Chem.*, **37**, 53–64.

70. (a) Franchetti, P., Cristalli, G., Grifantini, M., Cappellacci, L., Vittori, S., and Nocentini, G. (1990) *J. Med. Chem.*, **33**, 2849–2852; (b) Goldstein, B.M., Takusagawa, F., Berman, H.M., Srivastava, P.C., and Robins, R.K. (1985) *J. Am. Chem. Soc.*, **107**, 1394–1400.

71. (a) Hobbs, S.H., Johnson, S.J., Kesten, S.R., Pavia, M.R., Davis, R.E., Schwarz, R.D., Coughenour, L.L., Myers, S.L., Dudley, D.T., and Moos, W.H. (1991) *Bioorg. Med. Chem. Lett.*, **1**, 147–150; (b) Sauerberg, P., Chen, J., WoldeMussie, E., and Rapoport, H. (1989) *J. Med. Chem.*, **32**, 1322–1326.

72. Pelletier, J.C., Hesson, D.P., Jones, K.A., and Costa, A.-M. (1996) *J. Med. Chem.*, **39**, 343–346.

73. (a) Babczinski, P., Blunck, M., Sandmann, G., Shiokawa, K., and Yasui, K. (1995) *Pestic. Biochem. Physiol.*, **52**, 45–59; (b) Babczinski, P., Sandmann, G., Schmidt, R.R., Shiokawa, K., and Yasui, K. (1995) *Pestic. Biochem. Physiol.*, **52**, 33–44; (c) Babczinski, P., Blunck, M., Sandmann, G., Schmidt, R.R., Shiokawa, K., and Yasui, K. (1990) *Pestic. Sci.*, **30**, 339–342.

74. (a) Anziani, O.S., Guglielmone, A.A., and Schmid, H. (1998) *Vet. Parasitol.*, **76**, 229–232; (b) Hart, R.J., Cavey, W.A., Ryan, K.J., Strong, M.B., Moore, B., Thomas, P.L., Boray, J.C., and von Orelli, M. (1982) *Aust. Vet. J.*, **59**, 104–109.

75. (a) Maienfisch, P., Angst, M., Brandl, F., Fischer, W., Hofer, D., Kayser, H., Kobel, W., Rindlisbacher, A., Senn, R., Steinemann, A., and Widmer, H. (2001) *Pest Manag. Sci.*, **57**, 906–913; (b) Maienfisch, P., Huerlimann, H., Rindlisbacher, A., Gsell, L., Dettwiler, H., Haettenschwiler, J., Sieger, E., and Walti, M. (2001) *Pest Manag. Sci.*, **57**, 165–176.

76. Wilkie, J.S. and Winzenberg, K.N. (1992) *Aust. J. Chem.*, **45**, 457–461.

Part I
Neurological Disorders

Bioactive Heterocyclic Compound Classes: Pharmaceuticals,
First Edition. Edited by Jürgen Dinges and Clemens Lamberth.

2
Tropane-Based Alkaloids as Muscarinic Antagonists for the Treatment of Asthma, Obstructive Pulmonary Disease, and Motion Sickness

Michael L. Schulte and Craig W. Lindsley

2.1
Introduction

Tropane is the generic name for a group of over 200 nitrogen-containing bicyclic alkaloids possessing an *N*-methyl-8-azabicyclo[3.2.1]octane core, which can be drawn in two different representations: **1**, a flat one-dimensional view, and **2**, a three-dimensional view (Figure 2.1) [1–10]. The nitrogen bridge is between C1 and C5, and despite two asymmetric carbons, tropane is not optically active due to a plane of symmetry. Tropane congeners lacking the *N*-Me moiety, azabicyclo[3.2.1]octanes (**3**), are generically referred to as nortropanes. Virtually all naturally occurring tropanes possess a hydroxyl group at C3, designated β-tropine (**4**) and α-tropine (**5**), equatorial and axial, respectively, derived from the ketone congener commonly referred to as *tropinone* (**6**) [1–10].

Tropane alkaloids obtained as extracts from various plants of the family Solanaceae have been widely used by man for over 2000 years to treat numerous diseases and elicit hallucinations and as deadly poisons via their pharmacological actions as either anticholinergics or stimulants [1–13]. Synthetic tropane alkaloids are frontline therapeutic agents for the treatment of asthma, obstructive pulmonary disease, and motion sickness by virtue of more selective muscarinic acetylcholine receptor (mAChR) inhibition and reduced systemic exposure resulting from quarternization of the amine [1–17]. This chapter focuses on the history, synthesis, mechanism of action, and therapeutic benefit of tropane alkaloids.

2.2
History

Tropane alkaloids are secondary metabolites extracted from flowering plants of the family Solanaceae, such as deadly nightshade (*Atropa belladonna*), jimsonweed (*Datura stramonium*), and mandrake (*Mandragora officinarum*) [1–13]. The name tropane is derived from the plant in which they were first discovered, *A. belladonna*. Extracts of plants of Solanaceae for medical purposes, toxins, and hallucinogens

Bioactive Heterocyclic Compound Classes: Pharmaceuticals,
First Edition. Edited by Jürgen Dinges and Clemens Lamberth.

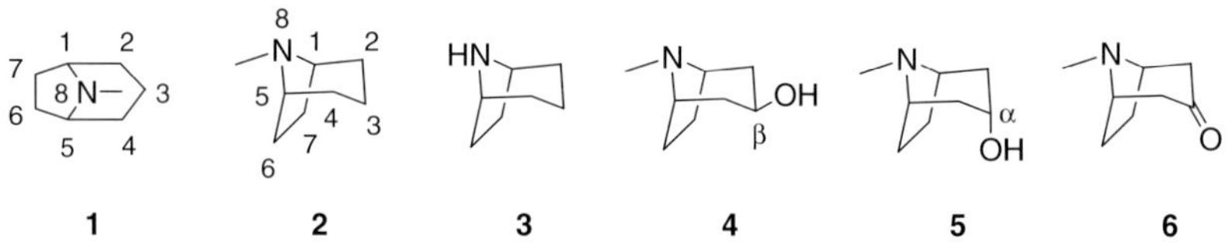

Figure 2.1 Structures and numbering for tropanes (**1**–**6**).

date back to the fourth century BC, and their use was commonplace during the Roman, Islamic, and Egyptian empires.

Cleopatra used the extracts to dilate her pupils (mydriatic action) to appear more alluring. During the renaissance, juice from the berries of *A. belladonna* was used to enlarge the pupils for beauty; interestingly, "bella donna" means "beautiful lady" in Italian [1–13]. Ancient Egyptians with airway diseases, such as asthma, burnt distillates of *Hyoscyamius* and inhaled the smoke, and in Europe, similar traditional medicine used the deadly nightshade shrub [17]. As mentioned above, extracts can also be quite toxic, and the genus *Atropa* is named after one of the Greek Fates, Atropos, who chose how a person was to die (the fate that cut the thread of life); thus, the genus was chosen to signify the lethality of these compounds and their classical use as poisons [18]. An overdose of tropane alkaloids elicits hallucinations and paranoia (similar to the positive symptoms of schizophrenia), severe cognitive disturbances, coma, and death [1–18].

Early naturally occurring tropane alkoids that still have medical applications as anticholinergics and stimulants for the parasympathetic nervous system in the present day include atropine (**7**), scopolamine (**8**), and cocaine (**9**) (Figure 2.2) [1–17]. Atropine (**7**) is a racemic mixture of D-hyoscyamine (**10**) and L-hyoscyamine

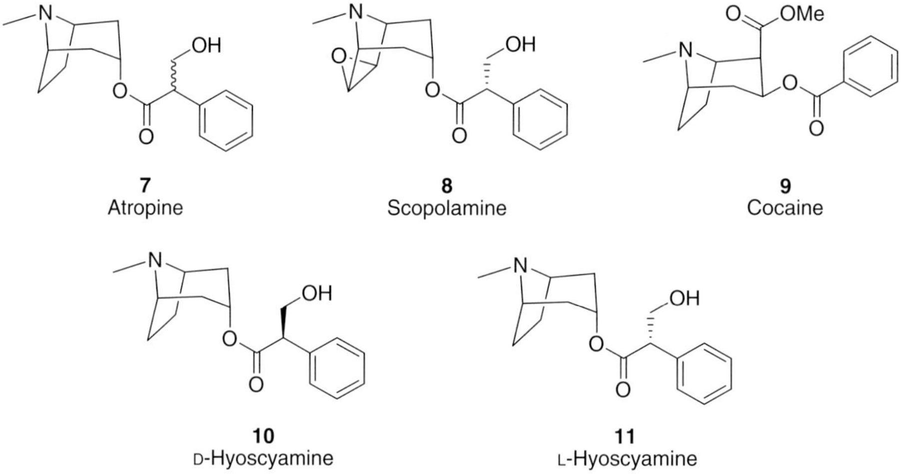

Figure 2.2 Naturally occurring tropane alkaloids with historical and present day medical value as anticholinergics and stimulants: atropine (**7**) [19], scopolamine (**8**) [20] and cocaine (**9**) [23], and the active enantiomer of racemic atropine, L-hyoscyamine (**11**) [1–10].

(**11**), with the therapeutic anticholinergic effects due to L-hyoscyamine (**11**). Extracts containing atropine were the standard of treatment until 1831 when Mein isolated pure crystalline atropine (**7**) [19]. Atropine (**7**) is widely used in ophthalmics as a pupil-dilating agent. Atropine (**7**) has numerous actions on the parasympathetic nervous system, and, as such, it is used for resuscitation to accelerate heart rate and to inhibit secretions from the salivary, sweat, and mucous glands. In fact, atropine (**7**) was commonly used with anesthesia to limit mucous accumulation and excessive salivation [1–19]. Scopolamine (**8**) has also been used since antiquity as extracts, until the German chemist Ladenburg isolated it in the pure form in 1880 [20]. Scopolamine (**8**), an anticholinergic agent, is also referred to as both L-*duboisine* and *hyoscine*. Scopolamine (**8**) was used from the 1940s to the 1960s to put women in labor into a "twilight sleep" that eliminated the memory of the pain, while not preventing the pain. During the same time period, **8** was an active ingredient in Asthmador, an over-the-counter medication to treat asthma and bronchitis [1–17, 21]. At present, **8** is widely used in very small doses in a transdermal patch as a frontline therapy for motion sickness [16]. At higher doses, **8** elicits significant memory impairment and prevents memory consolidation, for which it has received severe negative attention in the popular press as a date rape drug [22]. Cocaine (**9**) was a more recent discovery. In 1860, **9** was isolated by Niemann [23] from the Peruvian *Erythroxylon coca* plant, well known for anesthetic properties as well as a powerful stimulant of the CNS leading to improved physical performance. These properties led to widespread use in multiple medications and products until the addictive properties of **9** were understood; as a result, cocaine is now an illegal narcotic around the world.

In addition to the naturally occurring tropane alkaloids **7** and **8**, several synthetic congeners (Figure 2.3), such as ipratropium (**12**, Atrovent®), oxitropium (**13**, Oxivent®), tiotropium (**14**, Spiriva®), and trospium (**15**, Sanctura® and Trosec®), are marketed therapeutics for the treatment of asthma, chronic obstructive pulmonary disease (COPD), and motion sickness via anticholinergic mechanisms (see below) [1–16]. The biosynthesis of scopolamine (**8**), and the origin of the epoxide moiety, has long been of interest, and it is also found in oxitropium (**13**, Oxivent) and tiotropium (**14**, Spiriva). Interestingly, it has been shown that the epoxide is generated by the direct oxidation of L-hyoscyamine (**11**) and not through the intermediacy of an olefin (Scheme 2.1) [24, 25]. The enzyme from *Hyoscyamus niger* has been purified and shown to require both Fe^{2+} and 2-oxoglutarate for the two-step oxidation process. The first step involves oxidation at the 6-position to install a 6β-hydroxyl moiety to produce **16**, and then the same enzyme facilitates epoxide ring closure to provide scopolamine (**8**) [24, 25].

2.3 Synthesis

The first effort toward structural elucidation of atropine (**7**) was led by Kraut in 1863 [10, 26]. By boiling **7** in $BaSO_4$ solution, Kraut recovered tropine (**5**) and

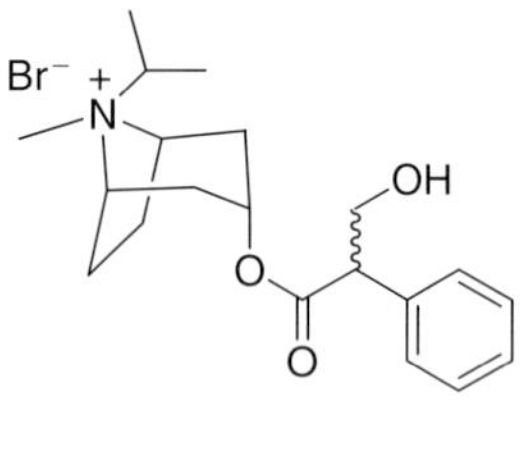

12
Ipratropium

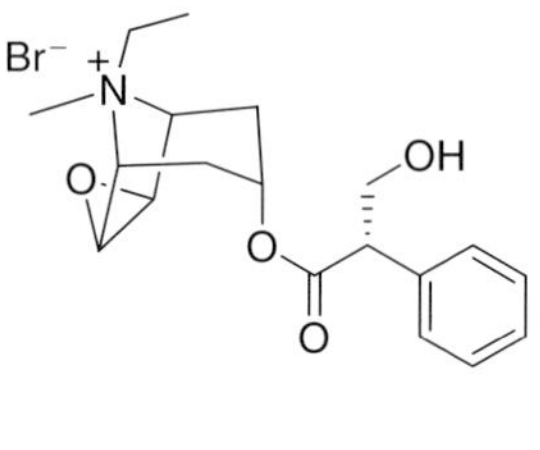

13
Oxitropium

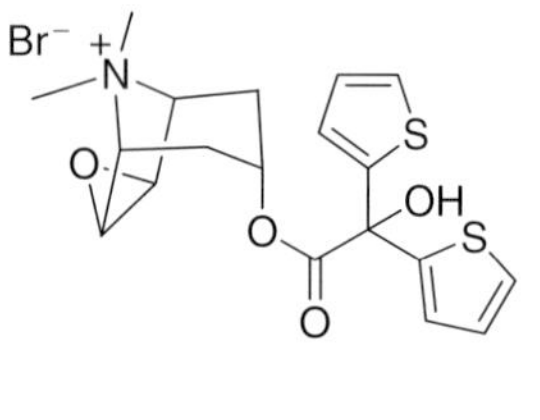

14
Tiotropium

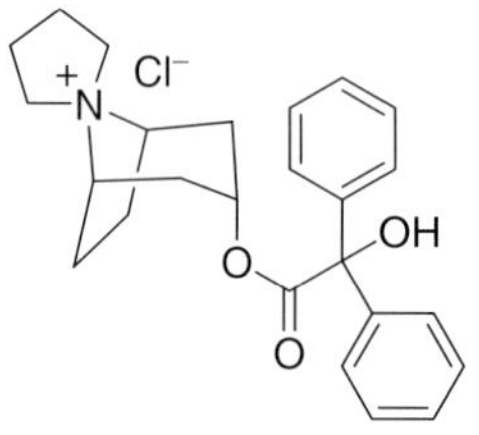

15
Trospium

Figure 2.3 Structures of synthetic, marketed tropane alkaloids: ipratropium (**12**) [14–16], oxitropium (**13**) [14–16], tiotropium (**14**) [14–16], and trospium (**15**) [14–16].

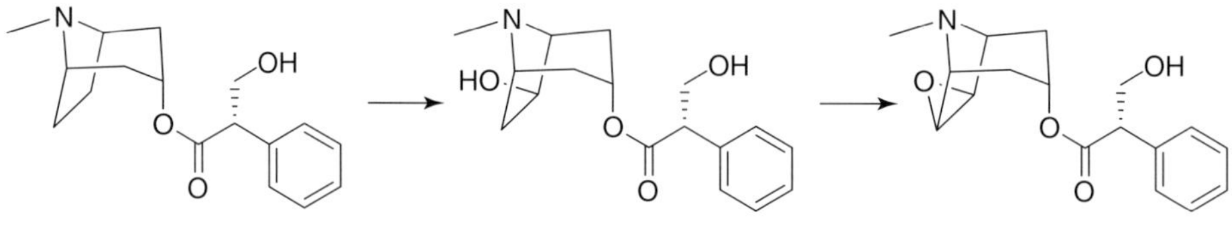

11 **16** **8**

Scheme 2.1 Biosynthesis of scopolamine (**8**) from L-hyoscyamine (**11**).

atropic acid (**17**). In 1864, Lossen showed that the initial product of hydrolysis of **7** was tropic acid (**18**), which rapidly dehydrated to afford **17** [10]. Almost 20 years later, Ladenberg demonstrated that the process could be reversed and atropine (**7**) could be prepared by the esterification of **5** with **18** in the presence of dry HCl (Scheme 2.2) [27].

A similar hydrolysis of cocaine (**9**) produced ecgonine (**19**) and benzoic acid (**20**) (Scheme 2.3); however, the structures of these fragments and the relative connectivities were unknown at the time. The same is true for the reaction shown in Scheme 2.2.

The earliest synthetic efforts toward **7** were by Ladenburg and Rugheimer [28], who completed the first total synthesis of tropic acid (**18**) in 1880; however, the absolute stereochemistry of the (*S*)-enantiomer was not deduced until 1961 [29]. Shortly thereafter, Willstätter conducted degradation studies on **5** and **19** with

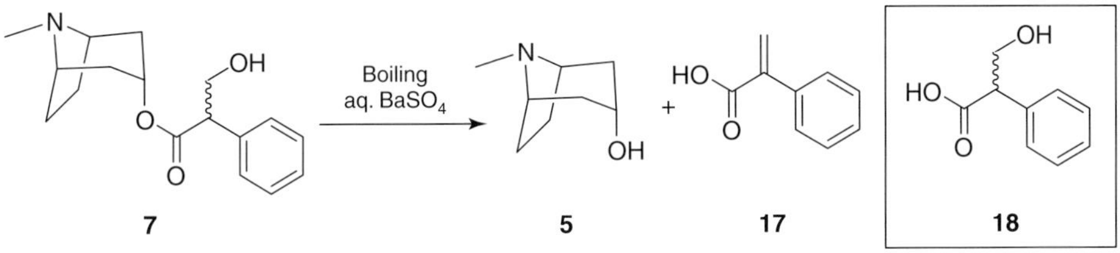

7 5 17 18

Scheme 2.2 Hydrolysis of atropine (7) to produce tropine (5) and tropic acid (18), which rapidly dehydrates to form atropic acid (17).

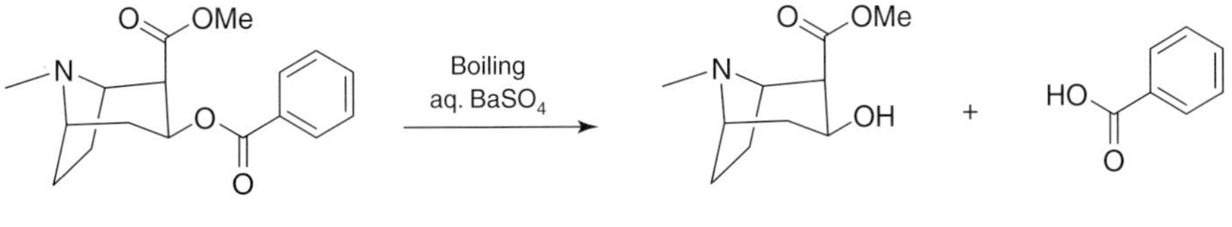

8 19 20

Scheme 2.3 Hydrolysis of cocaine (9) to produce ecgonine (19) and benzoic acid (20).

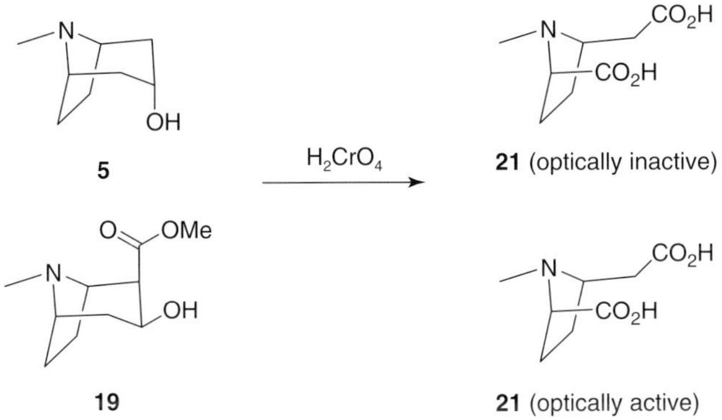

Scheme 2.4 Degradation of tropine (5) and ecgonine (19) to afford tropinic acid (21).

chromic acid and discovered that both produced tropinic acid (21). However, the compound 21 when obtained from 5 was optically inactive (5 is *meso*), while when derived from 19 was optically active (Scheme 2.4) [30].

These degradation studies laid the foundation for a true classic in total synthesis, the first preparation of the tropane ring system by Willstätter [31] in 1889 (Scheme 2.5). While a lengthy route, the successful synthesis of the tropane ring system is outstanding considering the tools available to organic chemists in the nineteenth century. Willstätter's synthesis began with cycloheptanone, which was converted in eight steps to cycloheptatriene (22). Addition of bromine and dimethylamine provided 24, followed by monoreduction to deliver 25. The key reaction in the sequence involved the addition of bromine across the olefin to produce 26, followed by intramolecular nucleophilic displacement to form, for the first time,

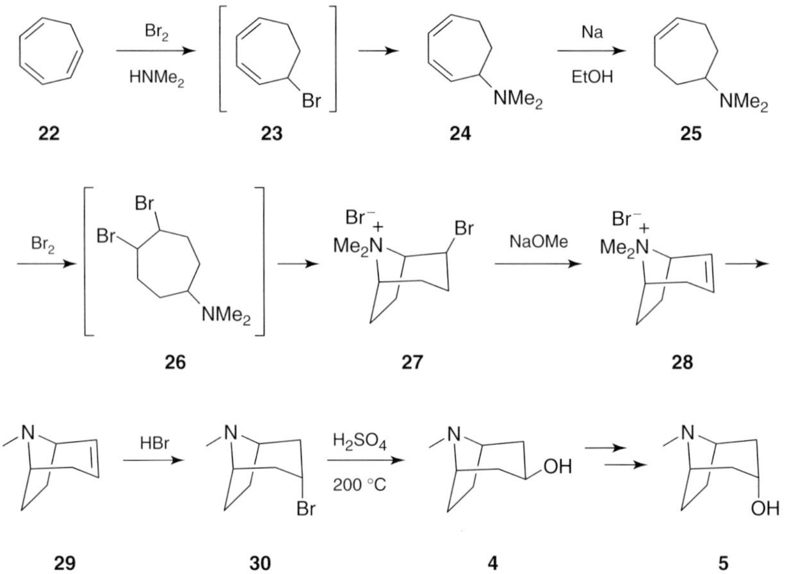

Scheme 2.5 Willstätter's classic, as well as the first, synthesis of the tropane ring system.

the core tropane ring system as the tetraalkylammonium salt (**27**). Alkali metal treatment first formed **28**, via elimination, and then **29**. Addition of HBr led to the isolation of what was assumed to be the α-bromide (**30**), which on treatment with H_2SO_4 in a sealed tube at 200 °C produced β-**4**, with the undesired stereochemistry at C3. Oxidation of the hydroxyl group to afford tropinone (**6**) and a reduction sequence delivered the desired β-tropine (**5**) [31]. Starting from **6** generated in this route, Willstätter also completed the first total synthesis and the structural elucidation of cocaine (**9**) [32]. From these efforts and for solving the structure of chlorophyll, Willstätter won the 1915 Nobel prize for Chemistry [10].

In 1917, Robinson published a one-pot, three-component biomimetic synthesis of tropinone (**6**) that continues to influence tropane syntheses to this day [33]. In the optimized sequence, succinaldehyde (**31**) is added to an aqueous solution of methylamine and acetonedicarboxylic acid (**32**) (early efforts used acetone in place of **32**, but the yields were very low) at physiological pH to afford a 42% yield of **6** (Scheme 2.6) [33]. Willstätter's [34, 35] second-generation tropine synthesis, as

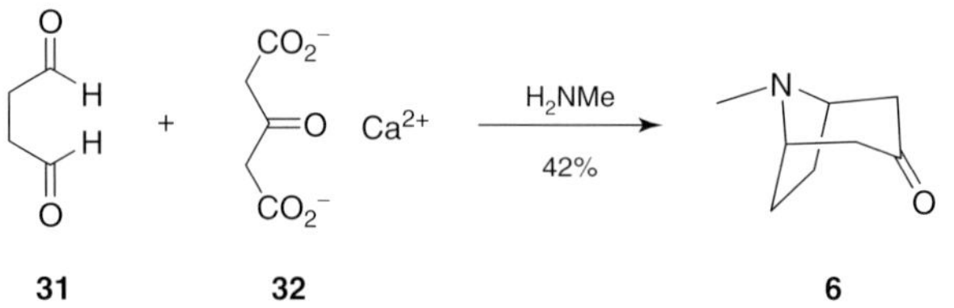

Scheme 2.6 Robinson's concise, biomimetic synthesis of the tropane ring system.

well as Elming's [36] tropine synthesis, uses a variation of Robinson's route, as do most routes today. This led Robinson [37] to a propose a biosynthetic route for **6** in 1995, which was later validated by Abraham and Lette [38] in 1995. On the basis of seminal contributions to alkaloid chemistry, Robinson won the Nobel prize in Chemistry in 1947 [10].

Over the years, alternative routes (Scheme 2.7) to the tropane nucleus have been reported based on, for example, (i) a [4 + 2] nitroso cycloaddition with 1,3-cycloheptadienes [39], (ii) a reductive [3 + 4] cycloaddition of a polybromo ketone-iron complex with 1-methylpyrrole (**38**) [40], and (iii) enyne ring-closing metathesis strategies [41, 42]. Despite the synthetic innovation of these, and other, approaches, virtually all clinically relevant tropane alkaloids are prepared using the Robinson approach or directly from atropine (**7**) or scopolamine (**8**) [10]. For example, Schemes 2.8 and 2.9 highlight the two commercial synthetic routes used to access ipratropium (**12**, Atrovent) [43–45]. Following the classical Robinson approach, 2,5-dimethoxytetrahydrofuran (**43**) is treated with HCl to generate succinaldehyde (**31**), which is then treated with aqueous solution of isopropylamine and acetonedicarboxylic acid (**32**) to afford *N*-isopropylnortropinone (**44**) (Scheme 2.8).

Raney Ni reduction delivers *N*-isopropylnortropine (**45**), which then undergoes a transesterification with **46** to deliver **47**. $NaBH_4$ reduction and N-alkylation with methyl bromide leads to ipratropium (**12**, Atrovent) [43–45]. Alternatively, heating tropine (**5**) with tropic acid (**18**) and HCl effects a Fischer esterification to provide atropine (**7**) (Scheme 2.9). Subsequent alkylation with isopropyl bromide leads to quarternization of atropine (**7**), delivering ipratropium (**12**, Atrovent) [43–45].

2.4 Mode of Action

As mentioned previously, tropane alkaloids are among the most potent anticholinergic agents (i.e., mAChR antagonists) known, and their biologic effects in man are driven by this mechanism of action [1–15]. The mAChRs are members of the class A G-protein-coupled receptors (GPCRs) that mediate a broad range of actions of the neurotransmitter acetylcholine (ACh) in both the central and peripheral nervous systems. To date, five distinct subtypes of mAChRs (M_1–M_5) have been cloned and sequenced. M_1, M_3, and M_5 activate phospholipase C through G_q to stimulate calcium release, whereas M_2 and M_4 couple to inhibition of adenylyl cyclase and other effector systems through activation of $G_{i/o}$ (Figure 2.4) [46–48]. Both atropine (**7**) and scopolamine (**8**) are the prototypical pan-muscarinic antagonists, inhibiting M_1–M_5 with equal potency by competing for binding at the ACh binding site, and are invaluable basic pharmacological research tools [1, 15, 46–48]. As mentioned previously, both still have clinical use, with atropine (**7**) on the World Health Organization's "Essential Drug List" [49] and utilized in ophthalmics as a pupil-dilating agent, to accelerate heart rate and to inhibit secretions from the salivary, sweat, and mucus glands during anesthesia [14, 17, 50]. Scopolamine (**8**),

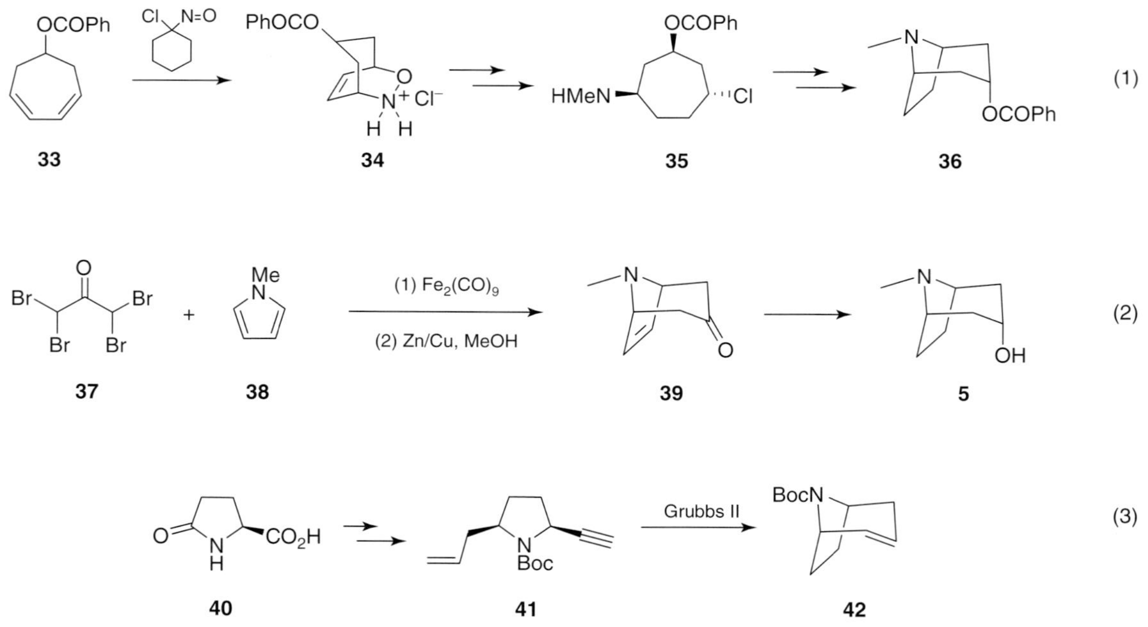

Scheme 2.7 Alternate synthetic approaches to the tropane ring system.

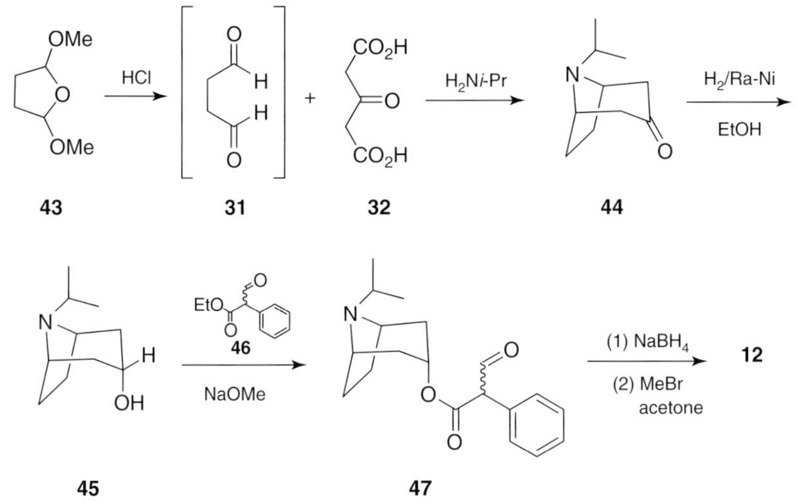

Scheme 2.8 Classical Robinson approach to ipratropium (**12**, Atrovent®).

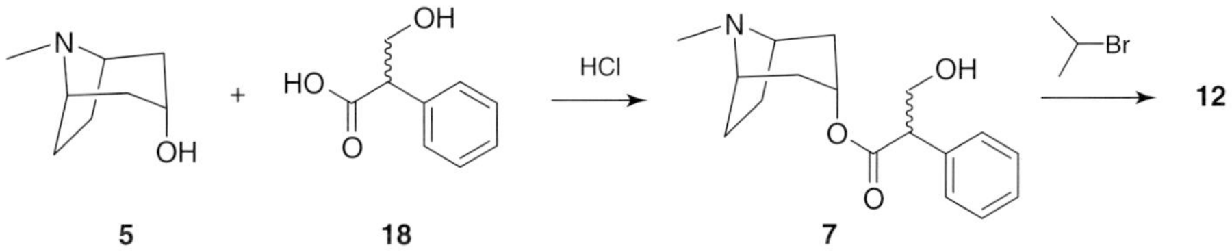

Scheme 2.9 Alternative to ipratropium (**12**, Atrovent®), via alkylation of atropine (**7**).

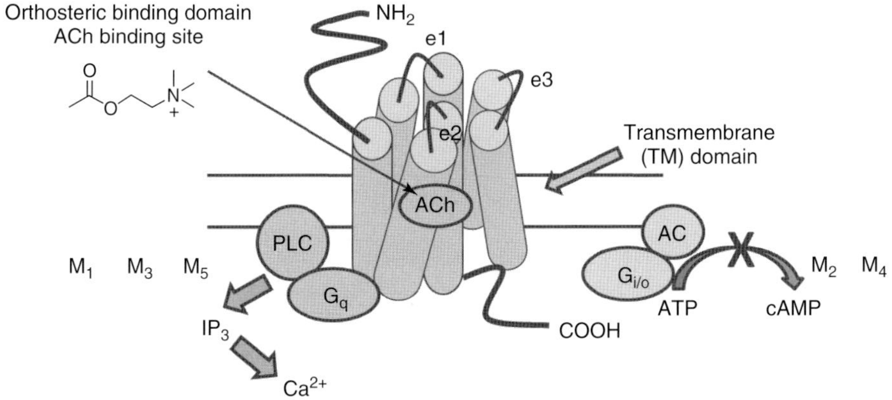

Figure 2.4 Cartoon representation of the muscarinic acetylcholine receptors (M_1–M_5).

owing to antiemetic properties, is used in very small doses (330 $\mu g\,day^{-1}$) in a transdermal patch (Transderm Scōp) as a frontline therapy for motion sickness [50]. Interestingly, atropine (**7**) has very limited CNS exposure, while scopolamine (**8**) achieves high CNS exposure; consequently, scopolamine (**8**) elicits more severe adverse events (AEs) due to inhibition of neuronal mAChRs. Both are available as oral, intravenous, and ocular formulations as well, and efficacy for different indications depends on the route of administration used [1–15].

A major therapeutic area for anticholinergic tropane alkaloids has been in the treatment of respiratory diseases such as asthma and COPD [14, 17]. In fact, blockade of muscarinic receptors is one of the oldest treatments for asthma using plant extracts for over 2000 years [1–17]. Airway tone is mediated by parasympathetic nerves that release ACh onto the mAChRs (M_1–M_5), and patients with asthma or COPD display increased airway tone (also referred to as *airway hyperreactivity*) [1–17, 51]. Therapy for these diseases has been driven by the presumption that the increased cholinergic tone may be the only reversible event. As mAChR expression and function became understood throughout the 1960s to 1980s, discovery paradigms shifted to address targeting specific mAChR subtypes and/or tissues. Of relevance to asthma and COPD, M_1, M_2, and M_3 are found in, and exert physiological effects on, the lungs. Within the lung tissue, mAChRs are found in parasympathetic ganglia (M_1), submucosal glands (M_3 and modest M_1), airway smooth muscle (M_2 and M_3), and postganglionic cholinergic nerves (M_2) [14, 17, 52–54]. The presynaptic M_2 receptor affords a negative feedback loop wherein its stimulation inhibits the release of ACh; therefore, inhibition of M_2 with pan-muscarinic antagonists will increase cholinergic tone and enhance bronchoconstriction, leading to more severe asthma or COPD [14, 17, 52–54]. M_3 receptors, however, mediate contraction of smooth muscle and mucous secretion in the respiratory tract. Therefore, the ideal therapeutic for asthma or COPD would target M_3, or both M_3 and M_1, without inhibiting M_2. While both **7** and **8** have been used for respiratory disease indications, both suffer from undesirable AEs due to exposure outside the lungs and/or inhibition of M_2 [14, 17, 52–54]. The most common AEs are tachycardia (M_2), dry mouth, urinary retention, and constipation (all M_3). Thus, drug discovery efforts were focused on synthetic tropane alkaloids to limit exposure outside the respiratory tract and/or limit inhibition of M_2.

2.5
Structure–Activity Relationships

Attempts to develop highly selective M_3 antagonists have largely failed due to the highly conserved orthosteric (ACh) binding site across the mAChRs and the promiscuous binding of the tropane "privileged structure"; however, there have been no reports of M_3 negative allosteric modulators (NAMs), which should afford high M_3 selectivity and novel chemotypes. Efforts to limit exposure have been very fruitful. By quaternizing the nitrogen bridge, as in **12–15**, CNS exposure is eliminated (the charged species do not cross the blood–brain barrier), systemic absorption through inhalation formulation is limited (~5–7% of dose), and plasma stability is poor, thereby avoiding the systemic and CNS side effects observed with **7** and **8** [14, 17, 52–54]. Currently, ipratropium (**12**, Atrovent), oxitropium (**13**, Oxivent), and tiotropium (**14**, Spiriva) are marketed anticholinergics for the treatment of asthma and COPD [14–16].

Ipratropium (**12**, Atrovent, also marketed as Apovent® and Aerovent®: Boehringer Ingelheim) is a quaternary ammonium analog (*N*-*i*-Pr) of atropine (**7**) and has been the most widely used anticholinergic therapeutic for asthma or COPD since its introduction in the 1980s [14, 17, 52–54]. Ipratropium (**12**) is considered a short-acting muscarinic antagonist with a duration of action in the 4–6 h range, thus requiring administration by inhalation four times a day. Typically, ipratropium (**12**) is coadministered with a short-acting β-agonist, such as salbuterol/albuterol, and marketed as Combivent® or Duoned® (for the treatment of COPD) or coadministered with fenoterol and marketed as Duovent® (for the treatment of asthma). The N-ethyl variant, oxitropium (**13**, Oxivent), is marketed in Europe (Nippon Boehringer Ingelheim) and is pharmacologically equivalent in most respects to **12** [14, 17, 52–54]. The most common AE is dry mouth.

While **12** and **13** are short-acting muscarinic antagonists (4–6 h duration of action), tiotropium (**14**, Spiriva: Boehringer Ingelheim and Pfizer) was the first long-acting muscarinic antagonist (LAMA) [14, 17, 52–54]. Tiotropium (**14**) is also marketed by Cipla under the trade name Tiova®. Although **14** binds to M_1–M_5 with equal affinity, it exhibits kinetic selectivity for M_1 and M_3 over M_2. The residence time has been measured for the dissociation of **14** from M_1 (~15 h), M_2 (~4 h), and M_3 (~35 h), therefore conferring functional selectivity for M_3 inhibition with a very long duration of action (~24 h). Swinney [55] has suggested that **14** may bind irreversibly to M_3 in the lungs, and Price [56] believes that the two thiophene rings are the key to the M_3 functional selectivity. Tiotropium (**14**) was the first ever anticholinergic drug that was shown to be effective in patients with poorly controlled asthma and is more effective than **12** and **13** in COPD patients. Tiotropium (**14**) is administered once daily from a dry powder inhaler with trade name the Spiriva and HandiHaler® [14, 17, 52–54]. The oral bioavailability of **14** is 2–3%, providing low systemic exposure. On inhalation, 80% of the dose is swallowed and eliminated in the feces, with 19.5% reaching the lungs (i.e., almost 100% of available dose reaches the target). Multiple studies have demonstrated that **14** is superior to **12** and **13** in asthma and COPD patients, due to the functional selectivity for M_3 [14, 17, 52–54]. The most common AE is dry mouth; however, in 2008, a study reported that both **12** and **14** increased the risk of heart attack and stroke [57]. In 2010, the FDA announced the results of a long-term impact trial and found no significant increase in the risk of heart attack and stroke with **14** versus placebo [58]. Trospium (**15**, ALKS27, Sanctura and Trosec) was initially developed for overactive bladder by Endo and Speracor, but later repositioned by Alkernes and Indevus for COPD. An inhaled formulation showed efficacy in phase II trials, but in 2011, ALKS27 is no longer listed in the Alkernes pipeline [59].

Current efforts are focused on the development of additional LAMAs with high M_3 selectivity, and there are numerous phase II and phase III clinical trials underway

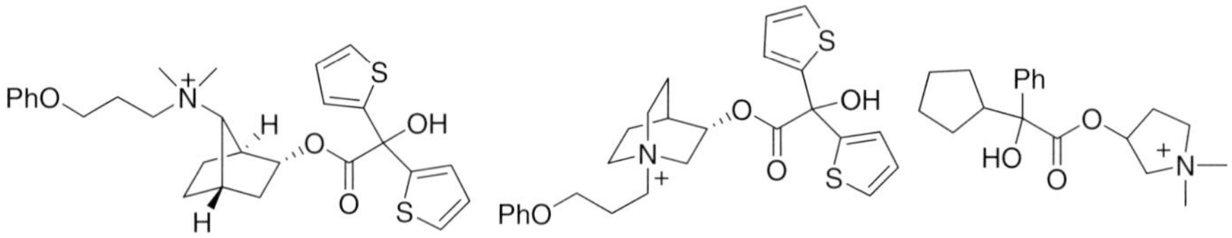

48

49
Acilidinium

50
Glycopyrrolate

Figure 2.5 Third-generation anticholinergic scaffolds for the treatment of asthma and COPD.

Table 2.1 Muscarinic antagonists and key properties.

Compound	M_1 pK_i	M_2 pK_i	M_3 pK_i	M_4 pK_i	M_5 pK_i	hM_3 Upreg pEC_{50} (%)	$M_3 > M_2$ affinity	M_3 $t_{1/2}$ (h)
7	9.77	9.47	9.68	9.97	9.50	9.1 (54%)	None	3.5
8	10.3	10.2	10.3	10.4	9.5	9.0 (19)	None	3.6
12	9.4	9.53	9.58	9.65	9.0	9.0 (41)	None	3.2
14	10.8	10.6	11.02	11.02	9.9	10.3 (18)	FS	34.7
49	10.78	10.68	10.74	10.84	10.26	9.5 (40)	FS	29
50	10.1	9.6	10.7	10.8	9.7	9.4 (43)	5x	3.7

FS, functional selectivity.

as well as an extensive amount of patent filings. Interestingly, many companies are moving away from the tropane core and are basing new therapeutics on bicyclo[2.2.1]heptane (**48**) [60], quinuclidinol (**49**, aclidinium bromide, Eklira®) [61], or quaternary pyrroldinyl (**50**, glycopyrrolate, NVA237) [62] scaffolds (Figure 2.5).

In summary, anticholinergic agents based on the tropane core represent not only the oldest known but also the current standard of care for the treatment of asthma, COPD, and motion sickness. Newer synthetic congeners display improved affinity for M_3, selectivity for M_3 versus M_2, hM_3 upregulation, and longer residence time at M_3, leading to more effective and safer therapeutic agents (Table 2.1) [17, 63, 64]; however, there is still a major unmet medical need for more selective agents.

References

1. Fordor, G. (1955) *Experientia*, **11**, 129–140.
2. Gyermek, L. and Nador, K. (1957) *J. Pharm. Pharmacol.*, **9**, 209–229.
3. Fodor, G., Soti, F., Dobo, P., Uresch, F., and Romeike, A. (1965) *Beitr. Biochem. Physiol. Naturstioffen Festschr.*, 135–142.
4. Lounasamaa, M. and Tamminen, T. (1993) Tropane alkaloids, in *The Alkaloids*. Chemistry and Pharmacology, Vol. 44, Academic Press, London, pp. 1–103.
5. Griffin, W.L. and Linn, D.G. (2000) *Phytochemistry*, **53**, 632–637.

6. Grynkiewicz, G. and Gadzikowska, M. (2008) *Pharmacol. Rep.*, **60**, 439–463.
7. Bruce, N.C. (2009) Alkaloids, in *Biotransformations I*, Biotechnology, Vol. 8A (ed. P.R. Kelly), Wiley-VCH Verlag GmbH, pp. 326–361.
8. Peretto, I., Petrillo, P., and Imbimbo, B.P. (2009) *Med. Res. Rev.*, **29**, 867–902.
9. Fordor, G. and Dharanipragada, R. (1988) *Nat. Prod. Rep.*, **5**, 67–72.
10. Humphrey, A.J. and O'Hagan, D. (2001) *Nat. Prod. Rep.*, **18**, 494–502.
11. Hanson, B. (2005) *Understanding Medicinal Plants: their Chemistry and Therapeutic Action*, Haworth Herbal Press, Binghampton.
12. Pengelly, A. (2004) *The Constituents of Medicinal Plants: An Introduction to the Chemistry and Therapeutics of Herbal Medicine*, CABI Publishing, Cambridge, MA.
13. Mann, J. (1992) *Murder, Magic, Medicine*, Oxford, University Press, Oxford.
14. Lee, A.M., Jacoby, D.B., and Fryer, A.D. (2001) *Curr. Opin. Pharmacol.*, **1**, 223–229.
15. Joss, G.F. (2010) *Expert Opin. Invest. Drugs*, **19**, 257–264.
16. Clissold, S.P. and Heel, R.C. (1985) *Drugs*, **29**, 189–207.
17. Moulton, B.C. and Fryer, A.D. (2011) *Br. J. Pharmacol.*, **163**, 44–52.
18. Goodman, L.S., Gilman, A., Brunton, L.I., Lazo, J.S., and Parker, K.L. (2006) *Goodman & Gillman's the Pharmacological Basis of Therapeutics*, 11th edn., McGraw-Hill, New York.
19. Sneader, W. (ed.) (1985) *Drug Discovery:the Evolution of Modern Medicines*, John Wiley & Sons, Ltd, Chichester, pp. 120–126.
20. Ladenburg, A. (1880) *Annalen*, **206**, 299.
21. Putcha, L., Cintron, N.M., Tsui, J., Vanderploeg, J.M., and Kramer, W.G. (1989) *Pharm. Res.*, **6**, 481–485.
22. Jones, D.M., Jones, M.E., Lewis, M.J., and Spriggs, T.L. (1979) *Br. J. Pharmacol.*, **7**, 479–483.
23. Niemann, A. (1860) *Arch. Pharm.*, **153**, 129–256.
24. Hashimoto, H. and Yamada, Y. (1987) *Eur. J. Biochem.*, **164**, 277–280.
25. Hashimoto, H., Matsuda, J., and Yamada, Y. (1993) *FEBS Lett.*, **329**, 35–41.
26. Kraut, A. (1863) *Ann. Chem. Pharm.*, 280.
27. Ladenberg, A. (1879) *Ber. Dtsch. Chem. Ges.*, **12**, 941.
28. Ladenberg, A. and Rugheimer, L. (1880) *Ber. Dtsch. Chem. Ges.*, **13**, 2041.
29. Fodor, G. and Csepreghy, G. (1961) *J. Chem. Soc.*, 3222.
30. Willstätter, R.M. (1889) *Ber. Dtsch. Chem. Ges.*, **31**, 1534.
31. Willstätter, R.M. (1903) *Annalen*, **317**, 204.
32. Willstätter, R.M. and Mueller, W. (1889) *Ber. Dtsch. Chem. Ges.*, **31**, 2665.
33. Robinson, R. (1917) *J. Chem. Soc. Trans.*, **111**, 762.
34. Willstätter, R.M. and Pfannenstiel, A. (1921) *Annalen*, **422**, 1.
35. Willstätter, R.M. and Bommer, M. (1921) *Annalen*, **422**, 15.
36. Elming, N. (1960) *Advances in Organic Chemistry*, Vol. II, Interscience, New York, p. 67.
37. Robinson, R. (1955) *The Structural Relations of Natural Products*, Clarendon Press, Oxford.
38. Abraham, T.W. and Leete, E. (1955) *J. Am. Chem. Soc.*, **117**, 8100.
39. Iida, H., Watanabe, Y., and Kibayashi, C. (1995) *J. Org. Chem.*, **50**, 1818–1825.
40. Noyori, R. and Hayakawa, Y. (1985) *Tetrahedron*, **41**, 5879–5886.
41. Aggarwal, V.K., Astle, C.J., and Rogers-Evans, M. (2004) *Org. Lett.*, **6**, 1469–1471.
42. Cheng, G., Wang, X., Zhu, R., Shao, C., Xu, J., and Hu, Y. (2011) *J. Org. Chem.*, **76**, 2694–2700.
43. Abdine, H.H., Belal, F., and Al-Badr, A.A. (2003) *Profiles of Drug Substances, Excipients and Related Methodology*, Vol. 30, Elsevier, pp. 85–99.
44. Tsyskovskaia, I., Kandil, M., and Beaumier, Y. (2007) *Synth. Commun.*, **37**, 439–446.
45. Issa, F., Kassiou, M., Chan, H.-K., and McLeod, M.D. (2006) *Aust. J. Chem.*, **59**, 53–58.
46. Levey, A.I. (1993) *Life Sci.*, **52**, 441–448.
47. Abrams, P., Andersson, K.E., Buccafusco, J.J., Chapple, C.,

Chet de Groat, W., Fryer, A.D., Kay, G., Laties, A., Nathanson, N.M., Pasricha, P.J., and Wein, A. (2006) *Br. J. Pharmacol.*, **148**, 565–578.
48. Wess, J., Eglen, R.M., and Gautam, D. (2007) *Nat. Rev. Drug Discovery*, **6**, 721–733.
49. WHO Model Lists of Essential Medicines *http://www.who.int/medicines/publications/essentialmedicines/en/* (accessed 2011).
50. White, P.F., Tang, J., and Song, D. (2007) *Anesth. Analg.*, **104**, 92–96.
51. Gross, N.J. (1988) *New Engl. J. Med.*, **319**, 486–494.
52. Barnes, P.J. (1989) *Thorax*, **44**, 161–167.
53. Whitsett, J.A. and Hollinger, B. (1984) *Pediatr. Res.*, **18**, 1136–1140.
54. Barnes, P.J. (1993) *Life Sci.*, **52**, 521–527.
55. Swinney, D.C. (2004) *Nat. Rev. Drug Discovery*, **3**, 801–808.
56. Price, D., Sharma, A., and Cerasoli, F. (2009) *Expert Opin. Drug Metab. Toxicol.*, **5**, 417–424.
57. Singh, S., Loke, Y.K., and Furberg, C.D. (2008) *J. Am. Med. Assoc.*, **300**, 1439–1450.
58. Follow-up to the October 2008 updated early communication about an ongoing safety review of tiotropium (marketed as Spiriva HandiHaler) *http://www.fda.gov/Drugs/DrugSafety/PostmarketDrugSafetyInformationforPatientsandProviders/DrugDrugSafetyInformationforHeathcareProfessio/ucm197429.htm* (accessed 2011).
59. Oleson, L., Turncliff, R.Z. and Silverman, B. (2010) *Am. J. Respir. Crit. Care Med.*, **181**, A4457.
60. Argenta Discover Ltd. (2008) WO2008149053.
61. Sentellas, S., Ramos, I., Alberti, J., Salva, M., Anton, F., Miralpeix, M., Beleta, J., and Gavalda, A. (2010) *Eur. J. Pharm. Sci.*, **39**, 283–290.
62. Viletti, G., Bergamaschi, M., Bassani, F., Bolzoni, P.T., Harrison, S., Gigli, P.M., Janni, A., Geppetti, P., Civelli, M., and Patacchini, R. (2006) *Br. J. Pharmacol.*, **148**, 291–298.
63. Casarosa, P., Bouyssou, T., Germeyer, S., Schnapp, A., Gantner, F., and Pieper, M. (2009) *J. Pharmacol. Exp. Ther.*, **330**, 660–668.
64. Casarosa, P., Kiechle, T., Sieger, P., Pieper, M., and Gantner, F. (2010) *J. Pharmacol. Exp. Ther.*, **333**, 201–209.

3
Morphinone-Based Opioid Receptor Agonist Analgesics

Stephanie M. Ng

3.1
Introduction

The active agent of opium, which is obtained from the readily grown poppy *Papaver somniferum*, is morphine (**1**), a potent analgesic medication that has been known as the *original alkaloid*, the first true drug, an illicit narcotic, and a compound that has arguably spawned the entire field of medicinal chemistry [1]. Although the structure of morphine (**1**) was neither defined until 1925 nor confirmed by total synthesis until 1952 [2], by then a number of chemical modifications had been made in the search for an improved nonaddicting opioid, including the synthesis of heroin and hydromorphone (**2b**).

Opioid analgesics can be categorized into eight related chemical classes [3], one of them being the morphone, morphin-6-one, or morphinone family (Figure 3.1).[1] This family includes, but is not limited to, compounds such as hydrocodone (**2a**), hydromorphone (**2b**), oxycodone (**3a**), and oxymorphone (**3b**) (Figure 3.2). Compounds of this family with a methylated nitrogen are known to be selective μ-type receptor agonists having a strong analgesic effect, similar to the biological mode of action of morphine (**1**) [4].[2] This chapter focuses on compounds with the morphin-6-one core structure, which are selective μ-type receptor agonists.

3.2
History

Long before, morphine (**1**) was isolated by Sertürner in the early nineteenth century, the opium poppy was widely used and recognized as an analgesic of unrivaled power [5]. Sertürner's isolation of the purified alkaloid improved the biological

1) Morphinanone, morphone, and morphinone are synonymous for the parent morphin-6-one for the purposes of this review.
2) *N*-substituted morphinanones, while demonstrating some analgesic potency, generate a range of activities from partial agonism to pure antagonism, and thus are beyond the scope of this chapter, information about these compounds can be found elsewhere.

Bioactive Heterocyclic Compound Classes: Pharmaceuticals,
First Edition. Edited by Jürgen Dinges and Clemens Lamberth.

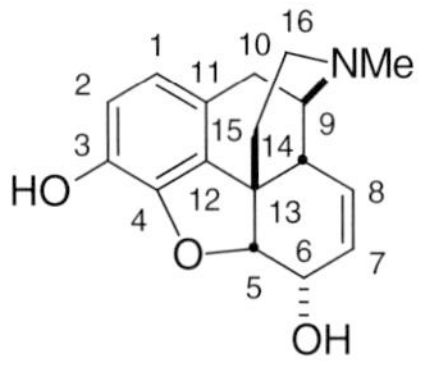

1
Morphine

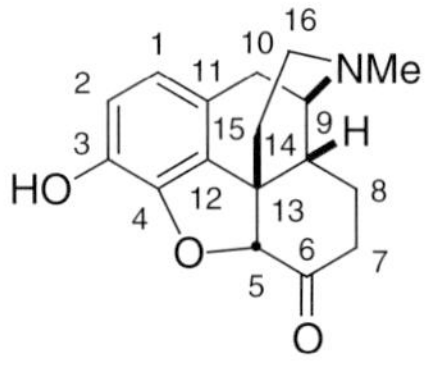

2
Parent morphin-6-one

Figure 3.1 Numbering conventions of morphine (**1**) and the morphone parent structure morphone (**2**).

effects compared with opium-based concoctions. Since the addictive properties and toxicity of morphine (**1**) were recognized early on, chemical adaptations of the parent molecule were carried out to improve pharmacological properties, even before the correct structure was known. With the development of the hypodermic syringe in 1853 [6], the use and problems of abuse and addiction to morphine (**1**) proliferated. Introduction of this superior method of administration further spurred the search for "nonaddictive" opioids, motivating not only a large part of opioid research but also a large amount of drug development for the next century. This research occurred well before significant advances were made on opioid receptors and on understanding the molecular biology behind specific receptor binding. The common mechanism for these agonists is now known to be interaction with μ-opioid receptors [3]. Nonetheless, with the goal of finding an analgesic medication with lower abuse potential but retaining the efficacy of morphine (**1**), synthetic and medicinal chemists created a number of semisynthetic opioids by simple modifications of the parent molecule. This early structure–activity relationship (SAR) exploration led to the discovery of many potent analgesics with the morphin-6-one parent core.

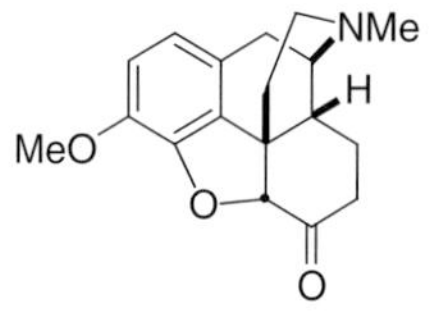

2a
Hydrocodone

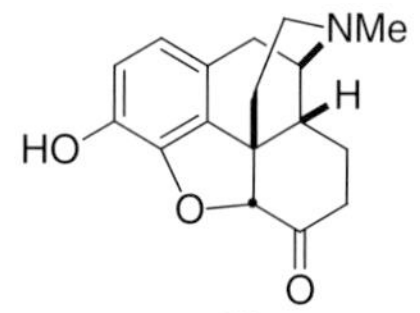

2b
Hydromorphone

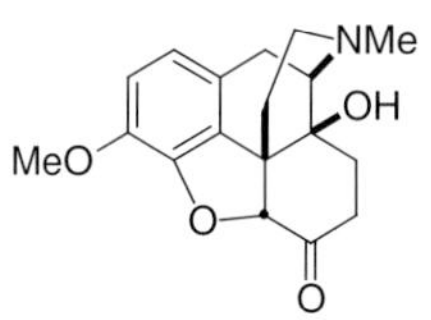

3a
Oxycodone

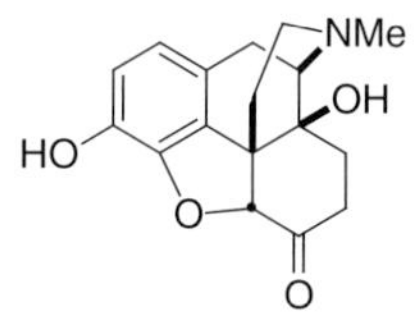

3b
Oxymorphone

Figure 3.2 Examples of the morphin-6-one family: hydrocodone (**2a**), hydromorphone (**2b**), oxycodone (**3a**), and oxymorphone (**3b**).

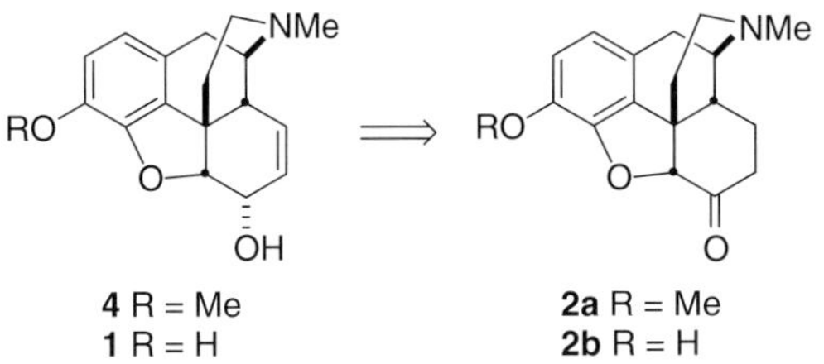

Figure 3.3 Historic development of **2a** and **2b** from morphine (**1**) and codeine (**4**).

Simple manipulations of readily available morphine (**1**) and codeine (**4**) led to the discovery of **2a** and **2b** (Figure 3.3). Hydrocodone (**2a**) was first synthesized by Mannich and Löwenheim [7] in 1923. In the same decade, hydrocodone (**2a**) and hydromorphone (**2b**) were patented in Germany under the name "Dicodid®" and "Dilaudid®," respectively, and were used especially for the treatment of cough [8]. Hydromorphone (**2b**) was eventually introduced in the United States in 1932 and was distributed to the mass market by the Bilhuber-Knoll Corporation [9]. The compound is still marketed today under the trade name Dilaudid (**2b**) as an immediate-release hydromorphone source (other immediate-release forms include Hydrostat IR®). Hydromorphone (**2b**) plays a key role in chronic and acute pain relief as an alternative to morphine (**1**) and has been studied as an analgesic for postoperative pain. Hydromorphone (**2b**) can be administered orally, parenterally, and epidurally. There are two major extended-release preparations of **2b**, including a modified-release capsule preparation (Palladone®, Hydromorph Contin®, Palladone SR®, and Palladone XL®) and the OROS® Push-Pull preparation using an osmotic technology (Jurnista®, Exalgo®). Despite two different drug delivery vehicles, the pharmacodynamics and clinical outcome of the two extended-release preparations appear to be similar [10]. The extended-release formulations of hydromorphone (**2b**) or any other opioid were created with the intent of decreasing the rate of entry into the brain by producing a slower rise in plasma levels, thus significantly lowering the abuse potential of the drug [11]. Hydrocodone (**2a**) also has analgesic properties, but its main metabolite hydromorphone (**2b**) is thought to have the greatest analgesic effect when **2a** is taken [12]. Interestingly, the trade name for hydrocodone (**2a**) in the United States was originally not "Dicodid," but rather "Hycodan®" [13]. At present, Hycodan (**2a**) is marketed as an antitussive containing both hydrocodone (**2a**) and homatropine, supplied by Endo Pharmaceuticals [14]. Hydrocodone (**2a**) as a stand-alone opioid is unavailable in the United States but instead is found in combination with paracetamol/acetaminophen (Lortab®, Norco®, and Vicodin®), aspirin (Damason-P®, Lortab ASA®), and ibuprofen (Ibudone®, Reprexain®, and Vicoprofen®), to name a few, for pain management.

One of the earliest manipulations of the morphine core was the installation of a 14-hydroxyl group into thebaine (**5**), a biogenetic precursor to codeine (**4**) and morphine (**1**) and naturally occurring in the poppy plant as well (Figure 3.4). Oxycodone (**3a**) was introduced as "Eukodal®" in Germany, after German doctors tested the drug in animals and on themselves, and by 1917, was in clinical use mainly

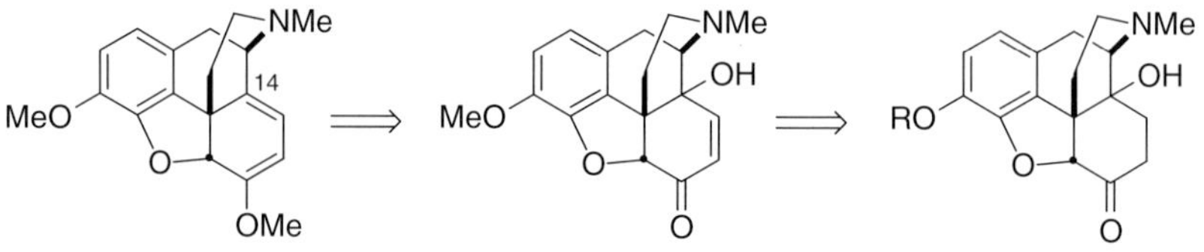

5
Thebaine

6

3a R = Me
3b R = H

Figure 3.4 Historic development of **3a** and **3b** from thebaine (**5**).

for acute pain [15]. At present, oxycodone (**3a**) has been shown to demonstrate efficacy in acute postoperative pain, cancer-related pain, and chronic noncancer pain and has been administered intravenously, intramuscularly, intranasally, subcutaneously, rectally, epidurally, and orally [16]. The controlled-release formulation of oxycodone hydrochloride is marketed by Purdue Pharmaceutical as OxyContin®, and immediate-release oxycodone (**3a**) is also available as OxyIR® by the same manufacturer; both are in tablet forms. Other brand names for oxycodone (**3a**) include Dazidox®, ETH-Oxydose®, Endocone®, Oxyfast®, Percolone®, and Roxicodone®. Similar to hydrocodone (**2a**), oxycodone (**3a**) is available in various combinations with other drugs, such as paracetamol/acetaminophen (Percocet®, Endocet®, Lynox®, Magnacet®, Perloxx®, Primalev®, Roxicet®, Taxadone®, Tylox®, and Xolox®), aspirin (Percodan®, Roxiprin®), and ibuprofen (Combunox®) [17]. Oxymorphone (**3b**), synthesized from oxycodone (**3a**) and patented in the United States by Endo Pharmaceutical in 1957 [18], has been available in parenteral formulation since 1959 [19]. Oxymorphone (**3b**) is marketed as an injection or a suppository (Nuomorphan®). At present, oxymorphone hydrochloride is sold by the same manufacturer as an immediate-release tablet (Opana IR®) or as a controlled-release tablet (Opana ER®) using an extended-release technology called *TIMERx*® developed by Penwest Pharmaceutical [20].

3.3 Synthesis

Morphine (**1**) and its many derivatives continue to be useful today for medicinal purposes, derived from the poppy or semisynthesized from an isolated natural compound. There is yet no practical synthetic route to morphine (**1**) that can compete with the natural supply. Since the intermediates in the biosynthesis of morphine (**1**) play a key role in the synthesis of the 6-morphinone scaffold, it is important to understand the biosynthesis of natural morphine (**1**), which has almost completely been elucidated.

Scheme 3.1 shows the beginning of the biosynthetic pathway toward morphine (**1**) [21]. The dimerization of L-tyrosine (**7**) accounts for all the nonmethyl carbon atoms incorporated into the core structure. The first crucial intermediate is (*S*)-reticuline (**8**), which is converted into its (*R*)-enantiomer **9** via an oxidation–reduction

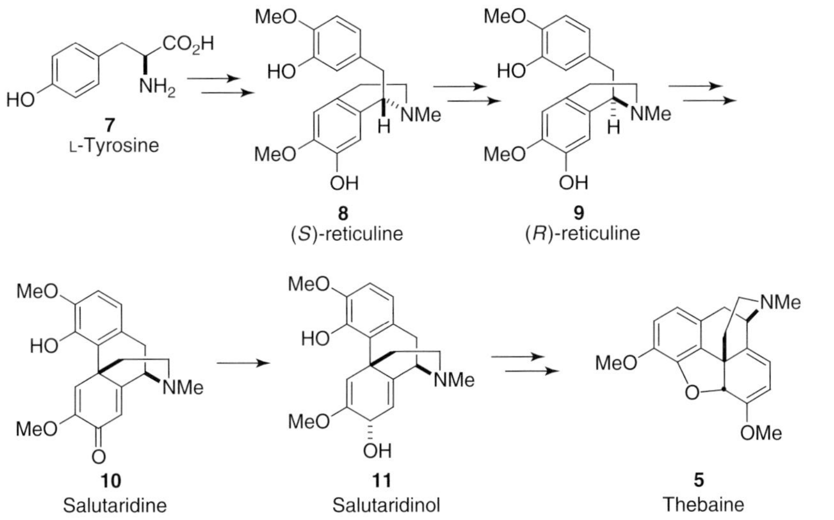

Scheme 3.1 Biosynthetic pathway to thebaine (**5**).

sequence with the NADPH-dependent enzyme 1,2-dehydroreticulinium reductase. Oxidative phenolic coupling occurs next, catalyzed by salutaridine synthase, which functions as an oxidase rather than a mono-oxidase, to establish the characteristic stereochemistry of the morphine skeleton. Stereospecific reduction of salutaridine (**10**) yields salutaridinol (**11**), which is acetylated, resulting in an allylic acetate that undergoes a spontaneous nonenzymatic *syn* S_N2' displacement by the phenolic hydroxyl group to create the final ring of the morphine skeleton, affording thebaine (**5**).

Having established the core scaffold of morphine (**1**), the biotransformation of thebaine (**5**) to morphine (**1**) is described in Scheme 3.2. Thebaine (**5**) is demethylated to neopinone (**12**), which readily isomerizes to codeinone (**13**), and reduction of **13** affords codeine (**4**). Final demethylation of **4** yields morphine (**1**), which itself is subject to further metabolic transformation [22].

Interestingly, Gates' first total synthesis of morphine (**1**) reported in 1955 was not biomimetic [23] but did follow Rapoport's initial reports in the final demethylation from codeine (**4**) to morphine (**1**) [24]. Rapoport was one of the many scientists chemically manipulating natural derivatives of morphine (**1**) in the laboratory well before the first total synthesis, the full understanding of the biosynthesis of the drug, and the correct structure was known. As a result, many parent morphone (**2**) compounds were obtained via semisynthetic routes that are still used today.

German patents from 1935 showed that morphine (**1**) and codeine (**4**) could easily be transformed into hydromorphone (**2b**) and hydrocodone (**2a**), respectively, with catalytic or excess palladium or platinum, with or without the presence of hydrogen (Scheme 3.3) [25]. Rapoport and coworkers sought to improve these methods in 1950, since they were unable to obtain yields higher than 50% using this method. After optimization, they were able to reduce codeine (**4**) and morphine (**1**) in

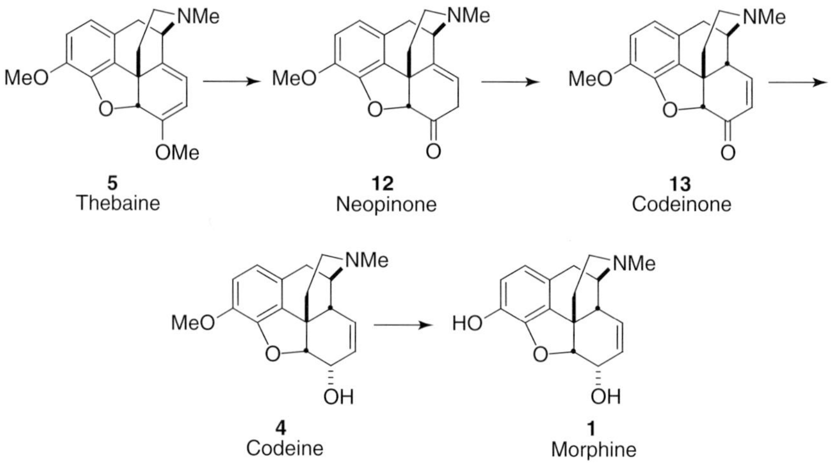

5 Thebaine; 12 Neopinone; 13 Codeinone; 4 Codeine; 1 Morphine

Scheme 3.2 Biosynthetic pathway to codeine (**4**) and morphine (**1**).

dilute acetic acid using a palladium on barium sulfate catalyst and hydrogen, followed by an Oppenauer oxidation using potassium *tert*-butoxide as a catalyst and benzophenone as an oxidant in benzene, improving their yield of **2a** and **2b** into the range of 70–80% [26].

The presence of the 3-methoxy group in the biosynthetic intermediate **5** made this alkaloid an attractive starting material for the synthesis of codeine (**4**) and other codeine-related compounds. Thebaine (**5**) is also naturally produced and isolable in poppy extracts, albeit in smaller quantities than codeine (**4**) or morphine (**1**). In fact, oxycodone (**3a**) was derived from thebaine (**5**) and was in clinical use in Germany by 1917 (Scheme 3.4) [15]. To obtain hydroxycodeinone (**6**), a solution of **5** in glacial acetic acid solution can be treated with hydrogen peroxide or peracids involving a postulated 1,4-addition reaction [27]. Later studies suggested an alternative mechanism involving epoxidation of the 8,14-double bond in **5**, which will then undergo an acid-catalyzed ring opening that can eventually lead to a diol [28] or directly to **6** [29]. Hydroxycodeinone (**6**) can be readily reduced in the presence of palladium–barium sulfate to give oxycodone (**3a**) [30], similar to the method used with the hydrocodone/hydromorphone series. In 1955, Weiss [31] reported that treatment of oxycodone (**3a**) in hot hydrobromic acid afforded oxymorphone (**3b**) in a 35% yield.

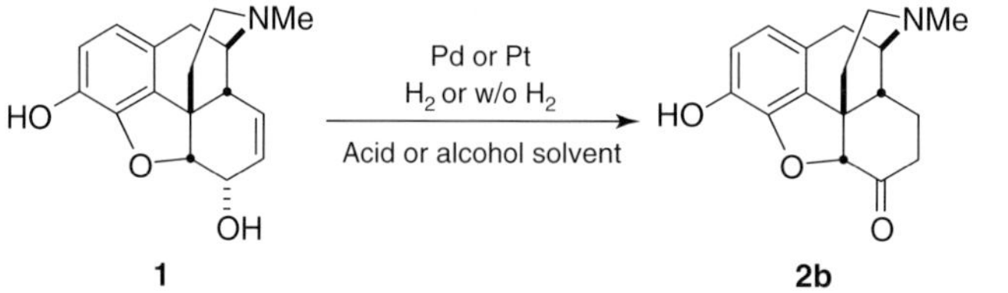

1; 2b

Scheme 3.3 Preparation of hydromorphone (**2b**) from morphine (**1**).

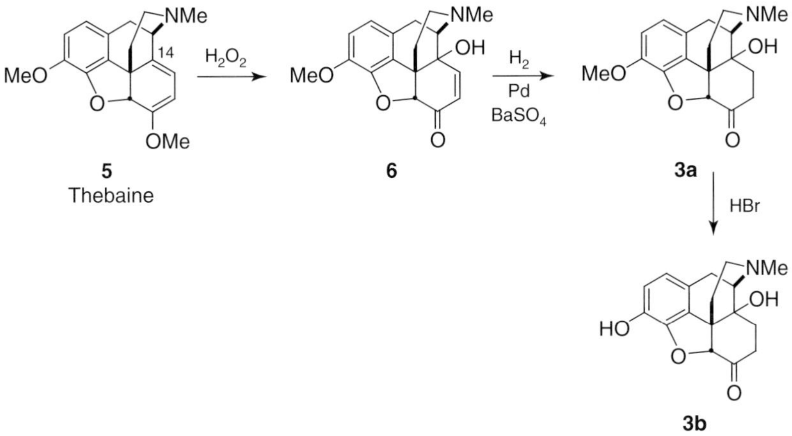

Scheme 3.4 Installing the 14-OH group into thebaine (**5**).

There are many reported total syntheses of morphine (**1**) [1, 21]. The most practical method of producing commercial quantities of morphine (**1**) and thus morphone-based analgesics, however, is still a semisynthetic route starting from naturally occurring compounds isolated from the poppy plant. Given the relative ease of poppy cultivation, production of opiates has had a worldwide impact as the most severe problem drugs. Perhaps, a commercially viable total synthesis of morphine (**1**) and its intermediates would help to alleviate some of the global, social, and economic implications that stem from illicit drug production [32].

3.4 Mode of Action

Opioid receptors are part of the larger family of G-protein-coupled receptors (GPCRs) and the subfamily of rhodopsin receptors. There are three major types of opioid receptors: μ, κ, and δ. These receptors share ~65% sequence homology, and the regions with the greatest homologies are the transmembrane domains and the intracellular loops. The N-terminus, extracellular loops and the C-terminus are the regions of least homology [3]. All three types of opioid receptors are located in the spinal cord tissues or human brain, and distribution can be categorized by gross region (i.e., supraspinal, spinal, etc.) and by function (i.e., nociception, respiration, etc.) [33].

All the opioid alkaloids and most of their synthetic derivatives are highly selective agonists for the μ-receptors. It is this selective interaction that gives these agonists similar analgesic effects and potencies, since injections of morphine (**1**) into μ-receptor knockout mice have demonstrated that all the major pharmacological responses such as analgesia, respiratory depression, tolerance, withdrawal symptoms, decreased gastric motility, and emesis occur by interactions with

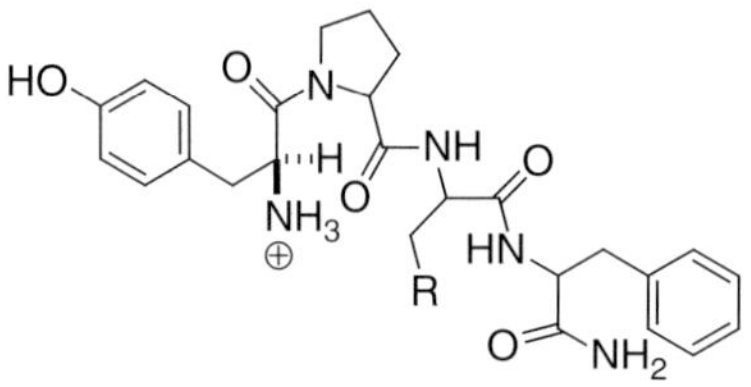

14 Endomorphin 1 (R = indolyl)
15 Endomorphin 2 (R = phenyl)

Figure 3.5 Endomorphins: endogenous opioid peptides that are highly selective for μ-receptors.

μ-receptors. Endogenous opioid peptides that are highly selective for μ-receptors include endomorphin 1 (**14**) and endomorphin 2 (**15**) (Figure 3.5) [34].

The signal transduction mechanism for all opioid receptors is through $G_{i/o}$ second messenger proteins. Agonist activity of opioid receptors activates $G_{i/o}$ proteins and initiates intracellular events that inhibit nerve firing. Such intracellular events include inhibition of adenylate cyclase activity, resulting in a decrease of cAMP levels. Combined with the efflux of K^+ ions and the closure of Ca^{2+} channels, these events cause hyperpolarization of postsynaptic neuronal membranes, inhibiting neuronal activity by increasing the membrane threshold. Initiation and propagation of action potentials are thus prevented [3, 34]. Repeated activation of μ-receptors can lead to a decrease in cAMP concentrations, which can be compensated in an upregulation of adenylate cyclase. This activity combined with receptor phosphorylation, G-protein uncoupling, high intracellular calcium concentration, and receptor internalization may all cause a decrease in the opioid response to a given dose of agonist [3]. The development of tolerance also leads to an imbalance of cAMP levels, and cessation of dosing the opioid increases cellular cAMP, ultimately leading to the unpleasant effects that are recognized as withdrawal symptoms (such as agitation, anxiety, sweating, insomnia, diarrhea, and nausea), lasting as long as required for cAMP to return to normal levels. The cellular adaptation may explain commonly seen side effects of frequent opioid use, such as tolerance to, withdrawal from, and even abuse and addiction to opioids. Thus, although μ-agonist activity is reliable in predicting analgesic potency, the correlation is just as good for predicting dependence liability [35]. The effects of morphinone opioids also depend highly on pharmacokinetic parameters, especially metabolism and absorption, and vary across species and from patient to patient. In general, the more lipophilic opioids have a faster rate of absorption and entry into the CNS [33].

3.5
Structure–Activity Relationship

The first systematic studies of SARs among derivatives of opium alkaloids began in 1929 under the direction of the Committee on Drug Addiction of the National

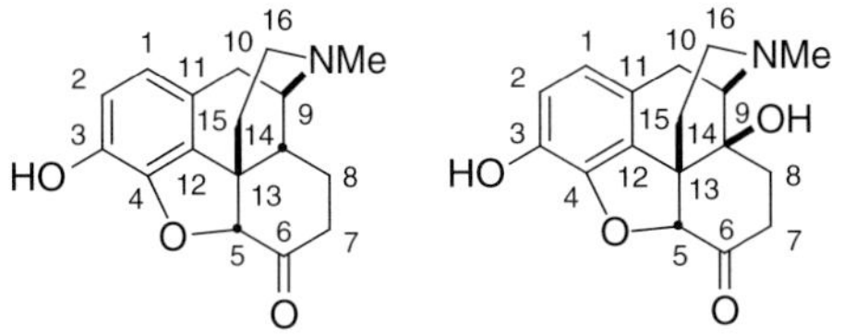

2
Parent hydromorphone structure

3
Parent oxymorphone structure

Figure 3.6 Numbering structure of parent motif.

Research Council. The committee was later named the Committee on Problems of Drug Dependence and became an independent organization in 1976 [36]. At present, it is a membership organization known as the College on Problems of Drug Dependence and continues to address problems of drug dependence and abuse [37]. The vast majority of SAR for manipulations of the 6-OH group on morphine was included in studies published from 1938 in an attempt to find safe, nonaddicting opioids and has since been extensively reviewed [38]. This section therefore does not list any specific enzyme potencies but rather briefly focuses on a few select SAR studies that were developed for activity relative to that of morphine and morphinone drugs (Figure 3.6). In addition to examining receptor binding, a wide variety of behavioral tests are used to evaluate potential efficacy of an alkaloid as an analgesic, including the writhing test, tail flick assay, hot plate test, flinch-jump test, and skin twitch response [39].

Introduction of substituents into nonfunctionalized portions of the morphinone skeleton has, in general, resulted in only quantitative differences in analgesic activity when compared to the parent morphinone (**2**) or morphine (**1**) itself. It is important to note here that replacement of the *N*-methyl group in the morphone structure brings about more qualitative changes, now producing compounds that are potent antagonists and mixed agonists–antagonists. Because these are no longer clean selective agonists, these structural changes are not covered here but are discussed elsewhere [4]. However, removal of the *N*-methyl group from oxymorphone (**3b**) afforded noroxymorphone (**3c**), a μ-receptor agonist that also demonstrated analgesic potency (Scheme 3.5). Both **3c** and **3d** are mainly formed via metabolism of **3a** and **3b** by specific CYP isoenzymes. Noroxymorphone (**3c**) was shown to have relatively high affinity for the μ-opioid receptor (three times that of oxycodone (**3a**) and 1.8 times lower compared to morphine (**1**)). Noroxymorphone (**3c**) was studied *in vivo* in rats and showed promise as an opioid for spinal analgesia [40].

In general, the introduction of the 14β-hydroxy group into the parent morphin-6-one (**2**) tends to maintain analgesic potency, if not enhance it, relative to morphine (**1**). For example, the 14β-hydroxy version of morphine (**1**) is equipotent in mice [42]. Oxycodone (**3a**) has >20 times less affinity to the μ-opioid receptor than morphine (**1**) but is equally potent or somewhat more potent than morphine (**1**) after systemic administration when factors such as bioavailability, metabolism,

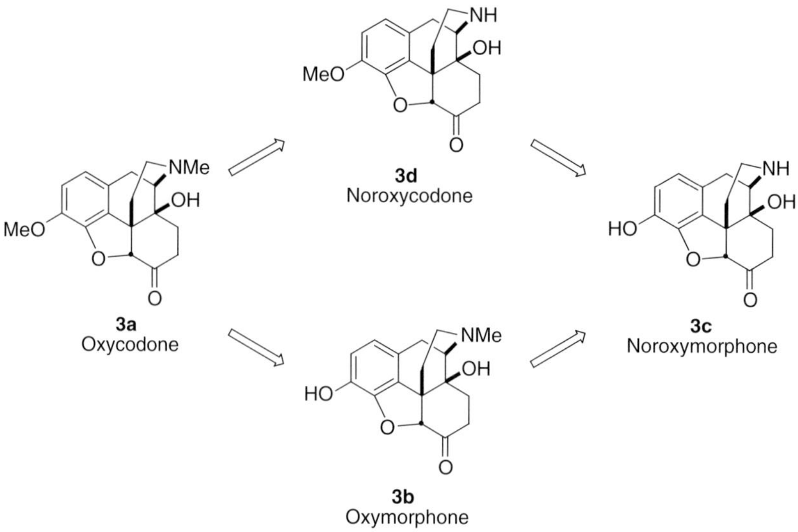

Scheme 3.5 Metabolic formation of noroxymorphone (**3c**). (Source: Modified from Lemberg *et al.* [40].)

and passage through the blood–brain barrier are considered [43]. Oxymorphone (**3b**) has a much higher binding affinity and is about 10 times as potent as morphine (**1**) after intravenous administration [41]. It is also more lipid soluble than oxycodone (**2a**) and morphine (**1**), crossing the blood–brain barrier at a faster rate. In an attempt to rationalize the enhancement of potency from the presence of the 14β-OH group, Zalucky and Hite [44] postulated that the strong hydrogen bond between the hydroxyl group and the amino nitrogen was responsible for enhancing metabolism to provide potency. However, esterification of the hydroxyl group will also produce a more potent opioid than morphine (**1**) and oxycodone (**2a**), but none are available commercially [19]. Hydroxycodeinone (**6**), an intermediate in the commercial production of **3a**, is known to demonstrate analgesic potency, but the α,β-unsaturated ketone motif designates it as a potential gene toxin due to its susceptibility for Michael additions with biological nucleophiles [45]. Studies were carried out on installing other substituents in the 14-position for SAR purposes (Figure 3.7). The nitro (**16**), chloro (**17**), and bromo derivatives (**18**) were synthesized but showed only weak potencies. However, it should be noted that codeine (**4**) was extremely weak in the reported assay studies [46]. Schwab published studies involving 1,4-cycloaddition of thebaine (**5**) with nitrosobenzene, which afforded analogs **19**–**21** with a phenylamino group in the 14β-position. Unfortunately, these compounds were less active than morphine (**1**) in mice [47].

Much like the oxycodone/oxymorphone series, hydrocodone (**2a**) displays a weaker affinity for the μ-receptor compared to morphine (**1**). In contrast, hydromorphone (**2b**) displays stronger binding than **1** [12]. Despite the different binding affinities, both compounds exhibit adequate pain relief in humans. Usage of these active compounds over morphine (**1**) highly depends on the administration routes

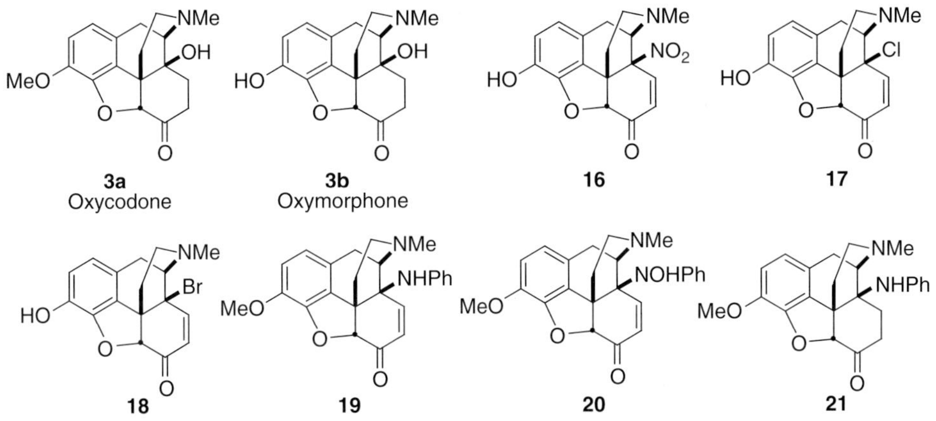

Figure 3.7 Examples of substituents in the 14-position.

and varies from patient to patient. Saturation of the 7,8-position in the morphinone systems therefore seems to retain analgesic potency [38]. In an attempt to install a double bond isostere and restrict the conformation of the parent core, Kotic synthesized the cyclopropane analog **2c** (Figure 3.8) of codeinone (**13**) and observed a 4-fold potency improvement over hydrocodone (**2a**) in a peripheral mouse writhing test and a 12-fold improvement in the central rat tail flick assay. Interestingly, the epoxide (**2d**) was deemed inactive [48]. Insertion of substituents into nonfunctionalized areas of the morphone core also demonstrated retention of analgesic activity. For example, the presence of a methyl group at the 5-position of the furan ring junction in hydromorphone (**2b**) gave metopon (**2e**), a drug that was three times more potent than morphine (**1**) when administered parenterally and

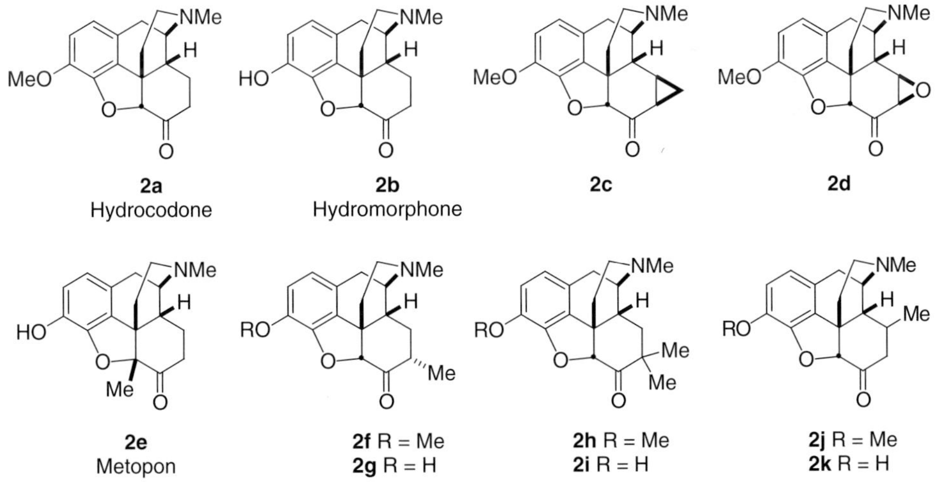

Figure 3.8 Examples of hydromorphone SAR.

less addicting [49]. Although further analogs were created for SAR purposes, the challenging synthesis [50] limited the use of metopon (**2e**) as a commercial drug. Introduction of a methyl group into the 7α-position of metopon (**2e**), **2f**, maintained some analgesic potency, but the compound was less active than hydrocodone (**2a**), hydromorphone (**2b**), and even oxycodone (**3a**). In contrast, the corresponding 3-hydroxy analog **2g** showed an improved analgesic effect. Increasing the size of the 7α-methyl group to a longer alkyl chain caused a drop in potency [51]. Incorporation of a 7-gem-dimethyl group in compounds **2h** and **2i** demonstrated analgesic activity but was also greatly affected by substitutions elsewhere in the molecule [52]. An 8α- or 8β-methyl or ethyl group such as in **2j** and **2k** seemed to afford relatively similar activity compared to hydrocodone (**2a**) and hydromorphone (**2b**), respectively, but larger side chains resulted in a drop in potency [53]. Decoration of the aromatic A ring has also been explored, but these analogs have not been extensively tested for opiate activity [54].

Despite the many years of extensive studies, a morphinelike analgesic with superior activity compared to morphine but without the undesirable side effects has yet to be discovered.

References

1. Blakemore, P.R. and White, J.D. (2002) *Chem. Commun.*, 1159–1168.
2. Gates, M. and Tschudi, G. (1956) *J. Am. Chem. Soc.*, **78**, 1380–1393.
3. Brenner, G.J., Mao, J., and Rosow, C. (2003) The opioid receptors, in *Contemporary Clinical Neurosciences: Neural Mechanisms of Anesthesia* (eds J.F. Antognini and E. Carstens), Humana Press, Totowa, pp. 413–426.
4. For select references on *N*-substituted narcotic antagonists: (a) Smits, S.E. and Takemori, A.E. (1970) *Br. J. Pharmacol.*, **39**, 627–638; (b) Fishman, J., Cotter, M.L., and Norton, B.I. (1973) *J. Med. Chem.*, **16**, 556–557; (c) Lewis, J.W., Bentley, K.W., and Cowan, A. (1971) *Ann. Rev. Pharmacol.*, **11**, 241–270.
5. Seppala, M.D. and Rose, M.E. (2010) *Prescription Painkillers: History, Pharmacology, and Treatment*, Hazelden, Center City, pp. 17–28.
6. Howard-Jones, N. (1971) *Sci. Am.*, **224**, 96–102.
7. Mannich, V.C. and Löwenheim, H. (1920) *Arch. Pharm.*, **258**, 295–316.
8. Eddy, N.B. and Reid, J.G. (1934) *J. Pharmacol. Exp. Ther.*, **52**, 468.
9. (a) Nathanson, I.T. and Daland, E.M. (1935) *N. Engl. J. Med.*, **213**, 741–746; (b) David, N.A. (1934) *J. Am. Med. Assoc.*, **103**, 474–478.
10. Murray, A. and Hagen, N.A. (2005) *J. Pain Symptom Manage.*, **29**, S57–S66.
11. Schuster, C.R. (2006) *Drug Alcohol Depend.*, **83S**, S8–S14.
12. Prommer, E. (2010) *J. Opioid Man.*, **6** (4), 295–299.
13. Drugs @FDA Home Page. *http://www.accessdata.fda.gov/scripts/cder/drugsatfda/index.cfm* (accessed 3 August 2011).
14. Drugs.com Web site. *http://www.drugs.com/pro/hycodan.html* (accessed 3 August 2011).
15. Falk, E. (1917) *Münch. Med. Wochenenschr.*, **20**, 381–384.
16. Kalso, E. (2005) *J. Pain Symptom Manage.*, **29**, S47–S56.
17. PubMed Health Web Site. *http://www.ncbi.nlm.nih.gov/pubmedhealth/PMH0000589* (accessed 3 August 2011).
18. Cox, D.P., Zhang, Y., Zhang, W.-C., and James, K.E. (2011) US Patent 2, 806,033, Mar. 15, 2011.

19. Sloan, P. (2008) *Ther. Clin. Risk Manag.*, 4, 777–787.
20. Zannou, E.A., Li, P., and Tong, W.-Q. (2009) in *Developing Solid Oral Dosage Forms* (eds Y. Qiu, Y. Chen, G.G.Z. Zhang, L. Liu, and W.R. Porter), Elsevier Inc, Oxford, pp. 911–921.
21. Novak, B.H., Hudlicky, T., Reed, J.W., Mulzer, J., and Trauner, D. (2000) *Curr. Org. Chem.*, **4**, 343–362.
22. For more on morphine biotransformation, see: Lenz, G.R., Evans, S.M., Walter, D.E., and Hopfinger, A.J. (1986) in *Opiates*, Academic Press Inc., Orlando, pp. 1–28.
23. Gates, M. and Tschudi, G. (1952) *J. Am. Chem. Soc.*, **76**, 1109–1110.
24. Rapoport, H., Lovell, C.H., and Tolbert, B.M. (1951) *J. Am. Chem. Soc.*, **73**, 5900.
25. (a) Knoll and Co. (1935) German Patent 617,238, Oct. 2, 1935; (b) Knoll and Co. (1935) German Patent 607,931, Jan. 11, 1935.
26. Rapoport, H., Naumann, R., Bissell, E.R., and Bonner, R.M. (1950) *J. Org. Chem.*, **15**, 1103–1107.
27. Hauser, F.M., Chen, T.-K., and Carroll, F.I. (1974) *J. Med. Chem.*, **17**, 1117.
28. Weiss, U. (1957) *J. Org. Chem.*, **77**, 1505–1508.
29. Ijima, I., Rice, K.C., and Brossi, A. (1977) *Helv. Chim. Acta*, **60**, 2135–2137.
30. Lutz, R.E. and Small, L. (1939) *J. Org. Chem.*, **4**, 220–233.
31. Weiss, U. (1955) *J. Am. Chem. Soc.*, **77**, 5891–5892.
32. For current drug trends and social implications of opiates, see: UNODC (2010) World Drug Report 2010, United Nations Publication, E.10.XI.13, United Nations, New York.
33. For more information on receptor subtypes, partial agonists/antagonists, and ADME profile, see: Gutstein, H.B. and Akil, H. (2006) in *Goodman & Gilman's The Pharmacological Basis of Therapeutics*, 11th edn (eds L.L. Brunton, J.S. Lazo, and K.L. Parker), McGraw-Hill, New York, pp. 547–590, and references therein, Also, see Refs. 3 and 34.
34. Fries, D. (2008) in *Foye's Principles of Medicinal Chemistry*, 6th edn (eds T.L. Lemke, D.A. William, V.F. Roche, and S.W. Zito), Williams and Wilkins, Baltimore, pp. 652–678.
35. Corbett, A.D., Henderson, G., McKnight, A.T., and Paterson, S.J. (2006) *Brit. J. Pharm.*, **147**, S152–S162.
36. May, E.L. and Jacobson, A.E. (1989) *Drug Alcohol Depend.*, **23**, 183–218.
37. The College on Problems of Drug Dependence Home Page. *http://www.cpdd.vcu.edu/#* (accessed 3 August 2011).
38. Lenz, G.R., Evans, S.M., Walter, D.E., and Hopfinger, A.J. (1986) *Opiates*, Academic Press Inc., Orlando, pp. 45–165, and references therein.
39. Hammond, D.L. (1986) *Opiates*, Academic Press Inc., Orlando, pp. 29–44, and references therein.
40. Lemberg, K.K., Siiskonen, A.O., Kontinen, V.K., Yli-Kauhaluoma, J.T., and Kalso, E.A. (2008) *Anesth. Analg.*, **106**, 463–470.
41. Mayyas, F., Fayers, P., Kaasa, S., and Dale, O. (2010) *J. Pain Symptom Manage.*, **39**, 296–308.
42. Weiss, U. and Daum, S.J. (1965) *J. Med. Chem.*, **8**, 123–125.
43. Kalso, E. (2007) *Pain*, **132**, 227–228.
44. Zalucky, T.B. and Hite, G. (1961) *J. Med. Chem.*, **3**, 615–616.
45. Cox, P.D., Zhang, Y., Zhang, W.-C., and James, K.E. (2011) US Patent 7, 906,647B2, Mar. 15, 2011.
46. Osei-Gyimah, P. and Archer, S. (1980) *J. Med. Chem.*, **23**, 162–166.
47. Schwab, L.S. (1980) *J. Med. Chem.*, **23**, 698–702.
48. Kotick, M.P. (1981) *J. Med. Chem.*, **24**, 722–726.
49. Sargent, L.J. and May, E.L. (1970) *J. Med. Chem.*, **13**, 1061–1063.
50. Stork, G. and Bauer, L. (1953) *J. Am. Chem. Soc.*, **75**, 4373–4374.
51. Kotick, M.P., Leland, D.L, Polazzi, J.O., Howes, J.F., and Bousquet, A.R. (1981) *J. Med. Chem.*, **24**, 1445–1450.
52. Leland, D.L. and Kotick, M.P. (1981) *J. Med. Chem.*, **24**, 717–721.
53. Kotick, M.P., Leland, D.L., Polazzi, J.O., and Schut, R.N. (1980) *J. Med. Chem.*, **23**, 166–174.
54. Wilson, M.L., Carroll, P.J., and Dalton, D.R. (2005) *J. Org. Chem.*, **70**, 6492–6495. Also, see Ref. 38.

4
Barbituric Acid-Based GABA(A) Receptor Modulators for the Treatment of Sleep Disorder and Epilepsy and as Anesthetics

Ingo Janser and Romy Janser

4.1
Introduction

Barbituric acid (**1**), IUPAC name pyrimidine-2,4,6(1*H*,3*H*,5*H*)-trione, is also known as *malonylurea* or *6-hydroxyuracil* (Figure 4.1). The structure of barbituric acid (**1**) is the basic structure of all barbiturates.

Barbituric acid (**1**) was first discovered by the German chemist Adolf von Baeyer in 1864 [1]. To celebrate the discovery, von Baeyer and his colleagues went to a tavern, where the town's artillery garrison was celebrating the day of Saint Barbara. Therefore, he chose the name "barbiturate" as a combination of St. Barbara and "urea" (one of the reagents he used for the synthesis of barbituric acid) [2]. This is only one of the hypotheses about the origin of the term "*barbiturates.*" A second one is that Baeyer used the name in honor of his friend Barbara [3]. A third reported possibility is that he was inspired by the "barbed" appearance of the crystals of the uric compounds [4]. Barbituric acid (**1**) is an odorless powder, soluble in water. It is the parent compound of several barbiturates (e.g., butobarbital (**2**), amobarbital (**3**), pentobarbital (**4**), quinalbarbital (**5**), phenobarbital (**6**), mephobarbital (**7**), and many others) (Figure 4.2). Barbiturates can be used for the treatment of anxiety (anxiolytic agent), epileptic seizure (anticonvulsant agent), and insomnia (hypnotic agent). The barbiturates act as central nervous system (CNS) depressants and thus possess a broad spectrum of effects, which range from mild sedation to total anesthesia. Barbiturates have a great addiction potential, physical as well as psychological. Especially for the treatment of anxiety and insomnia, barbiturates have been replaced over the past years by benzodiazepines, mostly because these drugs are less dangerous with respect to overdosage. However, barbiturates are still in use for the treatment of epilepsy and in general anesthesia.

First, this chapter focuses on the historical development, the discovery, and the use of various barbituric acid derivatives (barbiturates) especially with different substituents at position 5 of the pyrimidine skeleton. A general synthesis for barbiturates is introduced, followed by a closer look at their mode of action and the structure–activity relationship of various substitution patterns.

Bioactive Heterocyclic Compound Classes: Pharmaceuticals,
First Edition. Edited by Jürgen Dinges and Clemens Lamberth.

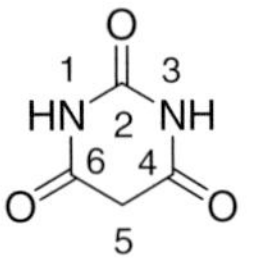

1
Barbituric acid

Figure 4.1 Structure and numbering convention for barbituric acid (**1**).

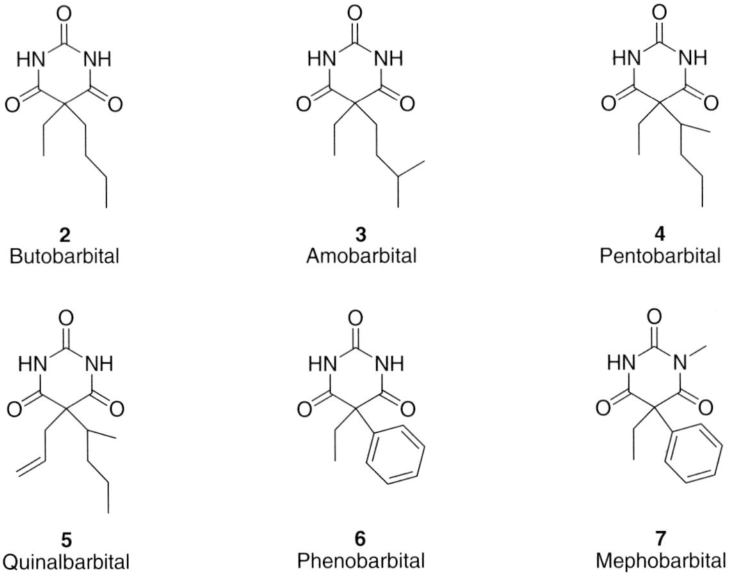

Figure 4.2 Some selected examples of barbiturates (derivatives of barbituric acid): butobarbital (**2**), amobarbital (**3**), pentobarbital (**4**), quinalbarbital (**5**), phenobarbital (**6**), and mephobarbital (**7**).

4.2
History

Barbituric acid (**1**) was first synthesized in 1864 by Adolf von Baeyer through a condensation reaction of urea (**8**) with malonic acid (**9**) (Scheme 4.1) [1].

Chemists of the late nineteenth and the early twentieth century had only a few tools available to determine the properties of chemical compounds. Very often, they used the one tool they all had available: the reaction of their body

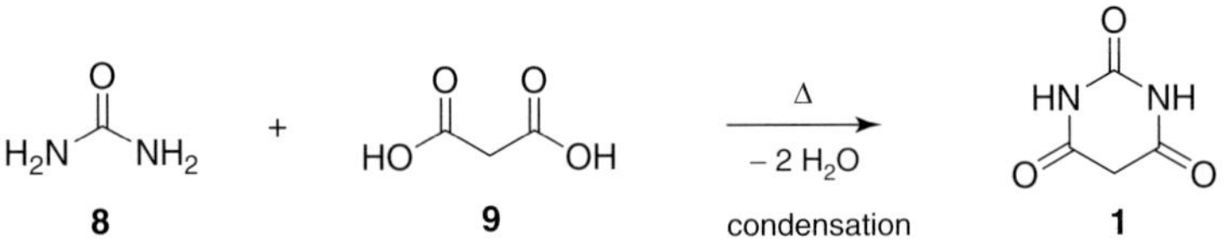

Scheme 4.1 First synthesis of barbituric acid (**1**) by the German chemist Adolf von Baeyer through a condensation reaction of urea (**8**) and malonic acid (**9**) [1].

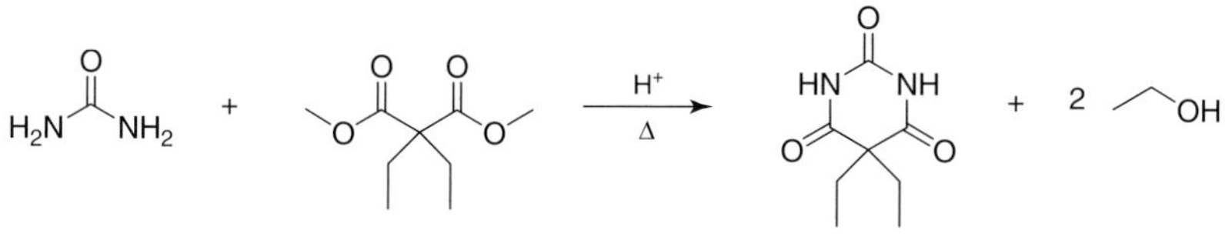

Scheme 4.2 Synthesis of barbital (11, Veronal®) from diethylmalonic acid diethyl ester (10) and urea (8) by Emil Fisher and Joseph von Mering [5].

to the substance synthesized. This way, it was determined that barbituric acid is pharmacologically inactive and therefore has no therapeutic significance. In 1903, almost 40 years later, 2 chemists, Emil Fisher and Joseph von Mering, synthesized a derivative of barbituric acid (**1**) with a similar structure and called it barbital (**11**) (5,5-diethylpyrimidine-2,4,6(1*H*,3*H*,5*H*)-trione) (Scheme 4.2) [5]. They discovered that this compound was very effective in putting dogs to sleep, and one year later, in 1904, it was marketed by the Bayer Company under the trade name Veronal® (**11**).

Veronal (**11**) was used as a sedative and anticonvulsant, and it was considered a great improvement over the existing hypnotics. Its taste was slightly bitter, but more importantly, it had only few side effects such as skin eruptions, which lasted from two days to three weeks. In addition, exfoliative dermatitis and pigmentation of the skin have been reported [6]. Although the therapeutic dose (0.65–0.97 g) was well below the lethal dose (3.5–4.4 g), prolonged usage resulted in tolerance to the drug, leading to a higher therapeutic dose necessary and therefore sometimes to a fatal overdose of the drug.

A soluble sodium salt of barbital (**11**) was marketed by the Schering company as Medinal® (sodium diethylbarbiturate).

After the discovery of the pharmacological activity of the barbiturates, chemists vigorously started to synthesize additional derivatives of barbituric acid (**1**), in order to discover novel drugs. Shortly after the synthesis of barbital (**11**), Emil Fisher synthesized phenobarbital (**6**) (5-ethyl-5-phenylpyrimidine-2,4,6(1*H*,3*H*,5*H*)-trione) (Figure 4.2), which was brought to the market by Bayer in 1912. It was marketed under the trade name Luminal® (**6**) and was used as a sedative and hypnotic agent. Both Veronal (**11**) and Luminal (**6**) were the first two representatives of the chemical class of barbiturates to be accepted by the international pharmacopoeia, such as the *United States Pharmacopoeia* (USP X) in 1926 and the *British Pharmacopoeia* in 1914 and 1932 [7]. The great advantage of these new drugs was their higher potency, longer duration of action, and wider therapeutic range in comparison with their classical predecessors such as alcohol or opiates [8]. In 1920, Roger Adams (Abbott Laboratories) synthesized the ester of 5-butyl-5-ethyl-malonic acid, an intermediate for the preparation of butobarbital (**2**) (5-butyl-5-ethyl-1,3-diazinane-2,4,6-trione) (Figure 4.2), which was synthesized in 1922 by Arthur Dox (Parke Davis and Company), and this was marketed one year later by Abbott Laboratories under the trade name Neonal® (**2**) [9]. This new drug was three times stronger than Veronal (**11**), and because of its higher lipophilicity the duration of pharmacological activity was much shorter, which prevented drowsiness the day after administration. During the following years, a number of new barbiturates continued to come on the market. In 1923, it was amobarbital (**3**) (5-ethyl-5-(3-methylbutyl)-1,3-diazinane-2,4,6-trione) (Figure 4.2), synthesized by Shonle and Moment at Eli Lilly and marketed by the same company under the trade name Amytal® (**3**). Butobarbital (**2**) and amobarbital (**3**) differ only by one additional methyl group on the *n*-butyl substituent in 5-position of the pyrimidine heterocycle. Six years later, Horace Abbott Shonle synthesized secobarbital (**5**) (5-[(2*R*)-pentan-2-yl]-5-prop-2-enyl-1,3-diazinane-2,4,6-trione), also known as quinalbarbital (**5**) (Figure 4.2). Both barbiturates, amobarbital (**3**) and secobarbital (**5**), possess anesthetic, anticonvulsant, sedative, and hypnotic properties [9]. Further important developments in the history of barbiturates were pentobarbital (**4**) (5-ethyl-5-(1-methylbutyl)-2,4,6(1H,3H,5H)-pyrimidinetrione) (Figures 4.2 and 4.3), synthesized by Volwiler and Tabern from Abbott Laboratories and marketed under the trade name Nembutal® (**4**), and thiopental (**12**) ((*RS*)-[5-ethyl-4,6-dioxo-5-(pentan-2-yl)-1,4,5,6-tetrahydropyrimidin-2-yl]sulfanide sodium) (Figure 4.3), better known as Pentothal® (**12**) (a trademark of Abbott Laboratories) [10, 11]. Thiopental, a sulfur derivative of pentobarbital (**4**) revolutionized intravenous anesthesia and is the only representative of the thiobarbiturate family to be officially recognized, being accepted first by the *British Pharmacopoeia* (1942, 7th Add) and subsequently by the *United States Pharmacopoeia* (1947, USP XIII) and the *Pharmacopoeia Internationalis* (1951, Volume 1). It is a rapid-onset short-acting anesthetic and is a core medicine in the World Health Organization's "Essential Drugs List," which is a list of minimum medical needs for a basic health care system [12].

Pentobarbital (**4**) is also a short-acting barbiturate and is approved for the treatment of seizures and preoperative (and other) sedations, and as a short-term hypnotic agent. In December 2010, it was also approved for use in executions in the United States in the states of Arizona, Oklahoma, and Texas [13]. In 2011, Lundbeck, Inc., the manufacturer of pentobarbital (**4**), opposed the use of its product for executions and announced that it would discontinue sales to US prisons that carried out the death penalty [14].

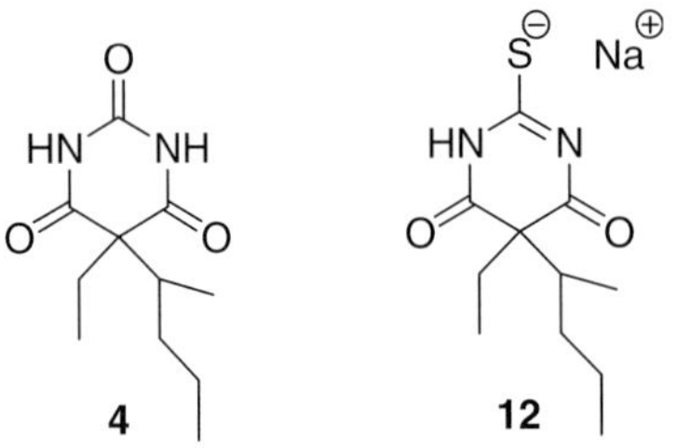

Figure 4.3 Structural similarity between pentobarbital (**4**) and thiopental (**12**).

4.2.1
Barbiturates in the Treatment of Sleep Disorders

Thanks to their induction of a state of deep and prolonged sleep, some barbiturates were rapidly applied to the treatment of psychotic patients. In 1915, the Italian psychiatrist Giuseppe Epifanio, working at the University Psychiatric Clinic in Torino, did some pioneering work in this area [15]. Unfortunately, he published his work solely in Italian journals, therefore, his work did not have a lot of impact, and the clinical introduction of his techniques is historically associated with Jakob Klaesi, a psychiatrist at the University Psychiatric Clinic in Zurich, Switzerland. In 1920, Klaesi proposed that his techniques for inducing deep hypnosis, taken from Epifanio, would facilitate communication between patient and psychotherapist in order "to achieve a better relationship between doctor and patient" [8]. The intravenous or subcutaneous administration of Somnifen®, a mixture of diethyl and dipropenyl barbituric acid and diethylamine (manufactured by the Hoffmann-LaRoche company), led to an improved cure of schizophrenic patients by 25–33% [16]. This treatment ("prolonged sleep therapy") acquired a great popularity during the 1920s [17]. However, it was noted that some patients died due to the drugs administered [16]. In 1927, it was reported that the mortality rate of Somnifen was around 5% [18]; therefore, in 1934, a new drug based on several active ingredients, among them isopropylallyl-barbituric acid (Cloettal®), was developed [19].

4.2.2
Barbiturates in the Treatment of Epilepsy

As mentioned above, phenobarbital (**6**) (Figure 4.2) was commercialized in 1912, shortly after the discovery of its anticonvulsant properties. This discovery is another example of serendipity in the field of psychopharmacology. It was already well known that phenobarbital (**6**) possesses sedative, soporific, and hypnotic properties, but nobody knew at the time that it was also an effective anticonvulsant. Alfred Hauptmann, a young doctor who gave it to his epilepsy patients as a tranquilizer, discovered that their epileptic attacks were susceptible to the drug. Hauptmann performed a careful study of his patients over an extended period of time. On phenobarbital (**6**), their epilepsy symptoms were greatly reduced: the worst patients suffered fewer and lighter seizures and some patients became seizure-free. Patients who had been institutionalized due to the severity of their epilepsy were able to leave and, in some cases, resume employment [20]. This way, the anticonvulsant properties of barbiturates were discovered, and phenobarbital (**6**) was the first truly effective drug for the treatment of epilepsy [21]. However, due to World War I, the international acceptance of phenobarbital (**6**) as an antiepileptic drug was seriously delayed [20]. For instance, it was not available in Great Britain until 1923, marketed by the Winthrop Chemical Company. Shortly after the introduction in England, Charles Brooks, Colony Medical Officer at the Chalfont Centre in London, reported the excellent efficacy of phenobarbital (**6**), particularly in very severe cases

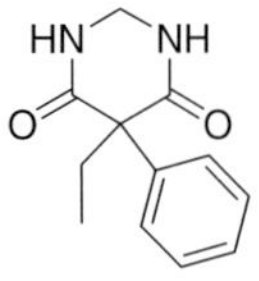

13

Figure 4.4 Structure of primidone (13).

of convulsions and in epileptic conditions associated with mental deficiency. He also noted that if the drug is not effective in the first months, it will be necessary to look for an alternative [22]. Right after the discovery of the antiepileptic properties of phenobarbital (**6**), many other barbiturates were synthesized and examined for their antiepileptic properties. It took 20 years until mephobarbital (**7**) (5-phenyl-5-ethyl-3-methylbarbituric acid) (Figure 4.2) was recognized as such a drug and was marketed under the trade name Prominal® (**7**) [23]. In 1952, the discovery of primidone (**13**) (5-ethyl-5-phenyl-hexahydropyrimidine-4,6-dione) (Figure 4.4) awoke great therapeutic interest, since this new drug showed high anticonvulsant effectiveness, greater than all other available barbiturates, and without sedative side effects [24]. This interest soon faded after it was shown that phenobarbital (**6**) was a metabolite of this drug [25]. Comparative clinical studies carried out with phenobarbital (**6**) and its prodrug primidone (**13**) showed no apparent differences between the two [26].

4.2.3
Barbiturates as Anesthetics

The first barbiturate to be used systematically in anesthesia was sodium *sec*-butyl-(2-bromo-allyl)-barbiturate (**14**, Pernocton®) (Figure 4.5), introduced in 1927 by the German obstetrician Bumm [27].

Subsequently, as new barbiturates were synthesized for their oral administration as sedatives, sodium salts of the same drugs were formulated, which could be administered intravenously and used as anesthetics [28]. Notable among the pioneers in this field is John Lundy of the Mayo Clinic (Rochester, USA). He introduced sodium amobarbital (1929) and sodium pentobarbital (1930) in anesthesia. In the early 1930s, the two chemists Kropp and Taub at Bayer (Leverkusen, Germany) synthesized hexobarbital (**15**) ((*RS*)-5-(cyclohex-1-en-1-yl)-

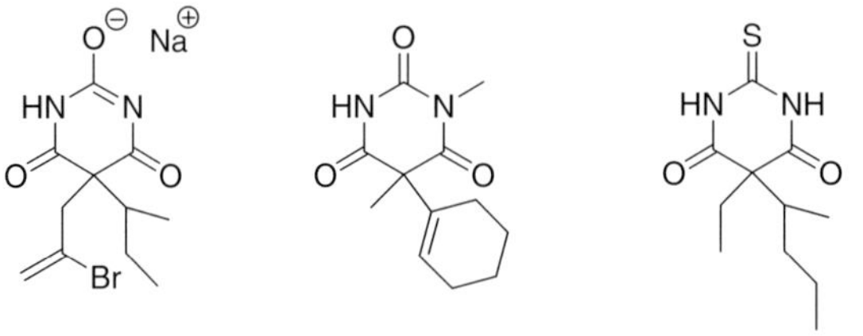

14 15 16

Figure 4.5 Structure of Pernocton® (**14**), hexobarbital (**15**), and thiopentobarbital (**16**).

1,5-dimethyl-1,3-diazinane-2,4,6-trione) (Figure 4.5), whose sodium salt (Evipal®) was introduced into clinical anesthesia in 1932 [29]. Sodium hexobarbital has a relatively fast onset of effects and a short duration of action. However, it can be difficult to control the depth of anesthesia with hexobarbital (**15**), which makes it quite dangerous.

Further developments led to the discovery of thiobarbiturates: compounds in which a sulfur atom replaces the carbonyl oxygen at position 2 of the pyrimidine skeleton. One very important thiobarbiturate is thiopentobarbital (**16**) (5-*sec*-pentyl-5-ethyl-2-thioxodihydropyrimidine-4,6(1*H*,5*H*)-dione) (Figure 4.5) derived from pentobarbital (**4**) (Figure 4.2). These agents revolutionized intravenous anesthesia. Most of the work was done by Volwiler and Tabern of Abbott Laboratories [11]. The properties of these agents were studied at the Mayo Foundation in Rochester by John Lundy's group, who gave the sulfur derivative of pentobarbital (**4**) the name Thionembutal® (**16**). The sodium salt of Thionembutal was marketed as Pentothal (**12**) [30]. This agent rapidly replaced the other barbiturates as an anesthetic, partly due to the swiftness of its onset and its short action period. Pentothal (**12**) currently remains the preferred intravenous anesthetic in many types of surgical intervention.

4.3 Synthesis

Adolf von Baeyer synthesized barbituric acid (**1**) for the first time via a condensation reaction of urea (**8**) and malonic acid (**9**), as illustrated in Scheme 4.1. Barbituric acid (**1**) is formed by nucleophilic attack of the urea nitrogens on the carboxyl carbons of malonic acid (**9**). However, due to the acidity of malonic acid (**9**) the carboxyl groups can easily be deprotonated and the carboxy anions no longer react with the nucleophilic nitrogens, resulting in poor yields. Therefore, barbituric acid (**1**) and its derivatives (**A**) are nowadays mainly prepared by the following two methods:

1) Reaction of urea (**8**) and malonyl dichloride (**17**) in the presence of phosphorus oxychloride ($POCl_3$) at 120 °C (Scheme 4.3).
2) Reaction of urea (**8**) and diethyl malonate (**18**) (Scheme 4.4).

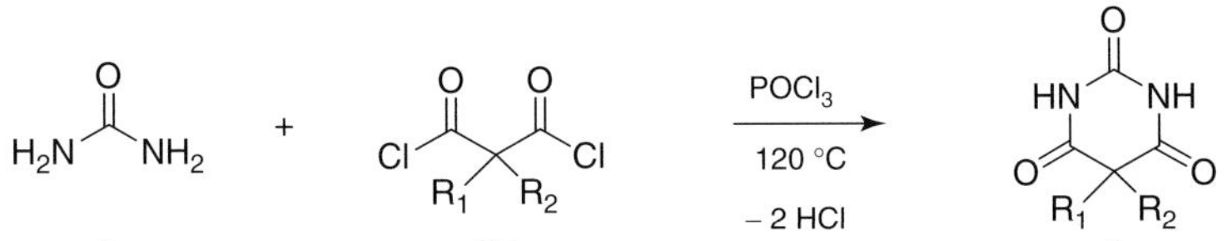

8 **17** **A**
1 $R_1 = R_2 = H$

Scheme 4.3 Reaction of urea (**8**) and malonyl dichloride (**17**) to yield barbituric acid (**1**) or its derivatives (**A**).

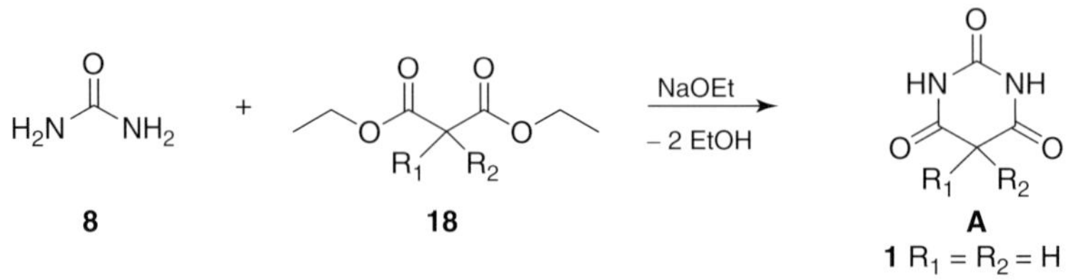

Scheme 4.4 Reaction of urea (**8**) and diethyl malonate (**18**) to yield barbituric acid (**1**) or its derivatives (**A**).

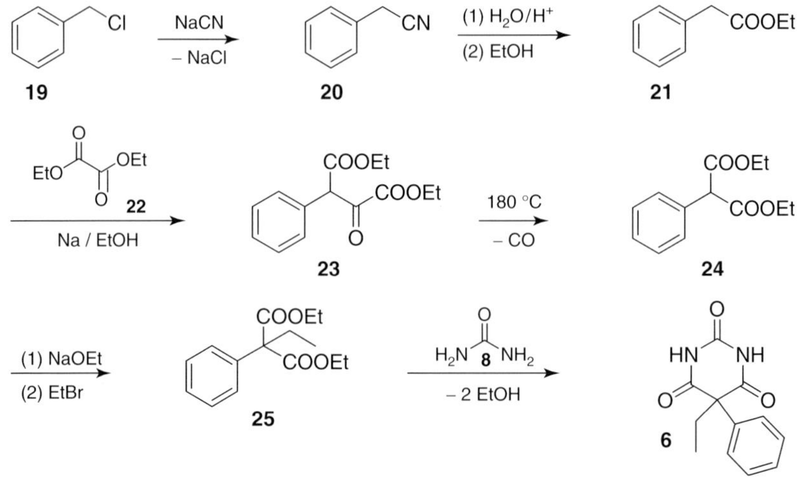

Scheme 4.5 Synthesis of phenobarbital (**6**) by method 1.

In the following discussion, additional synthetic routes to obtain specific barbiturates are described.

Phenobarbital (6) (method 1) (Scheme 4.5) is prepared by reacting benzyl chloride (**19**) with sodium cyanide (NaCN), forming 2-phenylacetonitrile (**20**). After acid-catalyzed hydrolysis, the corresponding carboxylic acid is treated with ethanol to form ethyl phenyl acetate (**21**). Reacting this ester with diethyl oxalate (**22**) in the presence of sodium ethoxide provides the diethyl phenyl oxaloacetate (**23**), which can be converted into phenyl malonic ester (**24**) through decarboxylation at 180 °C. Alkylation of **24** with ethyl bromide generates ethyl phenyl malonic ester (**25**). The last step of the synthesis is the condensation of **25** with urea (**8**) to afford the desired phenobarbital (**6**) [31].

Phenobarbital (6) (method 2) (Scheme 4.6) is prepared via a condensation reaction of 2-phenylacetonitrile (**20**) with diethylcarbonate (**26**) in the presence of sodium ethoxide to give α-phenylcyanoacetic ester (**27**). The following alkylation of the ester using ethyl bromide yields α-phenyl-α-ethylcyanoacetic ester (**28**), which is reacted with urea (**8**) to produce the 4-imino-derivative **29**. Acidic hydrolysis of **29** finally leads to the formation of phenobarbital (**6**) [32].

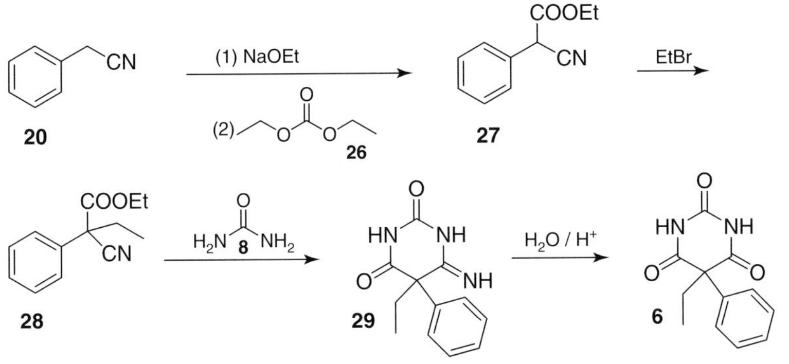

Scheme 4.6 Synthesis of phenobarbital (**6**) by method 2.

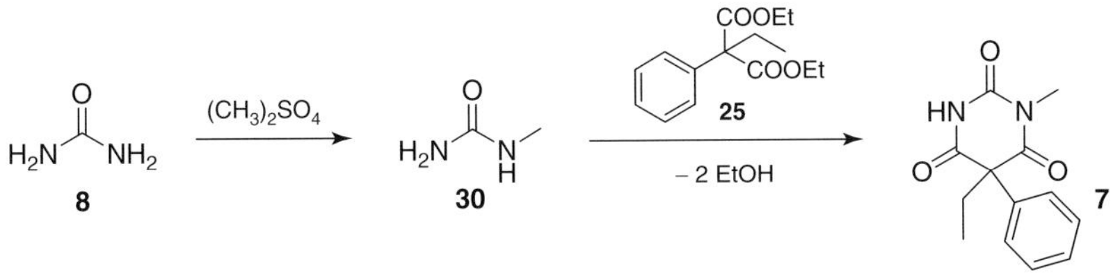

Scheme 4.7 Synthesis of methylphenobarbital (**7**).

Methylphenobarbital (7) (Scheme 4.7) can be generated by condensation of *N*-methyl urea (**30**), obtained through the treatment of urea (**8**) with dimethyl sulfate, and ethylphenyl malonic ester (**25**) (Scheme 4.5).

Amobarbital (3) (Scheme 4.8) is synthesized by reacting malonic ester (**19**) with 1 equivalent of ethyl bromide in the presence of sodium ethoxide. The resulting α-ethylmalonic ester (**31**) is alkylated a second time with 1-bromo-3-methylbutane to produce α-ethyl-α-isoamylmalonic ester (**32**). The desired amobarbital (**3**) is then obtained in a condensation reaction of the ester **32** with urea (**8**) [33].

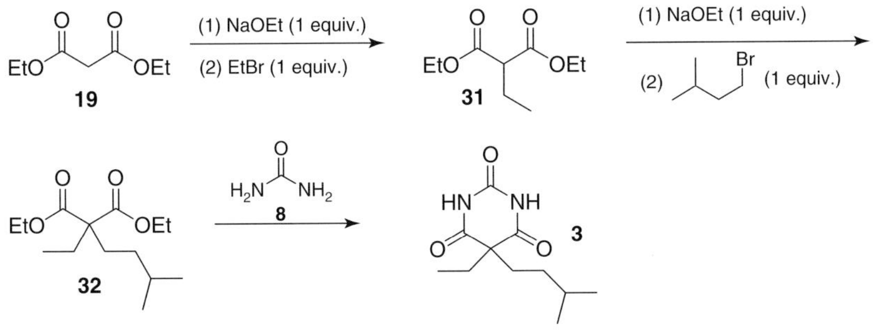

Scheme 4.8 Synthesis of amobarbital (**3**).

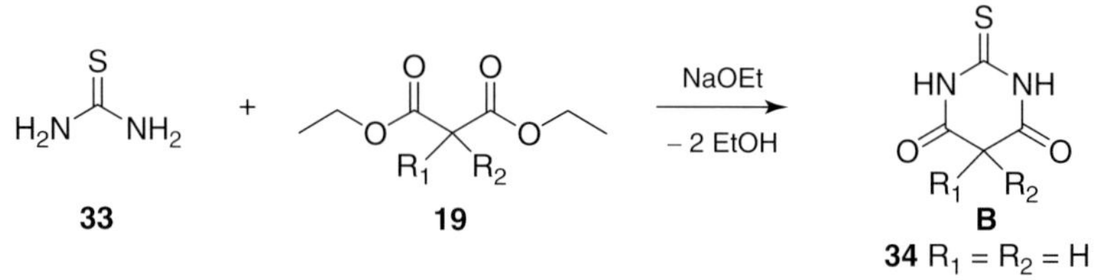

Scheme 4.9 Synthesis of thiobarbituric acid (**34**) and its derivatives (**B**) starting from thiourea (**33**).

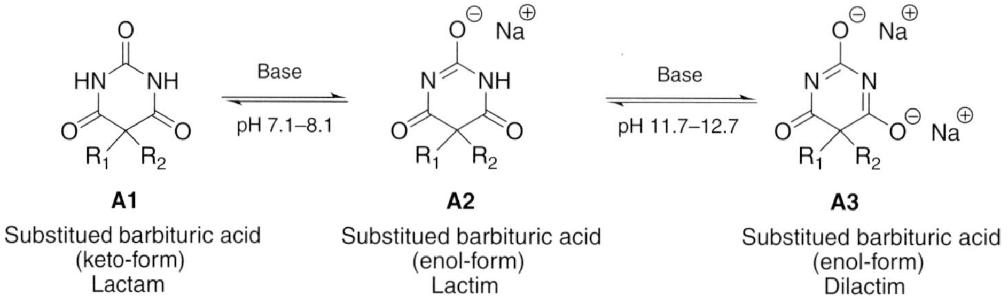

Scheme 4.10 pH-dependant lactam–lactim tautomerization of barbiturates.

Thiobarbituric acid (**34**) and its analogs (**B**) are derivatives of barbiturates in which a sulfur atom replaces the carbonyl oxygen in position 2. The synthesis of thiobarbituric acid (**34**) and its derivatives (**B**) starts with thiourea (**33**), instead of urea (**8**), which is condensed with diethyl malonate **19** (Scheme 4.9).

Barbituric acid (**1**) and substituted barbituric acids contain three enolizable carbonyl groups that can undergo pH-dependant lactam–lactim tautomerization as shown in Scheme 4.10 [33].

The acidity of barbiturates in aqueous solution depends on the number of substituents attached to barbituric acid (**1**). The 5,5-disubstituted barbituric acids, the 5,5-disubstituted thiobarbituric acids, and the 1,5,5-trisubstituted barbituric acids are relatively weak acids, and salts of these barbiturates are easily formed through treatment with base [33].

4.4 Mode of Action

Before the introduction of the benzodiazepines in the 1960s, barbiturates were widely used in clinics for the treatment of anxiety, insomnia, and seizures. In addition, they were used as muscle relaxants and anesthetic agents. Now, benzodiazepines and even newer nonbenzodiazepines have taken their place for most of these clinical uses because they possess a wider therapeutic index, tolerance develops more slowly, and their liability for abuse is significantly lower. However, barbiturates remain a very important class of drugs from a scientific point of view, because they played an important role in the characterization of the $GABA_A$ (γ-aminobutyric acid) receptor. The $GABA_A$ receptor is an inhibitory ligand-gated ion channel (also known as *ionotropic receptor*) that decreases neuronal activity. Barbiturates enhance the inhibitory action of the $GABA_A$ receptor. The following paragraph focuses on the understanding that scientists have thus far of the receptor actions of barbiturates [34].

It has to be mentioned that no drug has a single effect. Barbiturates are no exception; on the contrary, they have sedative-hypnotic, anticonvulsant, anesthetic, and respiratory depressant effects that are mediated by their actions at various target sites in the body. Since the discovery of the barbiturates, researchers have tried to identify their actions in the CNS. The most potent *in vitro* effect of barbiturates is the enhancement of GABA-mediated inhibition through a prolongation of the postsynaptic chloride conductance [35–41]. GABA is the principal inhibitory neurotransmitter in the mammalian CNS. Barbiturates bind to the $GABA_A$ receptor α-subunit at binding sites that are distinct from the GABA and the benzodiazepine binding sites. At slightly higher, but still relevant concentrations, the barbiturates mimic the action of GABA and directly activate the same chloride channels, leading to inhibition of neuronal activity [42]. A variety of neurochemical studies support the interpretation that barbiturates directly interact with the postsynaptic $GABA_A$ receptor – chloride ion channel complex [43]. These interactions at the tissue, cellular, and molecular level correlate fairly well with anesthetic activity. Like benzodiazepines, barbiturates potentiate the effect of GABA at this receptor. In addition to this GABA-ergic effect, barbiturates also block the α-amino-3-hydroxy-5-methyl-4-isoxazolepropionic acid (AMPA) receptor, a subtype of the glutamate receptor. Glutamate is the principal excitatory neurotransmitter in the mammalian CNS. Taken together, the findings that barbiturates potentiate inhibitory $GABA_A$ receptors and inhibit excitatory AMPA receptors can explain the CNS-depressant effects of these agents. The two distinct effects of barbiturates have their own pharmacological profiles. First, they are able to enhance GABA's effect, thereby shifting the GABA dose-response curve to the left, which means that in the presence of barbiturates, less GABA is necessary to produce the same inhibitory effect than without the presence of barbiturates. Second, barbiturates are capable of reversing the action of $GABA_A$ channel blocking drugs, such as picrotoxin [44]. It could be demonstrated that barbiturates inhibit the Ca^{2+}-dependent release of neurotransmitters at higher concentrations [45]. Barbiturates produce their pharmacological effects by increasing the lifetime of GABA-activated chloride ion channels (this increases the efficacy of GABA) [46]. Researchers suggest that barbiturates modify the kinetics of channel transitions (e.g., openings and closings) between multiple conformational states, resulting in a net increase in open time during activation of the $GABA_A$ receptor by agonists (e.g., GABA) [47]. In contrast, benzodiazepines increase the frequency of the chloride ion channel opening at the $GABA_A$ receptor (this increases the potency of GABA).

4.5 Structure–Activity Relationship

Despite the fact that barbituric acid (**1**) itself does not possess any pharmacological properties, many derivatives of **1** have been synthesized to discover possible pharmaceutics on a trial-and-error basis [48]. Throughout these studies, many structural features were recorded, which are required for their hypnotic activity. However, no clear correlation between structure and activity could be determined. In 1951, Sandberg made a fundamental postulation that in order to possess good hypnotic activity the drug must be a weak acid and must have a lipid–water partition coefficient within a certain range. It was found that the derivatives with the lowest (0–15) and highest values (>600) are ineffective as hypnotics. Therefore, only the 5,5-disubstituted barbituric acids, the 5,5-disubstituted thiobarbituric acids, and the 1,5,5-trisubstituted barbituric acids possess significant hypnotic, anticonvulsant, or anesthetic activities. All other substituted barbituric acids, such as 5-monosubstituted, 1,3-disubstituted, or 1,3,5,5-tetrasubstituted barbituric acids are either inactive or produce convulsions [49].

4.5.1 5,5-Disubstitution

With increasing carbon chain length of the substituents at 5-position, the lipophilic character of the substituted barbituric acids also increases. Branching, unsaturation, replacement of alicyclic or aromatic substituents by alkyl substituents, and introduction of halogen into the alkyl substituents increase the lipid solubility of the barbiturates. Although lipophilic character determines the ability of compounds to cross the blood–brain barrier, hydrophilic character is also important because it determines the solubility in biologic fluids and ensures that the compound reaches the blood–brain barrier [50].

The following principal points must be taken into consideration with respect to the structure–activity relationship amongst the barbiturates: [33]

1) For optimal therapeutic results, the total number of carbon atoms present in the substituents at position 5 must not be less than 4 or more than 10.
2) Only one of the substituents at position 5 may be a cyclic system.
3) The greater the branching in the substituents, the more potent is the drug (e.g., pentobarbital (**4**) is more potent than amobarbital (**3**)).
4) Double bonds in the alkyl substituent lead to compounds that are more readily vulnerable to tissue oxidation; hence, they are short acting (e.g., quinalbarbital (**5**)).
5) Stereoisomers do not differ noticeably in their potencies.
6) Aromatic and alicyclic moieties exert greater potency than the corresponding aliphatic moieties having the same number of carbon atoms.
7) Substituents at position 5 with a small amount of carbons resist oxidation and, therefore, are long acting. Long chains are readily oxidized and thus produce short-acting barbiturates.

8) Alkyl substituents at position 5 bearing a halogen atom enhance the activity of the barbiturate.
9) Substituents at position 5 possessing a polar group (e.g., OH, CO, COOH, NH_2, RNH, and SO_3H) reduce potency considerably.
10) Methylation of one of the imide hydrogens enhances onset and reduces duration of action (e.g., the transition from 5,5-disubstituted to 1,5,5-trisubstituted barbituric acid).
11) Replacement of the carbonyl oxygen at carbon 2 by sulfur results in thiobarbiturates, which exhibit rapid onset and short duration of action because they are readily detoxified.
12) Inclusion of more sulfur atoms (e.g., the 2,4-dithio or 2,4,6-trithio barbituric acid analogs) decreases activity. Likewise, introduction of imino group(s) into the barbiturate-scaffold reduces activity (e.g., the 2-imino, 4-imino, 2,4-diimino, and 2,4,6-triimino barbituric acid analogs).
13) Replacement of the hydrogen atom at carbon atoms 1 and 3 with an alkyl group increases the vulnerability of the molecule to tissue oxidation.

4.5.2 Substitution at the Nitrogens

Substitution of one imide hydrogen by alkyl groups increases lipid solubility. The result is a quicker onset and a shorter duration of activity. As the size of the *N*-alkyl substituent increases, the lipid solubility increases and the hydrophilic character decreases. Furthermore, attachment of large alkyl groups to the nitrogen imparts convulsant properties to barbiturates [50].

References

1. Baeyer, A. (1864) *Justus Liebigs Ann. Chem.*, **131**, 291–302.
2. Sharpless, S.K. (1970) in *The Pharmacological Basis of Therapeutics*, 4th edn (eds L.S. Goodman and A. Gilman), The MacMillan Company, New York, pp. 98–120.
3. Cohen, W.A.T. (1943) *Chem. Weekblad*, **40**, 176.
4. Fieser, L.F. and Fieser, M. (1944) *Organic Chemistry*, D.C. Heath and Company, Boston, p. 247.
5. Fischer, E. and von Mering, J. (1903) *Ther. Ggw.*, **44**, 97–101.
6. Poole, A.K. (1929) *Yale J. Biol. Med.*, **1**, 345–351.
7. López-Muños, F., Ucha-Udabe, R., and Alamo, C. (2005) *Neuropsychiatr. Dis. Treat.*, **1**, 329–343.
8. Shorter, E. (ed.) (1997) *A history of Psychiatry: from the Era of the Asylum to the Age of Prozac*, John Wiley & Sons, Inc., New York.
9. Sneader, W. (ed.) (1985) *Drug Discovery: the Evolution of Modern Medicines*, John Wiley & Sons, Inc., New York.
10. Volwiler, E.H. and Tabern, D.L. (1930) *J. Am. Chem. Soc.*, **52**, 1676–1679.
11. Tabern, D.L. and Volwiler, E.H. (1935) *J. Am. Chem. Soc.*, **57**, 1961–1963.
12. Essential Medicines (2005) WHO Model List, March, *http://whqlibdoc.who.int/hq/2005/a87017_eng.pdf* (accessed 22 June 2011).
13. Google News, Agence France-Presse (2010) US court okays use of animal drug for human executions, Nov 22, *http://www.google.com/hostednews/afp/article/ALeqM5gWB2tTONFG2O8FCzUQ3Zsnb-FonQ?docId=CNG.abc73085fe90435beeac2a10705c6628.601* (accessed 23 June 2011).

14. Jolly, D. (2011) Lundbeck Blocks Sale of Nembutal for U.S. Executions. The New York Times (July 1), *http://www.nytimes.com/2011/07/02/world/europe/02execute.html?_r=2&src=rechp* (accessed 1 October 2011).
15. Epifanio, G. (1915) *Riv. Patol. Nerv. Mentale*, **20**, 273–308.
16. Klaesi, J. (1922) *Z. Gesamte Neurol. Psychiatr.*, **74**, 557–592.
17. Windholz, G. and Witherspoon, L.H. (1993) *Hist. Psychiatry*, **4**, 83–93.
18. Müller, M. (1927) *Z. Gesamte Neurol. Psychiatr.*, **107**, 522–543.
19. Cloetta, M. and Maier, A.W. (1934) *Z. Gesamte Neurol. Psychiatr.*, **164**, 146–162.
20. Hauptmann, A. (1912) *Münch. Med. Wochenschr.*, **59**, 1907.
21. Iváñez, V. and Díez-Tejedor, E. (1998) in *Historia de la Neuropsicofarmacología: una Nueva Aportacíon a la Terepéutica Farmacológica de Los Trastornos del Sistema Nervioso Central* (eds F. López-Muños and C. Álamo), Ediciones Eurobook SL, Madrid, pp. 347–364.
22. Brooks, C. (1922) Report of the Medical Officer. The National Society for Epileptics, 30th Annual Report, p. 19.
23. Weese, H. and Scharpff, W. (1932) *Dtsch. Med. Wochenschr.*, **58**, 1205–1207.
24. Bogue, J.Y. and Carrington, H.C. (1953) *Br. J. Pharmacol.*, **8**, 230–235.
25. Butler, T.C. and Waddell, W.J. (1956) *Proc. Soc. Exp. Biol. N.Y.*, **93**, 544–566.
26. Oleson, O.V. and Dam, M. (1967) *Acta Neurol. Scand.*, **43**, 348–356.
27. Bumm, R. (1927) *Klin. Wochenschr.*, **6**, 725–726.
28. Dundee, J.W. and McIlroy, P.D.A. (1982) *Anaesthesia*, **37**, 726–734.
29. Weese, H. and Scharpff, W. (1932) *Dtsch. Med. Wochenschr.*, **58**, 1205–1207.
30. Pratt, T.W., Tatum, A.L., Hathaway, H.R., and Waters, R.M. (1936) *Am. J. Surg.*, **31**, 464–466.
31. Chamberlain, J.S., Chap, J.J., Doyle, J.E., and Spaulding, L.B. (1935) *J. Am. Chem. Soc.*, **57**, 352–354.
32. Pinhey, J.T. and Rowe, B.A. (1980) *Tetrahedron Lett.*, **21**, 965–968.
33. Kar, A. (ed.) (2007) *Medicinal Chemistry*, 4th edn, New Age International Limited Publishers, New Delhi.
34. Harrison, N., Mendelson, W.B., and de Wit, H. (2000) Neuropsychopharmacology: The Fifth Generation of Progress, *http://www.acnp.org/g4/GN401000173/Default.htm* (accessed 23 June 2011).
35. Eccles, J.C. and Malcolm, J.L. (1946) *J. Neurophysiol.*, **9**, 139–160.
36. Haefely, W., Pole, P., and Schaffner, R. (1979) in *GABA Neurotransmitters* (eds P. Krogsgaard-Larsen, J. Scheel-Kruger, and G. Kofod), Munksgaard, Copenhagen, pp. 357–375.
37. Nicoll, R.A., Eccles, J.C., Oshima, T., and Rubia, F. (1975) *Nature*, **258**, 625–627.
38. Evans, R.H. (1979) *Brain Res.*, **171**, 113–120.
39. Dunwiddie, T.V., Worth, T.S., and Olsen, R.W. (1986) *J. Pharmacol. Exp. Ther.*, **238**, 564–575.
40. Barker, J.L. and Mathers, D.A. (1981) *Trends Neurosci.*, **4**, 10–13.
41. Schulz, D.W. and Macdonald, R.L. (1981) *Brain Res.*, **209**, 177–188.
42. Mathers, D.A. and Barker, J.L. (1980) *Science*, **209**, 507–509.
43. Olsen, R.W. (1981) *J. Neurochem.*, **37**, 1–13.
44. Harrison, N.L. and Simmons, M.A. (1983) *Br. J. Pharmacol.*, **80**, 387–394.
45. Brunton, L.L., Lazo, J.S., and Parker, K.L. (eds) (2006) *Goodman & Gilman's Pharmacological Basis of Therapeutics*, 11th edn, McGraw-Hill, New York.
46. Study, R.E. and Barker, J.L. (1981) *Proc. Natl. Acad. Sci. U.S.A.*, **78**, 7180–7181.
47. Macdonald, R.L., Skerritt, J.H., and Werz, M.A. (1984) in *Actions and Interactions of GABA and Benzodiazepines* (ed. N.G. Bowety), Raven, New York, pp. 203–216.
48. Mautner, H.G. and Clemson, H.C. (1970) in *Medicinal Chemistry, Part II* (ed. A. Burger), Wiley-Interscience, New York, pp. 1365–1385.
49. Sandberg, F. (1951) *Acta Physiol. Scand.*, **24**, 7–26.
50. Vida, J.A. (1995) in *Principals of Medicinal Chemistry*, 4th edn (eds W.O. Foye, T.L. Lemke, and D.A. Williams), Williams & Wilkinson, Baltimore, pp. 154–180.

5 Phenothiazine-Based Dopamine D_2 Antagonists for the Treatment of Schizophrenia

Cristiana A. Zaharia

5.1 Introduction

Phenothiazines (IUPAC: 10*H*-phenothiazines) [1] with the general structure **1** are commonly found among the first generation of antipsychotic drugs, known as *typical* or *conventional antipsychotics* (Figure 5.1) [2–5]. These drugs, called *neuroleptics*, have been used since the 1950s for the treatment of patients with schizophrenia and other psychotic disorders. The term "*antipsychotic*" refers to the ability of these drugs to alleviate hallucinations and delusions, while the term "*neuroleptic*," although used interchangeably, also suggests neurological actions marked by adverse extrapyramidal effects [6, 7].

The first member of this class of drugs, chlorpromazine (**2a**), was discovered in 1952 and marketed as Largactil® in Europe and Thorazine® in the United States [4, 5]. Chlorpromazine (**2a**) marked an important moment in the history of psychiatry since before its introduction schizophrenic patients were permanently institutionalized [5, 8]. Shortly after the introduction of chlorpromazine (**2a**), several structurally related phenothiazines (**1**) were developed for use as antipsychotic agents [9]. It is generally believed that the observed antipsychotic activity of phenothiazines (**1**) is directly correlated with their antagonist activity against the dopamine D_2 receptor [10–12]. The first generation of antipsychotics also includes a chemically related class called *thioxanthenes* (for example, flupenthixol (**3**) also known as *Depixol®* or *Fluanxol®*) (Figure 5.2); however, only phenothiazine-based drugs are discussed in this chapter [13].

5.2 History

While working on developing new textile dyes at the Badische Anilin und Soda Fabrik (BASF) in Germany, Caro synthesized the first phenothiazine analog, methylene blue (**4**), in 1876 (Figure 5.3) [5, 14, 15].

Bioactive Heterocyclic Compound Classes: Pharmaceuticals,
First Edition. Edited by Jürgen Dinges and Clemens Lamberth.

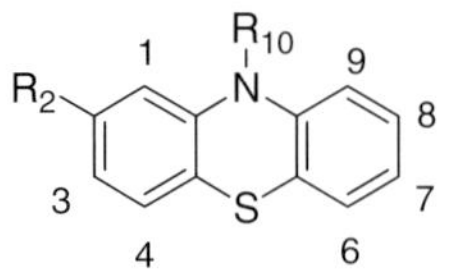

1

Figure 5.1 The basic structure and conventional numbering of phenothiazines (1).

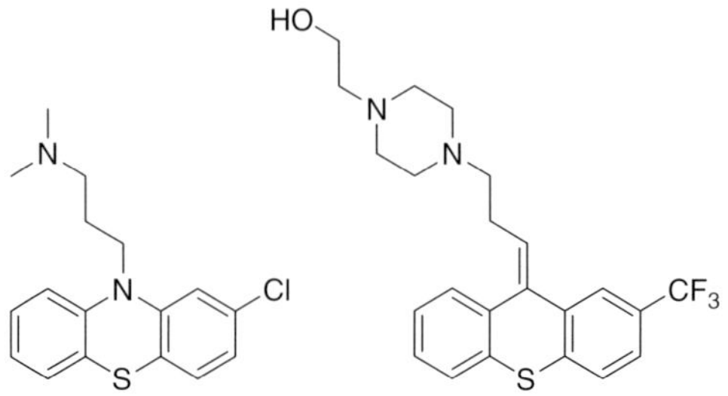

2a
Chlorpromazine

3
Flupenthixol

Figure 5.2 Different chemical entities of first-generation antipsychotics.

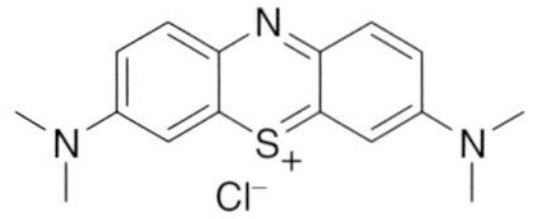

4

Figure 5.3 Structure of methylene blue (4).

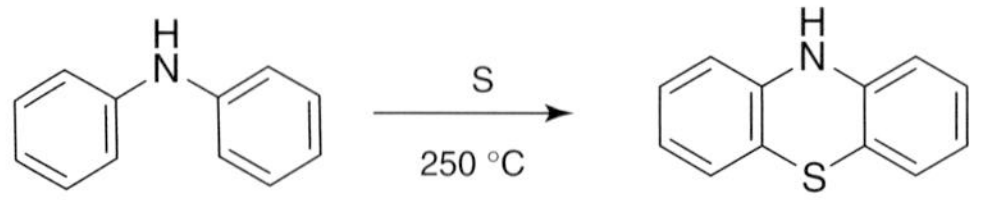

5

6

Scheme 5.1 Bernthsen's synthesis of the phenothiazine core (6).

The phenothiazine core (6) itself was synthesized in 1883 by August Bernthsen during his research on textile aniline dyes (Scheme 5.1) [16]. Treatment of diphenyl aniline (5) with elemental sulfur at 250 °C generated phenothiazine (6) [16].

In the 1930s, phenothiazine (6) was used extensively for its antimicrobial, insecticidal, and anthelmintic (parasitic worm expelling) properties [1, 5]. Before their utility as antipsychotic agents, phenothiazine derivatives were used as antipruritic (anti-itching), antispasmodic, anticonvulsant, antibacterial, anthelmintic, and antiemetic drugs [17]. Historically, the use of phenothiazines (1) as antihistamines originated from the discovery of 7 by Bovet and coworkers at the Rhône-Poulenc laboratories (Scheme 5.2) [18]. Bovet, Halpern, Ducrot, and their colleagues [18] modified benzodioxane (7) into ethanolamine (8), which was further transformed into either diphenhydramine (9) (Benadryl®) or ethylenediamine

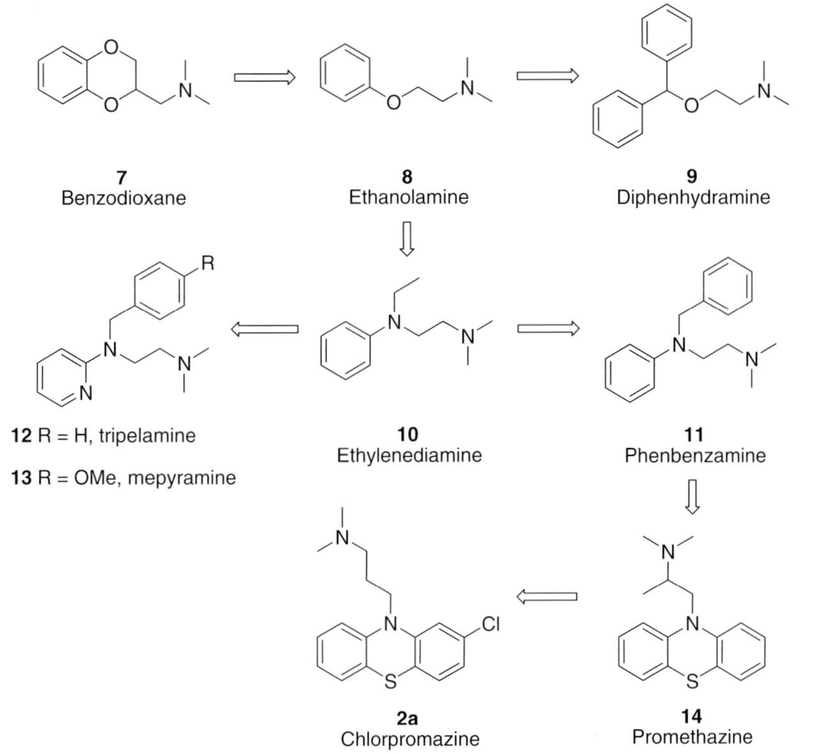

Scheme 5.2 The evolution of phenothiazine-based antipsychotics.

(**10**). Modifications of **10** led to the discovery of the allergy drug phenbenzamine (**11**) (Antergan®) [13, 19]. Replacing the phenyl in **10** with a pyridyl group and extending the ethyl substituent on the aniline nitrogen yielded mepyramine (**13**: Rhône-Poulenc) and tripelamine (**12**: Ciba) [11]. Meanwhile, the scientists at Rhône-Poulenc were also working on derivatives of methylene blue (**4**) as potential antimalarial treatments [2]. Although unsuccessful, they soon came to the realization that some of their phenothiazine analogs were actually analogs of phenbenzamine (**11**) and might therefore display antihistaminic activity [13]. While investigating *N*-substituted phenothiazines (**1**) as antihistaminic agents, French chemist Paul Charpentier and his team discovered the first commercial phenothiazine-based drug, promethazine (**14**) (Fenergan®) [11]. Guirard reported that promethazine (**14**) also had sedative effects and used it as a sleeping aid for some of his patients [5]. Further structure–activity relationship (SAR) studies on promethazine analogs led again by the Charpentier team revealed a promising chlorinated compound (RP-4560), later named *chlorpromazine* (**2a**) [5, 6, 10].

Animal studies by Courvoisier, a pharmacologist at Rhône-Poulenc, showed that rats injected with chlorpromazine (**2a**) would lose their previously conditioned reflexes, that is, rats that have been conditioned to climb a rope at the sound of a buzzer refused to do so after being administered **2a** [8]. Chlorpromazine's (**2a**) striking central nervous system (CNS) depressant effects prompted Henri Laborit to administer the drug to patients undergoing surgery. After giving a dosage of 50–100 mg intravenously, his patients became calmer and completely indifferent to their surroundings while their body temperature decreased in what Laborit called *a state of "artificial hibernation."* Thus, a team at Val de Grâce hospital in Paris administered chlorpromazine (**2a**) (50 mg) as part of a cocktail to a severely agitated psychotic patient and observed a calming state within a few hours of dosage. Repeated dosages totaling 855 mg of chlorpromazine (**2a**) showed a drastic improvement, enabling the release of the patient who was able to resume normal daily activities [5, 8, 19]. This result led to the first presentation and publication on the potential applications of chlorpromazine (**2a**) in psychiatry in 1952 [5, 8, 20]. Recognizing the potential of **2a** in psychiatry, Jean Delay and Pierre Deniker started conducting clinical investigations on chlorpromazine (**2a**) as a single agent at Saint Anne's hospital in Paris. They were able to demonstrate its effectiveness in multiple patients by slowing responses to external stimuli without a loss in intellectual functioning, thus producing a state of emotional indifference [5, 8, 21]. Following these studies, chlorpromazine (**2a**) was introduced to the French market by Rhône-Poulenc under the name of Largactil (large activity) for use in psychiatry and anesthesiology, among other indications. Within a few years, the international psychiatric community recognized the potential of chlorpromazine (**2a**) and began investigating the effects of the drug [5, 8]. The first randomized and placebo-controlled clinical study design in a psychiatric environment was conducted by Elkes and Elkes on patients with chronic psychoses. They were able to demonstrate efficacy of the drug by using a cross-over study with sequential placebo and chlorpromazine (**2a**) (250 mg day^{-1}) in each patient rather than a parallel drug design [5]. A chlorpromazine (**2a**) (up to 800 mg day^{-1}) clinical trial run on 71 patients by Lehman at Verdun hospital in Montreal revealed that long-term treatment of patients with various psychiatric conditions resulted in a positive response in 66% of the population. His results, published in the Archives of Neurobiology and Psychiatry, stimulated the interest of the North American clinicians in chlorpromazine (**2a**), and soon thereafter, Smith, Kline, & French licensed the drug from Rhône-Poulenc. In 1954, after a series of animal testing and clinical trials, the US Food and Drug Administration (FDA) approved the drug under the name of Thorazine (**2a**), for use in psychiatry. Within a few months, sales of Thorazine (**2a**) reached 75 million dollars while the number of patients in mental institutions decreased by 89% between 1954 and 1996 [5]. A study by Hogarty and Goldberg on patients who were discharged from the hospital and randomized to receive either chlorpromazine (**2a**) or placebo has shown the necessity of continuous treatment since in 68% of the patients treated with placebo the condition relapsed [22]. The discovery of phenothiazines (**1**) has drastically changed the world of psychiatry and the way schizophrenic patients are treated.

The introduction of chlorpromazine (**2a**) was a major breakthrough in the treatment of antipsychotic disorders, and immediately following its discovery, a series of other phenothiazine derivatives with similar properties came to the

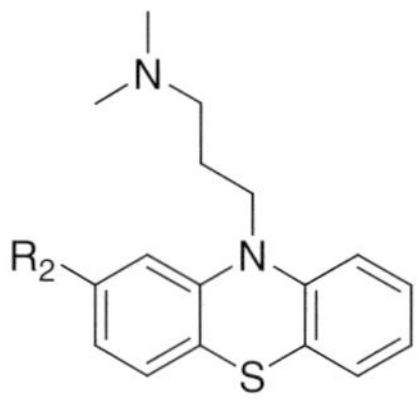

Aliphatics

2a R_2 = Cl chlorpromazine

2b R_2 = CF_3 triflupromazine

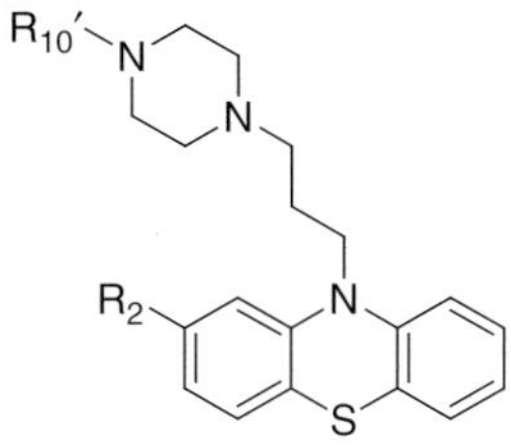

Piperazines

2c R_2 = Cl R_{10}' = Me prochlorperazine

2d R_2 = CF_3 R_{10}' = Me trifluoperazine

2e R_2 = Cl R_{10}' = $-CH_2CH_2OH$ perphenazine

2f R_2 = CF_3 R_{10}' = $-CH_2CH_2OH$ fluphenazine

2g R_2 = $COCH_3$ R_{10}' = $-CH_2CH_2OH$ acetophenazine

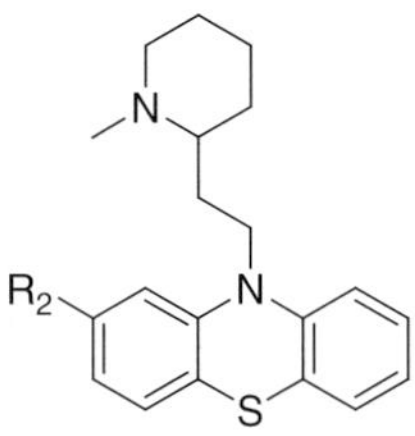

Piperidines

2h R_2 = SCH_3 thioridazine

2i R_2 = $SOCH_3$ mesoridazine

Figure 5.4 The three classes of phenothiazine-based antipyschotics.

market. Based on the nature of the nitrogen (N10) substituent (Figure 5.1), these structurally similar drugs can be grouped into three classes: (i) aliphatics, including chlorpromazine (**2a**, Thorazine) and triflupromazine (**2b**, Vesprin®); (ii) piperazines, including prochlorperazine (**2c**, Compazine®), trifluoperazine (**2d**, Stelazine®), perphenazine (**2e**, Trilafon®), fluphenazine (**2f**, Permitil®, Prolixin®), and acetophenazine (**2g**, Tindal®); and (iii) piperidines, including mesoridazine (**2h**, Serentil®) and thioridazine (**2i**, Mellaril®) (Figure 5.4) [11, 23].

Experiments by Carlson and Lindquist designed to elucidate the mechanism of action of chlorpromazine (**2a**) showed that administration of the drug resulted in a significant increase in dopamine (DA) turnover in the brain [24]. They found that chlorpromazine (**2a**) blocks the dopamine release from the presynaptic neurons thus acting as a neurotransmitter. *In vitro* receptor binding assays by Seeman showed selective antagonist activity of chlorpromazine (**2a**) against dopamine receptors and a correlation between the daily dose of antipsychotic and dopamine D_2 antagonism [25]. All these studies led to the dopamine hypothesis of schizophrenia, which suggests that there are more dopamine receptors in the brain of a schizophrenic patient and therefore more dopamine neurotransmitter [26–28].

Table 5.1 K_i values (nM) for antipsychotic drugs at various neurotransmitter receptors [6].

	Dopamine			Serotonin	Muscarinic cholinergic	Adrenergic		Histaminic
	D_2	D_1	D_4	5-HT_2		a_1	a_2	H_1
Fluphenazine	0.8	15	9	19	2000	9	1600	21
Perphenazine	1.4	–	–	5.6	1500	10	510	–
Thioridazine	2.3	22	12	41	10	1.1	–	–
Mesoridazine	5	–	13	6.3	–	–	–	–
Chlorpromazine	19	56	12	1.4	60	0.6	750	9.1

While the antipsychotic action of phenothiazines (**1**) results from the antagonist activity against the dopamine D_2 receptors in the mesolimbic pathways, blockage of the D_2 receptors within the nigrostriatal system results in significant extrapyramidal side effects (EPS) [11]. Examples of EPS include Parkinson's disease symptoms (tremors), akathisia (motor restlessness), acute dystonia, and neuroleptic malignant syndrome, and with long-term exposure, tardive dyskinesia and dystonia. D_2 receptor antagonism has also been linked to increased prolactic secretion due to blockage of the dopamine receptors in the pituitary gland. Phenothiazine-based drugs suffer from a variety of side effects mainly due to off target effects, that is, histamine (H_1) receptor antagonism causing sedation; cholinergic muscarinic (M_1) receptor antagonism resulting in dry mouth, urinary retention, and blurred vision; and adrenergic (α_1/α_2) receptor antagonism leading to orthostatic hypotension and in some cases sexual disturbance (Table 5.1) [6]. Snyder and Yamamura [29] suggested that a higher muscarinic receptor blockade might result in lower incidence of extrapyramidal effects. Some of the phenothiazine drugs also have complex effects on the cardiovascular system by prolonging the QTc and blunting the T waves [6, 30].

Phenothiazine drugs are highly lipophilic and are about 95–99% plasma protein bound. They tend to have high first-pass metabolism with elimination half-lives varying between 12 and 24 h [6]. Their low oral bioavailability is due to extensive liver metabolism, mostly by CYP2D6 [31, 32]. Major active chlorpromazine (**2a**) metabolites include aromatic hydroxylation at the 7-position followed by glucoronidation and oxidative *N*-dealkylation of both N10 (chlorophenothiazine (**17**)) and the side chain basic nitrogen (nor_1 and nor_2 chlorpromazine). Thioridazine (**2h**) undergoes sulfur oxidation to generate mesoridazine (**2i**) [11].

5.3 Synthesis

Early syntheses of the phenothiazine core relied on reactions of various diphenylamines with sulfur in the presence of catalysts [1, 15]. The synthesis of chlorpromazine (**2a**) as initially reported by Charpentier is described in Scheme 5.3.

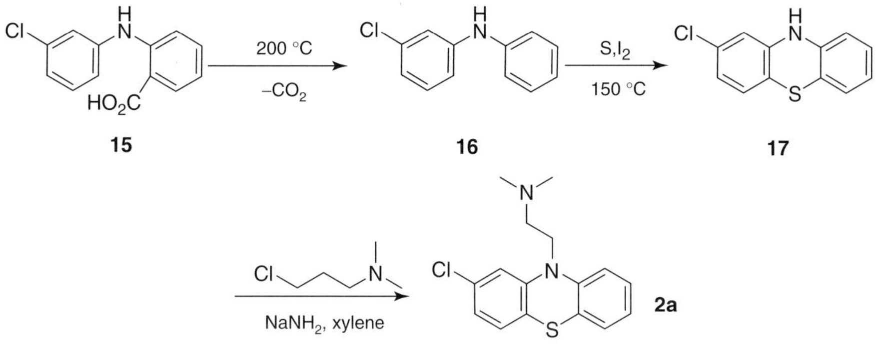

Scheme 5.3 Charpentier's synthesis of chlorpromazine (2a).

Decarboxylation of 2,3-chlorophenylaminobenzoic acid **15** at 200 °C followed by cyclization of 3-chlorodiphenyl amine (**16**) with sulfur in the presence of catalytic iodine generates a mixture of regioisomers, which are separated by fractional distillation. The addition of catalytic amounts of iodine not only lowers the reaction temperature but also increases the yields of the Bernthsen synthesis. Chlorpromazine (**2a**) is then obtained by treatment of phenothiazine (**17**) with 3-dimethylamino-1-chloropropane in the presence of sodium amide ($NaNH_2$) in xylene at reflux temperature [33–35].

Newer synthetic methodologies allow for the formation of the phenothiazine scaffold via cyclization reactions of substituted diphenyl sulfides [1]. Treatment of 2-aminobenzenethiol (**18**) with 2,4-dichloro-1-nitrobenzene (**19**) in the presence of a base and subsequent acylation affords thioether (**20**) (Scheme 5.4, Route 1). Smiles rearrangement of **20** results in the formation of 2-chlorophenothiazine (**17**) [36]. An alternate synthesis involves reaction of 2-bromothiophenol (**22**) with 2,4-dichloro-1-nitrobenzene (**19**) to provide the thioether (**23**) (Scheme 5.4, Route 2). Reduction of the nitro group to aniline and copper-mediated Ullmann reaction then yields 2-chlorophenothiazine (**17**) [1, 37].

There has been much synthetic effort in varying the basic side chain at the N10 position of the phenothiazines (**1**) (Scheme 5.5). Route A shows the direct alkylation of N10 with alkyl halides that already contain a basic functionality. For example, chlorpromazine (**2a**) [19–21] and triflupromazine (**2b**) [38–41] are synthesized via alkylations of the corresponding phenothiazine core with 3-dimethylaminopropylchloride in the presence of sodium amide. In analogy, thioridazine (**2h**) [42] and mesoridazine (**2i**) [43] are synthesized through alkylations with 2-(2-chloroethyl)-1-methylpiperidine. In contrast, prochlorperazine (**2c**) [44, 45] and trifluoperazine (**2d**) [39, 46, 47] are accessed via alkylation of the corresponding (3-chloropropyl)phenothiazine with methyl piperazine (Route B).

Intermediate **25** in the synthesis of fluphenazine (**2f**) is synthesized in an analogous manner to method B described in Scheme 5.6. Hydrolysis of the formyl group with sodium hydroxide, subsequent alkylation of **26** with 2-bromoethyl

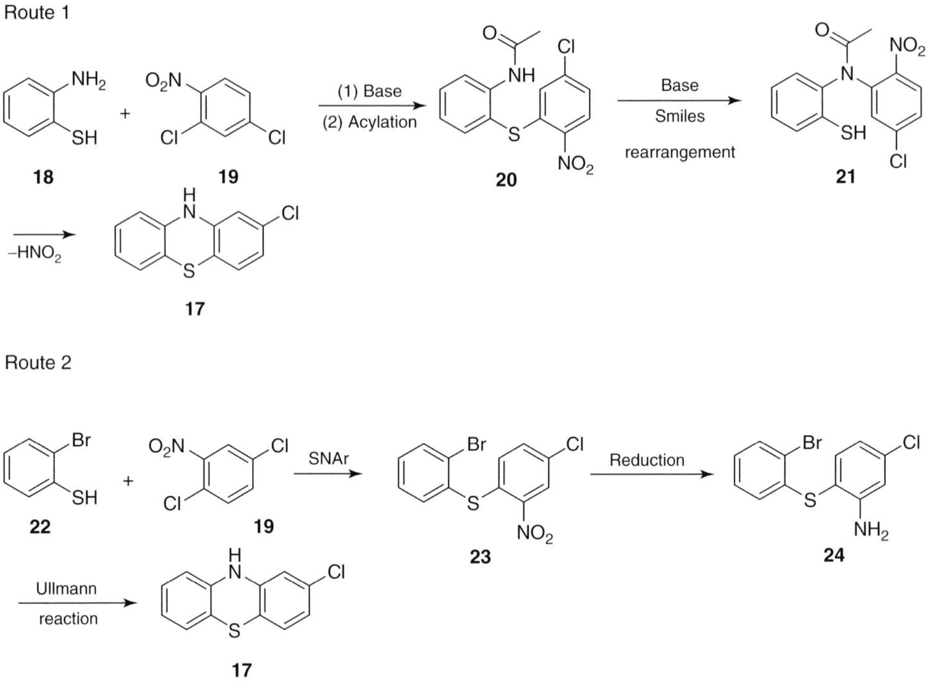

Scheme 5.4 Synthesis of 2-chlorophenothiazine (**17**) via Smiles rearrangement and Ullmann reaction.

acetate and removal of the acetate group under acidic conditions affords fluphenazine (**2f**) [48, 49].

Recently, Jorgensen and coworkers described a palladium-catalyzed three-component reaction of 2-bromothiophenol (**27**), amine (**28**), and the 4-substituted 1-bromo-2-iodobenzene (**29**) to generate the phenothiazine core of a variety of antipsychotic drugs such as chlorpromazine (**2a**) and triflupromazine (**2b**) (Scheme 5.7). The best yields are obtained with dipalladium tribenzylideneacetone (Pd_2dba_3) in the presence of 1,1'-bis(diphenylphosphino)ferrocene (dppf) and sodium *tert*-butoxide (NaO*t*Bu) as a base [50].

Ma and coworkers [51] described a new method for accessing *N*-substituted phenothiazines (**1**) from substituted 2-iodoanilines (**30**) and 2-bromobenzenethiol (**31**) in the presence of CuI/L-proline as a catalyst (Scheme 5.8).

5.4 Mode of Action

Schizophrenia is a chronic debilitating mental illness that affects ~1% of the world population. The discovery of the phenothiazine-based neuroleptic drug chlorpromazine (**2a**) has had a tremendous impact on the lives of patients with

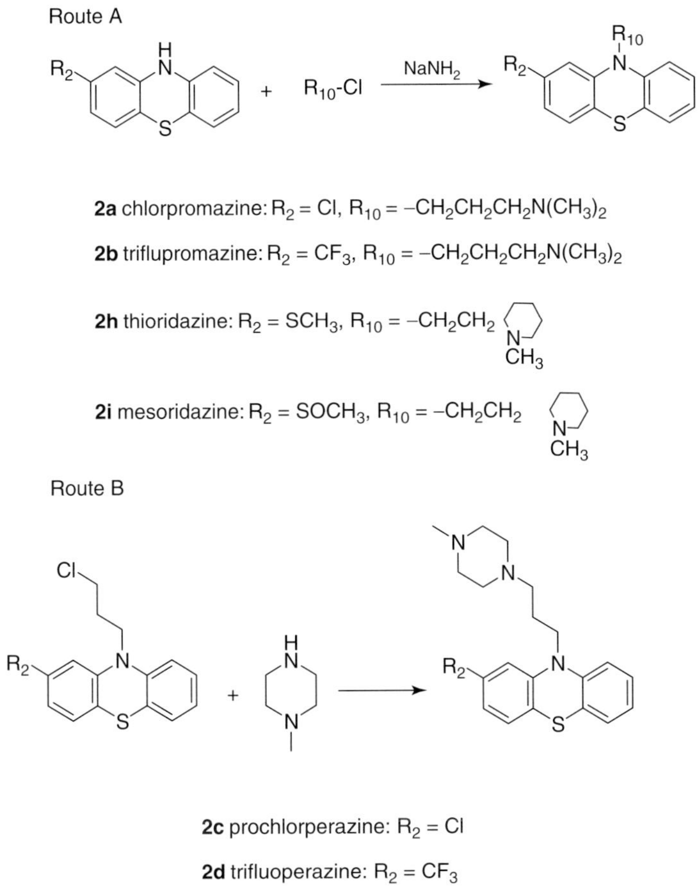

Scheme 5.5 Synthesis strategies for antipsychotic drugs.

schizophrenia by alleviating some of the symptoms associated with the disease. Auditory hallucinations, delusions, and disorganized patterns of behavior represent the positive symptoms of the disease. Apart from the negative symptoms of the disease that include social withdrawal, apathy, and blunting of emotions, there are other neurocognitive deficits (attention, memory, fluency) and mood symptoms (depression, hostility) [6]. While the underlying causes of the disease remain unclear, one of the oldest and most accepted neurochemical theories, entitled the dopamine hypothesis of schizophrenia, centers around the dysregulation of dopamine neurotransmission in the brain. According to this hypothesis, schizophrenia is associated with increased dopamine neurotransmission as well as an increase in monomeric dopamine D_2 receptors in the brain [52]. The ability of chlorpromazine (**2a**) and related phenothiazine drugs to exert their pharmacological action is believed to stem from their antagonistic activity against the dopamine D_2 receptors. According to early pharmacological studies, there are two types of dopamine receptors: (i) D_1,

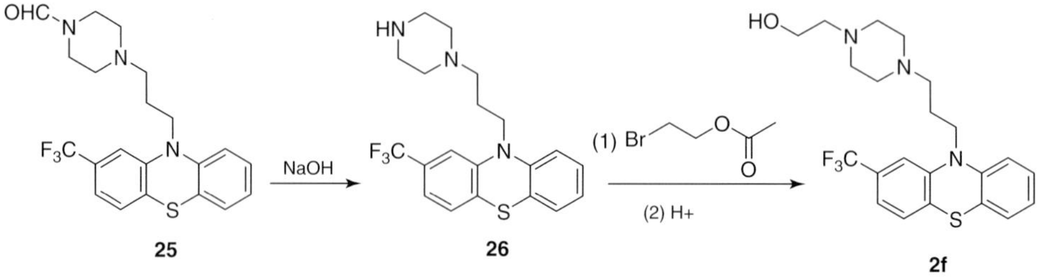

Scheme 5.6 Synthesis of fluphenazine (**2f**).

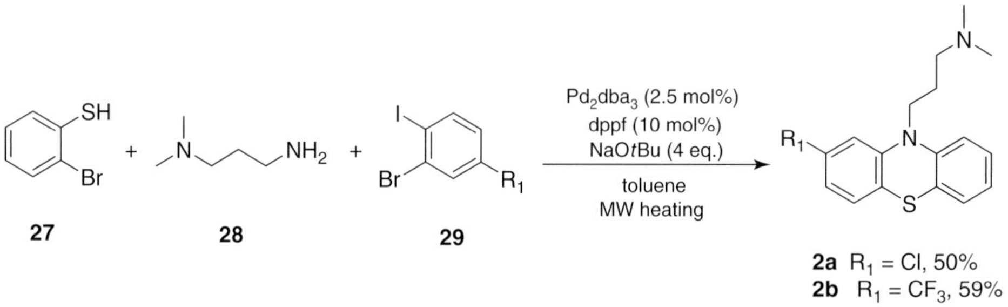

Scheme 5.7 Jorgensen's palladium-catalyzed three-component coupling reaction.

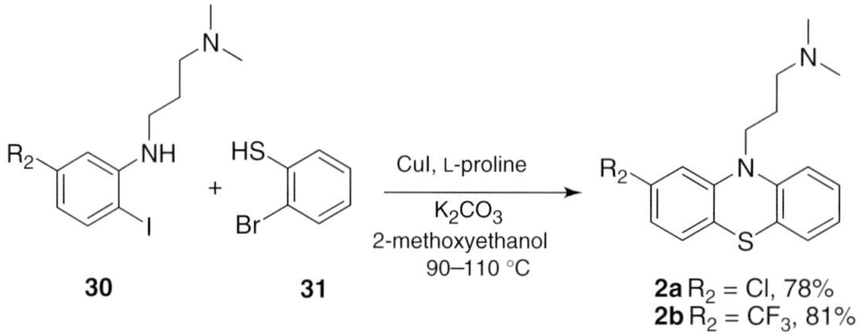

Scheme 5.8 Ma's synthesis of *N*-substituted phenothiazines (**1**).

which stimulate the synthesis of cAMP and (ii) D$_2$, which inhibit cAMP synthesis and modulate Ca^{2+} and K^+ currents [53]. Additional molecular cloning studies revealed five types of dopamine receptors in the human brain: D$_1$–D$_5$. Dopamine receptors can be further classified on the basis of their effect on adenylyl cyclase (AC): D$_1$-type receptors (D$_1$ and D$_5$) stimulate AC and D$_2$-type receptors (D$_2$, D$_3$, and D$_4$) inhibit AC and thus cause accumulation of cAMP as well as suppression of Ca^{2+} currents and activation of K^+ currents [54]. Dopamine receptors belong to the G-protein-coupled receptor (GPCR) family and contain an extracellular amino terminus and an intracellular carboxy terminus separated by seven hydrophobic transmembrane domains of approximately 24 amino acids each. The seventh transmembrane domain is believed to be responsible for the antagonist binding

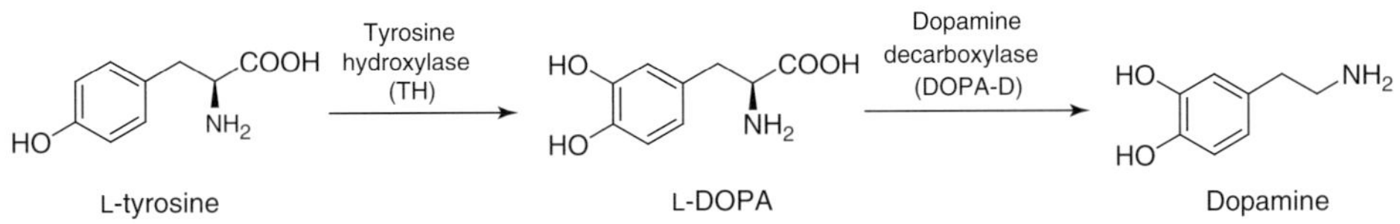

Scheme 5.9 Biosynthesis of dopamine (DA).

[55]. Dopamine receptors are also known as *slow* or *metabotropic receptors* since they act via a series of signaling cascades that involve G-proteins and adenylate cyclase [6]. Dopamine is a catecholamine neurotransmitter that is biosynthesized from tyrosine via two steps (Scheme 5.9). The first step is the rate-limiting step catalyzed by tyrosine hydroxylase (TH) and involves conversion of tyrosine to L-dihydroxylphenylalanine (L-DOPA). Decarboxylation by dopamine decarboxylase affords dopamine, which is stored in vesicles in the presynaptic neuron.

On depolarization of the presynaptic neuron, dopamine is released into the synaptic cleft and then binds to the dopamine receptors in the postsynaptic neuron. Some of the dopamine released in the cleft can also stimulate the presynaptic dopamine autoreceptors, which in turn inhibit tyrosine dehydroxylase, thus causing a decrease in the synthesis and release of dopamine. Cytoplasmic dopamine is metabolized via several steps catalyzed by monoamine oxidase (MAO) and catechol-O-methyltransferase (COMT) to homovanillinic acid [6, 11]. PET scan studies by Hietala have shown that presynaptic dopamine function is severely altered in schizophrenic patients. By measuring the uptake of fluorodopa, he was able to demonstrate that there was a higher intake of fluorodopa in sick than in healthy patients [56]. The major dopamine pathways in the human brain include the extrapyramidal nigrostriatal system, heavily involved in locomotor functions, and the mesolimbic and mesocortical systems, which control emotional behavior and cognitive symptoms. The antipsychotic therapeutic effects of phenothiazines (**1**) are associated with antagonist activity against the D_2 receptors in the mesolimbic pathway. However, during treatment with phenothiazine antipyschotics, the cognitive or negative symptoms of the disease remain unaltered. Antagonism of the nigrostriatal D_2 receptors causes extrapyramidal symptoms (e.g., parkinsonism, acute dystonia, and akathisia). Interestingly, the degradation of the neurons in the nigrostriatal pathway has also been associated with Parkinson's disease. Phenothiazine-based antipsychotics exert their effects on the D_2-type receptors located postsynaptically on neurons in the limbic and extrapyramidal pathways [11]. Recently, it has been demonstrated that activation of presynaptic D_2 receptors (autoreceptors) can also cause a decrease in dopamine synthesis and dopaminergic neurotransmission [57]. Aripiprazole (Abilify®) was approved in 2002 and is marketed as a partial D_2 agonist, which is believed to stabilize high or low dopamine levels in the brain [13]. As soon as the drug enters the synaptic space, it competes with endogenous dopamine for the dopamine receptor. Phenothiazine drugs have been shown to bind more tightly than dopamine to their respective D_2 receptor with lower dissociation constant values than dopamine ($K_i = 2.1$ nM) (e.g., $K_i = 0.99$ nM for chlorpromazine (**2a**), $K_i = 0.55$ nM for fluphenazine (**2f**),

$K_i = 0.27$ nM for perphenazine (**2e**), $K_i = 1.7$ nM for prochlorperazine). Studies by Seeman have shown that in order to achieve an antipsychotic effect, the occupancy of the D_2 receptors has to be between 60 and 80% [58]. In conclusion, the mechanism of action of phenothiazines (**1**) involves blocking the hyperactive dopamine neurons in the dopamine pathways of the brain. Antagonism of dopaminergic neurons in the mesolimbic pathway accounts for the relief of positive schizophrenia symptoms such as psychoses. However, blockage of the D_2 receptors in other dopamine pathways possibly results in major side effects. Phenothiazines (**1**) have very limited beneficial effects on the cognitive functions linked to schizophrenia [6]. Several atypical antipsychotics with dual antagonist activity against the D_2 and 5-hydroxytryptamine (5-HT) receptors seem to lack the undesirable EPS effects of phenothiazines (**1**); however, they have been associated with other serious metabolic effects [59]. Risperidone (Risperdal®, 1994), olanzapine (Zyprexa®, 1996), and ziprasidone (Geodon®, 2001) are among some of the atypical antipsychotic drugs, recently marketed for the treatment of schizophrenia [12]. Newer antipyschotics are currently in the clinic and they have been designed to address some of these issues, however, this is beyond the focus of this chapter. Treatment of schizophrenia remains a challenge, and first-generation antipsychotics are still prescribed as either first- or second-line therapy.

5.5 Structure–Activity Relationships

Molecular model and computer calculation studies by Snyder and coworkers revealed that the aromatic ring of dopamine superimposes over the substituted phenyl ring of the phenothiazines (**1**) in such a way that the meta hydroxyl group of dopamine is positioned to sit right above the sulfur atom. In this conformation, the primary amine of dopamine sits above the tertiary amine group of chlorpromazine (**2a**) [60]. Molecular modeling and computer calculations published later by Snyder show favorable van der Waal's interactions between the 2-substituent and the amine side chain. In this model, substituents at the 2-position of phenothiazines (**1**) are crucial for activity, with the optimal groups being electron withdrawing ($CF_3 > Cl$). For example, triflupromazine (**2d**) is more potent at blocking the dopamine receptors than chlorpromazine (**2a**) [61]. The absence of a 2-substituent (promazine) results in a compound that is less potent than its substituted analogs due to the inability of the core to assume the preferred conformation compared to dopamine. 3-Substitution is tolerated, while 4-substitution may interfere with the interaction of the phenothiazine sulfur atom and the dopamine receptor. Additional ring substitutions decrease activity by activating the cleavage of the central thiomorpholine ring. Maximum potency is achieved when the spacer group linking the phenothiazine core and the basic amine is a propyl group. A shorter or branched alkyl chain displays decreased neuroleptic activity. The highest activity against the dopamine receptor has been observed with aliphatic or cyclic tertiary amines. Diethylamine analogs have decreased potency, as do pyrrolidine and morpholine amines [62]. An interesting discovery was that replacement of the aliphatic chain with piperidines and *N*-substituted piperazines led to significantly more potent analogs. As predicted by the Snyder model, the piperazine-containing side chain affords more favorable van der Waal's interactions with the 2-substituent [61]. Also, these interactions should be greatly increased when the piperazine is further substituted with a hydroxyethyl group as exemplified by the fact that fluphenazine (**2f**) is more clinically efficacious than trifluoperazine (**2d**). The piperazine phenothiazines are the most potent phenothiazine antipsychotics because of their high affinity for the dopamine D_2 receptors. However, this also gives rise to an increased occurrence of EPS effects. On the other hand, the piperazine analogs have the lowest affinities for the histamine-1 receptor, hence their stimulatory effects when compared to the aliphatic amine compounds [6].

The presence of a basic amine allows phenothiazine-based drugs to exist either as mineral or organic acid salts (chlorpromazine hydrochloride, triflupromazine hydrochloride, thioridazine hydrochloride, fluphenazine hydrochloride, trifluoperazine hydrochloride, mesoridazine benzene sulfonate, prochlorperazine maleate). Fluphenazine (**2f**) and perphenazine (**2e**) contain a hydroxyl group that allows for the introduction of a prodrug functionality. The duration of action of these antipsychotics can be considerably increased by esterifying the hydroxyl group with a fatty acid ester [63]. A series of ester prodrugs of phenothiazines (**1**) have been marketed: fluphenazine decanoate (Prolixin decanoate®), fluphenazine enanthate (Prolixin enanthate), and perphenazine enanthate. These prodrugs are administered intramuscularly, and their effects last for a few weeks, thus alleviating some of the side effects as well as issues related to patient compliance. For example, fluphenazine hydrochloride is 100 times more potent than chlorpromazine (**2a**) and is administered in doses ranging from 5 to 40 mg day^{-1} with 24 h duration of action. On the other hand, the enanthate (heptanoic acid ester) is given in a dose range of 12–75 mg every one to two weeks via an intramuscular injection [6, 11].

References

1. Taurand, G. (2000) in *Ullmann's Encyclopedia of Industrial Chemistry* (ed. F. Ullmann), Wiley-VCH Verlag GmbH & Co. KGaA, Weinheim, pp. 1–16.
2. Healy, D. (2008) *Int. J. Neuropsychopharmacol.*, **1**, 191–194.
3. Swazey, J.P. (1974) *Chlorpromazine in Psychiatry – A Study of Therapeutic Innovation*, MIT Press, Cambridge, MA, p. 340.
4. Caldwell, A.E. (1978) in *Principles of Psychopharmacology* (eds W.G. Clark and J. del Giudice), Academic Press, New York, pp. 9–40.
5. López-Muñoz, F., Alamo, C., Cuenca, E., Shen, W.W., Clervoy, P., and Rubio, G. (2005) *Ann. Clin. Psychiatry*, **17**, 113–135.
6. Baldessarini, R.J. and Tarazi, F.I. (2005) in *Goodman and Gilman's the Pharmacological Basis of Therapeutics*, 11th edn, Chapter 18 (eds L.L. Brunton, J.S. Lazo, and K.L. Parker), McGraw-Hill, New York, pp. 461–500.
7. Marsden, C.D. and Jenner, P. (1980) *Psychol. Med.*, **10**, 55–72.
8. Ban, T. (2007) *Neuropsych. Dis. Treat.*, **4**, 495–500.

9. Sneader, W. (2005) *Drug Discovery: A History*, John Willey & Sons, Ltd, Essex, pp. 404–409.
10. Burns, M.J. and Linden, C.H. (2008) in *Intensive Care Medicine*, 6th edn (eds R.S. Irwin and J.M. Rippe), Lippincott, Williams, and Wilkins, Philadelphia, PA, pp. 1525–1536.
11. Booth, R.G. (2002) in *Foye's Principles of Medicinal Chemistry*, 6th edn (eds D.A. William, W.O. Foye, and T.L. Lemke), Lippincott, Williams, and Wilkins, Philadelphia, PA, pp. 601–630.
12. Kapur, S. and Mamo, D. (2003) *Prog. Neuropsychopharmacol. Biol. Psychiatry*, **27**, 1081.
13. Mailman, R.B. and Murthy, V. (2010) *Curr. Pharm. Des.*, **16**, 488–501.
14. Massie, S.P. (1954) *Chem. Rev.*, **54**, 797–833.
15. Bernthsen, A. (1912) *Ber. Dtsch. Chem. Ges.*, **45**, 2042.
16. Bernthsen, A. (1883) *Ber. Dtsch. Chem. Ges.*, **16**, 2896–2904.
17. Gordon, M., Craig, P.N., and Zirkle, C.L. (1964) *Advances in Chemistry*, Chapter 11, American Chemical Society, pp. 140–147.
18. Bovet, D. and Staup, A.M. (1937) *C.R. Soc. Biol.*, **124**, 547–549.
19. Laborit, H., Huguenard, P., and Alluaume, R. (1952) *Press Méd.*, **60**, 206–208.
20. Hanon, J., Paraire, J., and Velluz, J. (1952) *Ann. Méd. Psychol.*, **110**, 332–335.
21. Delay, J., Deniker, P., and Harl, J.M. (1952) *Ann. Méd. Psychol.*, **110**, 112–117.
22. Stroup, T.S., Alves, W.M., Hamer, R.M., and Lieberman, J.A. (2006) *Nat. Rev. Drug Disc.*, **5**, 133–202.
23. Vardanyan, R. and Hruby, V.J. (2006) *Synthesis of Essential Drugs*, Elsevier B.V., Amsterdam, pp. 83–88.
24. Carlsoon, A. and Lindqvist, M. (1963) *Acta Pharmacol. Toxicol.*, **20**, 140–144.
25. Seeman, P., Lee, T., and Chau-Wong, M. (1976) *Nature*, **261**, 717–719.
26. Van Rossum, J. (1967) in *Neuropsychopharmacology, Proceedings 5th Collegium Internationale Neuropsychopharmacologicum* (eds H. Brill, J. Cole, P. Deniker, H. Hippius, and P.B. Bradley), Excerpta Medica, Amsterdam, pp. 321–329.
27. Snyder, S.H., Banerjee, S.P., Yamamura, H.L., and Greenberg, D. (1974) *Science*, **184**, 1243–1253.
28. Seeman, P. (1987) *Synapse*, **1**, 133–152.
29. Snyder, S.H. and Yamamura, H.I. (1977) *Arch. Gen. Psychiatry*, **34**, 236–239.
30. Thomas, D., Wu, K., Kathofer, S., Katus, H.A., Scoels, W., Kiehn, J., and Karle, C.A. (2003) *Br. J. Pharmacol.*, **139**, 567–574.
31. Shin, J.-G., Soukhova, N., and Flockhart, D.A. (1999) *Drug Metab. Dispos.*, **27**, 1078–1084.
32. Michalets, E.M. (1998) *Pharmacotherapy*, **18**, 84–112.
33. Charpentier, P. (1953) U. S. Patent 2.645.640.
34. Charpentier, P. (1951) Ger. Patent 910.301.
35. Chapentier, P., Gailliot, P., Jacob, R., Gaudechon, J., and Buisson, P. (1952) *Compt. Rend.*, **235**, 59.
36. Evans, J. and Smiles, S. (1935) *J. Chem. Soc.*, 1263–1265.
37. Rhone-Poulenc (1960) BE. 556475.
38. Bernstein, J. and Sowinski, F. (1956) Ger. Patent 1.095.836.
39. Ulluot, G.E. (1960) U.S. Patent 2,921,069.
40. Smith, Kline & French Laboratories (1959) Brit. Patent 813.861.
41. Craig, P.N., Nodiff, E.A., Lassferty, J.J., and Ullyot, G.E. (1957) *J. Org. Chem.*, **22**, 709–711.
42. Bourquin, J.-P., Schwarb, G., Gamboni, G., Fischer, R., Ruesch, L., Guldimann, S., Theus, V., and Schenker, E. (1958) *Helv. Chim. Acta*, **41**, 1072–1108.
43. Renz, J. and Schwarb, G. (1960) U.S. Patent 3,084,161.
44. Horclois, R.J. (1959) U.S. Patent 2,902,484.
45. Horclois, R.J. (1957) GB 780193.
46. Ullyot, G.E. (1960) U.S. Patent 2,921,069.
47. Sharma, H., Banerjee, S., Sharma, V., and Mital, R. (1968) *J. Med. Chem.*, **11**, 1244–1245.
48. Anderson, E.L., Bellizona, G.B., Craig, P.N., Jaffe, G.E., Janewaes, K.P.,

Kaiser, C., Hester, B.M., Nikawitz, E.J., Pavloff, A., Reift, H.E., and Zirkle, Ch.L. (1962) *Arzneimittel-Forschung*, **12**, 937–942.

49. Cusic, J.W. (1956) U.S. Patent 2,766,235.
50. Dahl, T., Tornøe, C.W., Bang-Andersen, B., Nielsen, P., and Jørgensen, M. (2008) *Angew. Chem. Int. Ed.*, **47**, 1726–1728.
51. Ma, D., Geng, Q., Zhang, H., and Jiang, Y. (2010) *Angew. Chem. Int. Ed.*, 1291–1294.
52. Seeman, P. and Kapur, S. (2000) *Proc. Natl. Acad. Sci. U.S.A.*, **97**, 7673–7675.
53. Girault, J.A. and Greengard, P. (2004) *Arch. Neurol.*, 640–645.
54. Zhang, A., Neumeyer, J.L., and Baldessarini, R. (2007) *J. Chem. Rev.*, 274–302.
55. Hartman, D.S. and Civelli, O. (1997) *Prog. Drug Res.*, **48**, 174–194.
56. Hietala, J., Syvälahti, E., Vilkman, H., Vuorio, K., Räkköläinen, V., Bergman, J., Haaparanta, M., Solin, O., Kuoppamäki, M., Eronen, E., Ruotsalainen, U., and Salokangas, R.K. (1995) *Lancet*, **346**, 1130–1131.
57. Wolf, M.E. and Roth, R.H. (1990) *Ann. N. Y. Acad. Sci.*, **604**, 323–342.
58. Seeman, P. (2002) *Can. J. Psychol.*, **47**, 27–38.
59. Newcomer, J.W. (2005) *CNS Drugs*, **19** (Suppl. 1), 1–93.
60. Horn, A.S. and Synder, S.H. (1971) *Proc. Natl. Acad. Sci. U.S.A.*, **68**, 2325–2328.
61. Feinberg, A.P. and Snyder, S.H. (1975) *Proc. Natl. Acad. Sci. U.S.A.*, **72**, 1899–1903.
62. Sriram, D. and Yogeeswari, P. (2010) *Medicinal Chemistry*, 2nd edn, Chapter 9, Dorling Kindersley, India, pp. 173–181.
63. Groves, J.E. and Mandel, M.R. (1975) *Arch. Gen. Psychiatry*, **32**, 893–900.

6
Arylpiperazine-Based 5-HT_{1A} Receptor Partial Agonists and 5-HT_{2A} Antagonists for the Treatment of Autism, Depression, Anxiety, Psychosis, and Schizophrenia

Irini Akritopoulou-Zanze

6.1
Introduction

Arylpiperazines are a diverse group of compounds in which at least one of the nitrogens of a piperazine is substituted with an aryl group (Figure 6.1: **1**). Phenylpiperazines, a subset of arylpiperazines, refer to a class of compounds in which the aryl group is limited to phenyl analogs (**2**). The arylpiperazine moiety can be found in more than 50 launched drugs, which have been developed for such diverse indications as cancer; bacterial, viral, and fungal infections; urinary incontinence; glaucoma; cardiovascular diseases; and mental disorders. This chapter focuses on arylpiperazines that modulate serotonergic actions and pathways associated with mental disorders, such as autism, depression, anxiety, psychosis, and schizophrenia. These arylpiperazines are typically fused in larger molecules, such as **3**, or have substituents attached to the second piperazine nitrogen as in derivatives **4**.

Compounds with the general structures **3** and **4**, exhibit unique selectivity profiles that often, in addition to the serotonergic component, possess various degrees of dopaminergic, histaminergic, or adrenergic activities. Their actions on these receptors also vary from partial agonism to antagonism and numerous combinations thereof. As a result, the pharmacological profile and side effects of each arylpiperazine is unique with the serotonin component being the common thread across all molecules.

6.2
History

Arylpiperazine derivatives have been known since the late 1920s. A report in 1927 mentions the preparation of arylpiperazine analogs such as vanillyl piperamides (**5**) and the study of their pungency (Figure 6.2) [1].

In 1929, quipazine (**6**) was synthesized through substitution of α-chloroquinoline with piperazine [2]. It took, however, another 40 years for its pharmacological actions

Bioactive Heterocyclic Compound Classes: Pharmaceuticals,
First Edition. Edited by Jürgen Dinges and Clemens Lamberth.

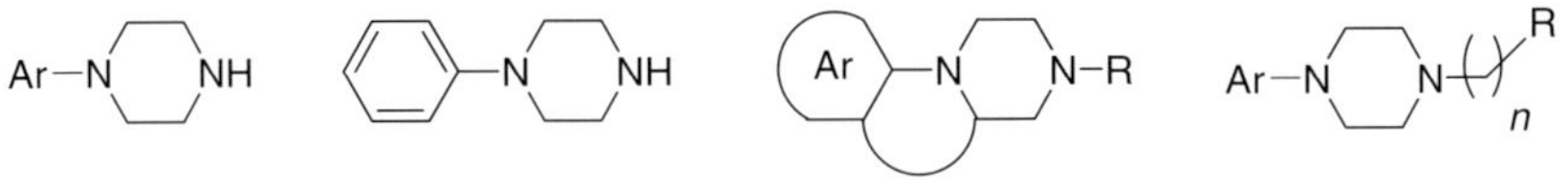

1 **2** **3** **4**

Figure 6.1 Representative arylpiperazine substructures.

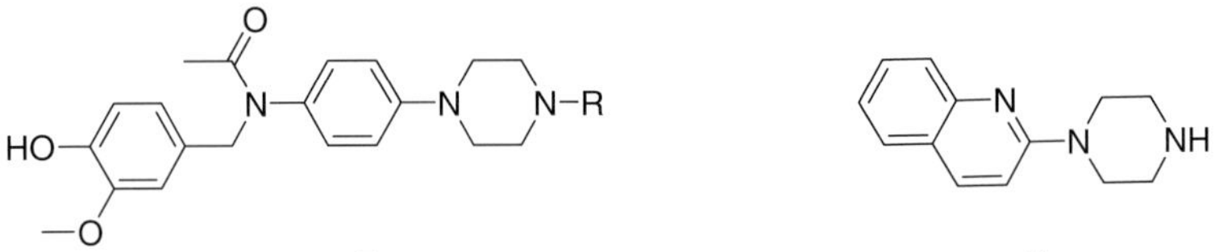

5 **6**

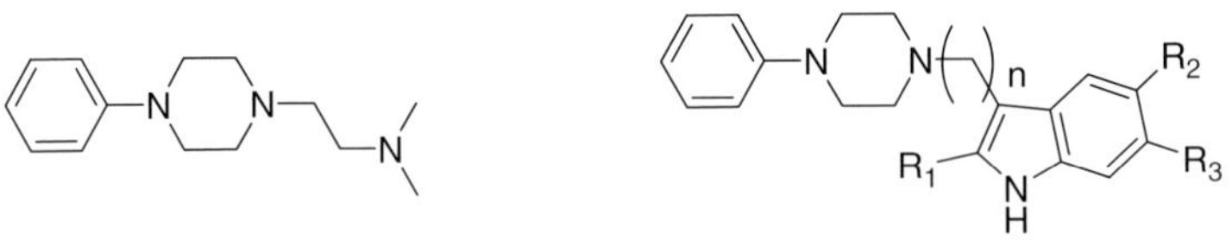

7 **8**

Figure 6.2 Early arylpiperazines.

[3] and antidepressant properties [4] to be understood. Plain phenylpiperazine (**2**) was prepared in 1933 from phenylamine and bis(2-chloroethyl)amine in MeOH under reflux conditions [5]. A decade later, the first hints of biological activity of this substructure came from a Croatian laboratory where numerous phenylpiperazine analogs, such as **7**, were tested in mice and guinea pigs and found to possess antihistaminic activities [6]. It was also noted that some compounds caused a narcotic effect in animals. By the late 1950s and early 1960s, the tranquilizing properties of arylpiperazines were studied by numerous laboratories. For example, phenylpiperazine derivatives **8** were found to block adrenergic action and reduce *d*-amphetamine-induced locomotor activity in mice [7, 8]. Several patents appeared claiming arylpiperazine substructures as tranquilizers [9] for the treatment of psychic disorders [10] and as cerebral depressants with antiemetic properties [11]. The 1950s and 1960s were also the times when the first antipsychotic drugs such as chlorpromazine, antidepressants such as imipramine (belonging to the tricyclic class), and anxiolytics such as the ones belonging to the benzodiazepine class emerged as major therapies for their respective indications. There was a great deal of enthusiasm and activity in the field toward the discovery of drugs for the treatment of neurological disorders. One of the first compounds containing the phenylpiperazine substructure **2** to be developed as a drug was trazodone (**9**) (Figure 6.3).

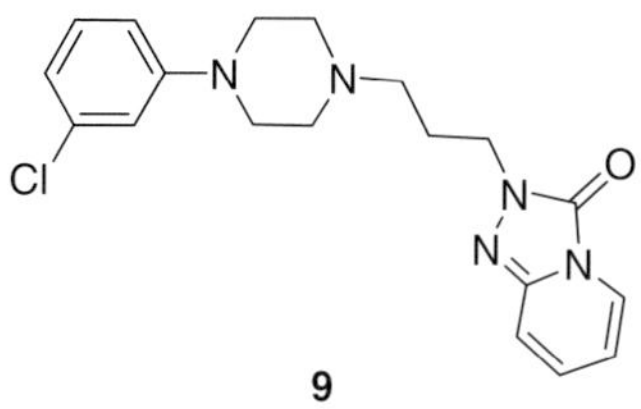

9
Trazodone

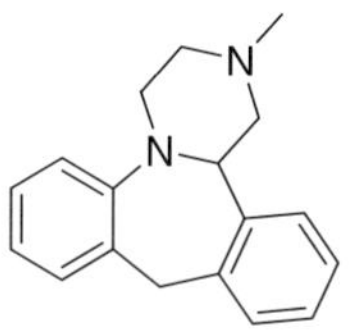

10
Mianserin

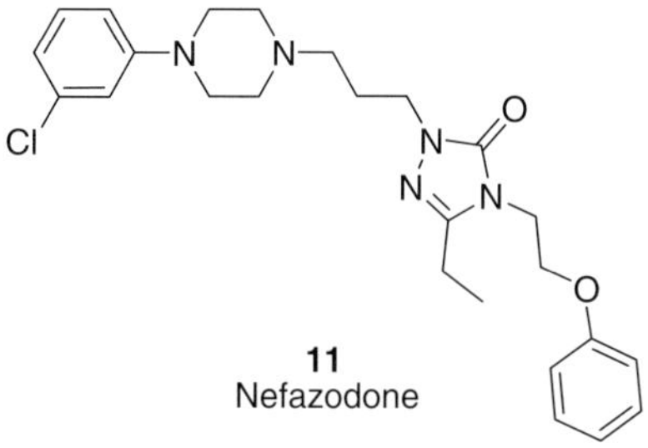

11
Nefazodone

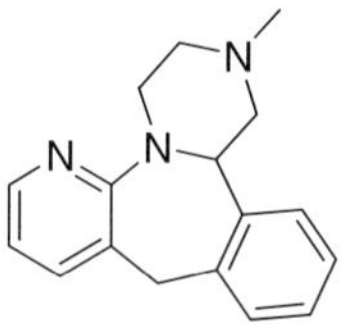

12
Mirtazepine

Figure 6.3 Drugs based on the arylpiperazine general structures **3** and **4** possessing antidepressant activity.

This compound was first disclosed in 1968 in a patent by an Italian company, F. Angelini Laboratories, claiming compounds with tranquilizing, hypotensive, and analgesic activities [12]. Several studies in the following years demonstrated the antiserotonergic and adrenolytic effects of this compound [13–15]. Clinical studies conducted in Italy indicated that trazodone (**9**) had an antidepressant profile similar to that of imipramine [16]; however, its effects on schizophrenia were less pronounced [17].

About the same time in the Netherlands, the scientists at Organon prepared mianserin (**10**) in a 14-step synthesis, a compound that shared the same phenylpiperazine substructure **2** with trazodone (**9**), placed, however, in a completely different type of molecule and made for a completely different purpose. At the time, it was hypothesized that antiallergic activity could have, in addition to antihistaminic, an antiserotonergic component; thus, mianserin (**10**) and several other analogs were synthesized as potential antiallergic agents [18]. Mianserin (**10**) was originally tested in rabbits and rats to confirm its antiserotonergic activity and in dogs to access its cardiovascular profile [19]. As the story was told by Organon scientists Peet and Behagel a decade later [20], several early human trials were carried out, first, for the treatment of migraine with disappointing results and then, for the treatment of asthma, where the results were more encouraging; however, it was found that mianserin (**10**) had sedative effects. In another unpublished study in Ireland, some of the patients treated with **10** showed a marked improvement in their mood. The effect of mianserin (**10**) was noted by the unit staff, who called it the "good humour pill." A year later, Organon scientists became aware of EEG studies being performed in the United States and saw these studies as the next step

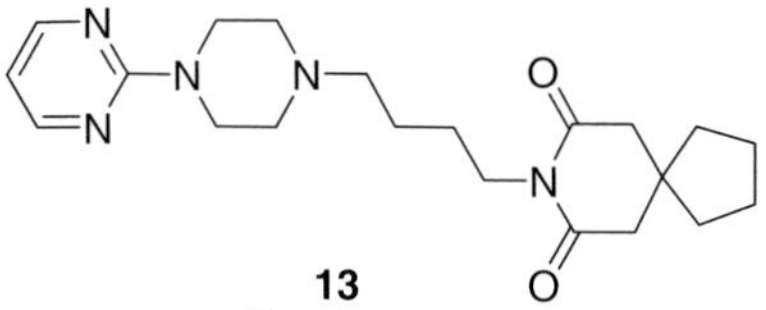

13
Buspirone

Figure 6.4 Buspirone (**13**), a drug possessing anxiolytic activity.

to the development of mianserin (**10**). Indeed, in 1970, Itil *et al.* [21] carried out EEG studies with **10** and noted that its EEG pattern was similar to that of tricyclic antidepressants. The compound was then assessed in multiple clinical trials for its antidepressant properties [22, 23]. While the efficacy of mianserin (**10**) was comparable to other antidepressants, it lacked the cardiovascular and anticholinergic side effects of the antidepressants known at the time [24, 25].

In the late 1960s, another company across the Atlantic was also hard at work to prepare numerous phenylpiperazine analogs as psychosedative agents. In the laboratories of Mead Johnson, various phenylpiperazines [26] and later pyrimidinopiperazines [27], including buspirone (**13**) (Figure 6.4), were prepared. Buspirone (**13**) was tested in a rhesus monkey aggressive behavior model, exhibiting inhibition of aggressive behavior [27, 28].

Meanwhile, trazodone (**9**) was marketed in Italy in 1972 for the treatment of depression. Mianserin (**10**) was launched in 1975 for the same indication. Mianserin (**10**) was later replaced by a second-generation tetracyclic compound mirtazapine (**12**), launched in 1994 and differing from mianserin (**10**) only by a C to N replacement on the aromatic ring. Trazodone (**9**) was marketed in the United States by Bristol Myers Squibb (BMS) in 1982 and was followed up by nefazodone (**11**), which was launched in 1994. Nefazodone (**11**) was expected to have fewer side effects; however, in 2004, BMS discontinued the production of this drug due to rare, but severe, hepatotoxicity. At present, trazodone (**9**) and mirtazapine (**12**) have become some of the most prescribed antidepressants.

Further testing of buspirone (**13**) in animal models as well as in human clinical trials demonstrated its beneficial profile for anxiety disorders [29]. Buspirone (**13**) was launched in 1985 by BMS, which had acquired Mead Johnson, for the treatment of anxiety. Currently, it is the only nonbenzodiazepine anxiolytic commonly prescribed by doctors.

The next major development in the history of arylpiperazines came in the early 1990s when the benzoisothiazolyl piperidine ziprasidone (**14**) (Figure 6.5) was developed by Pfizer for the treatment of schizophrenia. By that time, the majority of arylpiperazines were known to bind to serotonin receptors [30] and a series of compounds named "atypical antipsychotics" had already entered the market having a mixed dopamine/serotonergic component. The atypical antipsychotics were expected to have better safety profiles than older drugs that mainly targeted the dopamine receptors. Ziprasidone (**14**) was a dual dopamine/serotonin antagonist [31] that blocked brain dopamine receptors in a positron emission

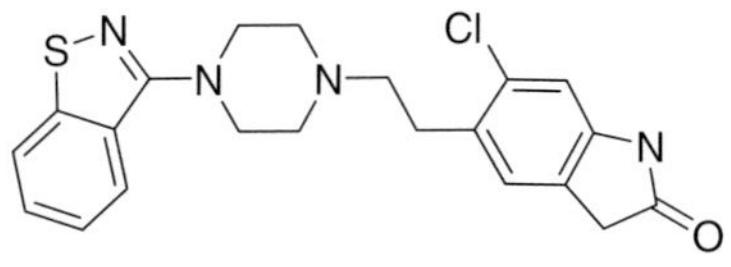

14
Ziprasidone

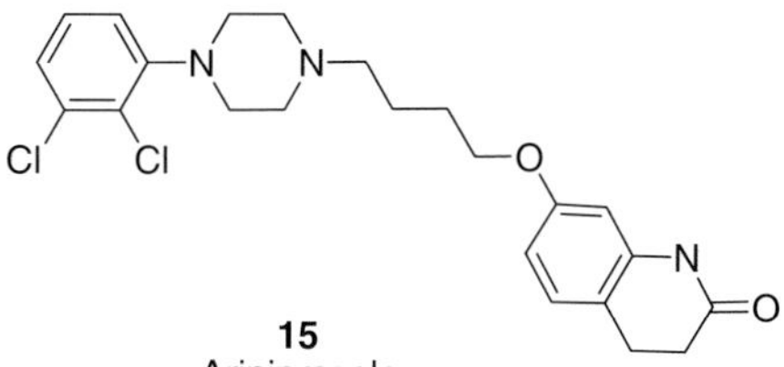

15
Aripiprazole

Figure 6.5 Drugs based on the arylpiperazine substructure **1**, possessing antipsychotic activity.

tomography dopamine receptor occupancy study in humans, with ^{11}C-raclopride as the radiotracer [32].

After a series of clinical studies that confirmed its efficacy and superior safety toward older antiphychotics, ziprasidone (**14**) was approved by the FDA in 2000 for the treatment of schizophrenia and in 2003 for bipolar disorder [33]. A couple of years later, another arylpiperazine, aripiprazole (**15**), was launched for the treatment of schizophrenia by Otsuka and BMS. Aripiprazole (**15**) was well tolerated in clinical trials addressing positive, negative, and cognitive symptoms of schizophrenia [34]. Aripiprazole (**15**) was also found to be effective in the treatment of irritability associated with autism [35]. Both ziprasidone (**14**) and aripiprazole (**15**) have low risk for metabolic or endocrine side effects, typically associated with older antipsychotics, and are among the most commonly prescribed antipsychotics today.

6.3 Synthesis

The construction of the basic arylpiperazine moieties can be accomplished by various routes, the most prominent of which are shown in Scheme 6.1. Aryl amines (**16**) can react with bis(2-chloroethyl)amine (**17**) to form the final arylpiperazines (**1**) (Route 1). In reversing this strategy, **17** can first be attached to an aryl group to form the bis(2-chloroethyl)arylamines (**18**) followed by reactions with amines (**19**) to close the piperazine rings (Route 2). Plain piperazine (**21**) also reacts with aryl halides (**20**), particularly heterocyclic halides, under basic conditions to form the final arylpiperazines (**1**) (Route 3).

The most common way of installing additional substituents on the free NH group of **1** is described in Scheme 6.2 (Route A). Arylpiperazines (**1**) react with aryl alkyl halides (**22**) under basic conditions in the presence of an accelerator such as an alkalimetal iodide. In most cases, heat is also necessary [36]. For example, buspirone (**13**), ziprasidone (**14**), and aripiprazole (**15**) have been constructed using this general methodology.

Alternatively, the linker can first be attached to the arylpiperazine ring to form compounds **24** (Route B), which react, typically under basic conditions, with compounds **25** to introduce the heterocycles. This methodology has been

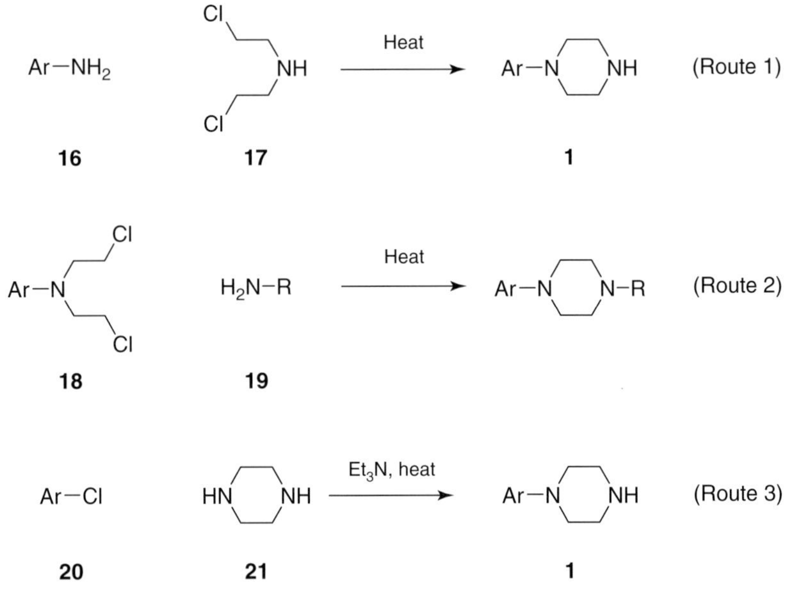

16 17 1 (Route 1)

18 19 (Route 2)

20 21 1 (Route 3)

Scheme 6.1 Most common synthesis routes for simple arylpiperidines.

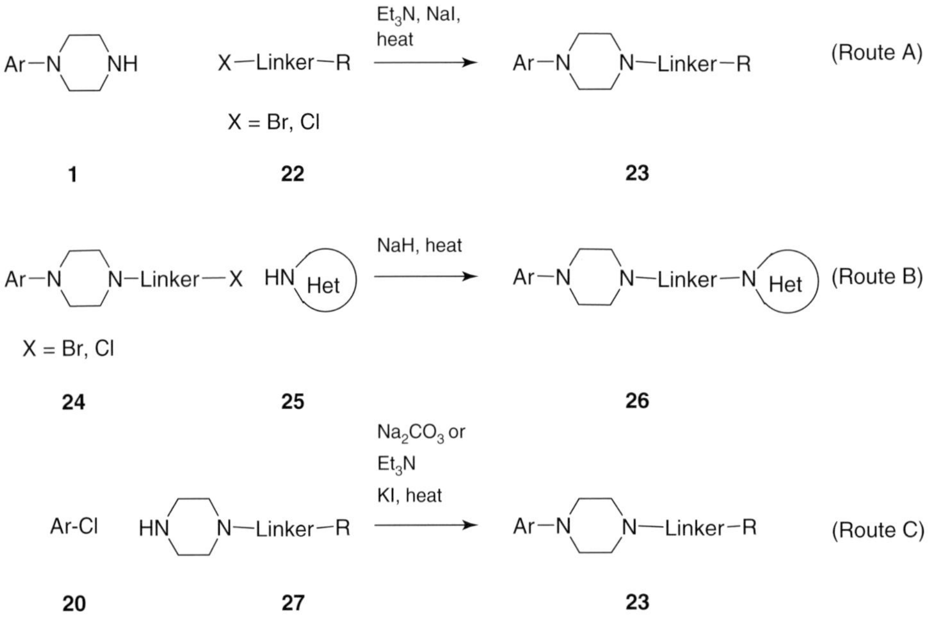

1 22 23 (Route A)

24 25 26 (Route B)

20 27 23 (Route C)

Scheme 6.2 Synthesis of *N*-substituted arylpiperidines.

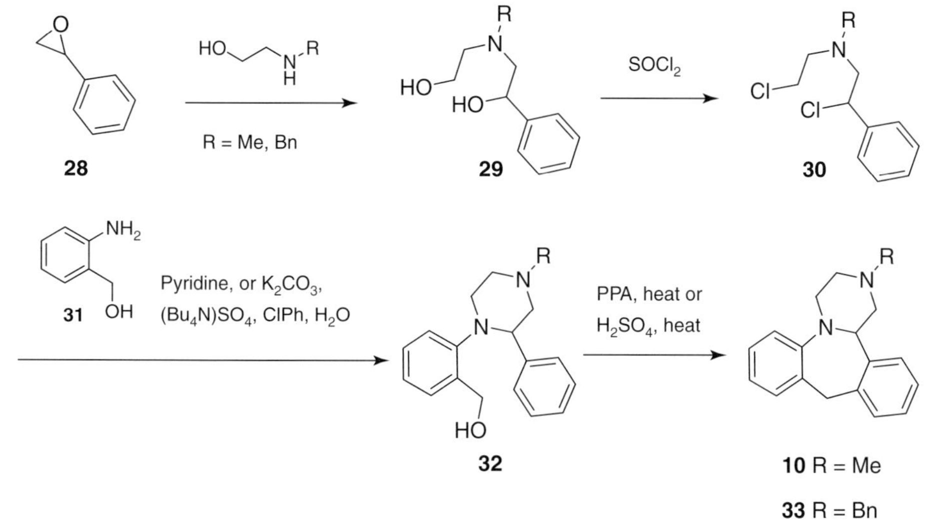

Scheme 6.3 Synthetic routes to mianserin (**10**) and analogs.

applied mostly in the preparation of compounds such as trazodone (**9**) and nefazodone (**11**) [12].

Another common preparation of arylpiperazine derivatives **23** is by formation of the arylpiperazine connection as shown in Route C. This reaction is typically carried out under basic conditions in the presence of a catalytic amount of NaI or KI and heat. A variety of solvents have been used with varying degrees of overall yields and ease of purification of the final products [37].

Tricyclic arylpiperazines such as mianserin (**10**) and mirtazapine (**12**) are prepared following slightly different routes. In the case of mianserin (**10**) (Scheme 6.3), the phenyl piperazine moiety **32** is constructed in a manner similar to Route 1 in Scheme 6.1, by reacting substituted bis(2-chloroethyl)amine **30** with aniline **31**. Compound **30** can easily be prepared from epoxide **28** via the dihydroxy compound **29**. The seven-membered ring is obtained under acidic conditions and heat [38, 39]. In cases where R is a protecting group, **33** can be deprotected allowing further manipulations of the free amino group.

In the case of mirtazapine (**12**) (Scheme 6.4), the seven-membered ring is constructed starting from the aminopyridine **34** either by reaction with 2,2-dimethoxy-*N*-methylethanamine **35** or the ester equivalent **40** under acidic conditions to provide intermediates **36** or **41**, respectively. In the case of **36**, the construction of the piperazine ring is then accomplished by reaction with bromoethanol (**37**), replacement of the hydroxyl group of **38** with chloride and ring closure of intermediate **39** under basic conditions. Alternatively, attachment of amine **42** to the intermediate **41** and subsequent replacement of the hydroxyl group with chloride yields **43**, which closes, in a manner similar to **39**, to the

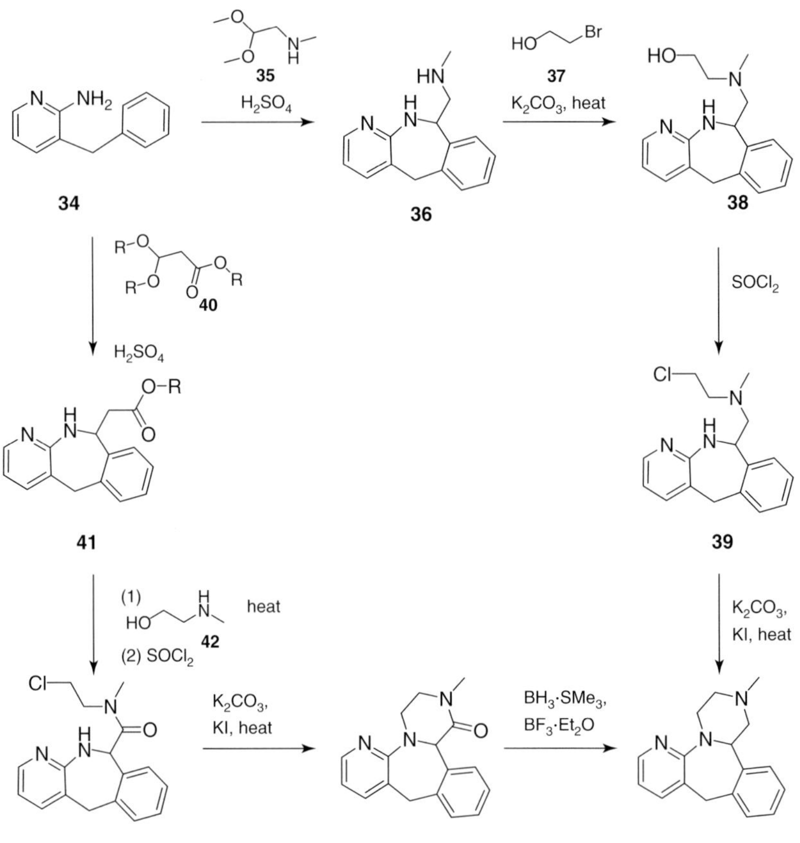

Scheme 6.4 Synthetic routes to mirtazapine (**12**).

piperazinone (**44**). Mirtazapine (**12**) is then obtained after borane reduction of the amide bond [40].

6.4 Mode of Action

While the therapeutic properties of compounds containing the arylpiperazine substructure **1** were discovered and capitalized on in the form of marketed drugs already in the 1970s, the precise mechanism of their action was examined decades later. The interactions of many of these compounds with multiple receptors in the brain made the interpretation of their pharmacological profiles difficult. Most arylpiperazines were known to have serotonergic action; however, the differentiation of serotonin receptors into the 5-hydroxytryptamine 1 5-HT_1 and 5-HT_2

types was not reported until 1979 [41]. In 1981, trazodone (**9**) and mianserin (**10**) were shown to bind to the 5-HT_2 receptor, and it was hypothesized that it was this interaction that was responsible for their antidepressant properties [42]. In 1983, buspirone (**13**) was reported to bind to the 5-HT_1 receptor, which led to the implication of this receptor to anxiolytic effects [43]. By the end of this decade, additional subtypes for both the 5-HT_1 [44, 45] and 5-HT_2 [46] receptors were reported. It was then found that trazodone (**9**) and mianserin (**10**) were targeting the 5-HT_{2A} receptor, while buspirone (**13**) was more selective for 5-HT_{1A} [47].

The discovery of the effects that various compounds had on multiple types and subtypes of serotonin receptors gave rise to a wide range of hypotheses regarding their antipsychotic [48] and antidepressant [49] properties that were investigated. Buspirone (**13**) and other arylpiperazines were found to be partial agonists for the 5-HT_{1A} receptor, and the target was implicated in psychiatric disorders such as anxiety and depression [50]. The antipsychotic actions of ziprasidone (**14**) and aripiprazole (**15**) were initially attributed to their mixed 5-HT_{2A}/D_2 antagonistic activities; however, the compounds were also 5-HT_{1A} receptor partial agonists [51].

Current hypotheses on modulation of mental disorders by arylpiperazine drugs are based on serotonin modulation and its effects on dopaminenergic [52] and glutaminergic systems [53]. Table 6.1 highlights some of the most common arylpiperazine drugs and their overall receptor profiles along with hypothesized mechanisms of action. Thus, the antidepressant effects of trazodone (**9**) and nefazodone (**11**) are explained partially by their 5-HT_{2A} antagonism and also possibly by serotonin uptake inhibition. Trazodone (**9**) also exhibits side effects such as hypertension and sedation, which have been attributed to adrenergic and histaminic activities, respectively [54].

Histaminic activity is the main driver of sedative side effects observed in mianserin (**10**) and mirtazapine (**12**), while the weight gain and increase in appetite associated with administration of these compounds is attributed to their 5-HT_{2C} antagonistic properties. Buspirone (**13**) is fairly selective as a partial 5-HT_{1A} agonist and exhibits no significant side effects. Ziprasidone (**14**) and aripiprazole (**15**) have antipsychotic properties attributed to their mixed 5-HT_{2A}/D_2 antagonism and anxiolytic/antidepressant properties attributed to 5-HT_{1A} agonism.

It should be noted that although there are many theories regarding mechanisms of action for the various mental disorders, these diseases are very complicated and most likely there are multiple factors contributing to their manifestation.

6.5 Structure–Activity Relationship

Arylpiperazines are divided into several subclasses exhibiting similar structural and biological characteristics. Mianserin (**10**) and mirtazapine (**12**) belong to a pharmacological class called *tetracyclic antidepressants* (Table 6.2) [61]. The majority

Table 6.1 Pharmacological profiles of marketed drugs.

Drug name (brand name)	Indication	Receptor and transporter binding affinities: K_i (nM)		Mechanism of action	References
Trazodone (**9**) (Desyrel®)	Depression	Alpha1	12	5-HT_{2A} antagonism	[55]
		5-HT_{2A}	20	SERT inhibitor	
		H1	29		
		5-HT_{1A}	42		
		Alpha2	106		
		5-HT_{2C}	244		
		SERT	252		
Nefazodone (**11**) (Serzone®)	Depression	Alpha1	6	5-HT_{2A} antagonism	
		5-HT_{2A}	7	SERT inhibitor	
		H1	30		[55]
		5-HT_{2C}	43		
		5-HT_{1A}	52		
		Alpha2	84		
		SERT	459		
		NET	618		
Mianserin (**10**) (Athymil®)	Depression	H1	1.7	5-HT_{2A} antagonism	[56]
		5-HT_{2A}	4	Alpha2 antagonism	
		Alpha2A	5		
		5-HT7	56		
		5-HT_{2C}	4		
		Alpha1	74		
		NET	101		
Mirtazapine (**12**) (Remeron®)	Depression Insomnia	H1	1.6	5-HT_{2A} antagonism	[56, 57]
		5-HT_3	8	5-HT_{2C} antagonism	
		Alpha2A	20	Alpha2 antagonism	
		5-HT_{2C}	39		
		5-HT_{2A}	69		
		5-HT_{2B}	199		
		5-HT_7	265		
		Alpha1	608		
Buspirone (**13**) (Buspar®)	Anxiety	5-HT_{1A}	20	5-HT_{1A} partial agonism	[58]
		Alpha1	367		
		5-HT_{2A}	482		
		D2	852		
Ziprasidone (**14**) (Geodon®)	Schizophrenia Mania Bipolar disorder	5-HT_{2A}	0.4	5-HT_{2A} antagonism	[58, 59]
		5-HT_{2C}	0.7	5-HT_{2C} antagonism	
		5-HT_{2B}	2	5-HT_{1A} agonism	
		$5\text{-HT}_{1B/1D}$	2	D2 antagonism	

Table 6.1 (*Continued*).

Drug name (brand name)	Indication	Receptor and transporter binding affinities: K_i (nM)		Mechanism of action	References
		5-HT_{1A}	2		
		D2	3		
		D3	7		
		5-HT_7	9		
		Alpha1	13		
		D_4	32		
		H1	47		
		NET	48		
		SERT	53		
		D1	130		
		5-HT_6	76		
		5-HT_3	398		
Aripiprazole (**15**) (Abilify®)	Schizophrenia Mania Agitation Bipolar disorder Autism Major depression	5-HT_{2B}	1	5-HT_{2A} antagonism D2/D3 partial agonism 5-HT_{1A} partial agonism	[60]
		D3	4		
		D2	5		
		5-HT_{1A}	8		
		5-HT_{2A}	16		
		5-HT_{2C}	50		
		5-HT_7	63		
		Alpha1	100		
		D4	100		
		H1	100		
		5-HT_6	199		
		5-HT_3	500		

Abbreviations: SERT, serotonin transporter; NET, norepinephrine transporter.

of these compounds contain a central seven-membered ring fused with two aromatic rings and a piperazine or piperidine. The compounds are racemic mixtures; however, the (*S*)-enantiomer is the most active one against serotonergic and adrenergic receptors. In the case of mirtazapine (**12**), the (*R*)-enantiomer blocks the 5-HT_3 receptor [62].

Compounds containing an arylpiperazine attached to an *n*-butyl side chain, linked to a dione-containing heterocycle, have been grouped together in a pharmacological class called *azapirones* (Table 6.3) [63].

Azapirones share agonistic activity for the 5-HT_{1A} receptor and exhibit antidepressant and anxiolytic properties. The structure–activity relationship (SAR) of azapirones toward 5-HT_{1A} activity has been extensively studied [66, 67]. Aryl groups have been modified to include substituted phenyl rings, in particular, with an ortho-methoxy substituent. Pyrimidines or isothiazolyl piperazines were

Table 6.2 Tetracyclic antidepressants.

Drug name (company)	Structure	Current status
Mianserin (**10**) (Organon/Merck)	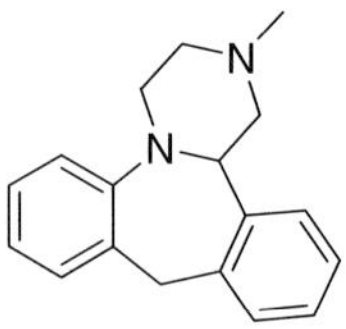	Approved in 1975; in most markets phased out and replaced by mirtazapine (**12**)
Mirtazapine (**12**) (Organon/Merck)	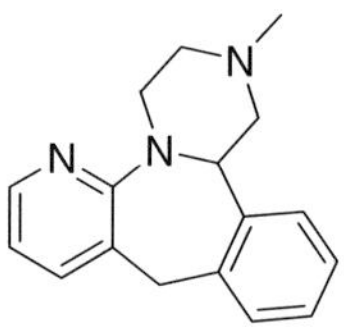	Approved in 1994; used for the treatment of depression/insomnia

ultimately the aryl groups that led to multiple compounds progressing to clinical evaluation and some of them becoming drugs. The ending "spirone" was introduced for buspirone (**13**) to describe the spirocyclic dione heterocycle at the other end of the side chain and persisted in the naming of compounds that did not contain a spirocycle but belonged to the azapirone class because of their overall pharmacological profile. The majority of compounds in this class are structurally similar. In most cases, the dione heterocycle is modulating the overall pharmacological profile and therefore the therapeutic properties and side effects of those compounds.

Compounds with an arylpiperazine attached to an *n*-propyl side chain, linked to a heterocycle and especially meta chloro-phenyl piperazines, have a different biological profile, most being mixed 5-HT_{2A} antagonists and serotonin reuptake inhibitors (Table 6.4).

Most of the compounds contain a triazolone heterocycle as in **9**, **11**, and **50** or in the case of mepiprazole (**51**), a pyrazole ring. The latest compound in this series is benzoxazole piperazine (**52**), a mixed 5-HT_{1A} partial agonist and serotonin reuptake inhibitor, which has been recently approved for the treatment of depression [68].

Another set of arylpiperazines distinguished for their antipsychotic effects are the isothiazolyl piperazines ziprasidone (**14**) and lurasidone (**53**) (Table 6.5).

It is interesting to note that these compounds bear extensive resemblance to compounds such as perospirone (**47**), containing, however, shorter side chains. As a result, the overall receptor binding profile of ziprasidone (**14**) and lurasidone (**53**) is different. The compounds are potent 5-HT_{2A} and D_2 antagonists, thus exhibiting antipsychotic properties. Similarly, aripiprazole (**15**) displays structural similarities

Table 6.3 Azapirones.

Drug name	Structure	Current status
Buspirone (**13**) (BMS)	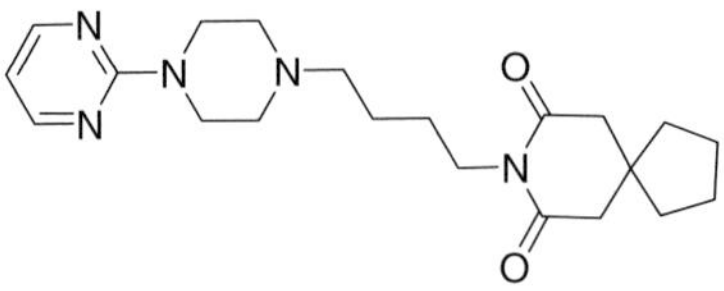	Approved in 1985; used for the treatment of anxiety[a]
Gepirone (**45**) (BMS/Fabre/Kramer)	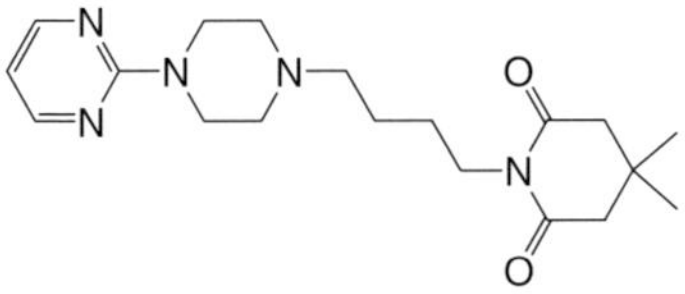	Rejected in 2004 and 2007 by FDA[a]
Tandospirone (**46**) (Dainippon Sumimoto)	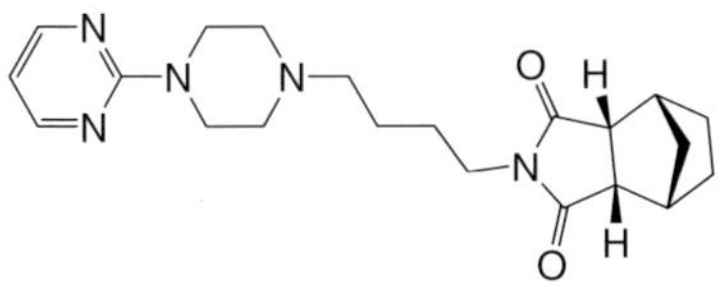	Approved in 1996, only in Japan and China; used for the treatment of depression, anxiety, and psychosis[a]
Perospirone (**47**) (Dainippon Sumimoto)	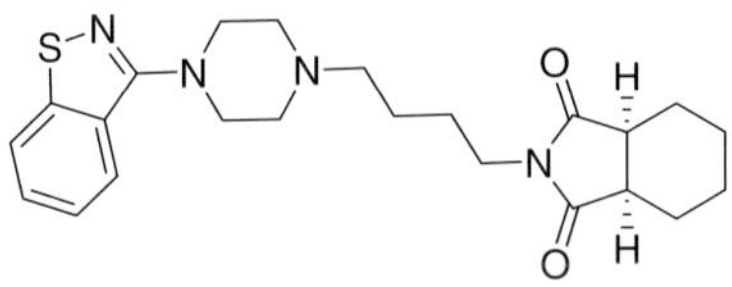	Approved in 2001, only in Japan; used for the treatment of schizophrenia[a]
Zalospirone (**48**) (Pfizer)	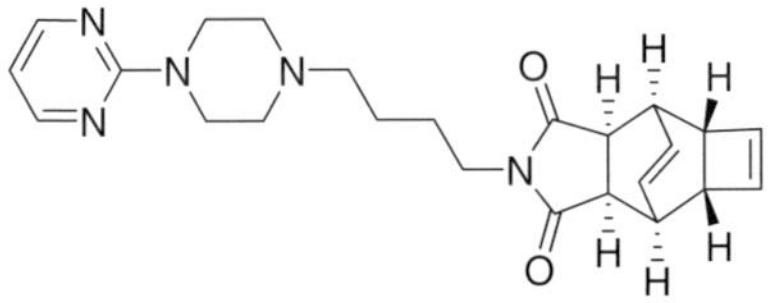	Reached phase II but was not further developed[a]. In clinical trials, nontolerated dizziness and nausea were observed at therapeutic doses [64]
Eptapirone (**49**) (Pierre Fabre)	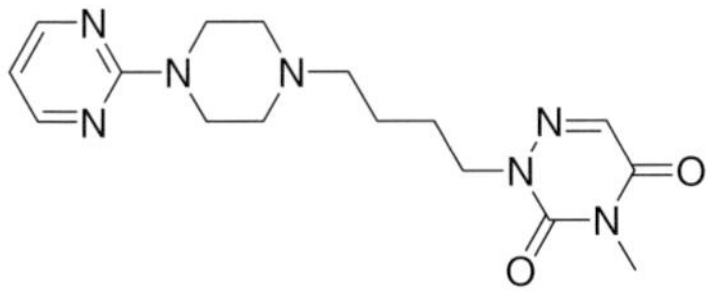	Reached phase II but was not further developed[a]. In clinical trials, dizziness, body temperature reduction, and hormonal increases were observed [65].

[a] Information obtained from the Thompson Reuter Integrity^{SM} database.

Table 6.4 Serotonin modulator antidepressants inhibiting serotonin reuptake.

Drug name	Structure	Current status
Trazodone (**9**) (Angelini/BMS)	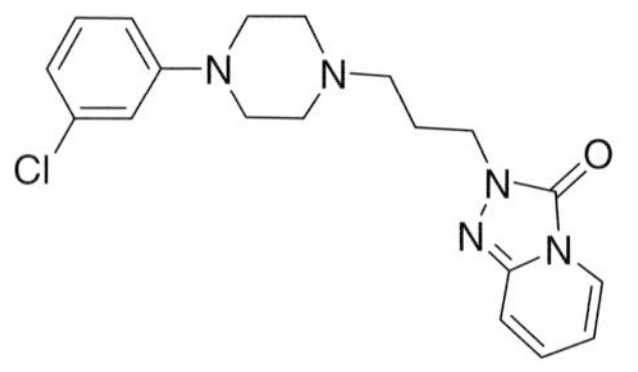	Approved in 1972 in Italy and in 1982 in the United States; used for the treatment of depression[a]
Etoperidone (**50**)	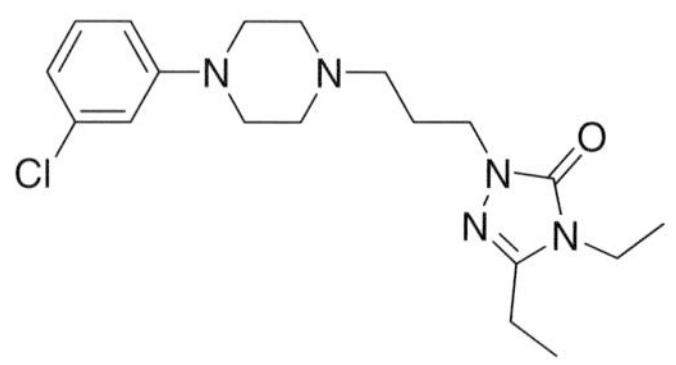	Launched in 1977 for the treatment of depression, discontinued[a]
Nefazodone (**11**) (BMS)	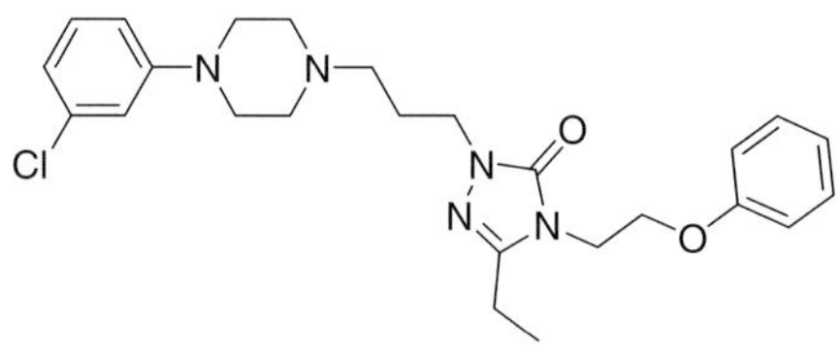	Approved in 1994 for the treatment of depression[a]
Mepiprazole (**51**)	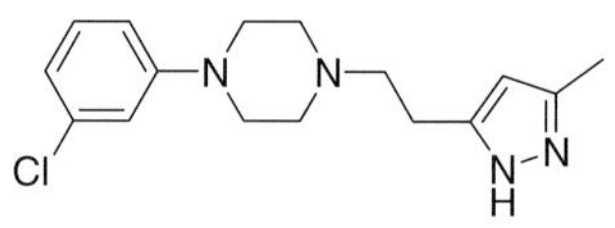	Approved only in Spain for the treatment of anxiety[a]
Vilazodone (**52**) (Merck KGaA/ Clinical Data, Inc.)	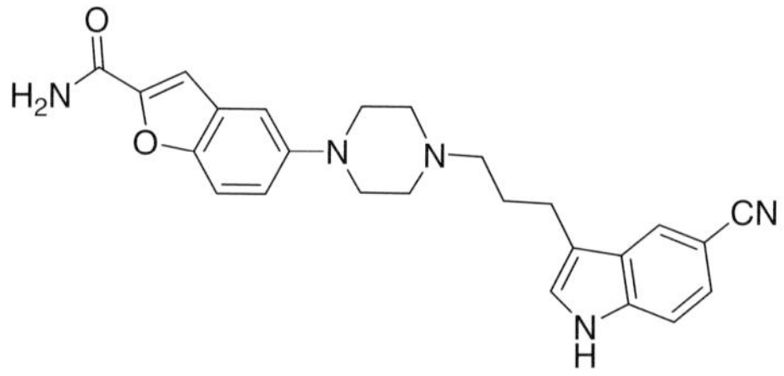	Registered in 2011 for the treatment of major depression[a]

[a]Information obtained from the Thompson Reuter Integrity^{SM} database.

to the compounds in Table 6.4, only containing a longer side chain and a different terminal heterocycle.

Compounds **54** and **55** are included in Table 6.5 to highlight the diversity of arylpiperazine analogs that possess a serotonin and dopaminergic component in their biological profiles, rendering them useful for the treatment of schizophrenia.

Table 6.5 Arylpiperazine antipsychotics.

Drug name	Structure	Current status
Ziprasidone (**14**) (Pfizer)	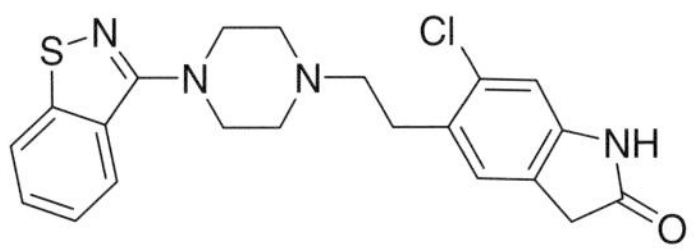	Approved in 2001; used for the treatment of schizophrenia, mania, and bipolar disorder. Received black box warning due to increased mortality in elderly patients with dementia-related psychosis[a]
Lurasidone (**53**) (Dainippon Sumimoto)	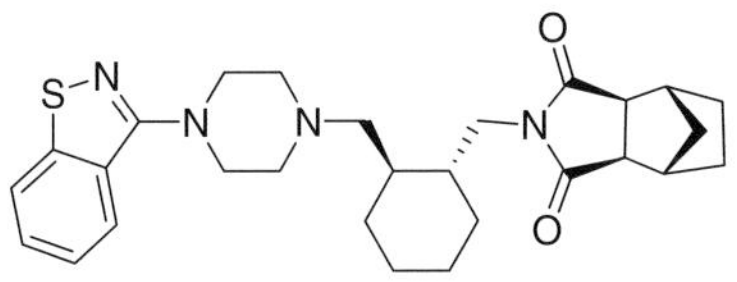	Approved in 2010 for the treatment of schizophrenia; pending approval in the United States for the treatment of bipolar disorder[a]
Aripiprazole (**15**) (BMS/Otsuka Pharmaceuticals)	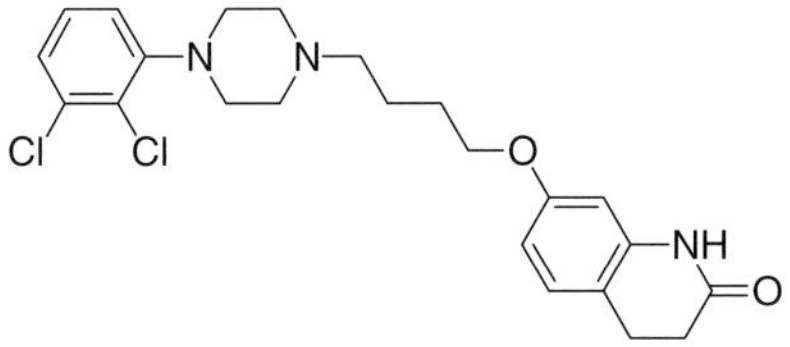	Approved in 2002; used for the treatment of schizophrenia, mania, bipolar disorder, autism, and major depression. Currently in multiple clinical trials for additional indications[a]
Blonanserin (**54**) (Dainippon Sumimoto)	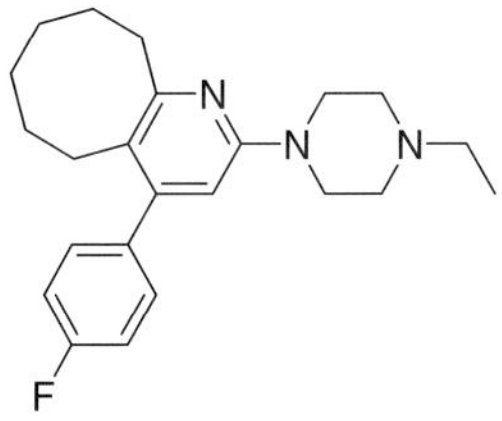	Approved in 2008, only in Japan; used for the treatment of schizophrenia[a]
Bifeprunox (**55**) (Solvay/ Lundbeck)	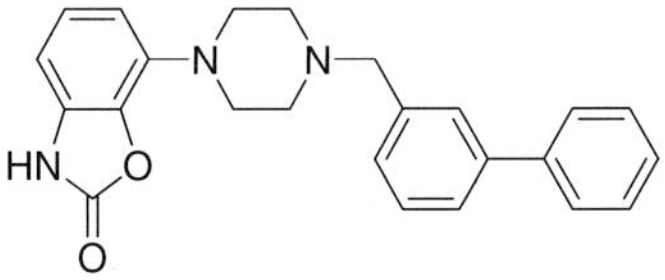	Phase III; used for the treatment of schizophrenia/bipolar disorder; efficacy data did not support further development for schizophrenia[a]

[a]Information obtained from the Thompson Reuter Integrity^{SM} database.

References

1. Kobayashi, S. (1927) *Sci. Pap. Inst. Phys. Chem. Res. (Jpn.)*, **6**, 166–184.
2. Fourneau, E. and Barrelet, C.E. (1929) *Bull. Soc. Chim., Fr.*, **45**, 1172–1189.
3. Hong, E., Sancilio, L.F., Vargas, R., and Pardo, E.G. (1969) *Eur. J. Pharmacol.*, **63**, 274–280.
4. Rodriguez, R. and Pardo, E.G. (1971) *Psychopharmacologia*, **21**, 89–100.
5. Prelog, V. and Driza, G.J. (1933) *Collect. Czech. Chem. Commun.*, **5**, 497–502.
6. Cerkovnikov, E. and Stern, P. (1946) *Arh. Kemiju*, **18**, 12–36.
7. Maxwell, D.R. (1959) Proceedings of International Congress of Neuro-Pharmacology, pp. 365–373, Volume Date 1958.
8. Wylie, D.W. and Archer, S. (1962) *J. Med. Pharm. Chem.*, **91**, 932–943.
9. Janssen, P.A.J. (1961) US2997472.
10. FR M208 19610320 (1961).
11. Parcell, R.F. (1962) US3028390.
12. Palazzo, G. and Silvestrini, B. (1968) US3381009.
13. Garattini, S. (1974) *Mod. Probl. Pharmacopsychiatry*, **9**, 29–46.
14. Yamatsu, K., Kaneko, T., Yamanishi, Y., Kitahara, A., and Ohtake, S. (1974) *Mod. Probl. Pharmacopsychiatry*, **9**, 11–17.
15. Angelucci, L. and Bolle, P. (1974) *Mod. Probl. Pharmacopsychiatry*, **9**, 65–75.
16. De Gregorio, M. and Dionisio, A. (1971) *Panminerva Med.*, **13**, 27–30.
17. Galvan, L., Nair, N.P., Kussin, D., and Ban, T.A. (1973) *Curr. Ther. Res., Clin. Exp.*, **15**, 776–780.
18. Van der Burg, W.J., Bonta, I.L., Delobelle, J., Ramon, C., and Vargaftig, B. (1970) *J. Med. Chem.*, **13**, 35–39.
19. Vargaftig, B.B., Coignet, J.L., de Vos, C.J., Grijsen, H., and Bonta, I.L. (1971) *Eur. J. Pharmacol.*, **16**, 336–346.
20. Peet, M. and Behagel, H. (1978) *Br. J. Clin. Pharmacol.*, **5** (Suppl. 1), S5–S9.
21. Itil, T.M., Polvan, N., Dincmen, K., Egilmez, S., and Sungerbey, K. (1974) *Dis. Nerv. Syst.*, **35** (7, Pt. 2), 10–17.
22. Vogel, H.P., Bente, D., Feder, J., Helmchen, H., Muller-Oerlinghausen, B., Bohacek, N., Mihovilovic, M., Brandle, A., Fleischhauer, J., and Walcher, W. (1975) *Act. Nerv. Super.*, **17**, 229–230.
23. Wheatley, D. (1975) *Curr. Ther. Res., Clin. Exp.*, **18**, 849–854.
24. Murphy, J.E. (1978) *Br. J. Clin. Pharmacol.*, **5** (Suppl. 1), S81–S85.
25. Brogden, R.N., Heel, R.C., Speight, T.M., and Avery, G.S. (1978) *Drugs*, **16**, 273–301.
26. Wu, Y.H., Smith, K.R., Rayburn, J.W., and Kissel, J.W. (1969) *J. Med. Chem.*, **12**, 876–881.
27. Wu, Y.H., Rayburn, J.W., Allen, L.E., Ferguson, H.C., and Kissel, J.W. (1972) *J. Med. Chem.*, **15**, 477–479.
28. Tompkins, E.C., Clemento, A.J., Taylor, D.P., and Perhach, J.L. Jr. (1980) *Res. Commun. Psychol., Psychiatr. Behav.*, **5**, 337–352.
29. Goldberg, H.L. and Finnerty, R.J. (1979) *Am. J. Psychiatry*, **136**, 1184–1187.
30. Fuller, R.W. (1986) *Phychopharmacol. Bull.*, **22**, 825–828.
31. Seeger, T. (1995) *J. Pharmacol. Exp. Ther.*, **275**, 101–113.
32. Bench, C.J., Lammertsma, A.A., Dolan, R.J., Grasby, P.M., Warrington, S.J., Gunn, K., Cuddigan, M., Turton, D.J., Osman, S., and Frackowiak, R.S. (1993) *Psychopharmacology*, **112**, 308–314.
33. Rosa, A.R., Franco, C., Torrent, C., Comes, M., Cruz, N., Horga, G., Benabarre, A., and Vieta, E. (2008) *CNS Neurosci. Ther.*, **14**, 278–286.
34. Stip, E. and Tourjman, V. (2010) *Clin. Ther.*, **32** (Suppl. A), S3–S20.
35. Farmer, C.A. and Aman, M.G. (2011) *Exp. Opin. Pharmacother.*, **12**, 635–640.
36. Laitinen, I. WO2007118923, (Fermion Oy) Oct. 25, 2007.
37. Tyagi, O.D., Srivastava, T.K., Nalam, V.K., and Patil, S.N. (2008) WO2010010566.
38. Olivie, J. (1980) US4217452.
39. Kisielowski-Ruppert, L., Moersdorf, J.P., Grafe, I., and Ahrens, K.-H. (1994) DE4305659.
40. Kemperman, G.J. (2008) WO2008125577.
41. Peroutka, S.J. and Snyder, S.H. (1979) *Mol. Pharmacol.*, **16**, 687–699.

42. Enna, S.J. and Kendall, D.A. (1981) *J. Clin. Psychopharmacol.*, **1** (Suppl. 6), S12–S16.
43. Glaser, T. and Traber, J. (1983) *Eur. J. Pharmacol.*, **88**, 137–138.
44. Pedigo, N.W., Yamamura, H.I., and Nelson, D.L. (1981) *J. Neurochem.*, **36**, 220–226.
45. Schnellmann, R.G., Waters, S.J., and Nelson, D.L. (1984) *J. Neurochem.*, **42**, 65–70.
46. McKenna, D.J. and Peroutka, S.J. (1989) *J. Neurosci.*, **9**, 3482–3490.
47. Peroutka, S.J. (1985) *Brain Res.*, **344**, 167–171.
48. Meltzer, H.Y. (1995) *Clin. Neurosci.*, **3**, 64–75.
49. Oegren, S.O., Ross, S., Hall, H., and Archer, T. (1983) *Drugs Psychiatry*, **1**, 13–34 (Antidepressants).
50. Murasaki, M. and Miura, S. (1992) *Prog. Neuro-Psychopharmacol. Biol. Psychiatry*, **16**, 833–845.
51. Wood, M.D., Scott, C., Clarke, K., Cato, K.J., Patel, N., Heath, J., Worby, A., Gordon, L., Campbell, L., Riley, G., Davies, C.H., Gribble, A., and Jones, D.N.C. (2006) *CNS Neurol. Dis.: Drug Targets*, **5**, 445–452.
52. Di Giovanni, G., Esposito, E., and Di Matteo, V. (2010) *CNS Neurosci. Ther.*, **16**, 179–194.
53. Ciranna, L. (2006) *Curr. Neuropharmacol.*, **4**, 101–114.
54. He, H. and Richardson, S.J. (1997) *CNS Drug Rev.*, **3**, 34–48.
55. Owens, M.J., Morgan, W.N., Plott, S.J., and Nemeroff, C.B. (1997) *J. Pharmacol. Exp. Ther.*, **283**, 1305–1322.
56. Fernández, J., Alonso, J.M., Andrés, J.I., Cid, J.M., Díaz, A., Iturrino, L., Gil, P., Megens, A., Sipido, V.K., and Trabanco, A.A. (2005) *J. Med. Chem.*, **48**, 1709–1712.
57. de Boer, T. (1996) *J. Clin. Psychiatry*, **57** (Suppl. 4), S19–S25.
58. Lopez-Rodriguez, M.L., Rosado, M.L., Benhamu, B., Morcillo, M.J., Sanz, A.M., Orensanz, L., Beneitez, M.E., Fuentes, J.A., and Manzanares, J. (1996) *J. Med. Chem.*, **39**, 4439–4450.
59. Schmidt, A.W., Lebel, L.A., Howard, H.R., and Zorn, S.H. (2001) *Eur. J. Pharmacol.*, **425**, 197–201.
60. Lange, J.H.M., Reinders, J.-H., Tolboom, J.T.B.M., Glennon, J.C., Coolen, H.K.A.C., and Kruse, C.G. (2007) *J. Med. Chem.*, **50**, 5103–5108.
61. Sperling, W. and Demling, J. (1997) *Drugs Today*, **33**, 95–102.
62. Antilla, S.A.K. and Leinonen, E.V.J. (2001) *CNS Drug Rev.*, **7**, 249–264.
63. Eison, A.S. (1990) *J. Clin. Psychopharmacol.*, **10** (Suppl. 3), S2–S5.
64. Rickels, K., Derivan, A., Kunz, N., Pallay, A., and Schweizer, E. (1996) *J. Clin. Psychopharmacol.*, **16**, 212–217.
65. Wilson, S.J., Bailey, J.E., and Nutt, D.J. (2005) *Psychopharmacology*, **179**, 895–896.
66. Lopez-Rodriguez, M.L., Ayala, D., Benhamu, B., Morcillo, M.J., and Viso, A. (2002) *Curr. Med. Chem.*, **9**, 443–469.
67. Caliendo, G., Santagada, V., Perissutti, E., and Fiorino, F. (2005) *Curr. Med. Chem.*, **12**, 1721–1753.
68. Kate, T. (2011) *Am. J. Health-Syst. Pharm.*, **68**, 366.

7
Arylpiperidine-Based Dopamine D_2 Antagonists/5-HT_{2A} Antagonists for the Treatment of Autism, Depression, Schizophrenia, and Bipolar Disorder

Ying Wang

7.1
Introduction

Arylpiperidines are chemical compounds with aryl groups directly attached to piperidine rings. Arylpiperidines, especially 4-arylpiperidines (**1**) (Figure 7.1), are considered privileged substructures in medicinal chemistry for drug design and development [1]. Many launched drugs, in particular, analgesics and antipsychotics, contain 4-arylpiperidines (**1**) as structural elements. This chapter focuses on drugs that contain 4-arylpiperidine substructures for the treatment of autism, depression, schizophrenia, and bipolar disorder. The shared mechanism of action, that is, dopamine D_2 antagonism/serotonin 5-HT_{2A} antagonism, suggests that the 4-arylpiperidine moieties are key pharmacophores in these drugs.

7.2
History

Compounds containing 4-phenylpiperidine substructures have been known in the literature since the late 1800s. From the early to mid-1900s, there are more than 300 reports describing the syntheses and the pharmacological properties of 4-phenylpiperidines [2]. The interest in this type of compound is mainly due to the discovery that many of them possess analgesic activities, especially meperidine (**2**) (Figure 7.2) [3]. Indeed, to date, 4-phenylpiperidine derivatives still play a predominant role in anesthesia [4].

In 1954, Dr Paul A. J. Janssen, at Janssen Laboratories in Belgium, was studying an extended series of 4-phenylpiperidines related to meperidine (**2**) in an effort to find new analgesics. His group were intrigued to find that by elongating the N-Me side chain, they were able to produce compounds possessing both morphinelike analgesic and chlorpromazinelike (**3**) (Figure 7.2) neuroleptic properties [5]. After preparing and testing hundreds of analogs, they were able to produce compounds with typical neuroleptic, chlorpromazinelike properties devoid of any morphinelike, analgesic effects. In 1959, their study led to the

Bioactive Heterocyclic Compound Classes: Pharmaceuticals,
First Edition. Edited by Jürgen Dinges and Clemens Lamberth.

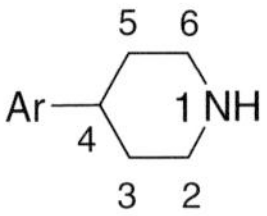

1

Figure 7.1 Basic structure of 4-arylpiperidine (**1**).

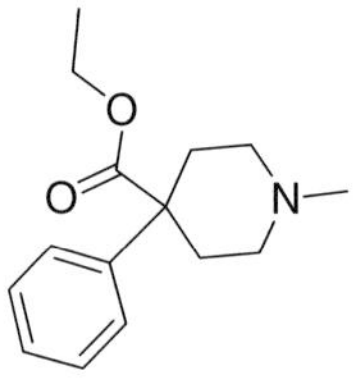

2

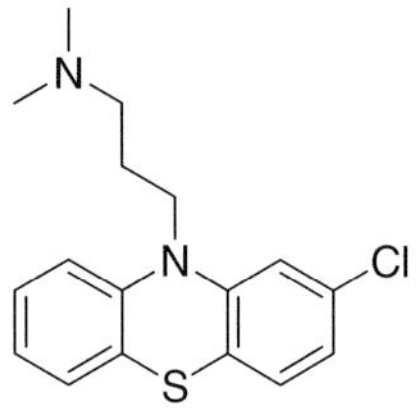

3

Figure 7.2 Chemical structure of meperidine (**2**) and chlorpromazine (**3**).

discovery of haloperidol (**4**) (Table 7.1), the first antipsychotic drug containing a 4-phenylpiperidine substructure [6]. On a weight basis, haloperidol (**4**) is ~50 times more potent than chlorpromazine (**3**), the first marketed antipsychotic drug. The discovery of haloperidol (**4**) represents a significant milestone in the development of antipsychotics because it introduced a new chemotype [7]. The chemical structures of the antipsychotic drugs before haloperidol (**4**) were primarily based on the tricyclic phenothiazines such as chlorpromazine (**3**). To date, a significant number of marketed antipsychotic drugs are also based on the arylamine motif of haloperidol (**4**). The discovery of haloperidol (**4**) led to a flurry of research activities with this class of compounds in the 1960s and 1970s. More than 5000 haloperidol analogs were synthesized and evaluated at Janssen Laboratories at the time, which led to the discovery of several antipsychotic drugs with structures closely related to haloperidol (**4**) (Table 7.1) but with a variety of pharmacological profiles.

Trifluperidol (**5**) (Table 7.1) was launched in 1959 by Janssen Laboratories [8]. Although its pharmacological properties are similar to those of haloperidol (**4**), it is considerably more potent on a weight basis. However, it also causes more severe side effects such as tardive dyskinesia, characterized by repetitive and involuntary movements [9, 10].

In 1963, droperidol (**6**) was approved in Europe for the treatment of psychosis [11]. The structure of **6** features an unsaturated double bond within the piperidine ring. Droperidol (**6**) was approved in the United States in 1970 but was primarily being used as a first-line antiemetic. In 2001, 30 years after the approval of droperidol (**6**), the US Food and Drug Administration (FDA) issued a "black box warning," indicating that there was a significant risk of arrhythmias even at low doses. This came as a surprise for many pharmacists and anesthesiologists who have been using low doses of droperidol (**6**) for many years without any significant complications. Many clinicians have questioned the FDA's decision since then. They felt that there was no convincing evidence to support the FDA's statement and more definitive studies would be needed [12]. Currently, the principal use of

Table 7.1 Drugs containing a 4-arylpiperidine moiety for the treatment of schizophrenia, bipolar disorder, autism, and depression.

Year launched	Generic name (brand name)	Company	Structure	Indication
1959	Haloperidol (**4**, Haldol®)	Janssen	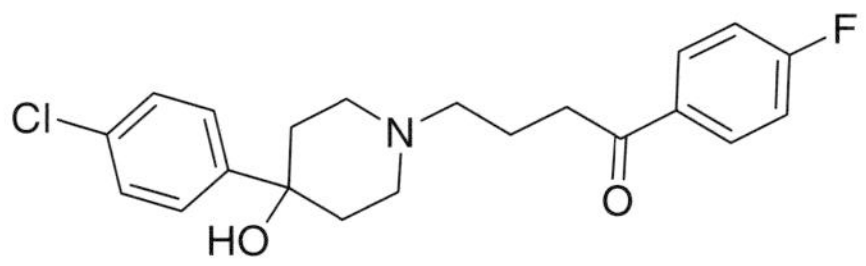	Schizophrenia
1959	Trifluperidol (**5**, Triperidol®)	Janssen	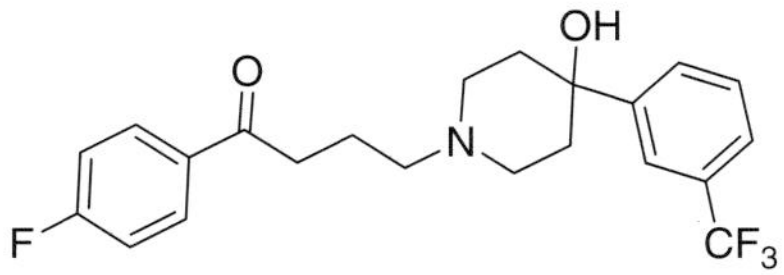	Schizophrenia
1963	Droperidol (**6**, Droleptan®, Inapsine®, Xomolix®)	Janssen	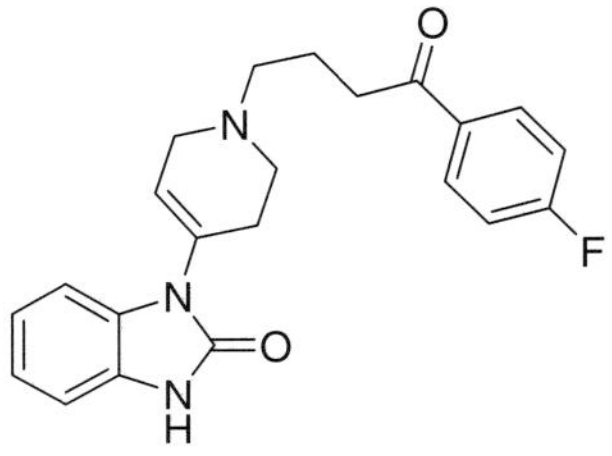	Schizophrenia and bipolar disorder

(continued overleaf)

Table 7.1 (*Continued*)

Year launched	Generic name (brand name)	Company	Structure	Indication
1966	Benperidol (**7**, Anquil®, Glianimon®)	Janssen	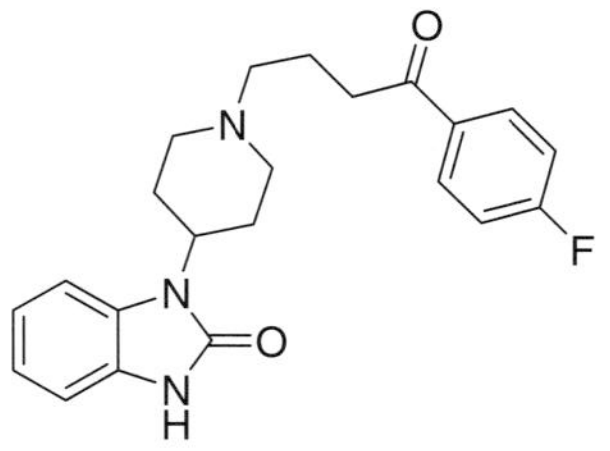	Schizophrenia
1969	Pimozide (**8**, Orap®)	Janssen	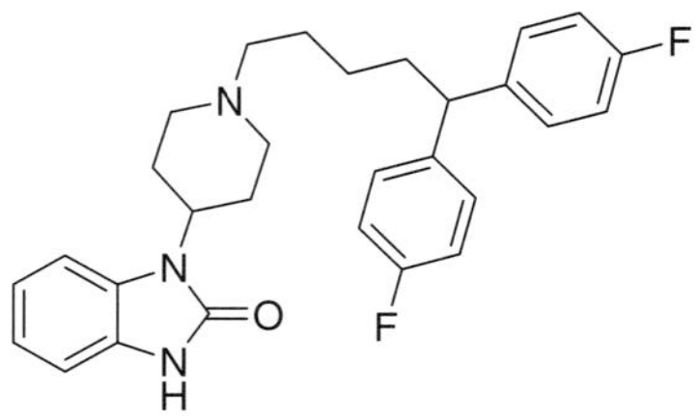	Schizophrenia

1981	Bromperidol (**9**, Impromen®)	Janssen	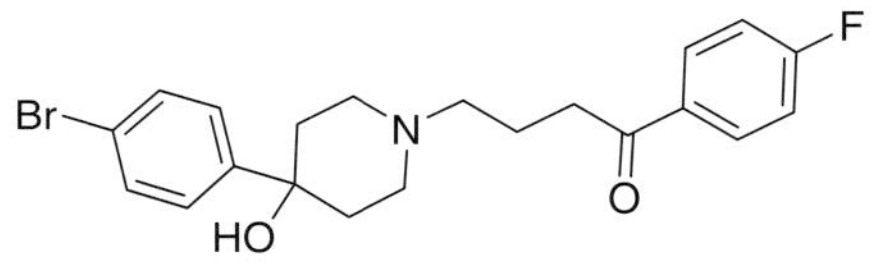	Schizophrenia
1993	Risperidone (**10**, Risperdal®, Belivon®)	Janssen	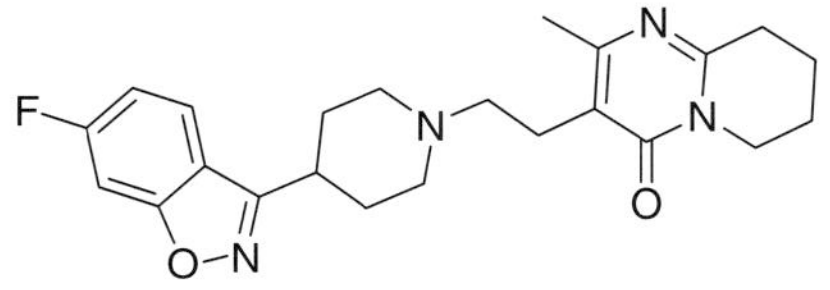	Depression schizophrenia, bipolar disorder, and autism
2006	Sertindole (**11**, Serdolect®, Serlect®)	Lundbeck	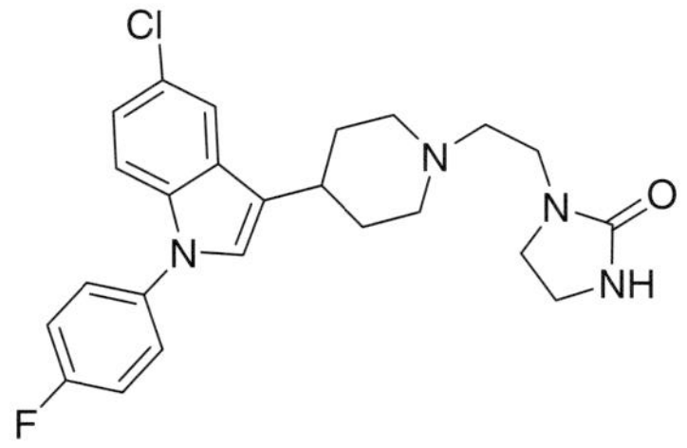	Schizophrenia

(continued overleaf)

Table 7.1 (*Continued*)

Year launched	Generic name (brand name)	Company	Structure	Indication
2007	Paliperidone (**12**, Invega®)	Johnson&Johnson	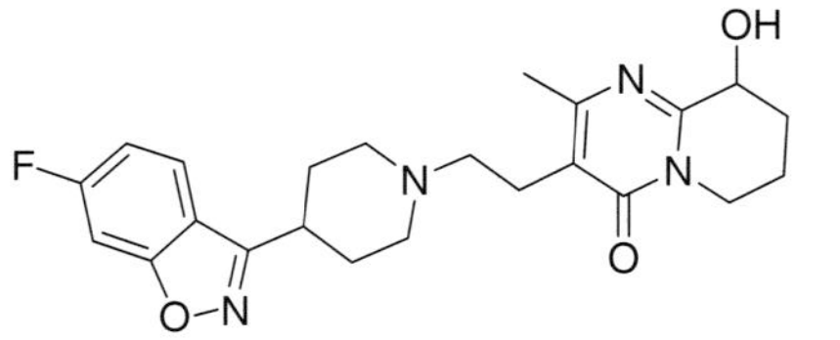	Schizophrenia, bipolar disorder, and autism
2010	Iloperidone (**13**, Fanapt®, Zomaril®)	Vanda Pharmaceuticals	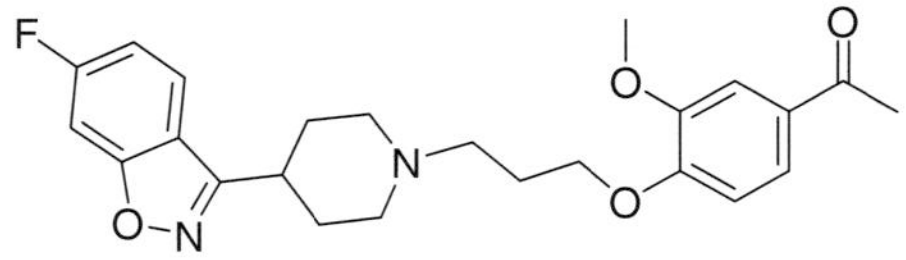	Schizophrenia, bipolar disorder, and depression

this drug is in anesthesiology for neuroleptanesthesia in combination with fentanyl [13].

Benperidol (**7**), which was derived by saturation of the double bond in droperidol (**6**), was marketed in 1966. It is one of the most potent antipsychotics known to date [14]. Modification of the side chain of droperidol (**6**) resulted in pimozide (**8**), which was approved by the FDA in 1969. Pimozide (**8**) is generally prescribed to patients with chronic schizophrenia [15].

After the launch of haloperidol (**4**) and its analogs, the development of new antipsychotic drugs slowed down between the mid-1970s and the late 1980s. Only bromperidol (**9**) was launched in 1981 [16]. It took another 12 years for the next 4-arylpiperidine-based antipsychotic drug, risperidone (**10**), to be developed [17]. Although risperidone (**10**) was derived from the same basic chemical scaffold as haloperidol (**4**), its mechanism of action is thought to be very different. It acts not only as a dopamine D_2 antagonist but also as a serotonin 5-HT_2 antagonist. Unlike haloperidol (**4**) and its analogs, which are considered as the first-generation antipsychotics, or "typical" antipsychotics [18], risperidone (**10**), belongs to the family of second-generation, or "atypical" antipsychotics [19]. Since its approval in 1993 for the treatment of schizophrenia, risperidone (**10**) has also been approved for additional indications, such as the treatment of bipolar mania, irritability in children, and autism in adolescents [20]. It is one of the few antipsychotics approved for the treatment of schizophrenia in youths aged 13–17 and bipolar disorder in youths aged 10–17 [21]. The drug is also used off-label to treat tourette syndrome and anxiety disorder [22]. Risperidone (**10**) is considered as one of the first-line therapeutics for psychosis.

In 1996, sertindole (**11**), with a phenylindole ring attached to the piperidine ring, was developed by Lundbeck. The drug was first launched in the United Kingdom for the treatment of schizophrenia [23]. It was subsequently withdrawn from the market in late 1998, following the observation of a high ratio of serious cardiac arrhythmias and sudden cardiac deaths. Since then, Lundbeck conducted additional studies, which demonstrated that the mortality rates among patients treated with sertindole (**11**) are comparable to those of other antipsychotics and that the product can be prescribed safely [24]. Sertindole (**11**) was approved in the European Union for the treatment of schizophrenia in 2006. The FDA has accepted a new drug application (NDA) filed by Lundbeck for the treatment of schizophrenia in 2008. However, although the FDA advisory committee recommended the approval one year later, sertindole (**11**) still has not been officially approved in the United States.

Following the discovery of risperidone (**10**) and paliperidone (**12**), the active metabolite of **10**, was launched in the United States in 2007 for the treatment of schizophrenia [25]. Structurally, paliperidone (**12**) only differs from risperidone (**10**) by one hydroxyl group. Thus, it is also known as *9-hydroxyrisperidone* (**12**). The extended-release formulation of **12** was developed with the OROS® drug delivery system and designed to deliver paliperidone (**12**) over a 24-h period [26]. In 2009, the US FDA approved paliperidone palmitate, an extended-release injectable suspension, for the acute and maintenance treatment of schizophrenia in adults.

As such, paliperidone (**12**) was the first once-a-month, long-acting injectable atypical antipsychotic approved in the United States.

Another recently approved antipsychotic drug with a benzisoxazolylpiperidine moiety is iloperidone (**13**) [27]. The development of iloperidone (**13**) was a rather long journey. The compound was originally discovered by Hoechst Marion Roussel Inc., now part of Aventis, in the early 1990s. Then it was handed to Titan Pharmaceuticals in 1997, which sold the rights to Novartis in 1998. Vanda Pharmaceuticals acquired the phase III development rights in 2004 and announced in 2007 that the FDA had accepted their NDA for iloperidone (**13**). However, in 2008, the FDA issued a "not approvable" letter to Vanda Pharmaceuticals. The agency questioned iloperidone's (**13**) efficacy compared with risperidone (**10**) in the phase III study and asked Vanda Pharmaceuticals to conduct an additional clinical trial to compare iloperidone (**13**) with either risperidone (**10**) or olanzapine. Vanda Pharmaceuticals appealed the FDA's decision and resubmitted an updated NDA. This time the FDA decided that the existing data were sufficient and approved iloperidone (**13**) in 2009 for the acute treatment of schizophrenia in adults.

7.3 Synthesis

As stated earlier in this chapter, syntheses of 4-arylpiperidines (**1**) trace back to the late 1800s. Over time, numerous synthetic methods and routes have been developed to construct this class of compounds. This is also reflected in the syntheses of the drugs that contain 4-arylpiperidine moieties. This chapter discusses some of the routes reported for the syntheses of haloperidol (**4**) and risperidone (**10**).

The original route for the synthesis of haloperidol (**4**), reported by Dr Janssen and his colleagues [28] in 1959, is illustrated in Scheme 7.1. The styryl derivative **14** was reacted with formaldehyde and ammonium chloride to give the intermediate **15**.

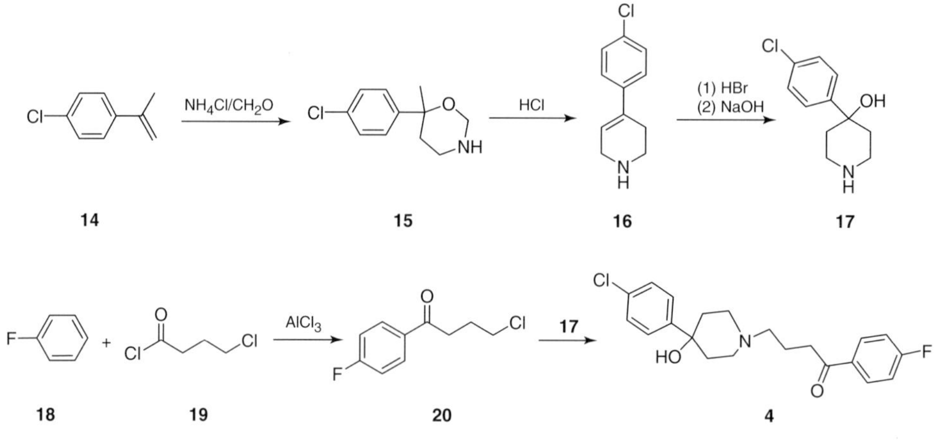

Scheme 7.1 Representative synthesis of haloperidol (**4**) (Route I).

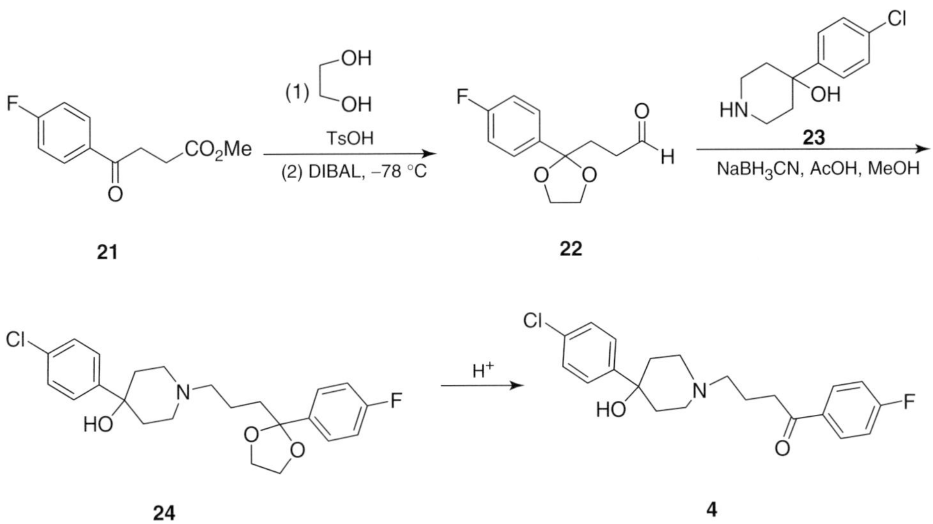

Scheme 7.2 Representative synthesis of haloperidol (**4**) (Route II).

When treated with hydrochloric acid, **15** was converted to the tetrahydropyridine (**16**), probably through opening the hydrogenated 1,3-oxazine ring followed by dehydration and cyclization. Compound **16** then underwent addition of HBr and subsequent hydrolysis to form the alcohol **17**. The second part of haloperidol (**4**), the chloride **20**, was easily synthesized by Friedel–Crafts reaction between fluorobenzene **18** and the acid chloride **19**. Alkylation of **17** with **20** produced the desired target compound **4**.

A more recent synthesis of haloperidol (**4**) is illustrated in Scheme 7.2 [29]. The ketone **21** was synthesized by 1,4-addition of *p*-fluorobenzaldehyde to methyl acrylate. Protection of the carbonyl group in **22** as a dioxolane, followed by reduction of the ester with diisobutylaluminum hydride (DIBAL) yielded the aldehyde **22**. Reductive amination of **22** with the now commercially available piperidine **23** afforded **24**, which was deprotected to give haloperidol (**4**) in 55% overall yield from compound **21**.

Similar to haloperidol (**4**), there are many reported routes for the synthesis of risperidone (**10**). Three representative routes are given below. In the first route (Scheme 7.3) [30], the bicyclic heterocycle **27** is obtained by refluxing acetylbutyrolactone (**25**) with 2-aminopyridine (**26**) in the presence of phosphorus oxychloride ($POCl_3$), followed by hydrogenation of the resulting intermediate in acetic acid. The synthesis of the arylpiperidine portion of **10** starts with the Friedel–Crafts acylation of 1,3-difluorobenzene with the acid chloride **28** to afford the ketone **29**. The *N*-acetyl group is then hydrolyzed with aqueous HCl. Formation of the oxime with hydroxylamine followed by ring closure under basic conditions gives the benzisoxazole **30**. In the final step, **30** is alkylated with the chloride **27** in the presence of sodium carbonate and potassium iodide to yield risperidone (**10**).

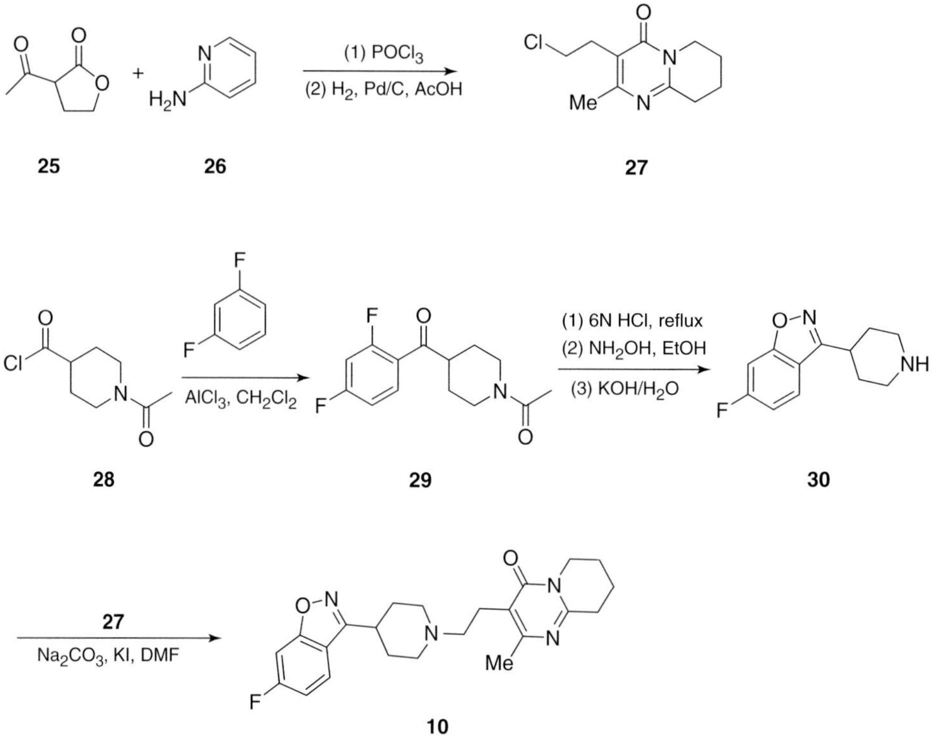

Scheme 7.3 Representative synthesis of risperidone (**10**) (Route I).

Alternatively, the piperidine ring can be constructed in the final steps to join the two fragments of risperidone (**10**) together. A representative synthesis is illustrated below (Scheme 7.4) [31]. The synthesis starts with the Friedel–Crafts acylation of 1,3-difluorobenzene with the acyl chloride **31**. The resulting ketone **32** is converted to the benzisoxazole **33** as described above. A combination of NaI and AcCl is used to open the tetrahydropyran ring in **33**. *In situ* formation of AcI cleaves the ether under mild conditions and without the need of a Lewis acid. The subsequently formed acetate is then hydrolyzed using potassium carbonate in refluxing acetonitrile to give the iodopentanol derivative **34**. Activation of the hydroxyl group in **34** via mesylation sets the stage for the formation of the piperidine ring. Refluxing a mixture of **35**, the amine **36**, and sodium bicarbonate in acetonitrile affords risperidone (**10**).

Scheme 7.5 illustrates a third possibility to synthesize risperidone (**10**), in which the benzisoxazole ring is constructed last [32]. Isonipecotic acid (**37**) is protected as an ethyl chloroformate **38** and is then converted to the acyl chloride **39**. Friedel–Crafts reaction with 1,3-difluorobenzene gives the ketone **40**. The protecting group is hydrolyzed to afford **41**, which is alkylated with compound **27** (Scheme 7.3) to produce the ketone **42**. Oxime formation and subsequent ring closure yields risperidone (**10**).

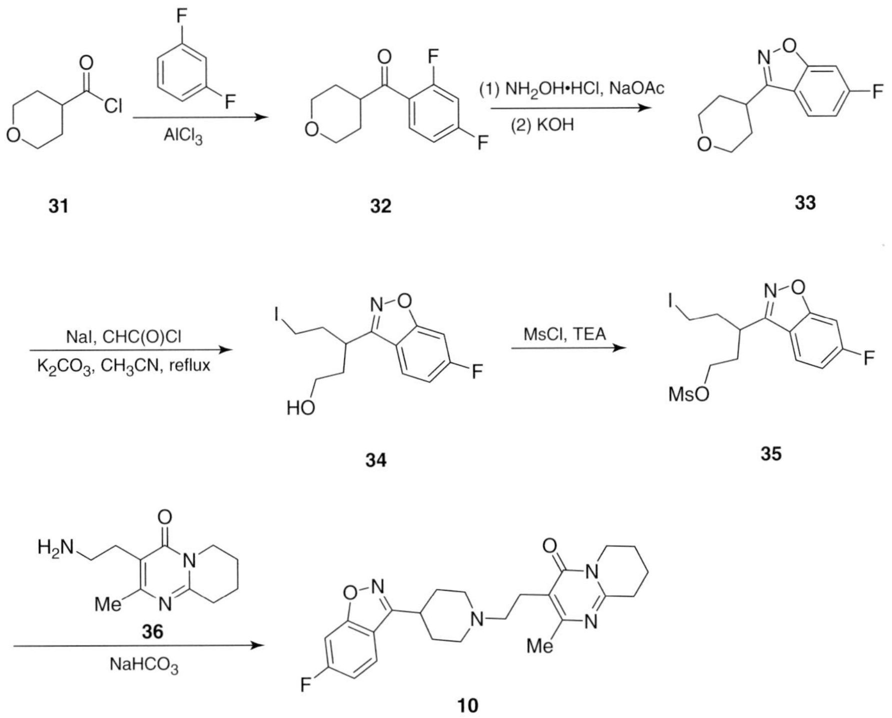

Scheme 7.4 Representative synthesis of risperidone (**10**) (Route II).

7.4
Mode of Action

Although haloperidol (**4**), the first 4-phenylpiperidine structurally based antipsychotic drug, was discovered in the 1950s, its exact mechanism of action is still not fully understood. As for most of the marketed antipsychotics, this is in part due to its complex pharmacological profile [33]. On the other hand, the etiology and mechanistic basis for diseases such as schizophrenia are also not yet completely understood. Haloperidol (**4**), along with the first antipsychotic drug chlorpromazine (**3**), was shown to be potent dopamine D_2 antagonist. In 1963, Carlsson and Lindqvist [34] demonstrated that haloperidol (**4**), along with other antipsychotic drugs, increased the turnover of dopamine, suggesting that dopamine and its receptors were involved in the antipsychotic drug actions. Their findings resulted in additional efforts to identify selective dopamine antagonists. In the late 1970s and early 1980s, two dopamine pathways in the brain, namely, the mesolimbic and mesocortical pathways, were implicated in psychosis [35]. Other dopamine pathways such as the nigrostriatal pathway (implicated in extrapyramidal side effects (EPSs) resulting from the dopamine blockade by antipsychotics) and the tuberoinfundibular pathway (associated with elevated prolactin levels) have also been proposed [36]. Haloperidol (**4**) is quite effective against the positive

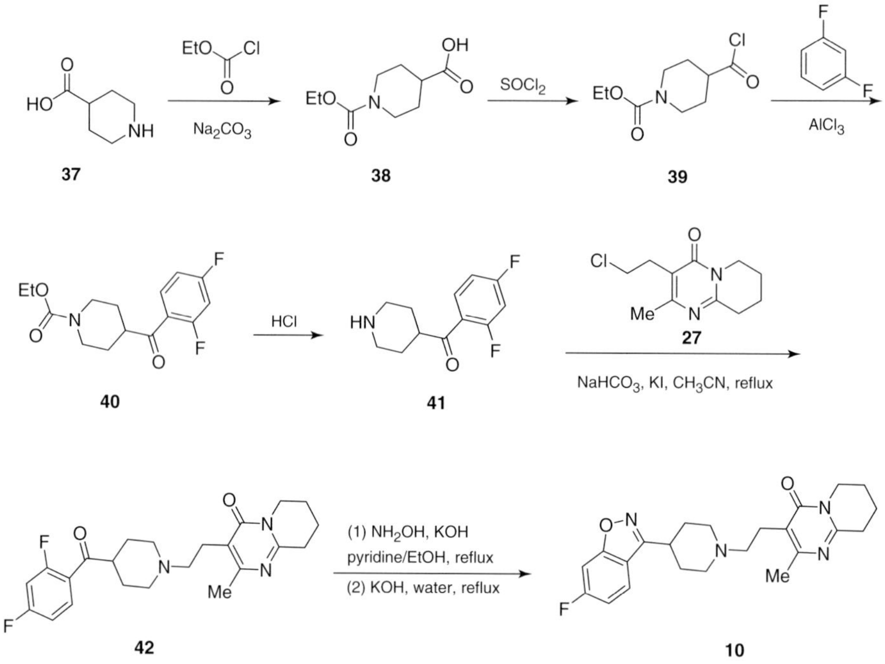

Scheme 7.5 Representative synthesis of risperidone (**10**) (Route III).

symptoms of schizophrenia (delusions, hallucinations, extreme emotions, excited motor activity, and incoherent speech), which mechanistically is believed to be due to its blockade of the mesolimbic pathway. However, **4** is ineffective against the negative symptoms of schizophrenia (poverty of speech, social withdrawal, avolition, anhedonia, and affective blunting). In addition, haloperidol (**4**) blocks the dopaminergic action in the nigrostriatal pathway, which is hypothesized to be the reason for its EPSs. EPSs are physical symptoms that include tremor, slurred speech, akathisia, dystonia, anxiety, distress, paranoia, and bradyphrenia, which are primarily associated with improper dosing of or unusual reactions to antipsychotic medications [37]. Haloperidol (**4**), along with its close analogs, is considered to be a first generation or typical antipsychotic drug. In general, first-generation or typical antipsychotic drugs are very effective in the treatment of positive symptoms of schizophrenia, do little for the negative symptoms, and cause EPS.

In a 1961 report, the prevalence of EPS in patients treated with antipsychotic drugs was estimated to be as high as 39% [19]. The majority of the clinicians and pharmacologists at the time believed that there is an absolute connection between EPS and the clinical effectiveness of antipsychotic drugs such as chlorpromazine (**3**) and haloperidol (**4**). However, later in the 1960s, the introduction of the second-generation "atypical" antipsychotics, led by the discovery of clozapine, demonstrated that these drugs can be efficacious not only against the positive symptoms of schizophrenia but also to some extent against the negative and cognitive symptoms. More importantly, these drugs, in general, have minimal EPS liabilities [38]. Although the second generation of antipsychotic drugs, similar to the first generation, has rich and complex pharmacological profiles, it is generally accepted that the dual antagonism at the D_2 and the serotonin 5-HT_{2A} receptors are mainly responsible for their efficacies with fewer side effects. In general, the second generation of antipsychotic drugs can be characterized by the following: no association with catalepsy in animal models, placebo-level EPS in humans, a high 5-HT_2/D_2 ratio and minimal association with elevated prolactin levels.

Risperidone (**10**) belongs to the second-generation (atypical) antipsychotics. Although the compound showed comparable affinity for the D_2 receptor compared to haloperidol (**4**), it has a much higher affinity for the 5-HT_{2A} receptor than **4** [39]. Clinical studies confirmed its combination of antipsychotic efficacy in treating positive and negative symptoms of schizophrenia with a reduced propensity to induce EPS. Doses of 4 mg or less per day resulted in negligible incidence of EPS. However, a dose above 6 mg per day caused EPS in a dose-dependent manner, implying a relatively narrow therapeutic window. Among all the atypical antipsychotics, risperidone (**10**) is most likely to cause prolactin increases, but it has a lower weight gain liability compared to other atypical antipsychotics such as olanzapine and quetiapine [40].

Paliperidone (**12**, Invega®) was developed as a follow-on product to risperidone (**10**). It is the 9-hydroxy metabolite of risperidone (**10**). The compound is a 5-HT_{2A} antagonist in the prefrontal cortex and a D_2 antagonist in the mesolimbic region [41].

Sertindole (**11**) is also an atypical antipsychotic drug. Its activity appears to be specific to the D_2, 5-HT_{2A}, and α1A adrenergic receptors and is largely confined to the limbic dopamine system of the brain. Sertindole (**11**) retains similar 5-HT_{2A}/D_2 pharmacology compared to other atypical antipsychotic drugs. Its affinity for the 5-HT_{1A}, H1, and muscarinic M1–M5 receptors is extremely low, a property thought to contribute to the low sedating potential and the lack of anticholinergic side effects of this drug [33, 38].

Overall, the pharmacological actions of all the antipsychotic drugs are very complicated. Most of these drugs have a number of central and peripheral effects. In addition, each drug has a unique pharmacological and clinical profile. Therefore, one must balance the benefits and risk factors for each patient in determining which drug to prescribe.

7.5 Structure–Activity Relationship

The drugs listed in Table 7.1 share a very common structural motif in that the molecules are composed of 4-arylpiperidine rings, simple alkyl chains, and aryl or heterocyclic groups R^2 (**43**) (Figure 7.3). In particular, haloperidol (**4**) and its close analogs (compounds **4**–**7** and **9** in Table 7.1) belong to a class of pharmaceutical

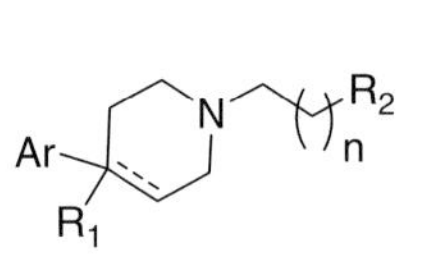

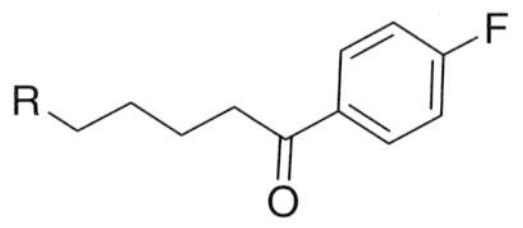

43 **44**

Figure 7.3 Common structure motifs for drugs in Table 7.1.

drugs known as *butyrophenones* [29], with a 4-fluorobutyrophenone group substituted at the piperidine nitrogen (**44**) (Figure 7.3). The diphenyl-butyl-arylpiperidine drug pimozide (**8**) (Table 7.1) can be considered as resulting from the modification of the butyrophenone structure in benperidol (**6**). Although haloperidol (**4**) and its analogs are structurally very closely related, there are differences in their clinical indications as illustrated earlier in this chapter. Risperidone (**10**) is considered to be a "masked" butyrophenone derivative, derived from the same chemical scaffold as haloperidol (**4**) [42].

The medicinal chemistry approach for the structure design of iloperidone (**13**) is an interesting story [43]. Researchers at Hoechst Marion Roussel have proposed that 4-substituted benzoisoxazoles **46** (Figure 7.4) could be considered as bioisosteres of 4-aroylpiperidines **45**. They have also noted in the literature that compounds with the general structures **47** were described as possessing antipsychotic activity in animal models. By replacing the 4-aroylpiperidine part in **47** with 4-substituted benzoisoxazoles, they were able to generate a group of compounds **48** with potent antipsychotic activities in animal models. Many of the compounds thus generated displayed moderate binding affinity for the D2 receptor coupled with a significantly greater affinity for the 5-HT_{2A} receptor, as mentioned previously, a property that has

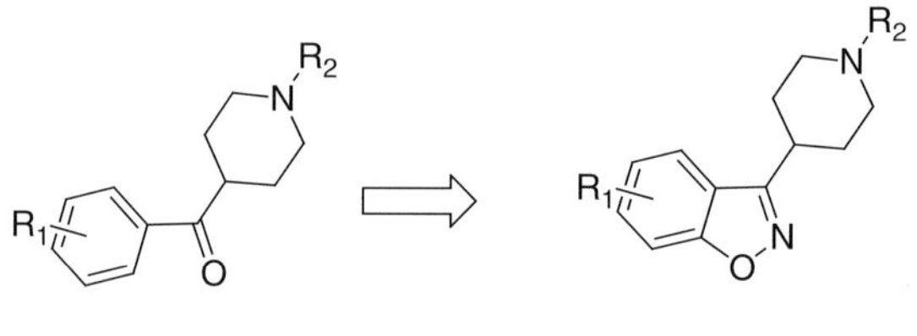

45 **46**

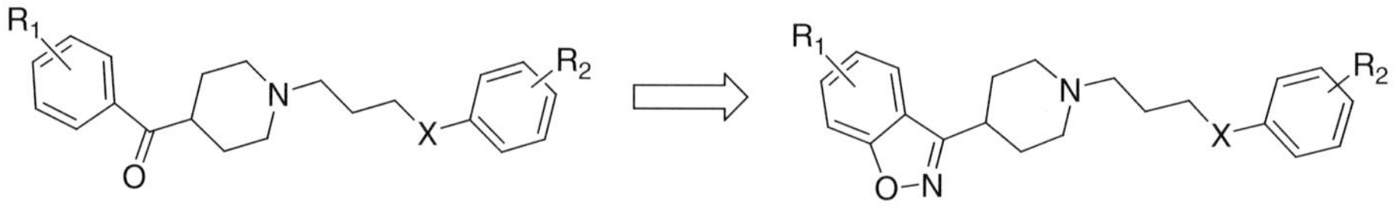

47 **48**

13 R_1 = 6-F, R_2 = 2-MeO and 4-$COCH_3$, X = O

Figure 7.4 The discovery of iloperidone (**13**).

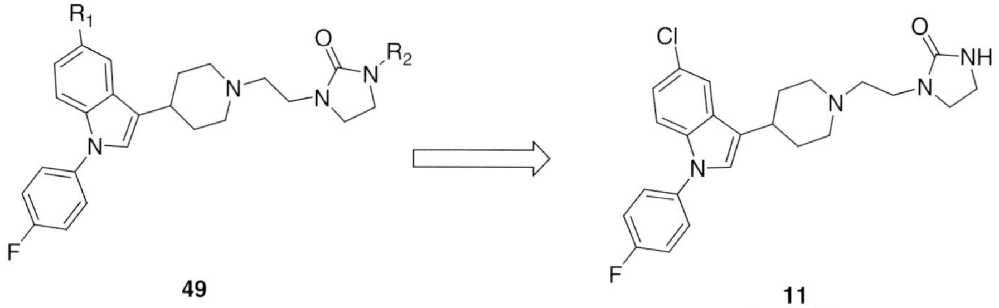

Figure 7.5 The discovery of sertindole (**11**).

been postulated as necessary for atypicality. Their structure–activity relationship (SAR) study eventually led to the identification of iloperidone (**13**). It has been reported that the incorporation of the three-carbon linker and the introduction of a fluorine at the 6-position of the benzoisoxazole were crucial to obtain the desired receptor profile.

Sertindole (**11**) was discovered from SAR studies on a series of *N*-phenylindolylpiperidines [44]. Among all the substituents R_1 studied at the 5-position of the indole nucleus of **49** (Figure 7.5), chlorine was found to have the best desired receptor profile. Substituents larger than hydrogen at the end of the imidazolone ring (R_2) decreased receptor affinity. Eventually, sertindole (**11**) was chosen because of its overall favorable receptor profile.

References

1. Welsch, M.E., Snyder, S.A., and Stockwell, B.R. (2010) *Curr. Opin. Chem. Biol.*, **14**, 347–361.
2. Walters, L.A. and McElvain, S.M. (1933) *J. Am. Chem. Soc.*, **55**, 4625–4629.
3. Christensen, E.M. and Gross, E.G. (1948) *J. Am. Med. Soc.*, **137**, 594–599.
4. Omoigui, S. (1999) *Anesthesia Drugs Handbook*, Blackwell, Oxford.
5. Janssen, P.A. (1965) *Intern. Rev. Neurobiol.*, **8**, 221–263, and references therein.
6. Janssen, P.A. (1967) *Intern. Rev. Neurobiol.*, **3**, 10–18.
7. Bernard, G. and Simona, A. (2005) *Ann. Clin. Psychiatry*, **17**, 137–140.
8. Janssen, P.A. (1965) *Intern. Rev. Neurobiol.*, **8**, 221–263.
9. Pratt, J.P., Bishop, M.P., and Gallant, D.M. (1963) *Curr. Ther. Res., Clin. Exp.*, **5**, 562–571.
10. Gallant, D.M., Bishop, M.P., Timmons, E., and Steele, C.A. (1963) *Am. J. Psychiatry*, **120**, 485–487.
11. Yelnosky, J., Katz, R., and Dietrich, E.V. (1964) *Toxicol. Appl. Pharmacol.*, **5**, 37–47.
12. Kao, L.W., Kirk, M.A., Evers, S.J., and Rosenfeld, S.H. (2003) *Ann. Emerg. Med.*, **41**, 546–558.
13. Henschel, W.F. (1973) *Int. Anesth. Clin.*, **11**, 37–49.
14. Leucht, S. and Hartung, B. (2002) Benperidol for schizophrenia. *Cochrane Database of Systematic Reviews* (Online) 1 (Art. No.: CD003083). Journal code: 100909747. E-ISSN:1469-493X. PubMed ID 11869652 AN 2002134519 MEDLINE.
15. Stirling, G.S. (1979) *Curr. Med. Res. Opin.*, **6**, 331–337.

16. Poldinger, W., Bures, E., and Haage, H. (1977) *Int. Pharmacopsychiatry*, **12**, 20–24.
17. Livingston, M.G. (1994) *Lancet*, **343**, 457–460.
18. Werkman, T.R., Glennon, J.C., Wadman, W.J., and McCreary, A.C. (2006) *CNS Neurol. Dis.: Drug Targets*, **5**, 3–23.
19. Shen, W.W. (1999) *Compr. Psychiatry*, **40**, 407–414.
20. Posey, D.J. and McDougle, C. (2002) *J. Curr. Opin. Invest. Drugs*, **3**, 1212–1216.
21. Davanzo, P.A. (2002) *Med. Psychiatry*, **18**, 355–414.
22. Haw, C. and Stubbs, J. (2007) *Expert Opin. Pharmacother.*, **6**, 533–545.
23. Azorin, J., Kaladjian, A., Fakra, E., and Adida, M. (2010) *Expert Opin. Pharmacother.*, **11**, 3053–3064.
24. Muscatello, M.R.A., Bruno, A., Pandolfo, G., Mico, U., Settineri, S., and Zoccali, R. (2010) *Drug Des., Dev. Ther.*, **4**, 187–201.
25. Green, B. (2009) *Curr. Drug Ther.*, **4**, 7–11.
26. Chwieduk, C.M. and Keating, G.M. (2010) *Drugs*, **70**, 1295–1317.
27. Marino, J. and Caballero, J. (2010) *Ann. Pharmacother.*, **44**, 863–870.
28. Janssen, P.A.J., Van de Westeringh, C., Jageneau, A.H.M., Demoen, P.J.A., Hermans, B.K.F., Van Daele, G.H.P., Schellekens, K.H.L., Van der Eycken, C.A.M., and Niemegeers, C.J.E. (1959) *J. Med. Pharm. Chem.*, **1**, 281–297.
29. Anjaiah, S., Chandrasekhar, S., and Gree, R. (2004) *Adv. Synth. Catal.*, **346**, 329–1334.
30. Srinivasa, R.G., Prasanna K., Basavapatna N., Manjunatha, S.G., and Kulkarni, A.K. (2005) WO 2005030772, PCT Int. Appl.
31. Dalmases Barjoan, P., Marquillas Olondriz, F., Huguet Clotet, J., Bosh Rovira, A., Del Castillo Nieto, J., and Caldero Ges, J. (1995) ES 2074966, Span.
32. Marquillas Olondriz, F., Bosch Rovira, A., Dalmases Barjoan, P., and Caldero Ges, J.M. (1994) ES 2050069, Span.
33. Nikam, S.S. and Awasthi, A.K. (2008) *Curr. Opin. Invest. Drugs*, **9**, 37–46.
34. Carlsson, A. and Lindqvist, M. (1963) *Acta Pharmacol. Toxicol.*, **20**, 140–144.
35. Lieberman, J.A. (2004) *CNS Drugs*, **18**, 251–267.
36. Goldstein, J.M. (1999) *Emerging Drugs*, **4**, 127–151.
37. Cheronis, J.C., Erinoff, L., Heller, A., and Hoffmann, P.C. (1979) *Brain Res.*, **169**, 545–560.
38. Rasmussen, J.G.C. (1997) *Int. J. Psychiatry Clin. Pract.*, **1**, 227–230.
39. Jack, J., Johnson, D.S., Sliskovic, D.R., and Roth, B.D. (2004) *Contemporary Drug Synthesis*, John Wiley & Sons, Ltd, Chichester, pp. 89–96.
40. Volavka, J., Czobor, P., Cooper, T.B., Sheitman, B., Lindenmayer, J., Citrome, L., McEvoy, J.P., and Lieberman, J.A. (2004) *J. Clin. Psychiatry*, **65**, 57–61.
41. Revill, P., Serradell, N., and Bolos, J. (2006) *Drugs Future*, **31**, 579–584.
42. MacDonald, G.J. and Bartolome, J.M. (2010) *Prog. Med. Chem.*, **49**, 37–80.
43. Strupczewski, J.T., Bordeau, K.J., Chiang, Y., Glamkowski, E.J., Conway, P.G., Corbett, R., Hartman, H.B., Szewczak, M.R., Wilmot, C.A., and Helsley, G.C. (1995) *J. Med. Chem.*, **38**, 1119–1131.
44. Perregaard, J., Arnt, J., Boegesoe, K.P., Hyttel, J., and Sanchez, C. (1992) *J. Med. Chem.*, **35**, 1092–1101.

8
Dibenzazepine-Based Sodium Channel Blockers for the Treatment of Neuropathic Pain

Derek W. Nelson

8.1
Introduction

Compounds containing three fused rings and various combinations of heteroatoms belong to the broad structural class known as *fused tricycles*. A number of tricycles have found use as drugs for the treatment of different conditions and disease states, particularly as tricyclic antidepressants (TCAs) [1] and antiepileptic drugs (AEDs) [2]. Compounds with a central seven-membered, nitrogen-containing azepine ring fused to two aromatic rings constitute a subset of the tricycles known as *dibenzazepines* (**1**) (Figure 8.1). Selected dibenzazepines, **1–4**, can be grouped based on the pharmacology attributable to the blockade of voltage-gated sodium channels (VGSCs) and the use for the treatment of diseases associated with sodium channel function. Dibenzazepine sodium channel blockers have been used as anticonvulsants, for the treatment of epileptic syndromes, and other diseases of the neurological system [3]. There are also examples of dibenzazepines that produce analgesic effects. The relationships between structure and function as sodium channel blockers and the utility as agents for the management of neuropathic pain are presented. The specific structures of the compounds covered are shown in Figure 8.1.

8.2
History

The dibenzazepine ring system (**1**, R = H) was prepared by Thiele and Holzinger [4] in the nineteenth century [4]. The synthesis was difficult and lacked generality, so the dibenzazepine derivatives were not well studied for several decades. The Swiss firm J.R. Geigy, which grew into the pharmaceutical company Novartis, played a pivotal role in the development of the dibenzazepine compounds. In the 1950s, chemists at Geigy prepared functionalized dibenzazepines and explored the pharmacology of these compounds. Carbamazepine (**5**, CBZ) (Figure 8.2) emerged from those efforts, and the patent disclosing the synthesis of CBZ (**5**) described

Bioactive Heterocyclic Compound Classes: Pharmaceuticals,
First Edition. Edited by Jürgen Dinges and Clemens Lamberth.

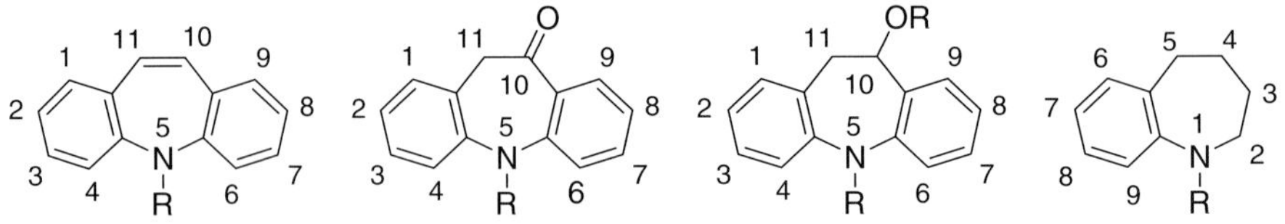

Figure 8.1 Structures and numbering conventions of dibenzazepine derivatives.

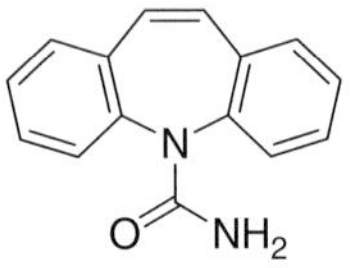

Figure 8.2 The structure of carbamazepine (**5**, CBZ, Tegretol®).

anticonvulsant activity suitable for use as "antiepileptica" [5]. CBZ (**5**) was originally developed under the brand name Tegretol®, and it still serves as an anticonvulsant.

The neurological disorder epilepsy manifests in spontaneous and recurring seizures resulting from abnormal, excessive, or hypersynchronous neuronal activity in the brain [6]. More than 40 epilepsy syndromes exist, and each presents a unique combination of symptoms. The pathology of epilepsy is complex and parallels the range of symptoms. Ion channels have been directly implicated in epilepsy, and genes that code for specific proteins in both ligand- and voltage-gated ion channels have been identified [7, 8]. The development of CBZ (**5**) as an analgesic and an anticonvulsant occurred concomitantly, although the function of the compound as an ion channel blocker was not elucidated until later.

After the disclosure of CBZ (**5**), reports described its capacity to provide relief from a specific type of facial pain called *trigeminal neuralgia* [9–11]. Also known as *tic douloureux*, trigeminal neuralgia originates from the trigeminal nerve and produces intense, recurring facial pain [12]. Mechanistically, trigeminal neuralgia causes nerve demyelination, and the disease symptoms constitute a specific type of neuropathic pain. The initial clinical studies demonstrated that CBZ (**5**) decreased pain and lessened the number of paroxysms. However, adverse events including rash formation were noted. Controlled clinical trials confirmed the analgesic efficacy of Tegretol (**5**) as a treatment for trigeminal neuralgia [13]. Additional clinical trials to evaluate the efficacy of CBZ (**5**) in the treatment of neuropathic pain, including diabetic neuropathy, have been conducted [14]. To date, the US Food and Drug Administration (FDA) has approved CBZ (**5**) only for the treatment of trigeminal neuralgia [15]. The extensive profile of adverse events resulting from CBZ (**5**) limits the utility of the drug. At present, CBZ (**5**) is sold as a generic equivalent under a variety of different brand names, including Microtrol®, Carbitrol®, and Equetro® [16].

The identification of new dibenzazepine derivatives that successfully treated neuropathic pain followed the introduction of CBZ (**5**), albeit slowly. Several factors contributed to the pace of development of new medicines from this class. First, the dibenzazepine core structures proved difficult to synthesize and

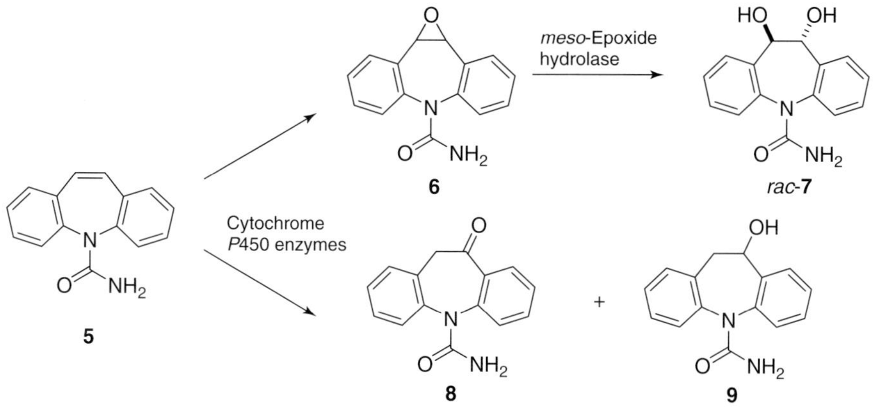

Scheme 8.1 Oxidative metabolism pathway of carbamazepine (**5**) in humans.

manipulate. Second, small structural changes, including heteroatom insertion in the rings or substituent addition, produced substantial changes in pharmacology. Dibenzazepine derivatives that retained significant function as sodium channel blockers and produced analgesic efficacy for neuropathic pain were related to the by-products of the oxidative metabolism of CBZ (**5**). In humans, CYP3A4 in the liver generates the majority of the metabolites (Scheme 8.1) [17]. Epoxide (**6**) forms initially, and subsequent hydrolysis of the epoxide leads to a racemic mixture of the vicinal diol (**7**) [18]. The 10-oxo (**8**) and 10-hydroxyl (**9**) derivatives are formed as minor metabolites. The epoxide (**6**) and diols (**7**) block sodium channels, but the general and neuronal toxicity of these compounds preclude their use as drugs [19].

The keto derivative oxcarbazepine (**8**, OXC) was the next dibenzazepine developed for the treatment of epilepsy. OXC (**8**) was prepared and patented as an anticonvulsant in 1969 by Ciba-Geigy [20, 21]. Similar to CBZ (**5**), the compound functions as a weak, nonspecific sodium channel blocker with complex pharmacology [22]. Following successful clinical trials for efficacy in a range of epileptic syndromes, Novartis launched OXC (**8**) under the name Trileptal® as an antiepileptic/anticonvulsant in Denmark in the early 1990s. Launch in the United States occurred nearly a decade later for use as monotherapy or adjunct therapy for the treatment of juvenile epilepsy. As with CBZ (**5**), adverse events occur with OXC (**8**) therapy depending on patient populations [23]. The incidence of adverse events was found to be lower for OXC (**8**) than for CBZ (**5**) [24]. Randomized, placebo-controlled clinical trials using OXC (**8**) as a potential treatment for painful diabetic neuropathies were supported by Novartis. The results of the clinical trials for Trileptal (**8**) were mixed. In one study, patients reported a statistically significant reduction in pain scores [25], but no effect versus placebo was reported in another study [26]. Case reports suggest the efficacy of OXC (**8**) for the treatment of trigeminal neuralgia [27–29]. The US FDA has not approved OXC (**8**) for the treatment of neuropathic pain, and use for indications involving neuropathy is off label.

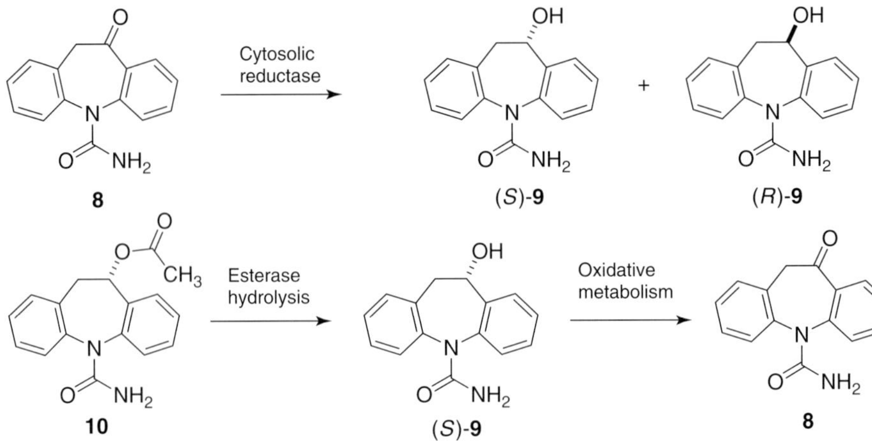

Scheme 8.2 Metabolic interconversion of oxcarbazepine (**8**) and eslicarbazepine (*S*-**9**).

Cytosolic ketoreductases reduce OXC (**8**) to form the enantiomers of hydroxyl derivative **9** (Scheme 8.2) [30]. Metabolite identification studies revealed that the (*S*)-isomer was formed in ~95% (vs. 5% of the (*R*)-isomer). The enantiomers, licarbazepine ((*R*)-**9**) and eslicarbazepine ((*S*)-**9**), were synthesized in enantiomerically pure form and evaluated [22]. Eslicarbazepine acetate (**10**), known as *BIA-2093* and developed by Bial, undergoes cleavage by nonspecific esterases in the first-pass metabolism process and serves as a prodrug of the active enantiomer of eslicarbazepine ((*S*)-**9**) [31]. Both CBZ (**5**) and OXC (**8**) interact with numerous receptors, but eslicarbazepine ((*S*)-**9**) does not interact significantly with calcium channels and glutamate transporters [32]. Metabolism of eslicarbazepine ((*S*)-**9**) does not produce the toxic epoxide and diol metabolites obtained from CBZ (**5**) [31]. Clinical trials for the use of BIA-2093 (**10**) for the treatment of epilepsy were successful [33]. The drug was approved in the European Union as an adjunctive therapy for seizures, and it is currently marketed under the brand names Exalief® and Zebinix® [34]. Eslicarbazepine acetate (**10**) produced efficacy in preclinical models of pain with some sedation noted [35].

Relatively few dibenzazepine derivatives function as analgesics via sodium channel blockade. The subset of these compounds that has been generally approved for use in the treatment of neuropathic pain is even smaller. The dibenzazepine core appears frequently in tricyclic antidepressants, and examples from this drug class have been used as analgesics [36]. The first TCA, imipramine (**11**) (Figure 8.3), was discovered at Geigy in the 1950s and has been widely used for the treatment of depression [37]. Other TCAs derived from the dibenzazepine core, including desipramine (**12**), doxepin (**13**), amitriptyline (**14**), nortriptyline (**15**), and many others, followed the discovery of imipramine (**11**). These compounds differ from CBZ (**5**) and its derivatives in the structure of the substituent on the nitrogen atom of the azepine ring. In some cases, the azepine ring is modified by the insertion or removal of heteroatoms. The tricyclic antidepressants frequently contain an

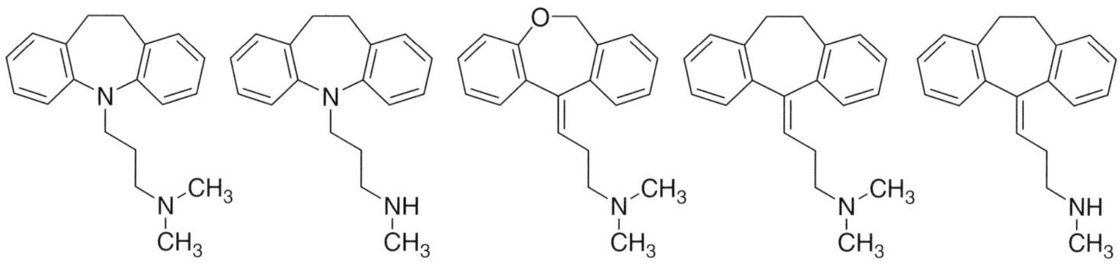

11 12 13 14 15

Figure 8.3 Examples of tricyclic antidepressants.

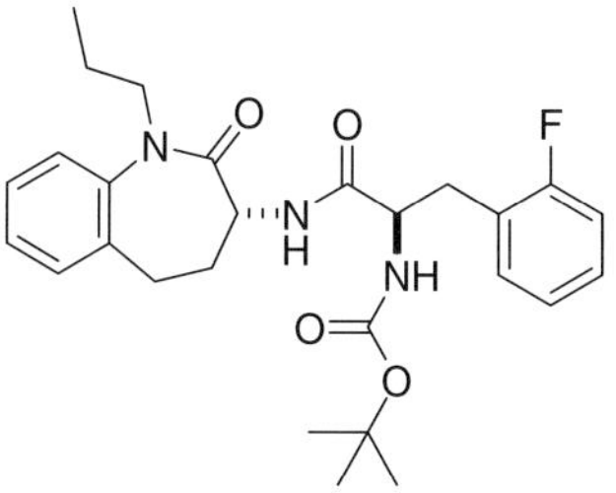

16

Figure 8.4 Selective sodium channel blocker with benzazepinone core.

aminopropyl moiety instead of the urea found in CBZ (**5**). Examples of the TCAs have found use in the treatment of neuropathic pain, but the complex pharmacology associated with these compounds distinguishes them from the dibenzazepine sodium channel blockers described here.

Current efforts to identify sodium channel blockers for therapeutic use focus on the identification of compounds that block specific sodium channel isoforms and thereby minimize the adverse event profiles of previous drugs [38]. Merck reported the discovery of a potent and selective sodium channel blocker, **16**, featuring the benzazepinone core (Figure 8.4) [39]. Replacement of one of the benzo groups of the dibenzazepine core with an amide moiety enabled selective blockade of sodium channels and provided a potential route to analgesia without associated adverse events.

8.3
Synthesis

The dibenzazepine core has been known for more than 100 years [5]. Early synthetic approaches to dibenzazepine analogs assembled the tricyclic core **19** and subsequently functionalized the nitrogen and the two-carbon bridge between the two phenyl rings. The sequence for the formation of the core is outlined in Scheme 8.3 [40]. Starting from diphenylethylene (**17**), 10,11-dihydrodibenzazepine (**18**) was

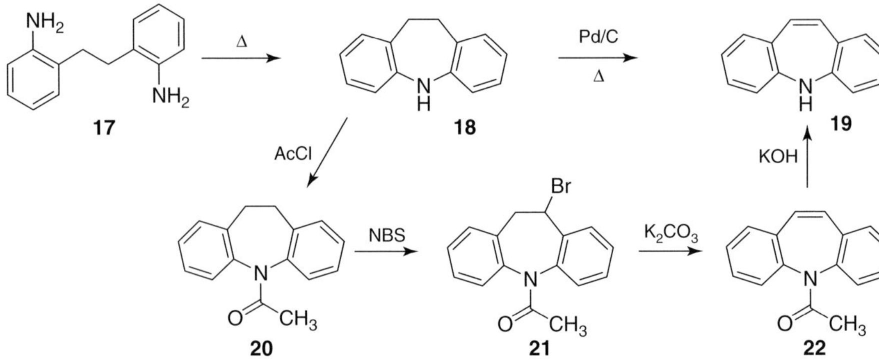

Scheme 8.3 Synthesis of the dibenzazepine core (**19**).

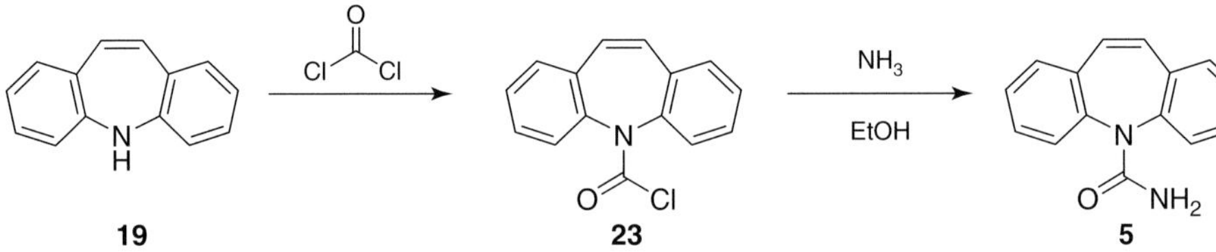

Scheme 8.4 Original sequence for the synthesis of CBZ (**5**).

assembled thermally. The two-carbon bridge was oxidized by two different routes, catalytic dehydrogenation or a multistep bromination/elimination protocol, to form **19**. This sequence enabled the synthesis of dibenzazepine-derived drugs, and numerous alternate routes were later developed [41]. The utility of the dibenzazepine core as a key component of pharmaceuticals [42] continues to spur the discovery of new and efficient routes to this material [43, 44].

Schindler [4] reported the first synthesis of CBZ (**5**) in 1960. The synthesis (Scheme 8.4) utilized **19** as the starting material, which was treated sequentially with phosgene and ammonia in ethanol to generate CBZ (**5**). Alternate preparations of CBZ (**5**) have appeared in the 50 years since the drug emerged, and most routes start with the assembled tricyclic core. Treatment of **19** with a variety of reagents, including metal cyanates followed by acid hydrolysis [45, 46], cyanogen halides followed by hydrolysis [47], and urea in combination with strong acid [48], are among the methods claimed.

OXC (**8**) was prepared in 1970 by chemists at Ciba-Geigy [49]. The synthesis was later optimized by Novartis (Scheme 8.5) [50]. The process generated cyanamide (**24**) from core **19** and used a selective nitration reaction to install a masked ketone (**25**). Reduction of the nitro group to form the enamine (**26**) occurred concomitantly with the oxidation of the cyanamide, and boric acid converted the enamine to the ketone (**8**). Additional syntheses of **8** have appeared, and the majority of synthetic routes involved functionalization of the core tricyclic system **19** to generate the oxidized central ring [51].

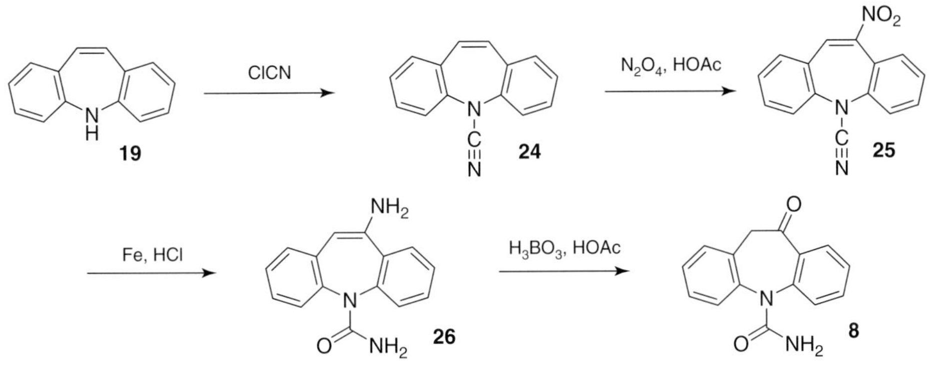

Scheme 8.5 Optimized synthesis of oxcarbazepine (OXC, **8**).

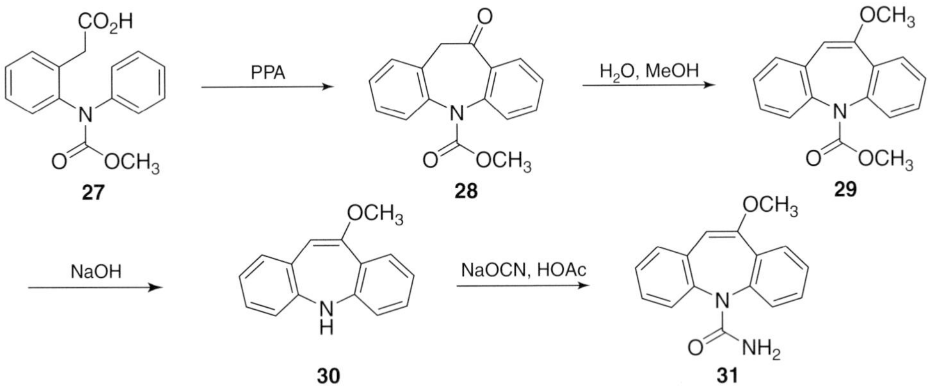

Scheme 8.6 Synthesis of OXC (**8**) precursor using an initial Friedel–Crafts cyclization.

The Novartis process group developed a new strategy involving a Friedel-Crafts cyclization to establish the fused tricyclic system with the 10-oxo group in place [52]. As shown in Scheme 8.6, the new synthetic sequence generates enol ether (**31**), an immediate precursor to OXC (**8**) from a more accessible starting material.

The tricyclic core **28** was created with the ketone in place, but protection as the enol ether was required to accommodate the urea formation. Additional improvements in the process continue to appear [53]. The preparation of the oxodibenzazepine core using a remote metalation/cyclization procedure using *N*-tolylanthranilic esters (**32**) has also been reported (Scheme 8.7) [54].

Palladium-catalyzed *C*- and *N*-arylation sequences for the preparation of OXC analogs were also described [55].

The synthesis of eslicarbazepine acetate (**10**) requires stereochemical control or a practical method for the resolution of the stereoisomers. Enantioselective reduction of OXC (**8**) has been used to generate eslicarbazepine ((*S*)-**9**)

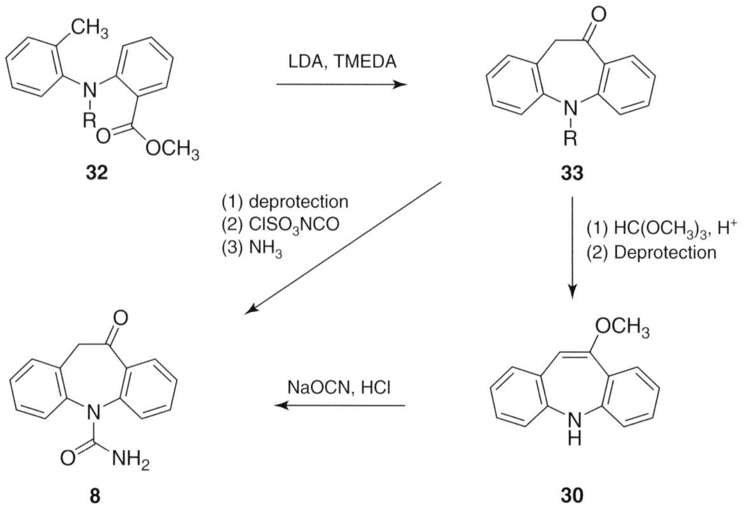

Scheme 8.7 Route to oxcarbazepine core via remote metalation/cyclization.

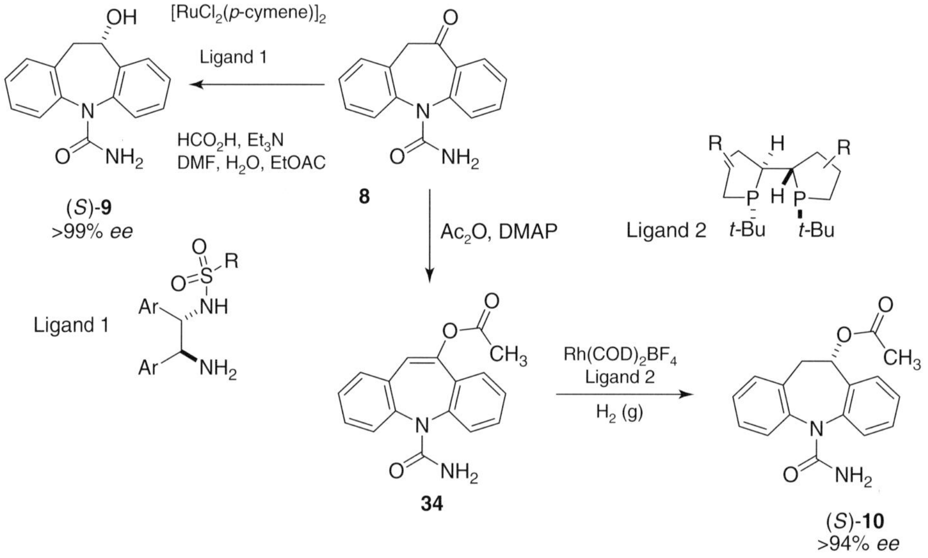

Scheme 8.8 Strategies for the preparation of chiral eslicarbazepine derivatives.

(Scheme 8.8). Asymmetric reduction of **8** was accomplished with a chiral ruthenium catalyst using the Noyori method [56, 57]. Alternately, reduction of the eno acetate using a rhodium catalyst sufficed, but the enantioselectivity was lower [58].

Two methods for the resolution via crystallization of diastereomeric salts have been reported (Scheme 8.9). Reduction of OXC (**8**) using sodium borohydride

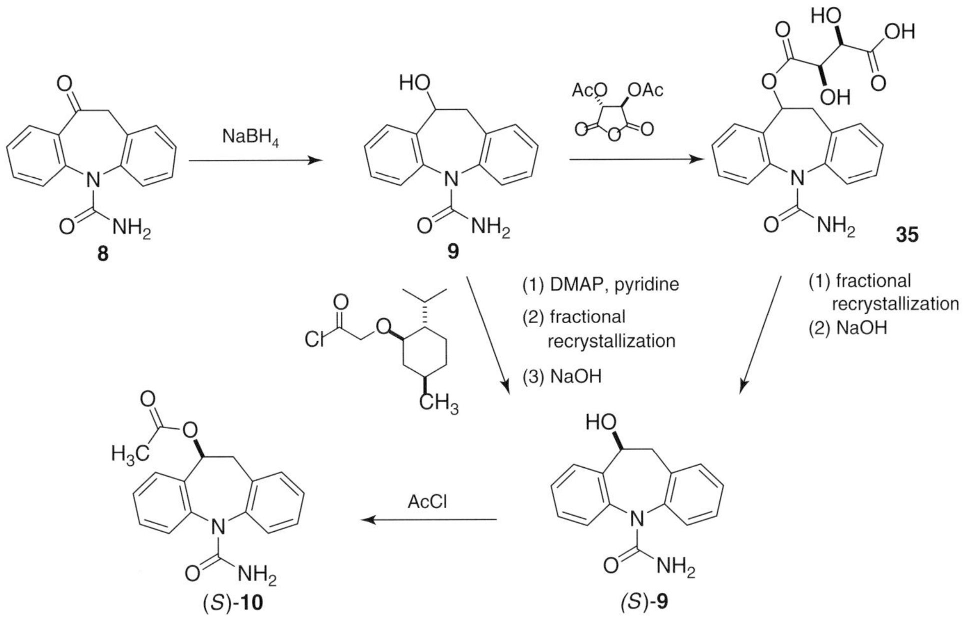

Scheme 8.9 Strategies for resolution of diastereomeric eslicarbazepine derivatives.

generated a racemic mixture of the hydroxy dibenzazepine (**9**). Treatment of **9** with menthoxyacetyl chloride in the presence of a base afforded diastereomeric esters. Fractional recrystallization provided the desired eslicarbazepine ((*S*)-**9**) in high enantiomeric excess [22]. A second method utilized the cyclic anhydride derived from *L*-tartaric acid [59]. The racemic alcohol was treated with the bisacetylated anhydride to give a mixture of diastereometic esters. Fractional recrystallization provided the enantiomerically enriched derivative of eslicarbazepine ((*S*)-**9**). For both recrystallization procedures, the chiral auxiliary was removed by saponification and standard methods generated the desired acetate ((*S*)-**10**). To minimize the loss of material represented by the undesired enantiomer, methods for mass recovery involving racemization [60] or chiral inversion [61] were developed.

The syntheses of the benzazepinone derivatives differ significantly from the preparations of the fused tricyclic systems. An outline of the sequence used to prepare BZP (**44**), an optimized selective sodium channel blocker that provided efficacy in preclinical pain models [62], is shown in Scheme 8.10 [63].

Following installation of the chloro group on the aromatic ring, alkylation of the nitrogen of lactam (**37**) was accomplished using sodium hydride and 2-iodopropane. The remainder of the structure was assembled by deprotection of selected nitrogen atoms as needed and amide bond formation using established EDC coupling techniques.

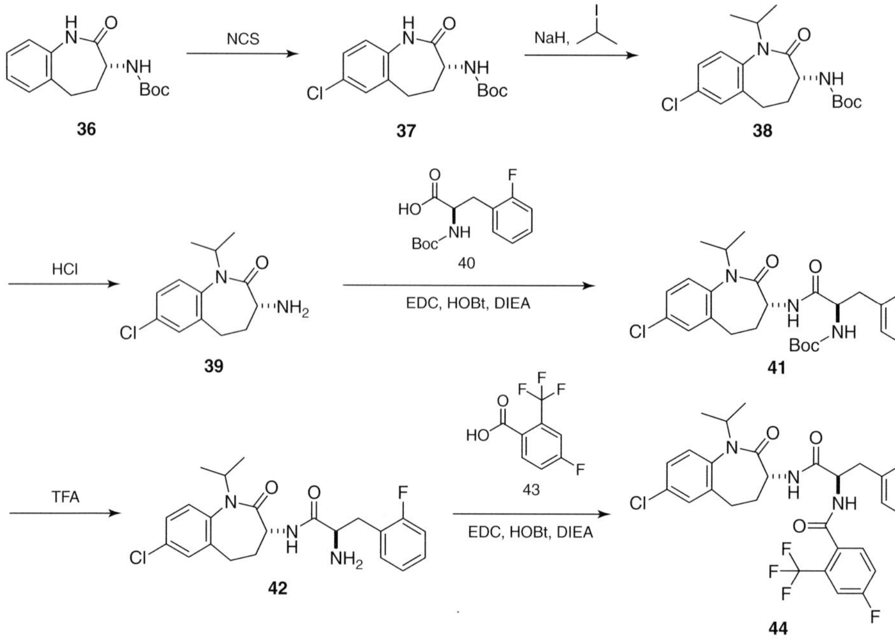

Scheme 8.10 Representative synthetic sequence for the preparation of benzazepinones.

8.4 Mode of Action

The family of VGSCs contains 10 proteins that control the transport of sodium ions through cell membranes [64–66]. Each distinct sodium channel protein consists of 1700–2000 amino acids, and the topologies of the channels have been studied. Functional sodium channels exist as heterodimeric combinations of 1 of the 10 alpha forms ($Na_V1.1Na_V1.9$, Na_Vx) with one or more beta forms ($\beta1-\beta4$) [67, 68]. The structures of the large (~260 kD) alpha subunits include four homologous domains (DI–DIV or D1–D4) consisting of six transmembrane segments (S1–S6) each (Figure 8.5).

The smaller beta subunits (33–45 kDa) participate in gating the function of the ion channel, and they are associated with the termini of the larger alpha subunits. Five of the transmembrane segments of the alpha subunits assemble in a bundle surrounding segment S4. The S4 segments in each domain serve as voltage sensors [69, 70]. The four domains DI–DIV aggregate, so segments S5 and S6 line the ion channel pore and facilitate ion transport [71].

The VGSCs exist in three distinct conformations: resting (closed, nonconducting), inactivated (closed), and open (Figure 8.6) [72]. The pore opens to allow sodium ion flux in response to membrane potentials. When cells containing VGSCs experience a negative resting potential, the ion channels exist in the resting state. When the membrane depolarizes, the sodium channels activate to the open state

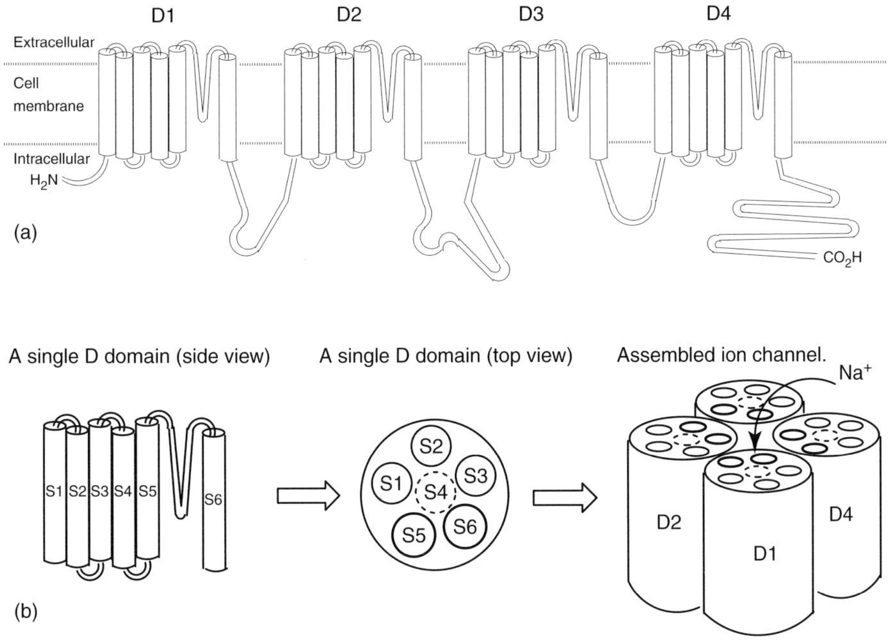

Figure 8.5 Schematic diagram of the sodium channel alpha subunit in extended form (a) and assembled in the conformation of a functional ion channel (b).

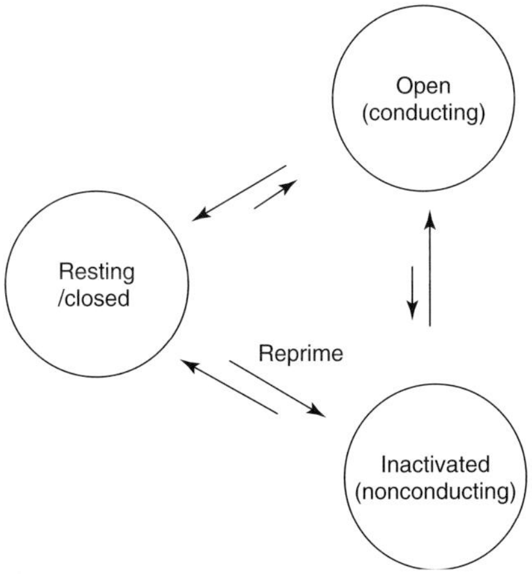

Figure 8.6 Representation of sodium channel functional states.

and allow the influx of sodium ions. Following ion flux, the channels change to the inactivated state, and a subsequent equilibrium to the closed state must occur before the ion channel can be primed for return to the open state [73]. The cycle through the states of the ion channel and the resulting sodium flux modulate action potentials that lead to neuronal conduction [74]. In disease states and conditions of

Table 8.1 Characteristics and sites of expression for the cloned Na_V sodium channels.

Subtype (alternate names)	TTX blockade (IC_{50})	Inactivation	Expression
$Na_V1.1$ (Brain Type I)	Potent	Fast	CNS, DRG
$Na_V1.2$ (Brain Type II)	Potent	Fast	CNS
$Na_V1.3$ (Brain Type III)	Potent	Fast	CNS, embryo
$Na_V1.4$ (Skm1–μ1)	Potent	Fast	Skeletal muscle
$Na_V1.5$ (Skm2, H1)	Weak	Slow	Cardiac muscle
$Na_V1.6$ (PN4)	Potent	Fast	CNS, DRG
$Na_V1.7$ (PN1, hNE-Na)	Potent	Fast	CNS, DRG
$Na_V1.8$ (PN3, SNS)	Weak	Slow	DRG
$Na_V1.9$ (NaN, SNS2)	Weak	Slow	DRG

neuronal injury, the cycle of the VGSCs through the activation states increases in rate.

The sodium channel isoforms exhibit different gating kinetics and voltage-dependent properties and have been categorized using several different criteria (Table 8.1) [3]. The channels were segregated into two groups based on sensitivity to tetrodotoxin (TTX, **45**) (Figure 8.7). TTX (**45**) efficiently blocks sodium currents in tissues of the central nervous system and skeletal tissues [75], but sodium currents in cardiac tissues are resistant to TTX blockade [76].

Functional VGSCs contribute to the initiation and propagation of action potentials in cells, and the activity of the sodium channels impacts motor activity, cognition, and nociception. Of the nine isoforms listed in Table 8.1, all but $Na_V1.4$ and $Na_V1.5$ are implicated in neurological function [77].

The mutation of the genes encoding for certain sodium channel isoforms, or some other etiology that impairs the function of these channels, can result in epileptic syndromes or other undesirable outcomes [78]. $Na_V1.1$ has been linked to epileptic syndromes, and mutations of the SCN1A gene that expresses $Na_V1.1$ are associated with increased ion channel function. The SCN2A gene encodes $Na_V1.2$, and mutations of SCN2A are also associated with epileptic disorders. Changes in the sites of expression of $Na_V1.2$ occur in multiple sclerosis and other diseases involving deterioration of the neuronal myelin sheaths. Blockade of $Na_V1.2$ contributes to unfavorable CNS side effects. Mutation of the SCN3A gene that expresses $Na_V1.3$ has been linked to epilepsy, but other physiological responses can occur on $Na_V1.3$ blockade. The $Na_V1.5$ sodium channel is expressed in cardiac tissues, and its

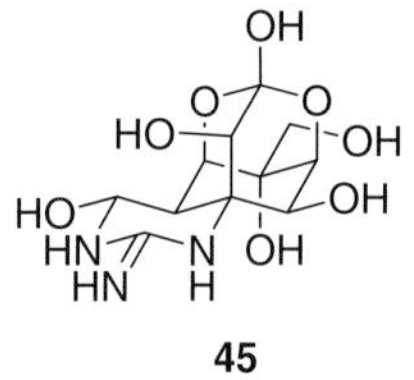

45

Figure 8.7 Marine toxin tetrodotoxin (**45**), used to categorize sodium channel activity.

blockade produces a range of undesirable cardiovascular liabilities. Disease states such as multiple sclerosis that are characterized by neuronal demyelination coincide with increased levels of $Na_V1.6$ expression. Sodium channel blockers may mediate inflammation and contribute to neuroprotection. Knockout of the SCN8A gene that encodes for $Na_V1.6$ produces mice with an unhealthy phenotype characterized by ataxia, tremors, muscular atrophy, and dystonia.

The list of sodium channels potentially involved in nociceptive and neuropathic pain includes $Na_V1.3$, $Na_V1.7$, $Na_V1.8$, and possibly $Na_V1.9$ [79–81]. Upregulation of $Na_V1.3$ in response to peripheral nerve injury occurs in different types of neurons involved in the transmission of pain signals. Increases in $Na_V1.3$ expression have been determined in preclinical pain models, including the spinal nerve ligation model (SNL) and the chronic constriction model (CCI). However, $Na_V1.3$ knockout mice develop conditions that mimic neuropathic pain. Clinical relevance of the potential of selective blockade of $Na_V1.7$ to produce analgesia has been established. Humans who lack functional $Na_V1.7$ fail to experience physical pain. Human phenotypes involving diminished or increased pain sensation are associated with mutations in the SCN9A gene that encodes for $Na_V1.7$. Expression of $Na_V1.8$ in the peripheral nervous system, upregulation, and increased function in neuropathies support the role of this ion channel in pain. Small interfering RNA and antisense oligonucleotides knocked down $Na_V1.8$ and reduced mechanical allodynia and thermal hyperalgesia in rat models of neuropathic pain. However, mice lacking $Na_V1.7$ and $Na_V1.8$ still develop conditions related to neuropathic pain. As with other sodium channels, the expression of $Na_V1.8$ changes in response to pain states. $Na_V1.9$ has not been studied as thoroughly as other sodium channels because of limitations in expression using recombinant systems. Mice that lack functional $Na_V1.9$ channels exhibit the normal pain phenotype. The use of antisense technology to inhibit the expression of $Na_V1.9$ did not alter the development of neuropathic pain in mice.

Small-molecule ligands of the voltage-gated sodium ion channels can interact with one or more of the conformations of the proteins that correspond to the open, closed, or inactivated states. Ligands may also bind to any number of the Na_V isoforms. The interactions of neurotoxins with sodium channels have been widely studied, and data regarding the binding site and state for the toxins have been generated [82]. Local anesthetics such as lidocaine function as nonselective, state-dependent sodium channel blockers, and numerous structural classes of blockers have been identified [3, 38]. Many small ligands can bind preferentially to one state of the sodium channels, and thereby exert influences on the equilibrium function by modulating this single state. CBZ (**5**) binds with low affinity ($>100\ \mu M$) to the resting state and with higher ligand affinity to the inactivated states of $Na_V1.3$, $Na_V1.7$, and $Na_V1.8$ with comparable IC_{50} values [83]. The interaction of **5** with the slow inactivation state of $Na_V1.8$-like sodium channels may influence the analgesic effects [84]. Homology models of the sodium channels have been developed, and the specific binding interactions of ligands including CBZ (**5**) have been proposed [85]. CBZ (**5**) and related compounds inhibit the sustained repetitive firing in spinal cord neurons [86].These sodium channel blockers also inhibit the sodium-induced release of the neurotransmitter glutamate in hippocampal nerve endings [87].

Many frequently used AEDs also function as calcium channel blockers [88]. Calcium channels contribute to depolarization-evoked neurotransmitter release [89]. Both CBZ (**5**) and OXC (**8**) exert antagonistic effects on the function of calcium channels [90, 91]. Very high doses that result in high plasma concentrations are required for blockade, and these exposure levels coincided with adverse events [92]. A report claims that BIA2-093 (**10**) inhibits the release of the neurotransmitter glutamate by blockade of sodium channels rather than calcium channels [93].

8.5
Structure–Activity Relationships

Phenytoin (**46**, Figure 8.8) was originally discovered in the early twentieth century [94], but it was not utilized as a drug until Dilantin® was introduced as an anticonvulsant in the 1950s [95].

The structure of **46** lacks the central fused tricyclic ring system of the dibenzazepines, but elements of the CBZ (**5**) and phenytoin (**46**) pharmacophores approximately overlap [85]. Phenytoin (**46**) functions as a sodium channel blocker with weak ligand affinity and state-dependent function [96]. Additional pharmacology has been determined for phenytoin (**46**) [97], including blockade of calcium channels comparable to that of CBZ (**5**). For both compounds, significant blockade of potassium-induced glutamate release occurs at high concentrations (> 500μM) [92]. Phenytoin (**46**) has been used for the treatment of trigeminal neuralgia [98], but it is not used frontline treatment for this condition.

Numerous structural variants of the TCAs were developed for the treatment of depression and related syndromes [12, 99]. Structurally, the imipramine analogs contain an aminoalkyl chain on the nitrogen of the azepine ring. They may also contain a carbon moiety in place of the azepine nitrogen. The pharmacology and mechanism of action of these compounds differ significantly from the dibenzazepine sodium channel blockers. The tricyclic antidepressants interact with several systems related to the transport mechanism of neurotransmitters including serotonin (5-HT) and noradrenaline (NA or norepinephrine) [100]. The pharmacology of the TCAs extends to interaction with the α2-adrenoreceptors of the noradrenergic system, binding to the glutamate receptors, blockade of potassium and calcium channels, and modification of opioid receptor density [36]. Examples

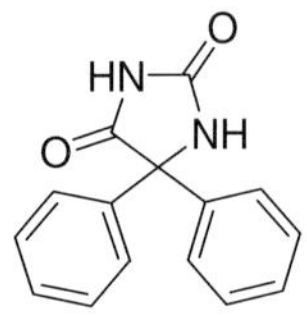

46

Figure 8.8 Structure of phenytoin (**46**).

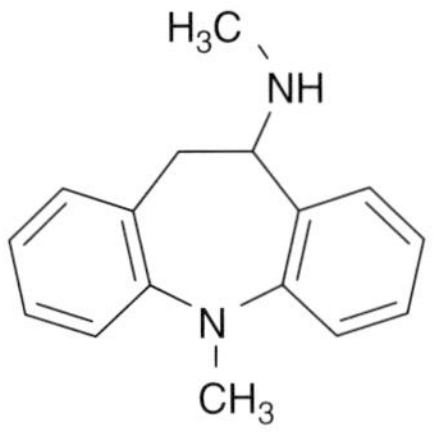

47

Figure 8.9 Structure of antidepressant metapramine (**47**).

of the tricyclic antidepressants exhibit sodium channel blockade, but this is only a component of their mechanism of action. Amitriptyline (**11**) blocked $Na_v1.7$ with an IC_{50} of 0.24 μM, and related compounds demonstrated similar potency [101]. The TCAs have been used successfully for the treatment of chronic pain, including postherapeutic neuralgia and diabetic neuropathy [102]. Efforts to develop selective sodium channel blockers for the treatment of pain have shifted away from the tricyclic cores of the anticonvulsants and antidepressants. However, a limited number of structure–activity relationship (SAR) studies on these cores have appeared. Compounds with fluorenyl or diphenylethylene cores provided ~50% blockade of $hNa_v1.2$ at concentrations of 10 μM [103]. Minor changes to the substituents on the dibenzazepine core can produce changes in pharmacology. The antidepressant metapramine (**47**) (Figure 8.9) functions as a low-affinity antagonist of the *N*-methyl-D-aspartic acid (NMDA) receptor [104].

Merck reported the discovery and investigation of a series of benzazepinone derivatives with selectivity as sodium channel blockers [39]. The efforts focused on the identification of blockers with selectivity for $Na_v1.7$. Two lead compounds emerged from these efforts. Benzodiazepinone (**48**) blocked human $Na_v1.7$ with an IC_{50} of 30 nM [105]. The potency in blockade function surpasses that of the simple tricyclic sodium channel blockers. Compound **48** demonstrated oral efficacy in a mouse model of epilepsy. The SAR studies at Merck produced benzazepinone compounds, including **44** (BZP), that effectively blocked $hNa_v1.7$ (Figure 8.10). The reported IC_{50} values for **44** and **49** were 30 and 232 nM, respectively. The oral

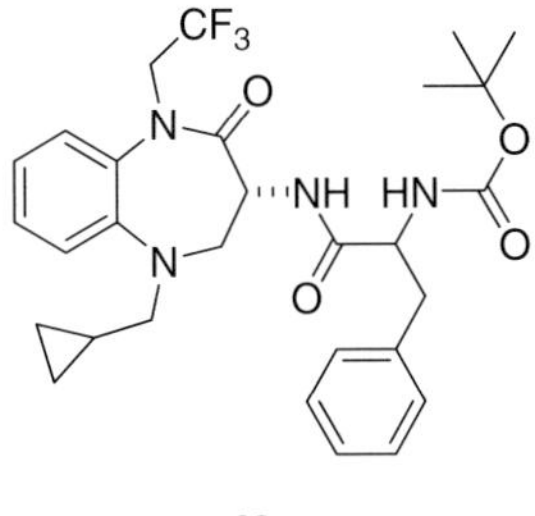

48

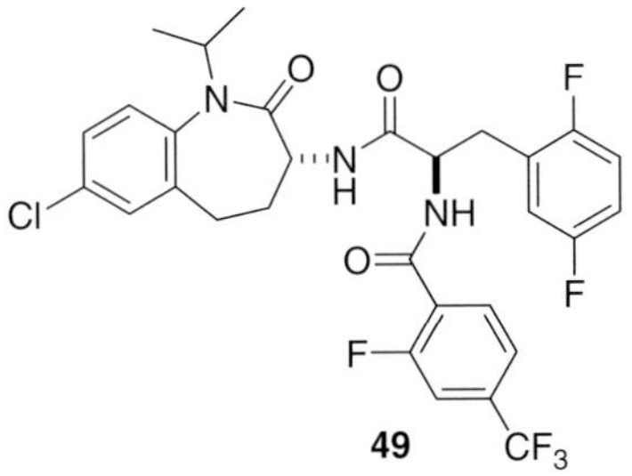

49

Figure 8.10 Structures of potent $hNa_v1.7$ sodium channel blockers.

bioavailability of **44** was higher than that reported for **49**, but both compounds significantly reversed allodynia in a rat model of neuropathic pain. Limited sodium channel isoform selectivity data were reported for **44** [62]. Modest levels of selectivity over $Na_V1.5$ and $Na_V1.8$ were determined.

Selected examples of dibenzazepine-derived sodium channel blockers may be used clinically for the treatment of specific types of neuropathic pain, but they do not provide analgesics of general utility. The clinical use of compounds such as CBZ (**5**) is limited by the side effect profile. Treatment of nociceptive pain resulting from stimuli that potentially or actively damage tissue and accompanying inflammation is currently not feasible with existing dibenzazepine sodium channel blockers. The applicability of selective sodium channel blockers for the treatment of neuropathic pain that originates from demonstrable neuronal injury or dysfunction of the tissues has been reported, but general use has not been approved. Significant efforts are currently directed toward the identification of new compounds to address this medical need.

References

1. Lader, M. (2004) *Handb. Exp. Pharmacol.*, **157**, 185–208.
2. Perucca, E. (2005) *Epilepsia*, **46S**, 31–37.
3. Anger, T., Madge, D.J., Mulla, M., and Riddall, D. (2001) *J. Med. Chem.*, **44**, 115–137.
4. Thiele, J. and Holzinger, O. (1899) *Justus Liebigs Ann. Chem.*, **305**, 96.
5. Schindler, W. (1960) U.S. 2,948,718.
6. Engel, J., Pedley, T.A., and Aicardi, J. (1998) *Epilepsia: A Comprehensive Textbook*, Lippinkott-Williams & Wilkins, Philadelphia, PA.
7. Khosravani, H. and Zamponi, G.W. (2006) *Physiol. Rev.*, **86**, 941–966.
8. Meisler, M.H., O'Brien, J.E., and Sharkey, L.M. (2010) *J. Physiol.*, **588**, 1841–1848.
9. Blom, S. (1962) *Lancet*, **1**, 839–840.
10. Spillane, J.D. (1964) *Practitioner*, **192**, 71–77.
11. Taylor, J.C. (1963) *J. Neurol. Neurosurg. Psychiatr.*, **26**, 553–554.
12. Zakzrzewska, J.M. (2010) *Expert Opin. Pharmacother.*, **11**, 1239–1254.
13. Campbell, F.G., Graham, J.G., and Zilkha, K.J. (1966) *J. Neurol. Neurosurg. Psychiatr.*, **29**, 265–267.
14. Bhattacharya, A., Wickenden, A.D., and Chaplan, S.R. (2009) *Neurotherapeutics*, **6**, 663–678.
15. (2011) US Food and Drug Administration Home Page Search: carbamazepine, *http://www.fda.gov/drugs* (accessed 30 June 2011).
16. (2011) Physician's Desk Reference Home Page Search: carbamazepine, *http://www.PDR3D* (accessed 30 June 2011).
17. Kerr, B.M., Thummel, K.E., Wurden, C.J., Klein, S.M., Kroetz, D.L., Gonzalez, F.J., and Levy, R.H. (1994) *Biochem. Pharmacol.*, **47**, 1969–1979.
18. Baker, K.M., Csetenyi, J., and Morselli, P.L. (1973) *J. Med. Chem.*, **16**, 703–705.
19. Kerr, B.M. and Levy, R.H. (1995) in *Antiepileptic Drugs* (eds R.H. Levy, R.H. Mattson, and M.S. Meldrum), Raven Press, New York, pp. 529–541.
20. Schindler, W. (1970) DE 2011087.
21. Schindler, W. (1972) US 3,642,775.
22. Benes, J., Parada, A., Figueirido, A.A., Alves, P.C., Freitas, A.P., Learmonth, D.A., Cunha, R.A., Garrett, J. and Soares-Da-Silva, P. (1999) *J. Med. Chem.*, **42**, 2582–2587.
23. (2011) US Food and Drug Administration Home Page Search: oxcarbazepine, *http://www.fda.gov/drugs* (accessed 30 June 2011).
24. Glauser, T.A. (2001) *Pharmacotherapy*, **21**, 904–919.

25. Dogra, S., Beydoun, S., Mazzola, J., Hopwood, M., and Wan, Y. (2005) *Eur. J. Pain*, **9**, 543–554.
26. Grosskopf, J., Mazzola, J., Wan, Y., and Hopwood, M.A. (2006) *Acta Neurol. Scand.*, **114**, 177–180.
27. Farago, F. (1987) *Eur. Neurol.*, **26**, 73–83.
28. Zakrzewska, J.M. and Passalos, P.N. (1989) *J. Neurol. Neurosurg. Psychiatr.*, **52**, 472–476.
29. Remillard, G. (1994) *Epilepsia*, **35**, s28–s29.
30. Theisohn, M. and Heimann, G. (1982) *Eur. J. Clin. Pharmacol.*, **22**, 545–551.
31. Hainzl, D., Parada, A., and Soares-Da-Silva, P. (2001) *Epilepsy Res.*, **44**, 197–206.
32. Ambrosio, A.F., Silva, A.P., Malva, J.O., Soares-Da-Silva, P., Cravalho, A.P., and Carvalho, C.M. (2001) *Biochem. Pharmacol.*, **61**, 1271–1275.
33. McCormack, P.L. and Robinson, D.M. (2009) *CNS Drugs*, **23**, 71–79.
34. (2011) US Food and Drug Administration Home Page Search: eslicarbazepine acetate, *http://www.fda.gov/drugs* (accessed 30 June 2011).
35. Soares-Da-Silva, P.M. (2007) WO 2007 094694 A1.
36. McCleane, G. (2008) *CNS Drugs*, **22**, 139–156.
37. Pletscher, A. (1991) *Experientia*, **47**, 4–8.
38. Matulenko, M.A., Scanio, M.J.C., and Kort, M.E. (2009) *Curr. Top. Med. Chem.*, **9**, 362–376.
39. Williams, B.S., Felix, J.P., Priest, B.T., Brochu, R.M., Dai, K., Hoyt, S.B., London, C., Tang, Y.S., Duffy, J.L., Parsons, W.H., Kaczorowski, G.J., and Garcia, M.L. (2007) *Biochemistry*, **46**, 14693–14703.
40. Schindler, W. and Hafliger, F. (1954) *Helv. Chim. Acta*, **59**, 472–483.
41. Kricka, L.J. and Ledwith, A. (1974) *Chem. Rev.*, **74**, 101–123.
42. Lednicer, D. (2009) *Strategies for Organic Drugs Synthesis and Design*, 2nd edn, Wiley-VCH Verlag GmbH, Hoboken, NJ.
43. Thansandote, P. and Lautens, M. (2009) *Chem. Eur. J.*, **15**, 5874.
44. Tsveliknovsky, D. and Buchwald, S.L. (2010) *J. Am. Chem. Soc.*, **132**, 14048–14051.
45. Gutman, D. and Baidossi, W. (2003) WO 106414 A2.
46. Eckhardt, R. and Jansch, H.-J. (2006) US 7,015,322 B1.
47. Citterio, A., Breviglieri, G., and Bruno, G. (2001) EP 1 127 877 A2.
48. Vyas, D.D., Jafri, W.S., and Kulkarni, A.K. (2001) US 6,245,908 B1.
49. Schindler, W. (1970) DE 2100087.
50. Aufderhaar, E. (1981) EP 28028.
51. Milanese, A. (2004) EP 1 600 443 A1.
52. Fuenfschilling, P.C., Zaugg, W., Beutler, U., Kaufman, D., Lohse, O., Mutz, J.-P., Onken, U., Reber, J.-L., and Shenton, D. (2005) *Org. Proc. Res. Dev.*, **9**, 272–277.
53. Singh, H., Gupta, N., Kumar, P., Dubey, S.K., and Sharma, P.K. (2009) *Org. Proc. Res. Dev.*, **13**, 870–874.
54. Lohse, O., Beutler, U., Funfschilling, P., Furet, P., France, J., Kaufmann, D., Penn, G., and Zaugg, W. (2001) *Tetrahedron Lett.*, **42**, 385–389.
55. Carril, M., SanMartin, R., Dominguez, E., and Tellitu, I. (2007) *Tetrahedron*, **63**, 690–702.
56. Learmonth, D.A., Grasa, G.A., and Zanotti-Gerosa, A. (2007) WO 2007 012793 A1.
57. Mathes, C., Sedelmeier, G., Blatter, F., Pfeffer, S., and Grimler, D. (2004) WO 2004 031155 A1.
58. Yu, B., Li, W., and Learmonth, D.A. (2007) WO 117166.
59. Learmonth, D.A. (2009) WO 092572.
60. Learmonth, D.A. (2004) WO 099153.
61. Learmonth, D.A., Weingaertner, G., and Kraemer, M. (2006) WO 005951.
62. McGowan, E., Hoyt, S.B., Li, X., Lyons, K.A., and Abbadie, C. (2009) *Anesth. Analg.*, **109**, 951–958.
63. Hoyt, S.B., London, C., Ok, H., Gonzalez, E., Duffy, J.L., Abbadie, C., Dean, B., Felix, J.P., Garcia, M.L., Jochnowitz, N., Karanam, B.V., Li, X., Lyons, K.A., McGowan, E., MacIntyre, D.E., Martin, W.J., Priest, B.T., Smith, M.M., Tschirret-Guth, R., Warren, V.A., Williams, B.S., Kaczorowski, G.J., and Parsons, W.H. (2007) *Biorg. Med. Chem. Lett.*, **17**, 6172–6177.

64. Caterall, W.A. (2000) *Neuron*, **26**, 13–25.
65. Goldin, A.L., Barchi, R.L., Caldwell, J.H., Hofmann, F., and Howe, J.R. (2000) *Neuron*, **26**, 365–368.
66. Goldin, A.L. (2001) *Annu. Rev. Physiol.*, **63**, 871–894.
67. Ko, S.H., Lenkowski, P.W., Hee, H.C., Mounsey, J.P., and Patel, M.K. (2005) *Pflugers Arch.*, **449**, 403–412.
68. Qu, Y., Rogers, J.C., Chen, S.F., McCormick, K.A., Scheuer, T., and Catterall, W.A. (1999) *J. Biol. Chem.*, **274**, 32647–32654.
69. Stuhmer, W., Conti, F., and Suzuki, H. (1989) *Nature*, **339**, 597–603.
70. Yang, N., George, A.L.J., and Horn, R. (1996) *Neuron*, **16**, 113–122.
71. Duclohier, H. (2009) *Biochem. Biophys. Acta*, **1788**, 2374–2379.
72. Kyle, D.J. and Ilyin, V.I. (2007) *J. Med. Chem.*, **50**, 2583–2588.
73. West, J.W., Patton, D.E., Scheuer, T., Wang, Y., Goldin, A.L., and Caterall, W.A. (1992) *Proc. Natl. Acad. Sci U.S.A.*, **89**, 10910–10914.
74. Gribkoff, V.K. and Winquist, R.J. (2005) *Expert Opin. Ther. Pat.*, **15**, 1751–1762.
75. Ritchie, J.M. and Rogart, R.B. (1977) *Rev. Physiol. Biochem. Pharmacol.*, **79**, 1–50.
76. Baer, M., Best, P.M., and Reuter, H. (1976) *Nature*, **263**, 344–345.
77. Termin, A., Martinborough, E., and Wilson, D. (2008) *Annu. Rep. Med. Chem.*, **43**, 43–60.
78. Kohling, R. (2002) *Epilepsia*, **43**, 1278–1295.
79. Cummins, T.R., Sheets, P.L., and Waxman, S.G. (2007) *Pain*, **131**, 243–257.
80. Hargus, N.J. and Patel, M.K. (2007) *Expert Opin. Invest. Drugs*, **16**, 635–646.
81. Momin, A. and Wood, J.N. (2008) *Curr. Opin. Neurobiol.*, **18**, 383–388.
82. Cestele, S. and Catterall, W.A. (2000) *Biochimie*, **82**, 883–892.
83. Kuo, C.C. (1998) *Mol. Pharmacol.*, **54**, 712–721.
84. Cardenas, C.A., Cardenas, C.G., de Armendi, A.J., and Scroggs, R.S. (2006) *Neurosci. Lett.*, **408**, 129–134.
85. Lipkind, G.M. and Fozzard, H.A. (2010) *Mol. Pharmacol.*, **78**, 631–638.
86. Macdonald, R.L. and McLean, M.J. (1986) *Adv. Neurol.*, **44**, 713–736.
87. Sitges, M., Chiu, L.M., Guarneros, A., and Nekressov, V. (2007) *Neuropharmacology*, **52**, 598–605.
88. Errington, A.C., Stohr, T., and Lees, G. (2005) *Curr. Top. Med. Chem.*, **5**, 15–30.
89. Spafford, J.D. and Zamponi, G.W. (2003) *Curr. Opin. Neurobiol.*, **13**, 308–314.
90. Schmidt, D. and Elger, C.E. (2004) *Epilepsy Behav.*, **5**, 627.
91. Ambrosio, A.F., Silva, A.P., Malva, J.O., Soares-Da-Silva, P., Carvalho, A.P., and Carvalho, C.M. (1999) *Neuropharmacology*, **38**, 1349–1359.
92. Sitges, M., Guarneros, A., and Nekrassov, V. (2007) *Neuropharmacology*, **53**, 854–862.
93. Ambrosio, A.F., Silva, A.P., Malva, J.O., Soares-Da-Silva, P., Carvalho, A.P., and Carvalho, C.M. (2001) *Biochem. Pharamacol.*, **61**, 1271–1275.
94. Biltz, H. (1908) *Berichte*, **41**, 1379–1393.
95. Friedlander, W.J. (1986) *Epilepsia*, **27** (Suppl. 3), S1–S21.
96. Xie, X.M. and Garthwaite, J. (1996) *Neuroscience*, **73**, 951–962.
97. Francis, J., Eubanks, J.H., and Burnham, W.M. (2000) *Brain Res.*, **876**, 131–140.
98. Tremont-Lukats, I.W., Megeff, C., and Backonja, M.-M. (2000) *Drugs*, **60**, 1029–1052.
99. Zettle, H., Schubert-Zsilavecz, M., and Siebert, C.D. (2008) *Pharm. Unserer Zeit*, **3**, 206–213.
100. Gillman, P.K. (2007) *Br. J. Pharmacol.*, **151**, 737–748.
101. Dick, I.E., Brochu, R.M., Purohit, Y., Kaczorowski, G.J., Martin, W.J., and Priest, B.T. (2007) *J. Pain*, **8**, 315–324.
102. Verdu, B., Decosterd, I., Buclin, T., Stiefel, F., and Berney, A. (2008) *Drugs*, **68**, 2611–2632.
103. Hudgens, D.P., Taylor, C.T., Batts, T.W., Patel, M.K., and Brown, M.L. (2006) *Bioorg. Med.Chem.*, **14**, 8366–8378.

104. Boireau, A., Bordier, F., Durand, G., and Doble, A. (1996) *Neuropharmacology*, **35**, 1703–1707.

105. Hoyt, S.B., London, C., Wyvratt, M.J., Fisher, M.H., Cashen, D.E., Felix, J.P., Garcia, M.L., Li, X., Lyons, K.A., MacIntyre, D.E., Martin, W.J., Priest, B.T., Smith, M.M., Warren, V.A., Williams, B.S., Kaczorowski, G.J., and Parsons, W.H. (2008) *Biorg. Med. Chem. Lett.*, **18**, 1963–1969.

Part II
Cardiovascular Diseases

Bioactive Heterocyclic Compound Classes: Pharmaceuticals,
First Edition. Edited by Jürgen Dinges and Clemens Lamberth.

9
Dihydropyridine-Based Calcium Channel Blockers for the Treatment of Angina Pectoris and Hypertension

Milan Bruncko

9.1
Introduction

The 1,4-dihydropyridine moiety (Figure 9.1) is a common core in many therapeutic agents including antibiotics [1], cytotoxics [2], and chelators [3]. However, 1,4-dihydropyridines (**1**, DHPs) are most commonly used as calcium channel blockers (CCBs) in the cardiovascular area. Drugs of this class are easily identified by the suffix "–dipine."

The DHPs (**1**) can be divided into several subgroups depending on the substitution pattern of the dihydropyridine ring. The typical CCBs are symmetrically substituted with a methyl group at the C2 and C6 positions, small alkyl carboxylates at C3 and C5, aryl at C4, and H at N1. Nifedipine (**2a**) [4, 5] is a classic representative for this subgroup that also includes the 4-nitroaryl DHPs *m*-nifedipine, dibudipine (**2b**) [6, 7], mebudipine (**2c**) [6], nisoldipine (**2d**) [8], and the nonsymmetrical analogs menidipine (**2e**) [9] and nitrendipine (**2f**) [10] (Figure 9.2).

The second subgroup of DHP analogs is characterized by more complex substituted polar neutral esters, for example, furnidipine (**3a**) [11, 12], nimodipine (**3b**) [13, 14], and cilnidipine (**3c**) [15]. These compounds possess a substituted phenyl moiety at the 4-position of the dihydropyridine, retain the 2,6-dimethyl groups, and have the small alkyl esters replaced with functionalized alkyl or cyclic ethers (Figure 9.3).

The third subgroup is DHPs (**1**) containing side chains with basic amines at the C2 position, like the best seller amlodipine (**4a**) [16] and the chiral levamlodipine (**4b**) [17], at the C3 position, for example, barnidipine (**4c**) [18] and nicardipine (**4d**) [19], at the C4 position, represented by labedipinedilol-A (**4e**) [20], or less often at N1, as in flordipine (**4f**) [21] (Figure 9.4).

This chapter focuses on dihydropyridine-based CCBs.

Bioactive Heterocyclic Compound Classes: Pharmaceuticals,
First Edition. Edited by Jürgen Dinges and Clemens Lamberth.

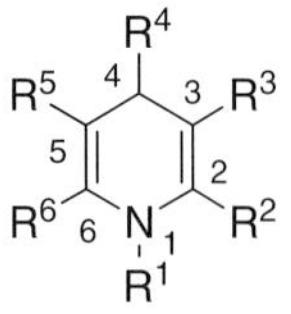

1

Figure 9.1 Basic structure and numbering convention for 1,4-dihydropyridines (**1**).

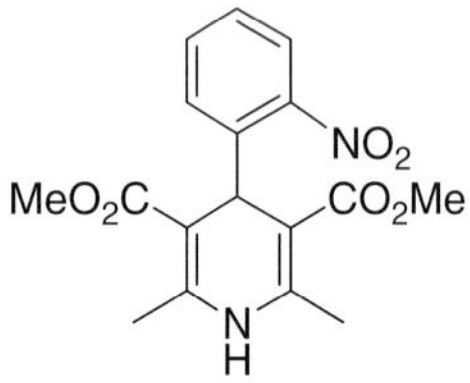

2a
Nifedipine

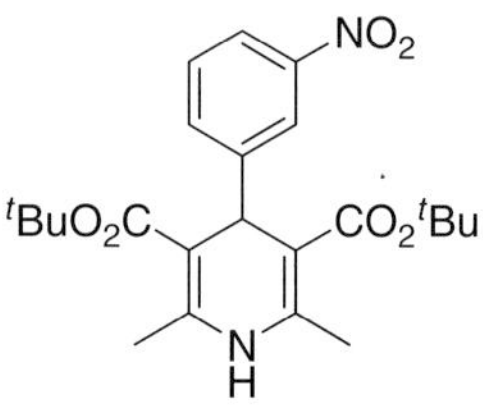

2b
Dibudipine

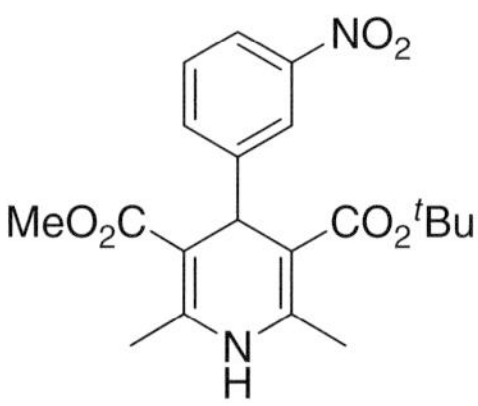

2c
Mebudipine

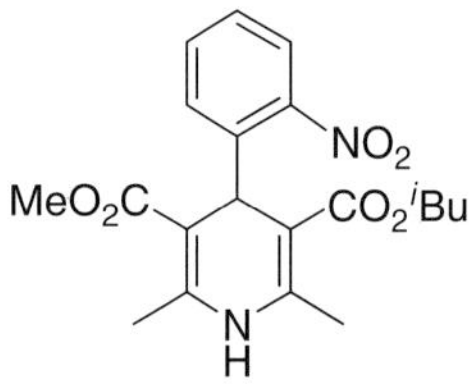

2d
Nisoldipine

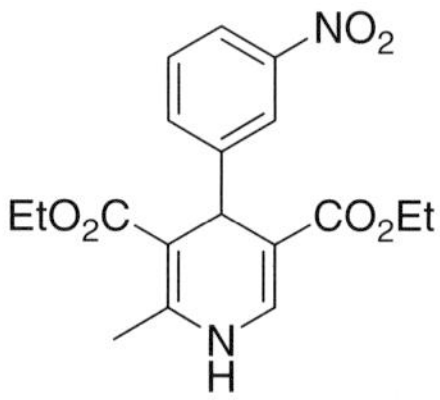

2e
Menidipine

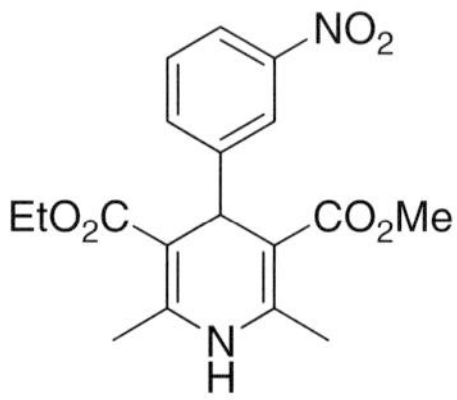

2f
Nitrendipine

Figure 9.2 Nifedipine (**2a**) and closely related dihydropyridines **2b**–**2f**.

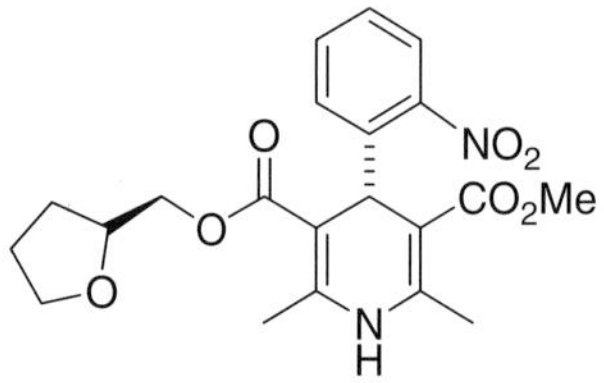

3a
Furnidipine

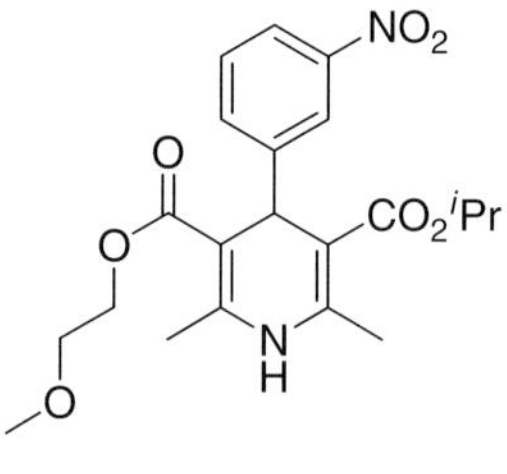

3b
Nimodipine

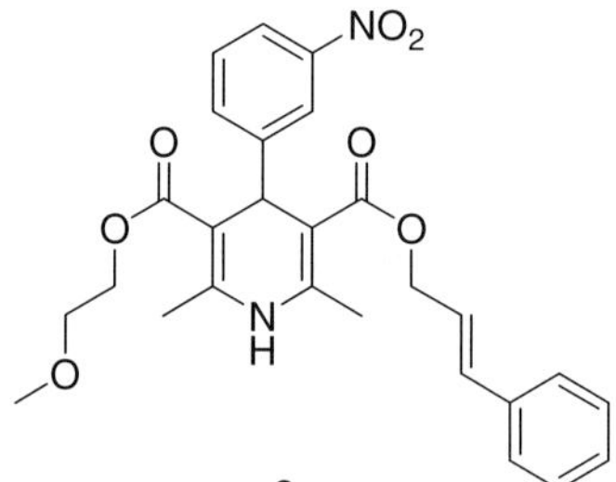

3c
Cilnidipine

Figure 9.3 Dihydropyridines (**1**) with more complex neutral ester substituents.

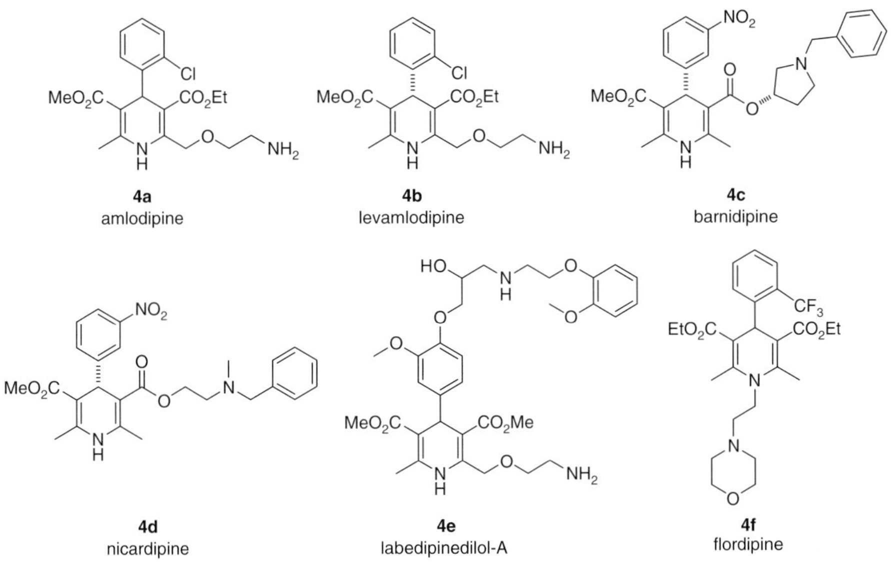

Figure 9.4 Dihydropyridines (**1**) with basic side chains.

9.2 History

The first synthesis of a dihydropyridine (**6**) was achieved by Arthur Hantzsch [22, 23] in 1882 by reacting 1 equivalent of formaldehyde, 1 equivalent of ammonium hydroxide, and 2 equivalents of the β-ketoester **5** (Scheme 9.1).

This methodology was further expanded to the synthesis of the asymmetrically substituted DHP **7** by Knoevenagel in 1898 [24]. Knoevenagel treated β-aminocrotonate (**8**) with the benzylidene analog **9**, providing the dihydropyridine **7** (Scheme 9.2).

The discovery of the "hydrogen transferring" properties of NADH, a dihydronicotinamide derivative, in 1930 prompted numerous studies on the biochemical properties of DHPs (**1**) [25, 26]. Besides the identification of mild analgesic and "curarelike" effects [27], more detailed pharmacological activities were not initially investigated. Loev *et al.* [5] were the first to fully explore those pharmacological properties and to develop structure–activity relationships (SARs) of the DHPs (**1**).

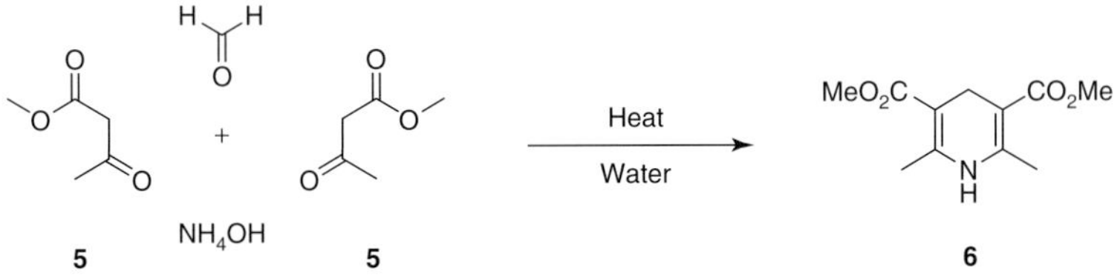

Scheme 9.1 Classical Hantzsch synthesis of DHP (**6**).

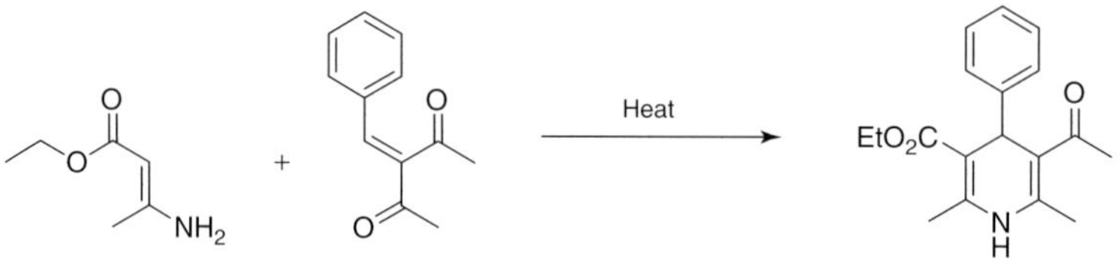

8 **9** **7**

Scheme 9.2 Knoevenagel synthesis of the asymmetrically substituted DHP (**7**).

In 1962, Hass and Hartfelder reported [28] that verapamil possessed negative inotropic and chronotropic effects that were not observed with other vasodilatory agents. Five years later, Fleckenstein [28] suggested that the inotropic effect was caused by the movement of calcium ions into cardiac myocytes. In 1969, Rougier *et al.* [28] presented evidence that the depolarization of atrial tissue is driven by two inwardly directed ionic currents. One of these currents is based on the influx of calcium ions into the cardiac cell through the "slow channel" or calcium channel.

Nifedipine (**2a**, BAY a1040, Adalat®, Nifediac®, Cordipin®, Nifedical®, and Procardia®) is the first dihydropyridine (**1**) developed by Bayer in the early 1970s that demonstrated antagonistic effects on calcium ion influx [29–31]. The discovery of nifedipine (**2a**) fueled extensive research in the area of DHP-based CCBs, yielding a vast number of DHP analogs with improved pharmacokinetic and pharmacodynamic properties, decreased adverse effects, and more convenient dosing.

In 1983, Schramm and coworkers [32] found that certain DHP analogs have the opposite effects on cardiac contractility and smooth muscle contraction when compared to nifedipine (**2a**). BAY K 8644 (**10**) (Figure 9.5) is one of these DHPs (**1**) with positive inotropic action functioning through the activation of calcium channels. BAY K 8644 (**10**) presumably binds at the same location or very close to the nifedipine (**2a**) binding site as was found from competition experiments. BAY K 8644 (**10**) is a noncardioselective agent that constricts the coronary arteries and therefore cannot serve as a therapeutic agent or as an antidote against CCB overdose.

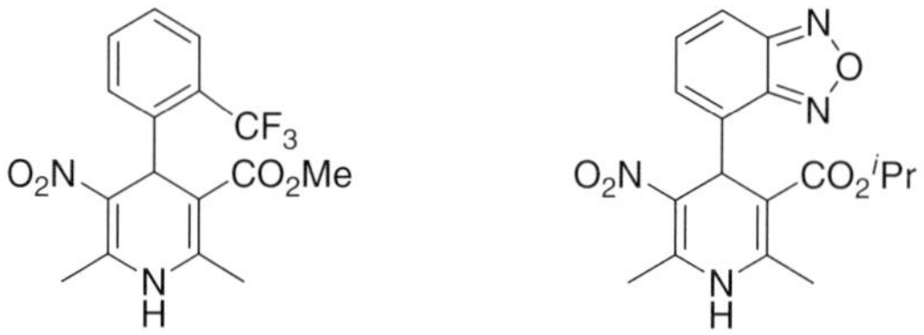

10
BAY K 8644

11
PN 202-791

Figure 9.5 The structures of BAY K 8644 (**10**), a calcium channel agonist, and PN 202-791 (**11**).

In 1985, Hof *et al.* [33] discovered that the enantiomers of PN 202-791 (**11**) have opposite pharmacological effects. The *R*-enantiomer inhibited contraction and calcium uptake, whereas the *S*-enantiomer enhanced contraction and calcium uptake. These results indicate the need for the development of enantiomerically pure compounds to avoid unnecessary adverse effects or to improve activities of racemates in the case that one functions as an agonist and the other as an antagonist.

9.3 Synthesis

Nifedipine (**2a**) and nifedipinelike analogs were easily prepared in one step by the standard Hantzsch synthesis. Methyl acetoacetate (**5**) heated with substituted benzaldehydes and ammonium hydroxide gave symmetrically substituted achiral DHPs (**1**) in moderate yields (Scheme 9.3) [5].

When DHPs (**1**) contain different ester groups at C3 and C5 or different alkyl groups at C2 and C6, the compounds possess a chiral center at the C4 position of the DHP ring. Racemic nitrendipine (**2f**) and several other similar analogs were synthesized via the Knoevenagel variation of the Hantzsch synthesis. First, the aldehyde **13** was treated with the β-ketoester **5** under acidic conditions, providing the benzylidene analog **14**. Subsequent coupling of **14** with beta-aminocrotonate (**8**) produced nitrendipine (**2f**) (Scheme 9.4) [34].

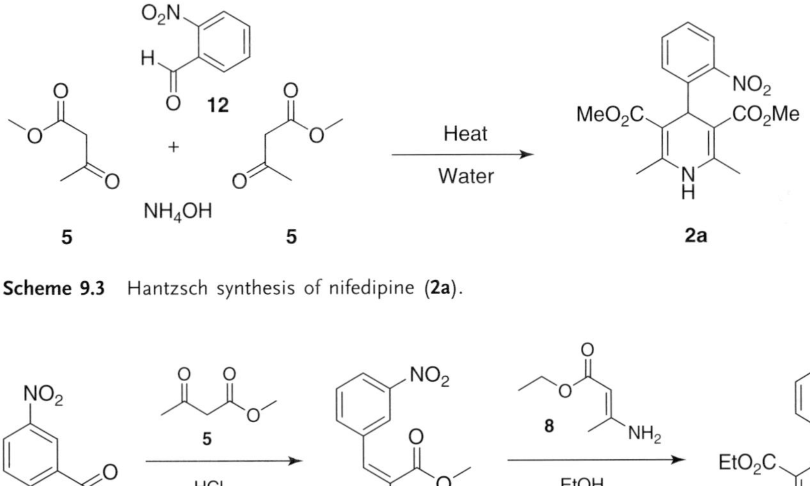

Scheme 9.3 Hantzsch synthesis of nifedipine (**2a**).

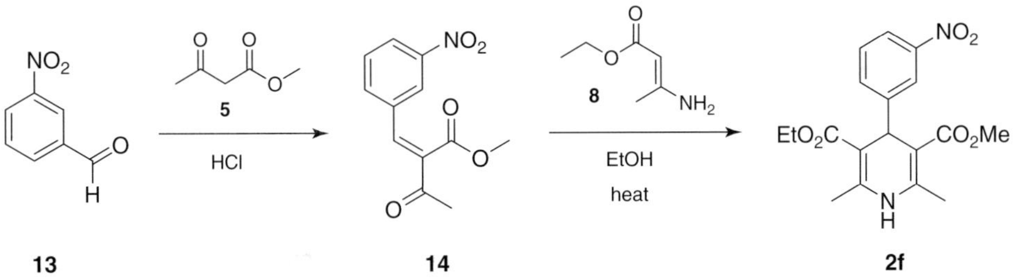

Scheme 9.4 Knoevenagel synthesis of nitrendipine (**2f**).

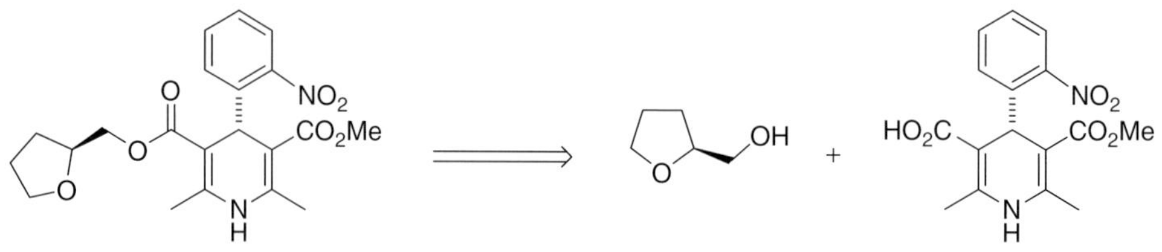

3a **15** **16**

Scheme 9.5 Failed retrosynthetic approach toward furnidipine (**3a**).

The majority of methods for the preparation of chiral DHPs are based on the chemical [35] or biological [36, 37] resolution of racemates or on the application of chiral auxiliaries [38–40].

Furnidipine (**3a**) is a dihydropyridine (**1**) with two chiral centers, which makes the compound and its diastereomers good examples to study the pharmacological effects of each individual diastereomer. However, early attempts for the asymmetric synthesis of chiral nifedipine analogs were abandoned because of their incompatibility with the 2-nitrophenyl substituent at C4 [40, 41]. Alajarin *et al.* studied a direct route (Scheme 9.5) in which both intermediates **15** and **16** would be prepared in enantiomerically pure form by chiral resolution. The chiral furnidipine (**3a**) would then be obtained following the esterification of the chiral DHP acid **16** or its N-protected analog **18** (Scheme 9.6) with the chiral tetrahydrofurfuryl alcohol **15** [12].

However, this promising attempt failed due to the low yielding (12%) hydrolysis of the diester **18** (Scheme 9.6). The formation of the nitroso compound **20** as the

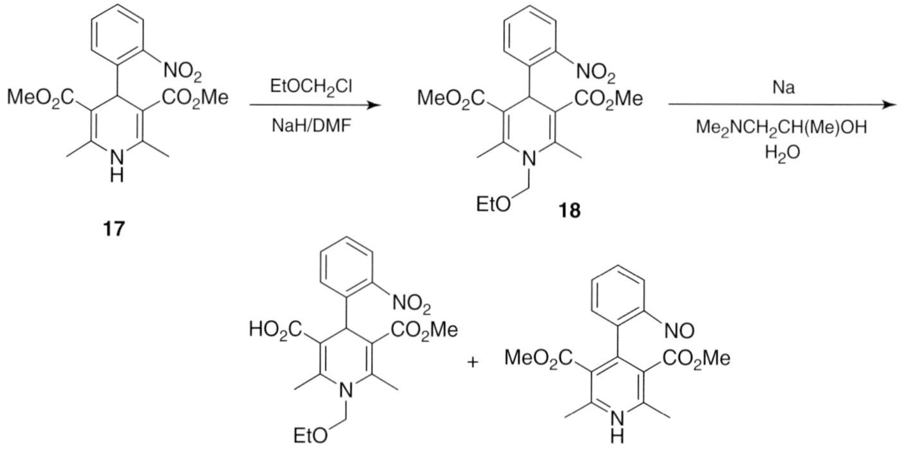

17 **18** **19** **20**

Scheme 9.6 Formation of the nitroso side product **20** during the ester hydrolysis of **18**.

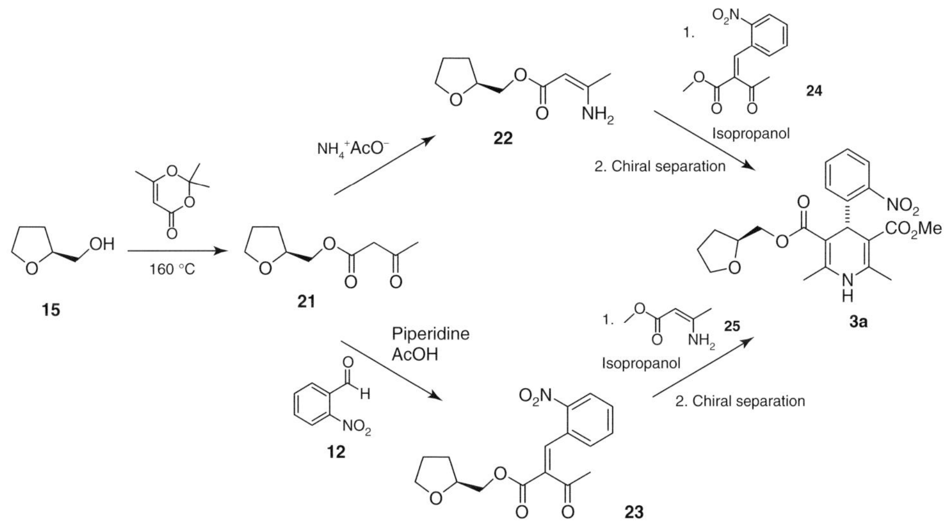

Scheme 9.7 Successful synthesis of enantiomerically pure furnidipine (**3a**).

major by-product was inevitable, and attempts to optimize the reaction conditions did not provide an improved yield of the acid **19**.

Alajarin *et al.* explored two alternate routes for the synthesis of furnidipine (**3a**) starting from the chiral tetrahydro furfuryl alcohol (**15**) (Scheme 9.7). They coupled either the chiral aminocrotonate **22** or the chiral benzylidene **23** with the corresponding achiral counterparts **24** or **25**, respectively, and then successfully separated obtained diastereomers.

Enders and coauthors [40] have developed an attractive chiral route toward the synthesis of enantiomerically pure DHPs (**1**) using the asymmetric Michael addition of metalated chiral alkyl acetoacetate hydrazones **27** to the Knoevenagel acceptors **28**. This reaction produced **30** in *enantiomeric excess* (*ee*) in the range of 84–98% (Scheme 9.8).

An elegant catalytic method was developed by Kanger *et al.* [42] using an organocatalytic asymmetric aza-ene-type cascade reaction and cocatalyzed by the (*S*)-diarylprolinol-trimethylsilyl ether **33** and benzoic acid (Scheme 9.9). The DHPs **34** were formed in moderate to good yields (45–86%) with *ee* of 62–96%. The limitation of this methodology is in the selection of aldehydes. The method only works well for aliphatic unsaturated aldehydes **31** (R^1 = alkyl), whereas the yields and selectivities are poor if aromatic unsaturated aldehydes with electron-donating substituents are used.

A similar approach was used by Takemoto *et al.* [43] (Scheme 9.10) who used a Brønsted acid (difluoroacetic acid) together with the novel thiourea derivative **37** to cocatalyze addition of β-enamino esters **35** to α, β-unsaturated aldehydes **36** (R = aryl), yielding chiral 1,4-DHPs **38** with moderate to good enantioselectivities (38–80%).

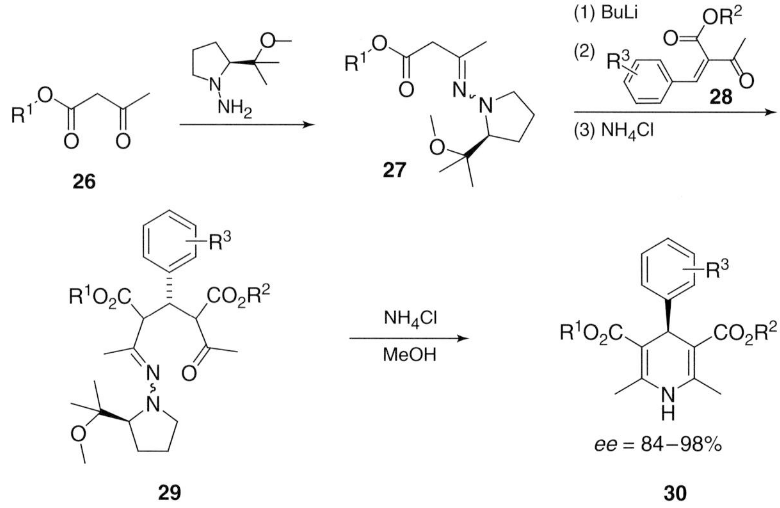

Scheme 9.8 Asymmetric synthesis of chiral dihydropyridines (**30**).

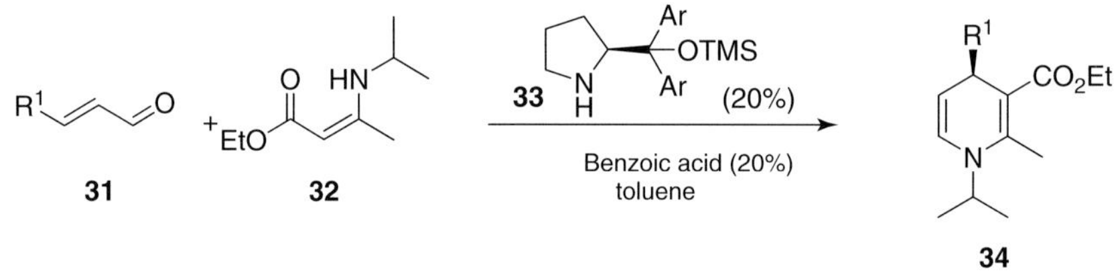

Scheme 9.9 Organocatalytic asymmetric aza-ene-type cascade reaction in the synthesis of dihydropyridines (**34**).

9.4
Mode of Action

Calcium (II) plays an essential role in the physiology of cells and organisms, where it has an important function as a second messenger in signal transduction pathways. The discovery of the role of calcium ions in the control of cardiac contractility goes back to the year 1883 when Ringer [44] performed the first experiments with various ionic solutions and studied their effect on heart muscle contractility.

Voltage-gated calcium channels are transmembrane proteins that control the influx of calcium ions in response to membrane depolarization [45]. These channels are important drug targets possessing specific binding sites for both antagonistic and agonistic ligands. CCBs are agents that disrupt the movement of calcium ions through voltage-dependent/voltage-gated calcium channels. There are several classes of voltage-gated calcium channels: L-type calcium channels (long lasting) are abundant in skeletal muscle, smooth muscle, and bone; T-type calcium channels (transient) are found in neurons, bone, and cells with pacemaker activities; P-type

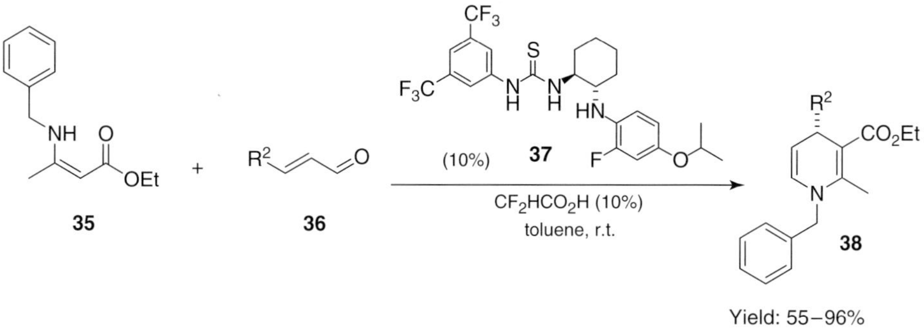

Scheme 9.10 Organocatalytic asymmetric addition of β-enamino esters to α, β-unsaturated aldehydes in the synthesis of dihydropyridines (**38**).

calcium channels (Purkinje) are found in Purkinje neurons; N-type calcium channels (neural) are located in the brain; and R-type calcium channels are found in neurons [46].

The CCBs, which bind to the L-type calcium channels, can be divided into three chemical classes: phenylalkylamines (verapamil) bind to the V binding site, benzothiazepines (diltiazem) bind to the D binding site, and DHPs (nifedipine (**2a**)) bind to the N binding site. These groups of compounds typically do not block other types of calcium channels; however, there are some DHP analogs with dual activity toward either L- or T-type and N-type calcium channels.

Calcium ions have an important function in regulating skeletal and smooth muscle contractility and thus the performance of the heart [47]. The inhibition of the calcium channels is achieved by the binding of CCBs to the L-type calcium channel, causing a decreased flow of calcium ions through the channel. This has a dilating effect on the coronary arteries, decreases cardiac contractility, and decreases hypertension (Figure 9.6).

The L-type calcium channel consists of several subunits such as the α_1 subunit, the $\alpha_2\delta$ subunit, the intracellular β_{1-4} subunits (Figure 9.7), and the transmembrane γ subunit (not shown in the figure).

The α1 subunit forms the pore through which calcium ions flow into the cell. The other subunits function as modulators of gating [46]. The DHP-based CCBs are selective antagonists that bind to the DHP binding site located on the surface of the α_1 subunit [48].

DHP antagonists have undergone optimizations at every position on the DHP core; currently, there are four generations of drugs characterized by their duration of action and their effect on heart rate and contraction.

The first generation of formulations/agents (nicardipine (**4d**), liquid nifedipine (**2a**) and clevidipine (**39**) [49]) (Figure 9.8), is characterized by a rapid onset of a vasodilatory effect but a short duration of action. This group of compounds caused a series of adverse reactions, such as cardiac depressant activity, and, therefore, the pharmaceutical industry rapidly developed the next generation of those compounds.

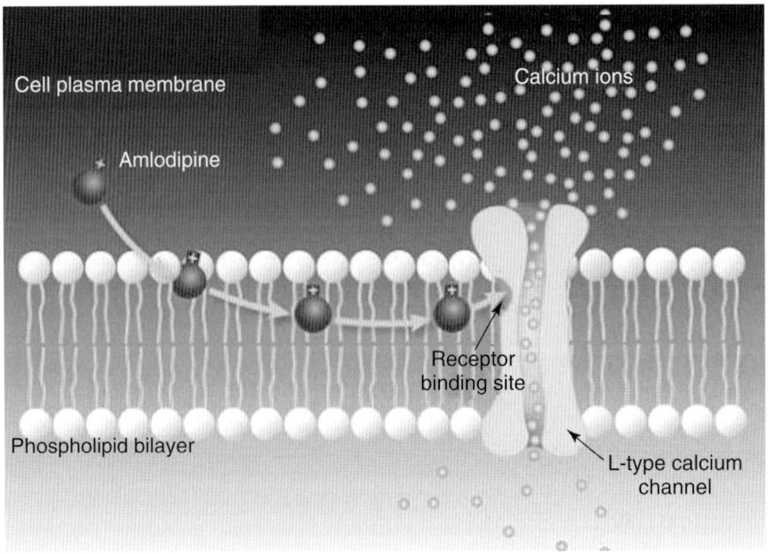

Figure 9.6 Mode of action of CCBs acting on an L-type calcium channel.

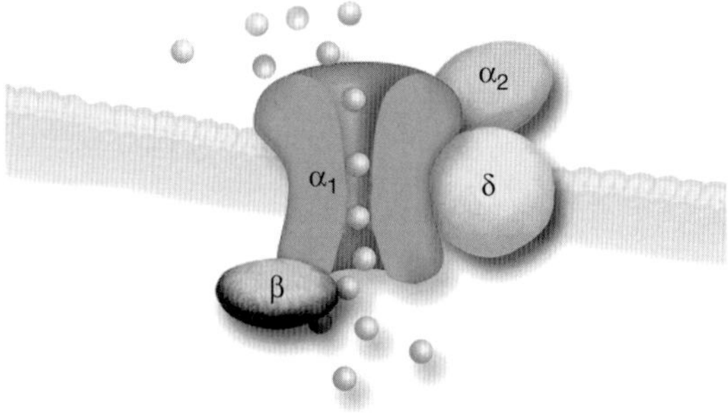

Figure 9.7 L-type calcium channel with subunits.

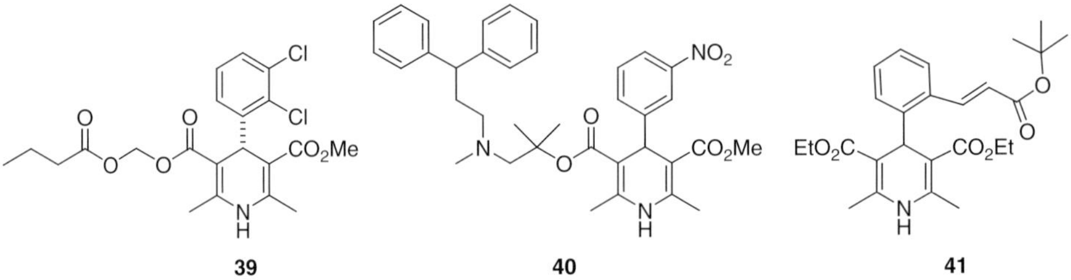

Figure 9.8 Structures of clevidipine (**39**), lercanidipine (**40**), and lacidipine (**41**).

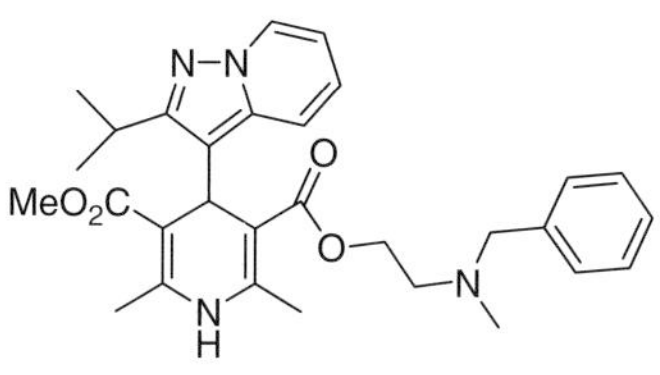

42
AHC-52

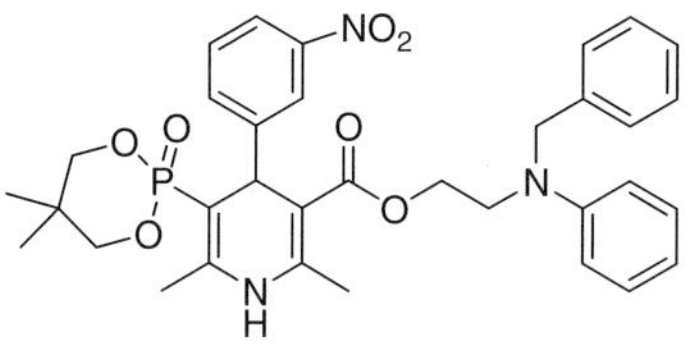

43
Efonidipine

Figure 9.9 Structures of AHC-52 (**42**) and efonidipine (**43**).

The second-generation DHPs are longer-acting slow-release formulations (nicardipine (**4d**), Carden SR®, nisoldipine (**2d**), and nifedipine GITS) with lower cardiac depressant activity. These agents allowed a better control of their efficacy with less adverse side effects.

The third generation of DHP analogs for the treatment of cardiovascular diseases includes cilnidipine (**3c**), amlodipine (**4a**), and nitrendipine (**2f**). These drugs are characterized by more stable pharmacokinetic profiles and no cardiac depressant activity, which makes them well tolerated in patients with heart failure.

The fourth generation of DHP-type CCB antagonists is represented by lercanidipine (**40**) and lacidipine (**41**) (Figure 9.8). These are highly lipophilic DHPs with stable activity, further reduction of adverse side effects, and a broad spectrum of activity, particularly in patients with congestive heart failure or myocardial ischemia [50].

Some DHPs (e.g., AHC-52 (**42**); Figure 9.9) do not act via antagonism of the L-type calcium channels, but instead, they block the cAMP-activated Cl^- channel [51, 52]. The Cl^- channel-blocking DHPs do not affect the myocardial contractile force during normoxia; however, they enhance the recovery of myocardial contractile force after experimental ischemia-reperfusion [53].

Efonidipine (**43**) [54, 55] (Figure 9.9) is a DHP with the ability to block both the L- and T-type calcium channels. Efonidipine (**43**) was developed as a CCB having a long-lasting effect similar to other DHPs; however, besides retaining a hypotensive effect, efonidipine (**43**) also showed a strong negative chronotropic effect similar to agents blocking the T-type calcium channels. Effect on bradycardiac activity through inhibition of the T-type Ca^{2+} current appears to be a promising strategy in the development of therapeutic agents for various cardiovascular disorders [56].

On the other hand, cilnidipine (**3c**) is a dual inhibitor, which is able to block calcium influx in the L-type and N-type calcium channels [57].

Recent studies have demonstrated that several DHPs are able to inhibit P-glycoproteins (Pgps) responsible for multidrug resistance in tumor cells. This inhibition does not correlate with the calcium-channel-blocking activity, and even DHPs with minimal CCB activity can be very potent Pgp inhibitors [58].

DHPs, such as many other drugs, can cause adverse side effects. The most common adverse effects of DHPs are due to vasodilation. In patients, vasodilation can lead to dizziness, hypotension, headache, nausea, constipation, coughing, wheezing, and pulmonary edema [28]. Aggravation of myocardial ischemia has also been observed with DHPs, caused by excessive hypotension and decreased coronary perfusion [59]. The use of nifedipine (**2a**) and nifedipine-like analogs was limited after 1995, when clinical trials exhibited increased mortality in patients with coronary artery disease [60].

9.5
Structure–Activity Relationship

Initial extensive SAR studies on the substituted Hantzsch DHPs ($R^2 = R^6 =$ Me and $R^3 = R^5 =$ alkyl ester) were performed by Loev *et al.* [5]. The group discovered that even analogs with small alkyl substituents at the 4-position have some hypotensive activity in anesthetized animals (Figure 9.10). Better activity, however, is generally observed with compounds containing cyclic substituents in the 4-position, particularly if those are aryl groups. Among the latter, the 4-heteroaryl and the 4-*ortho*-substituted phenyl derivatives showed the highest potencies. Electronic effects of the substituents on the phenyl ring do not play a substantial role for activity. Both electron-donating and electron-withdrawing substituents are nearly equipotent.

The position of substituents on the phenyl ring is crucial. For example, moving a substituent from the ortho position to the meta or the para positions has a negative effect on the binding of the DHPs to their binding site. It appears that a larger distortion of the plane on the phenyl ring from the plane of the DHP core has a positive influence on the potency of the compounds [5].

Triggle *et al.* [61] confirmed by X-ray diffraction that DHPs exist in a boatlike conformation with the C4 aryl substituent distorted in a pseudoaxial orientation. This arrangement was confirmed by Dagnino *et al.* [62]. Conformational studies on DHPs with a protonated nitrogen, utilizing AM1 and MM+ calculation methods, showed that the DHP ring does not exist in a clear boat conformation but that it is slightly twisted [63, 64]. The magnitude of the ring distortion, measured by the degree of the torsion angle between the aryl ring and the DHP ring, correlates well with the calcium channel antagonism activity [65]. The C4 and the C4-proton on the phenyl ring have a synperiplanar orientation, and the carboxylate groups at the C3 and C5 positions are not coplanar [66]. This orientation is associated with the antagonistic activity of the DHPs, while a coplanar orientation

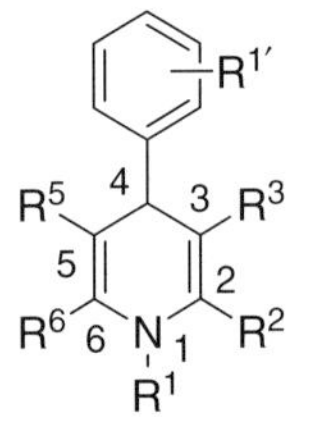

44

Figure 9.10 The general structure and numbering of DHPs **44**.

of the esters is associated with an agonistic character [67]. The phenyl group in 4-methyl-4-phenyl-substituted DHPs assumes an equatorial orientation, which results in a loss of activity [68].

Substitutions at the DHP nitrogen atom typically have detrimental effects on activity. However, flordipine (**4f**), substituted with a 2-morpholinoethyl substituent at the DHP nitrogen, is an effective calcium channel antagonist [21].

Loev *et al.* [5] also found that methyl groups at C2 and C6 are preferred when compared to other, bulkier substituents; however, it was later discovered that the next-generation DHPs tolerated substituents other than methyl and achieved good activities, for example, amlodipine (**4a**) and labedipinedilol-A (**4e**). Amlodipine (**4a**) has an alkyl amino group at the C2 position. This group significantly improves the pharmacokinetic profile of the drug by increasing the bioavailability and half-life. Amlodipine analogs have high calcium channel activities and prolonged duration of action when compared to nifedipine (**2a**). The C2 position of the DHP ring can be substituted with a variety of substituents and retain activity. The presence of a basic center at the C2 position is not necessary for the calcium channel antagonistic activity or selectivities [66].

The replacement of the ester groups at C3 and C5 with alternate electron-withdrawing substituents such as cyano or acetyl led to decreased potencies. Later, it was found that the replacement of one of the esters with a nitro substituent can reverse the antagonistic effect of the DHPs, and the compounds become calcium channel agonists (BAY K 8644 (**10**), PN 202-791 (**11**)).

The pharmaceutical industry continues to develop new DHPs. Numerous DHP analogs are in clinical trials in the hope that they will become the next blockbuster drugs for the treatment of cardiovascular disorders.

References

1. Sirisha, K., Bikshapathi, D., Achaiah, G., and Reddy, V.M. (2011) *Eur. J. Med. Chem.*, **46**, 1564–1571.
2. Abbas, H.-A.S., El Sayed, W.A., and Fathy, N.M. (2010) *Eur. J. Med. Chem.*, **45**, 973–982.
3. Karimian, K., Tam, T.F., Li, W., and Wang, Y. (2003) Can. Pat. Appl. CA 2379370 A1 20030928.
4. Bossert, F. and Vater, W.S. (1968) African, ZA 6801482 A 19680807.
5. Loev, B., Goodman, M.M., Snader, K.M., Tedeschi, R., and Macko, E. (1974) *J. Med. Chem.*, **17**, 956–965.
6. Mahmoudian, M., Mirkhani, H., Nehardani, Z., and Ghiaee, S. (1997) *J. Pharm. Pharmacol.*, **49**, 1229–1233.
7. Rowan, K.R. and Holt, E.M. (1996) *Acta Crystallogr., Sect. C: Cryst. Struct. Commun.*, **C52**, 2207–2212.
8. Kazda, S., Garthoff, B., Meyer, H., Schlossmann, K., Stoepel, K., Towart, R., Vater, W., and Wehinger, E. (1980) *Arzneim. Forsch.*, **30**, 2144–2162.
9. Xi, T. and Rao, M.-R. (1996) *Zhongguo Yaoli Xuebao*, **17**, 235–238.
10. Stoepel, K., Heise, A., and Kazda, S. (1981) *Arzneim. Forsch.*, **31**, 2056–2061.
11. Alajarin, R., Alvarez-Builla, J., Vaquero, J.J., Sunkel, C., Fau de Casa-Jauna, M., Statkow, P.R., and Sanz-Aparicio, J. (1993) *Tetrahedron Asymm.*, **4**, 617–620.
12. Alajarin, R., Vaquero, J.J., Alvarez-Builla, J., Pastor, M., Sunkel, C., Fau de Casa-Juana, M., Priego, J., Statkow, P.R., Sanz-Aparicio, J., and Fonseca, I. (1995) *J. Med. Chem.*, **38**, 2830–2841.

13. Meyer, H., Wehinger, E., Bossert, F., and Scherling, D. (1983) *Arzneim. Forsch.*, **33**, 106–112.
14. Bellemann, P., Schade, A., and Towart, R. (1983) *Proc. Natl. Acad. Sci. U.S.A.*, **80**, 2356–2360.
15. Lohn, M., Muzzulini, U., Essin, K., Tsang, S.-Y., Kirsch, T., Litteral, J., Waldron, P., Conrad, H., Klugbauer, N., Hofmann, F., Haller, H., Luft, F.C., Huang, Y., and Gollasch, M. (2002) *J. Hypertens.*, **20**, 885–893.
16. Arrowsmith, J.E., Campbell, S.F., Cross, P.E., Stubbs, J.K., Burges, R.A., Gardiner, D.G., and Blackburn, K.J. (1986) *J. Med. Chem.*, **29**, 1696–1702.
17. Liu, Z., Zheng, X., Yang, X., Wang, E., and Wang, J. (2009) *Biophys. J.*, **96**, 3917–3925.
18. Cummins, D.F. (1999) *Expert Opin. Invest. Drugs*, **8**, 1031–1042.
19. Higuchi, S. and Shiobara, Y. (1980) *Xenobiotica*, **10**, 447–454.
20. Liang, J., Chen, H., Chiu, C., Liou, S., Chen, I., and Yeh, J. (2006) *Life Sci.*, **79**, 1248–1256.
21. Rosenberg, M. and Choi, R.L. (1984) *J. Chromatogr.*, **308**, 382–386.
22. Hantzsch, A. (1882) *Justus Liebigs Ann. Chem.*, **215**, 1–82.
23. Hantzsch, A. (1884) *Ber. Dtsch. Chem. Ges.*, **17**, 1512–1522.
24. Knoevenagel, E. and Ruschhaupt, W. (1898) *Ber. Dtsch. Chem. Ges.*, **31**, 1025–1033.
25. Karrer, P., Schwarzenbach, G., and Utzinger, G.E. (1937) *Helv. Chim. Acta*, **20**, 72–79.
26. Karrer, P., Riniger, B.H., Buchi, J., Fritzsche, H., and Solmssen, U. (1937) *Helv. Chim. Acta*, **20**, 55–71.
27. Phillips, A.P. (1949) *J. Am. Chem. Soc.*, **71**, 4003–4007.
28. Robertson, R.M. and Robertson, D. (1995) in *Goodman, Gilman's, The Pharmacological Basis of Therapeutics*, 9th edn (eds J.G. Hardman, L.E. Limbird, P.B. Molinoff, R.W. Ruddon, and A.G. Gilman), McGraw Hill, pp. 767–779.
29. Bossert, F. and Vater, W. (1971) *Naturwissenschaften*, **58**, 578–585.
30. Vater, W., Kroneberg, G., Hoffmeister, F., Kaller, H., Meng, K., Oberdorf, A., Puls, W., Schlossmann, K., and Stoepel, K. (1972) *Arzneim. Forsch.*, **22**, 1–14.
31. Kohlhardt, M. and Fleckenstein, A. (1977) *Naunyn Schmiedebergs Arch. Pharmacol.*, **298**, 267–272.
32. Schramm, M., Thomas, G., Towart, R., and Franckowiak, G. (1983) *Nature*, **303**, 535–537.
33. Hof, R.P., Ruegg, U.T., Hof, A., and Vogel, A. (1985) *J. Cardiovasc. Pharmacol.*, **7**, 689–693.
34. Meyer, H., Bossert, F., Wehinger, E., Stoepel, K., and Vater, W. (1981) *Arzneim. Forsch.*, **31**, 407–409.
35. Marchalín, S., Cvopova, K., Kriz, M., Baran, P., Oulyadi, H., and Daïch, A. (2004) *J. Org. Chem.*, **69**, 4227–4237.
36. Sobolev, A., Franssen, M.C.R., Makarova, N., Duburs, G., and deGroot, A. (2000) *Tetrahedron Asym.*, **11**, 4559–4569.
37. Achiwa, K. and Kato, T. (1999) *Curr. Org. Chem.*, **3**, 77–106.
38. Fustero, S., Catalan, S., Sanchez-Rosello, M., Simon-Fuentes, A., and del Pozo, C. (2010) *Org. Lett.*, **12**, 3484–3487.
39. Yamamoto, T., Ohno, S., Niwa, S., Tokumasu, M., Hagihara, M., Koganei, H., Fujita, S., Takeda, T., Saitou, Y., Iwayama, S., Takahara, A., Iwata, S., and Shoji, M. (2011) *Bioorg. Med. Chem. Lett.*, **21**, 3317–3319.
40. Enders, D., Mueller, S., and Demir, A.S. (1988) *Tetrahedron Lett.*, **29**, 6437–6440.
41. Meyers, A.I. and Oppenlaender, T. (1986) *J. Chem. Soc., Chem. Commun.*, 920–921.
42. Noole, A., Borissova, M., Lopp, M., and Kanger, T. (2011) *J. Org. Chem.*, **76**, 1538–1545.
43. Yoshida, K., Inokuma, T., Takasu, K., and Takemoto, Y. (2010) *Synlett*, 1865–1869.
44. Ringer, S. (1883) *J. Physiol.*, **4**, 29–45.
45. Jones, S.W. (1998) *J. Bioenerg. Biomembr.*, **30**, 299–312.
46. Dolphin, A.C. (2006) *Br. J. Pharmacol.*, **147** (Suppl. 1), S56–S62.
47. Schwartz, A. (1989) *Am. J. Cardiol.*, **64**, 3I–9I.
48. Triggle, D.J. (1992) *Cleve. Clin. J. Med.*, **59**, 617–627.

49. Erickson, A.L., DeGrado, J.R., and Fanikos, J.R. (2010) *Pharmacotherapy*, **30**, 515–528.
50. Aouam, K. and Berdeaux, A. (2003) *Therapie*, **58**, 333–339.
51. Tanaka, H., Sekine, T., Terada, M., Kobayashi, Y., and Shigenobu, K. (1997) *Naunyn Schmiedebergs Arch. Pharmacol.*, **356**, 853–855.
52. Tanaka, H. and Shigenobu, K. (2000) *Cardiovasc. Drug Rev.*, **18**, 93–102.
53. Tanaka, H., Namekata, I., Komikado, Ch., Kawanishi, T., and Shigenobu, K. (2007) *Curr. Topics Pharmacol.*, **11**, 1–15.
54. Masuda, Y. and Tanaka, S. (1994) *Cardiovasc. Drug Rev.*, **12**, 123–135.
55. Suzuki, S., Ohkusa, T., Ono, K., Sato, T., Yoshida, M., Yano, M., Takebayahsi, S., and Matsuzaki, M. (2007) *Circ. J.*, **71**, 1970–1976.
56. Masumiya, H., Shijuku, T., Tanaka, H., and Shigenobu, K. (1998) *Eur. J. Pharmacol.*, **349**, 351–357.
57. Takahara, A., Fujita, S., Moki, K., Ono, Y., Koganei, H., Iwayama, S., and Yamamoto, H. (2003) *Hypertens. Res.*, **26**, 743–747.
58. Zhou, X., Zhang, L., Tseng, E., Scott-Ramsay, E., Schentag, J.J., Coburn, R.A., and Morris, M.E. (2005) *Drug Metab. Dispos.*, **33**, 321–328.
59. Egstrup, K. and Andersen, P.E. Jr. (1993) *Am. J. Cardiol.*, **71**, 177–183.
60. Furberg, C.D., Psaty, B.M., and Meyer, J.V. (1995) *Circulation*, **92**, 1326–1331.
61. Triggle, A.M., Shefter, E., and Triggle, D.J. (1980) *J. Med. Chem.*, **23**, 1442–1445.
62. Dagnino, L., Li-Kwong-Ken, M.C., Wynn, H., Wolowyk, M.W., Triggle, C.R., and Knaus, E.E. (1987) *J. Med. Chem.*, **30**, 640–646.
63. Hemmateenejad, B., Miri, R., Safarpour, M.A., Khoshneviszadeh, M., and Edraki, N. (2005) *J. Mol. Struct. Theochem*, **717**, 139–152.
64. Edraki, N., Mehdipour, A.R., Khoshneviszadeh, M., and Miri, R. (2009) *Drug Discov. Today*, **14**, 1058–1066.
65. Janis, R.A. and Triggle, D.J. (1983) *J. Med. Chem.*, **26**, 775–785.
66. Romero, M., Sanchez, I., and Pujol, M.D. (2003) *Curr. Med. Chem. Cardiovasc. Hematol. Agents*, **1**, 113–141.
67. Mahmoudian, M. and Richards, G.W. (1986) *J. Chem. Soc., Chem. Commun.*, **10**, 739–741.
68. Goldman, S., Born, L., Kazda, S., Pittel, B., and Schramm, M. (1991) *J. Med. Chem.*, **33**, 1413–1418.

10
Tetrazole-Based Angiotensin II Type 1 (AT_1) Antagonists for the Treatment of Heart Failure and Congestive Hypertension

Edward C. Lawson, Brian C. Shook, and James C. Lanter

10.1
Introduction

Tetrazoles are five-membered aromatic heterocycles consisting of one carbon and four nitrogen atoms. 5-Substituted tetrazoles may exist in two tautomeric forms, the 1*H*-tetrazole (**1**) and the 2*H*-tetrazole (**2**) (Figure 10.1).

The use of 5-substituted-tetrazoles as nonclassical isosteres for the carboxylic acid moiety in biologically active molecules (bioisosteres) has been reviewed in the literature [1–7]. Tetrazoles with a free NH are acidic and are ionized at physiological pH (7.4) like carboxylic acids. The ability of a tetrazole to stabilize a negative charge by electron delocalization is reflected in the p*K*as of both aromatic and aliphatic tetrazoles that are comparable to their corresponding carboxylic acids (4.5–4.9 vs 4.2–4.4, respectively) [4, 8–13]. The delocalization of charge by a tetrazole ring may lead to a more favorable receptor–substrate interaction depending on the local charge density at the interface [14]. Another important property for tetrazoles as carboxylic acid bioisosteres is the fact that they are 10 times more lipophilic than carboxylic acids [6], which has implications for a drug molecule to pass through a cell membrane. A clear advantage of tetrazolic acids over carboxylic acids is their resistance to many biological metabolic degradation pathways, although both tetrazoles and carboxylic acids may act as ligand-binding functionalities for the CYP450-derived oxidative metabolism. Aromatic and aliphatic tetrazoles do not undergo covalent bond formation with transferase enzymes, which often occurs with benzoic acid substrates; these then undergo subsequent conjugation transformation by a variety of pathways [5, 15]. Tetrazoles also do not degrade via β-oxidation, undergo glycine conjugation, or are incorporated into lipids [16]. However, tetrazoles are not completely devoid of metabolic degradation pathways as they can form *N*-β-glucuronides, in analogy to what often occurs with carboxylic acid *O*-β-glucuronic acid conjugates, followed by biliary excretion [16, 17]. Nevertheless, a number of orally administered tetrazolic-acid-containing drugs show long half-lives attributed to the enterohepatic recirculation mechanism (Figure 10.2) [18, 19].

Bioactive Heterocyclic Compound Classes: Pharmaceuticals,
First Edition. Edited by Jürgen Dinges and Clemens Lamberth.

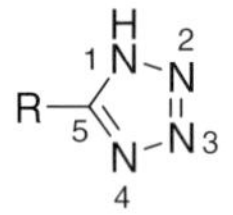

1

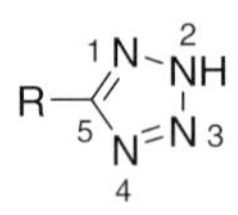

2

Figure 10.1 Basic structures and numbering conventions for the tautomeric tetrazoles **1** and **2**.

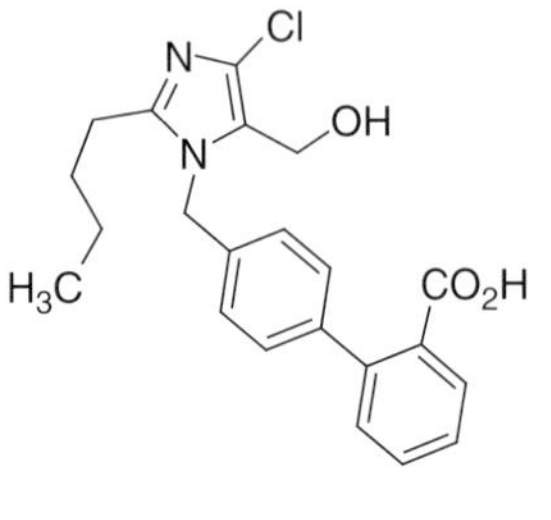

3

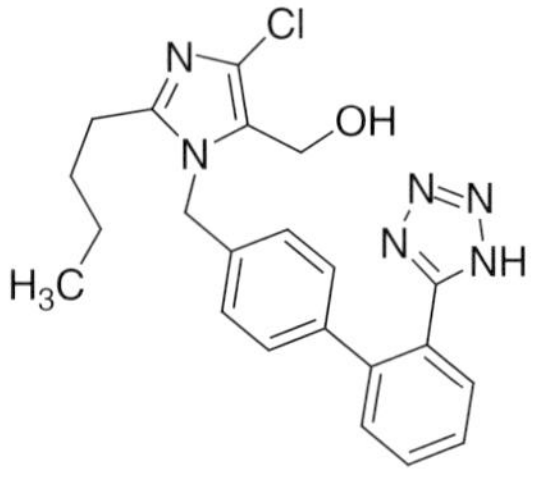

4
Losartan

Figure 10.2 DuPont angiotensin II antagonists **3** and losartan (**4**) [7, 20].

This rational was used successfully in the discovery and development of angiotensin II (Ang II) type 1 (AT_1) selective antagonists [7, 20, 21]. DuPont showed that the replacement of the carboxylic acid group in compound **3** with a tetrazole, losartan (**4**), increased oral bioavailability [20, 22].

This chapter focuses on the synthesis and history of tetrazoles as carboxylic acid bioisosteres specifically related to tetrazole-based AT_1 selective receptor antagonists. Additional information of the AT_1 pathway is also discussed in detail.

10.2 History

The earliest report of the synthesis of a tetrazole was in 1885 by J.A. Bladin at the University of Upsala [23]. Bladin observed that the reaction between dicyanophenylhydrazine and nitrous acid led to the formation of a compound with the chemical formula of $C_8H_5N_5$ and assigned the structure **6** shown in Figure 10.3. In 1886, he proposed the name tetrazole for the new ring system [24].

In 1893, it was shown by both Bamberger/De Grutter [25] and Widman [26] that Bladin had assigned the incorrect structure to the starting material. This structural error was carried over to the structural assignments of the corresponding tetrazoles derived from this material. Bladin's structural assignment of dicyanophenylhydrazine (**5**) versus the correct structural assignment (**7**) by Bamberger and De Gruyter are shown in Figures 10.3 and 10.4, along with the correct structure of the corresponding cyanophenyltetrazole (**8**).

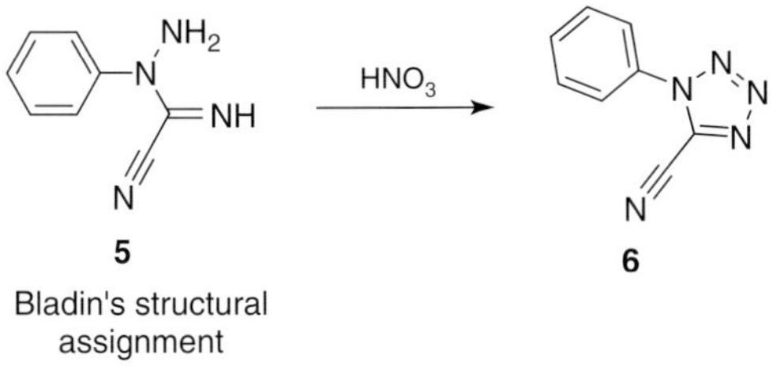

Figure 10.3 First tetrazole assigned by J.A. Bladin.

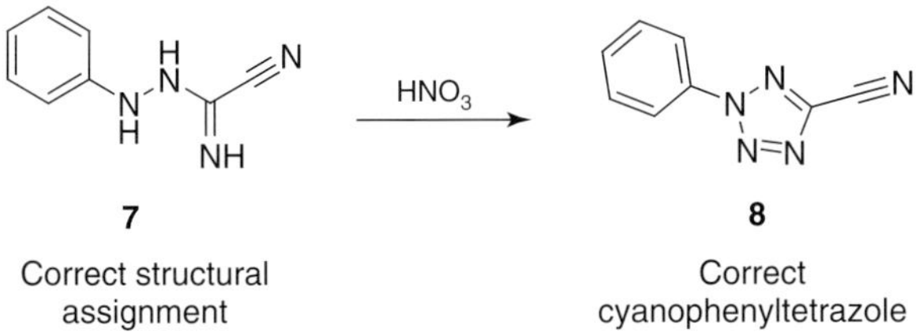

Figure 10.4 Correct assignment of dicyanophenyl hydrazine **7** and the corresponding tetrazole **8**.

The success of angiotensin-converting enzyme (ACE) inhibitors both clinically and commercially in the late 1970s for the treatment of congestive heart failure and hypertension has prompted interest in the exploration of additional targets and pathways to interfere with the renin–angiotensin system (RAS) [27–30]. Two specific side effects of ACE inhibitors, a dry cough and rare instances of angioedema, were attributed to the fact that ACE is a nonspecific protease that potentiates bradykinin [31–41]. In contrast, inhibition of the Ang II receptor in the RAS cascade offered the opportunity to block this system more specifically [40, 41]. Ang II is downstream of ACE, and therefore a potentiation of bradykinin would not be expected, eliminating the potential for the undesired side effects observed with the ACE inhibitors.

The first nonpeptidic AT_1-selective antagonists of Ang II, reported in 1982 by Takeda, were the 1-benzylimidazole-5-acetic acid derivatives S-8307 (**9**) and S-8308 (**10**) (Figure 10.5) [42, 43].

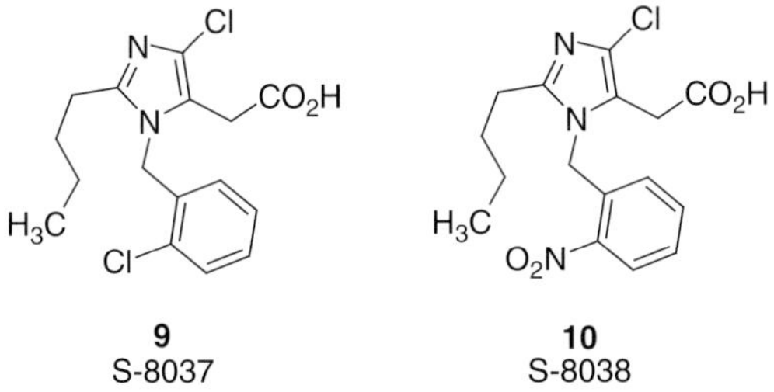

Figure 10.5 Takeda Chemical Co. angiotensin II antagonists S-8037 (**9**) and S-8038 (**10**).

Several analogs from this series were reported to inhibit the Ang II-induced response in rabbit aorta and the pressor response in the Ang II-infused rat. A group at DuPont was working on identifying a peptide fragment of Ang II and to use it in an analog synthesis effort toward identifying nonpeptidic antagonists [22]. When this strategy did not work out, the group decided to further explore the imidazole-based Takeda compounds. After the resynthesis of S-8037 and S-8038, the DuPont group confirmed that these compounds were weak but selective Ang II receptor antagonists [44, 45]. Timmermans further published *in vivo* results for additional analogs, but no information was given about the selectivity profiles of these compounds [46]. From this effort, losartan (**4**) (Figure 10.2) was identified as an AT_1-selective and competitive Ang II antagonist [47]. Instead of a benzyl substituent, this new generation of AT_1 antagonists now contained a 5-(4′-(methyl)biphenyl-2-yl)-1*H*-tetrazole group attached to the imidazole core. A direct comparison of the carboxylic-acid-substituted compound with the tetrazole replacement shows an increase in *in vitro* activity ($IC_{50} = 0.23$ vs $0.019\ \mu M$, respectively) [47]. After oral dosing, the tetrazole **4** was more efficacious than the corresponding carboxylic acid **3** in lowering of blood pressure in renal hypertensive rats (RHRs) ($ED_{30} = 0.59$ vs $11.0\ \mu M$, respectively) [47]. Cozaar® (**11**), the potassium salt of losartan (**4**), became the first marketed Ang II antagonist for the treatment of hypertension. In addition, it was found that this drug may also delay the progression of diabetic nephropathy, and it is also indicated for the reduction of renal disease progression in patients with type 2 diabetes [48].

Sankyo developed Benicar® (**12**), a prodrug containing the imidazole core of the active AT_1-selective Ang II inhibitors, to treat hypertension. This prodrug is rapidly converted to the active carboxylic acid by ester hydrolysis during absorption from the gastrointestinal tract [49]. Another prodrug, Atacand® (**13**) [50], which contains a benzimidazole core, was discovered by Takeda and jointly developed by AstraZeneca, Takeda, and Astra Merck. This drug was approved for both hypertension and heart failure. An Ang II antagonist with an acyclic core, Diovan® (**14**), was developed by Novartis. Here, the imidazole core was replaced by a valine *N*-pentanoic acid amide. Diovan (**14**) received FDA approval for the treatment of hypertension, heart failure, and postmyocardial infarction. Avapro® (**15**), an Ang II antagonist containing a spirocyclic core structure, was marketed by Sanofi. This drug was approved for hypertension and nephropathy in type 2 diabetic patients.

Besides the utilization as monotherapy, AT_1-selective Ang II antagonists are also sold as combination drugs. For example, Merck is marketing a combination drug of Cozaar (**11**, losartan potassium) with hydrochlorothiazide, a diuretic, as Hyzaar® for hypertension. Head-to-head clinical trials showed that this combination drug was more effective for patients with hypertension than Cozaar (**11**) alone [51]. Benicar (**12**, olmesartan medoxomil), Atacand (**13**, candesartan cilexetil), Diovan (**14**, valsartan), and Avapro (**15**, irbesartan) (Figure 10.6) were also combined with hydrochlorothiazide (Benicar HCT®, Atacand HCT®, Diovan HCT®, and Avalide®, respectively) to treat hypertensive patients more effectively.

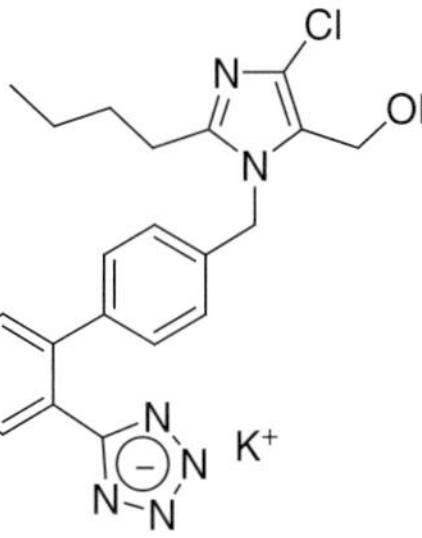

11
Losartan potassium

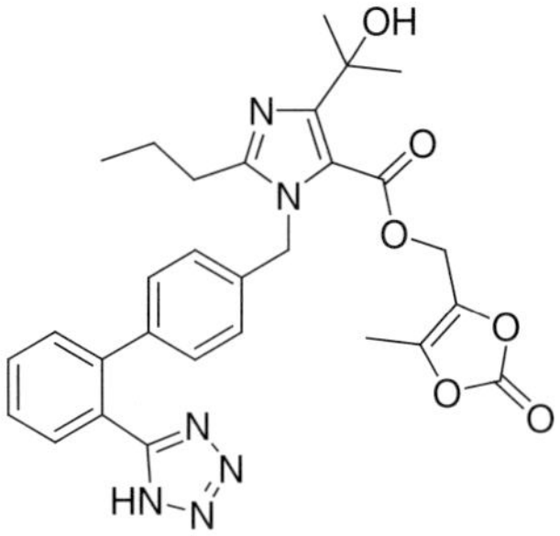

12
Olmesartan medoxomil

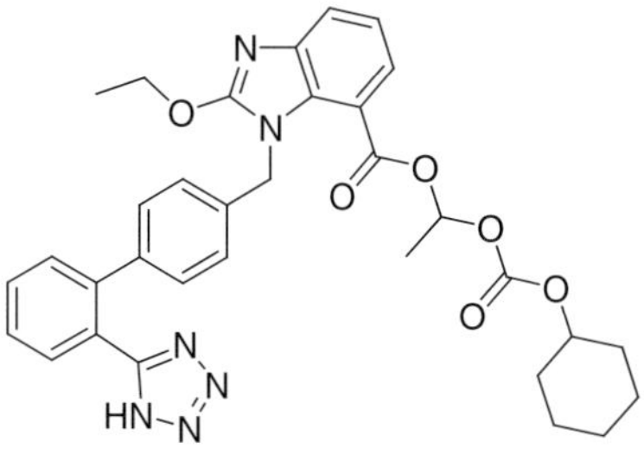

13
Candesartan cilexetil

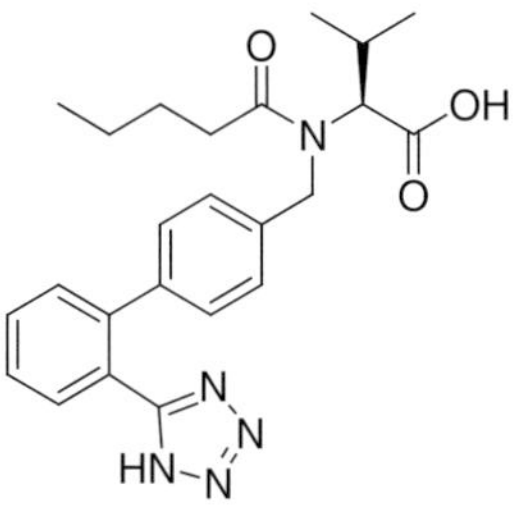

14
Valsartan

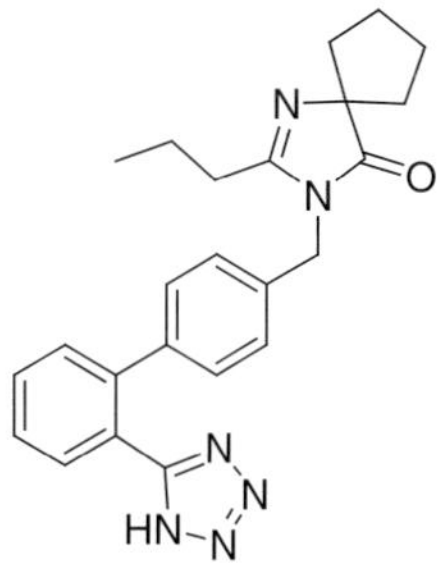

15
Irbesartan

Figure 10.6 Approved tetrazole-based angiotensin II antagonists: losartan potassium (**11**), olmesartan medoxomil (**12**), candesartan cilexetil (**13**), valsartan (**14**), and irbesartan (**15**).

10.3 Synthesis

As can be expected for a heterocyclic ring system identified more than a century ago, numerous methods exist for the preparation of every possible substitution pattern. Since this chapter focuses on 5-substituted therapeutic agents, synthetic

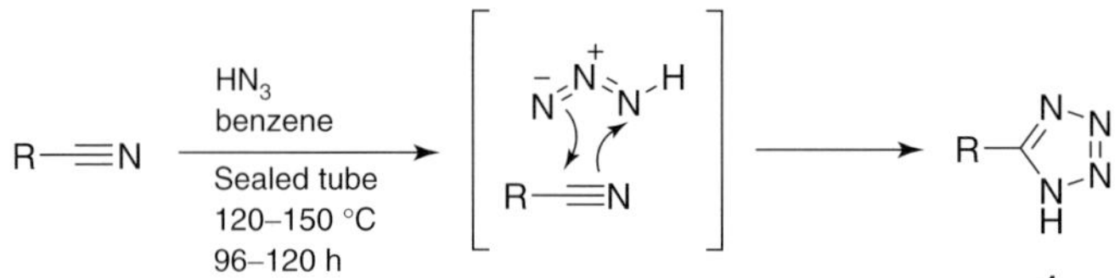

Scheme 10.1 Typical 5-substituted tetrazole synthesis via a [3 + 2]-cycloaddition.

discussion is limited to these specific systems. Mihina and Herbst [52] were the first to report 5-substituted tetrazole formation via the [3 + 2]-cycloaddition of nitriles with hydrazoic acid (Scheme 10.1).

To date, most 5-substituted tetrazole syntheses involve some version of the original approach, albeit with some variation to allow milder, safer, and more convenient reaction conditions. The most common of these include the *in situ* generation of hydrazoic acid by reaction of the nitrile with sodium azide in the presence of strong Lewis acids such as aluminum chloride in tetrahydrofuran (THF) [53], ammonium chloride in dimethylformamide (DMF) [54], or zinc bromide in water [55] at elevated temperatures. Alternatively, the authors note that it is possible to generate the hydrazoic acid *in situ* from sodium azide and acetic acid in isopropanol. Nitriles have also been reacted with sodium azide in ionic liquids in the presence of HCl [56]. One of the disadvantages of using sodium azide as a hydrazoic acid precursor is its limited solubility in organic media. To circumvent this problem, more soluble and "shelf-stable" alternatives such as tri-*n*-butylstannylazide (Bu_3SnN_3) [57] and trimethylsilylazide ($TMSN_3$) [58] can be used in the cycloaddition reaction, with removal of the metalloid–nitrogen bond

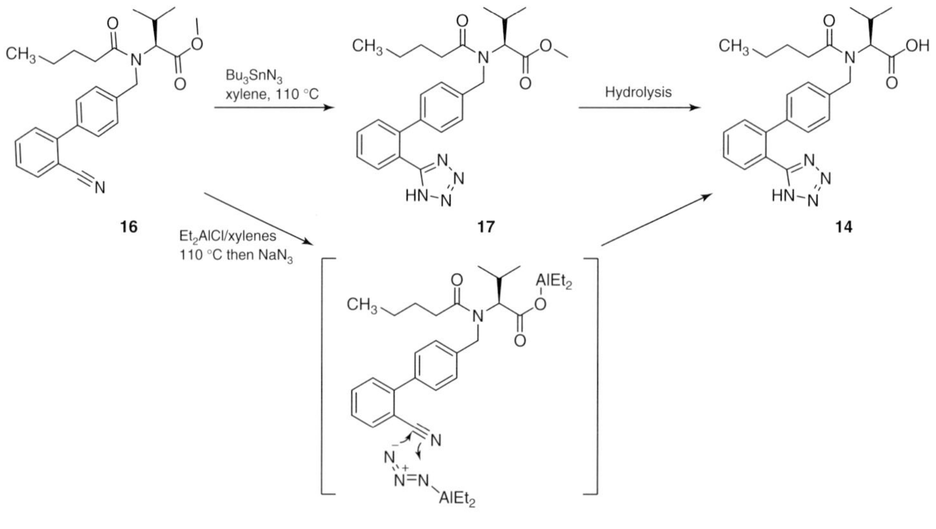

Scheme 10.2 Original (top) and an alternative (bottom) synthesis of Diovan® (**14**).

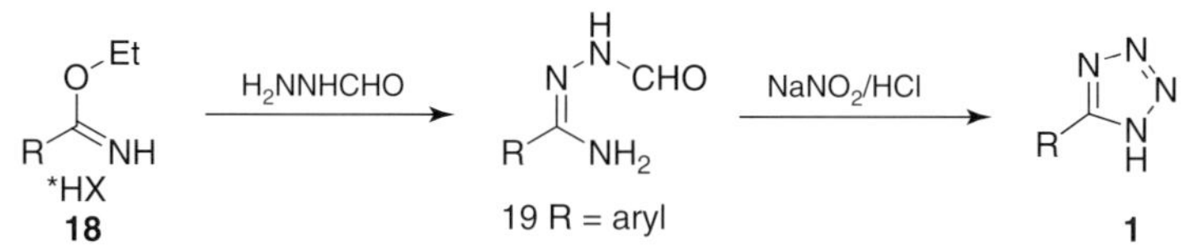

Scheme 10.3 The synthesis of tetrazoles via imidates.

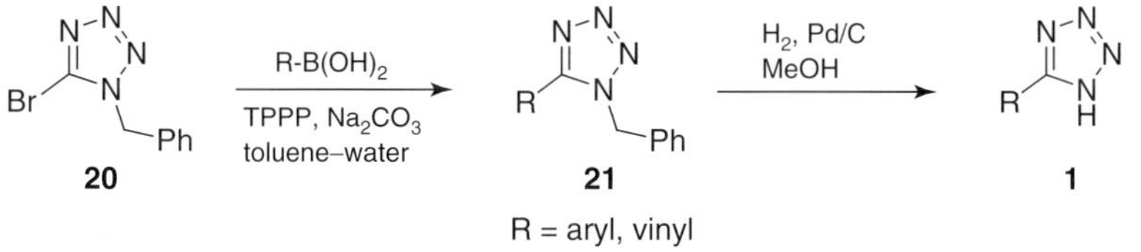

Scheme 10.4 The synthesis of tetrazoles (**1**) via Suzuki coupling.

during workup. In the case of the former, removal of the tin by-products can be problematic, an issue that has been addressed by fluorous phase technology [59]. Owing to this complication, the silyl reagent has been used more frequently in recent years, usually with additives such as dibutyltin oxide [60] or copper salts [61] to increase reactivity. Solventless reaction of $TMSN_3$ has also been reported in the presence of tetra-*n*-butylammonium fluoride (TBAF) [62]. Utilization of these methodologies is nicely illustrated in the original [63] (two-step) and a newer [64] (one-pot) syntheses of Diovan (**14**) (Scheme 10.2).

A few other methods to synthesize 5-substituted tetrazoles without using a cycloaddition reaction include the use of imidate precursors [65] (Scheme 10.3) and the Suzuki coupling of commercially available 5-bromo-1-benzyl tetrazole (**20**) [66] followed by debenzylation (Scheme 10.4).

10.4 Mode of Action

Hypertension is one of the leading risk factors for human health and has been attributed to a significant reduction in life expectancy [67]. In 2000, one-fourth of the world's adult population had been diagnosed with hypertension, and the number is continually on the rise [68]. *Hypertension* is defined as an elevated blood pressure and is influenced by the function and structure of blood vessels [29, 30]. When blood pressure is elevated, the blood vessels lose their elasticity, which can result in permanent damage of the arterial walls. The wall damage can give rise to the accumulation of fat and cholesterol, which, in turn, can eventually lead to the blockage of the vessels and subsequent organ damage. The vessel blockage causes the heart to overwork in trying to maintain its normal role, which leads to the development of cardiac hypertrophy [69].

Hypertension is a complex condition that occurs through multiple signaling pathways and is further complicated by its association with diabetes and renal

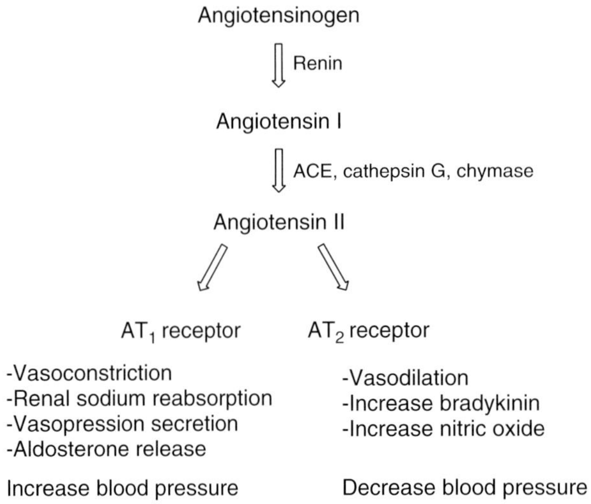

Figure 10.7 Renin–angiotensin system: generation of angiotensin II.

dysfunction [29, 30]. The RAS plays an important role in controlling blood pressure (Figure 10.7), and inhibition of the RAS effectively lowers blood pressure in patients with hypertension. The RAS is a coordinated proteolytic cascade that creates Ang peptides that operate in the circulatory system and within tissues such as blood vessels, heart, and kidneys. Ang II is the key mediator within the RAS that determines the pathophysiology of cardiovascular disease. Clinical and animal studies have shown that Ang II antagonists have protective effects on the heart, the vasculature, and the kidneys [70, 71].

Ang II is generated at both circulatory and tissue levels [27, 72]. At the circulating level, angiotensinogen from the liver is cleaved by the protease renin to give Ang I. Biologically inactive Ang I is rapidly converted to the octapeptide hormone Ang II by the metalloprotease ACE. Ang II is also secreted at the tissue level by various pathways in the arteries, heart, kidneys, and brain, while enzymes such as chymase and cathepsin G convert Ang I to Ang II. Ang II mediates biological responses through two receptor subtypes AT_1 receptor and AT_2 receptor. The circulatory AT_1 receptor induces vasoconstriction, aldosterone secretion, and nervous system activation, which all contribute to increase in blood pressure. Activation of AT_1 also decreases blood flow and together with aldosterone, increases the reabsorption of sodium leading to an increase in extracellular fluid volume and increased blood pressure [73]. Activity via AT_2 induces opposite effects to those seen with the AT_1 receptor, including vasodilation and natriuresis [74]. The AT_1 receptors predominate in most tissues, while the AT_2 receptors are much more limited in their distribution. The AT_1 receptor accounts for the majority of the cardiovascular effects evoked by Ang II and has been the primary target for the currently available agents that lower blood pressure.

10.5
Structure–Activity Relationship

The first Ang II receptor antagonist to be developed and enter the market was losartan (**4**). Since the discovery of losartan (**4**), the majority of the strategies to identify selective AT_1 receptor antagonists have been to replace or modify the imidazole ring (Figure 10.8, structure I) [22]. Losartan (**4**) is a selective and competitive ($IC_{50} = 6.1$, rabbit aorta) antagonist for the AT_1 receptor and showed antihypertensive effects when dosed orally ($ED_{30} = 0.59$ mg kg^{-1}) in RHRs. Numerous five- and six-membered heterocycles, fused heterocycles, and acyclic groups have been installed in place of the imidazole and resulted in potent, selective AT_1 antagonists [22]. Structure–activity relationships (SARs) of the imidazole ring showed that an alkyl chain (three to four carbon atoms) at the 2-position was essential for activity. A double bond within the alkyl chain was tolerated, but branched alkyl, cycloalkyl, and aromatics decreased binding affinity. Positions 4 and 5 were not critical for binding and therefore could be substituted with a variety of groups such as halides, alkyl groups, thiols, aromatics, and heteroaromatics while maintaining good activity. For example, substituting the 4-position with 2-chlorophenyl afforded Merck's L-158854, which showed potent *in vitro* activity ($IC_{50} = 0.55$ nM, rabbit aorta) and displayed 88% inhibition of the pressor response in normotensive rats at a dose of 0.1 mg kg^{-1}, p.o. [75]. Position 5, however, was substituted with hydroxyl methyl, carboxylic acid, carboxaldehyde, and carboxamide in the majority of the cases. The biphenyl tetrazole was important for activity, and the tetrazole was always unsubstituted. The biphenyl linker has been replaced with various linker groups, which had varying effects on activity [47]. Replacing either the central or the terminal phenyl rings with furan [76], thiophene [77], thiazole [76], pyrrole [76], or five- and six-membered cycloalkenyls [78] resulted in compounds having decreased binding affinities and decreased functional activities. For example, replacing the central phenyl ring in losartan (**4**) with thiophene led to a compound with an IC_{50} of 328 nM (rabbit aorta), more than 50-fold less potent than losartan (**4**). In contrast, replacing the central phenyl ring with benzofuran [79] or indole [80] gave compounds that

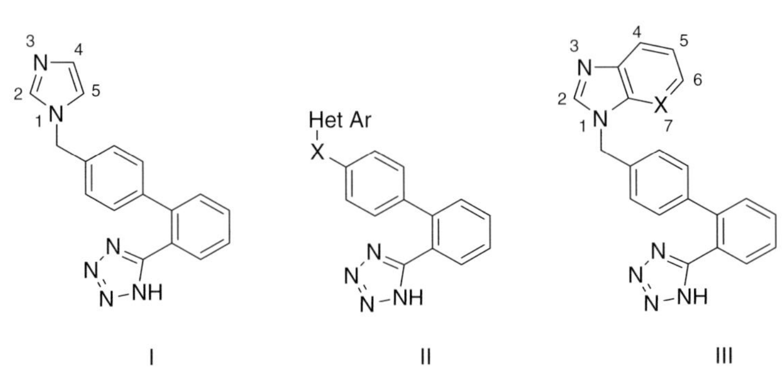

Figure 10.8 General structures for imidazole-based tetrazoles (I), heteroaryl-based tetrazoles (II), and benzimidazole- and imidazopyridine-based tetrazoles (III).

were highly potent with excellent *in vivo* activities. The indole-containing compound (BMS-180560) was a noncompetitive antagonist ($K_B = 0.068$ nM, rabbit aorta) and antagonized the pressor response in rats with an ED_{50} of 0.22 μmol kg^{-1} when dosed intravenously [80]. Chlorine and fluorine substituents on the central phenyl group were also well tolerated and often boosted activity [81].

There have been numerous efforts to replace the imidazole group with various heterocycles (Figure 10.8, structure II), but a very limited number of attempts to identify acyclic replacements. The most successful acyclic replacement of the imidazole (X-Het Ar is a substituted valine in structure II) resulted in the discovery of Diovan (**14**) [63]. Diovan (**14**) is a selective AT_1 antagonist ($IC_{50} = 2.7$ nM, rat aorta) and dose dependently decreased systolic blood pressure in RHR at doses of 3 and 10 mg kg^{-1}, p.o. [63]. The majority of heterocyclic replacements were attached via a methylene (X is CH_2 in structure II). Replacement of the methylene with S or SO_2 resulted in decreased activity [82]. Extending the linker by one atom (X is CH_2O [83] or CH_2N-alkyl [84] in structure II) generally gave potent compounds that maintained good *in vivo* activity. For example, replacing the imidazole with pyridine and using the CH_2N-alkyl linker to attach the biphenyl tetrazole resulted in a compound that significantly reduced blood pressure in furosemide-treated spontaneously hypertensive rats (SHRs), untreated SHR, and in renal-artery-ligated hypertensive rats at doses of 0.03, 0.1, and 3 mg kg^{-1}, p.o., respectively [84]. Replacing the imidazole with pyrazole [85], pyrrole [86], triazole [84], triazolone [87], dihydroimidazol-4-one [88], pyrazolidine-3,5-dione [89], pyrrolidin-2-one [90], thiazole [84], thiadiazole [84], pyridine [91], pyridinone [92], pyrimidine [84], pyrimidinone [93], piperidinone [94], and morpholine [95] gave compounds with good antagonist activity as long as there was an alkyl chain present analogous to the alkyl group in the 2-position of the imidazole in structure I. Fused heterocycles such as benzimidazole [96], imidazopyridine [97], quinazolinone [98], quinoline [99], quinoxaline [100], and naphthyridine [81] have also been used as viable replacements of the imidazole. The majority of replacements were screened by keeping the biphenyl tetrazole in place. Replacing either of the phenyl rings with pyridine gave compounds with comparable potency [101]. Fluorine substitution of the central phenyl ring was well tolerated [81].

Benzimidazoles [96] (X is CH in structure III) and imidazopyridines [97] (X is N in structure III) represent a large number of AT_1 antagonists [13]. Benzimidazole CV-11194 was potent ($IC_{50} = 0.06$, rabbit aorta) and dose dependently inhibited the pressor response in rats and dogs at 0.3–10 mg kg^{-1}, p.o. [96]. Like the imidazoles (structure I), an alkyl substituent at the 2-position of the benzimidazole was essential for activity. An ethoxy group in the 2-position was also effective for maintaining potent *in vitro* and *in vivo* activity [102]. The benzimidazoles that were unsubstituted in positions 4–7 had good activities. Carboxylic acid substitution at the 7-position was a common motif and also gave compounds with good potencies [102, 103]. Larger groups such as sulfonamides and benzimidazole were well tolerated at the 5- and 6-positions, respectively [104, 105]. The tetrazole-bearing phenyl ring could be replaced with pyrrole without losing activity [106]. The pyrrole-containing compound was equipotent to losartan (**4**) *in vitro* but showed greater inhibitory effects and longer duration of action in SHR at 1 mg kg^{-1}, p.o. than losartan (**4**) at the same dose [106]. The imidazopyridine (X is N in structure III) also required the two to four carbon alkyl chain at the 2-position [97]. The imidazopyridine L-158, 809 (Merck) was a potent ($IC_{50} = 0.3$ nM, rabbit aorta) and competitive antagonist that effectively inhibited the pressor response for 24 h in conscious normotensive rats when dosed at 0.1 mg kg^{-1}, i.v., and 0.3 mg kg^{-1}, p.o. [97]. Simple alkyl substituents were tolerated at positions 4–6 for both mono- and disubstituted analogs. Like the imidazoles, substitution of the imidazopyridine could be very liberal, as it had very little effect on the activity of the compounds. Replacements of the central phenyl ring with thiophene resulted in a 1000-fold decrease in potency [107], but replacing the tetrazole-bearing phenyl ring with thiophene [107], pyridine [108], and quinoline [109] gave compounds that had comparable activity to the corresponding biphenyl tetrazole. Replacing the central phenyl with dihydroindane [110], indole [111], or dihydroindole [111] also maintained potency.

References

1. Thornber, C.W. (1979) *J. Chem. Soc. Rev.*, **8**, 563.
2. Butler, R.N. (1977) *Adv. Heterocycl. Chem.*, **21**, 323.
3. Berger, A. (1991) *Prog. Drug Res.*, **37**, 287.
4. Butler, R.N. (1996) in *Comprehensive Heterocyclic Chemistry II*, vol. **4** (eds A.R. Katritzky, C.W. Rees, and E.F.V. Scriven), Pergamon, Oxford, p. 621.
5. Lipinski, C.A. (1986) *Annu. Rep. Med. Chem.*, **27**, 283.
6. Hansch, C. and Leo, L. (1995) *Exploring QSAR. Fundamentals and Applications in Chemistry and Biology*, Chapter 13, American Chemical Society, Washington, DC.
7. Herr, R.J. (2002) *Bioorg. Med. Chem.*, **10**, 3379–3393.
8. Herbst, R.M. and Wilson, K.R. (1957) *J. Org. Chem.*, **22**, 1142–1145.
9. McManus, J.M. and Herbst, R.M. (1959) *J. Org. Chem.*, **24**, 1643–1649.
10. Schaaf, T.K. and Hess, H.-J. (1979) *J. Med. Chem.*, **22**, 1340–1346.
11. Kaczmarek, J., Smagowski, H., and Grzonka, Z. (1979) *J. Chem. Soc., Perkin Trans. 2*, 1670–1674.
12. Albert, A. (1966) *J. Chem. Soc. B*, 427–433.
13. Ostrovskii, V.A. and Koren, A.O. (2000) *Heterocycles*, **53**, 1421–1448.
14. Kraus, J.L. (1983) *Pharmacol. Res. Commun.*, **15**, 183–189.
15. Cannon, J.G. (1995) in *Burger's Medicinal Chemistry and Drug Discovery*, 5th edn, vol. **1** (ed. M.E. Wolf), John Wiley & Sons, Inc., New York, p. 783.
16. Green, M.D. and Tephly, T.R. (1998) *Drug Metab. Dispos.*, **26**, 860–867.
17. Burchell, B., Brierley, C.H., and Rance, D. (1995) *Life Sci.*, **57**, 1819–1831.
18. Perrier, L., Bourrie, M., Marti, E., Tronquet, C., Masse, D., and Berger, Y. (1994) *J. Pharmacol. Exp. Ther.*, **271**, 91–99.
19. Chando, T.J., Everett, D.W., Kahle, A.D., and Starrett, A.M. (1998) *Drug Metab. Dispos.*, **26**, 408–417.
20. Wexler, R.R., Greenlee, W.J., Irvin, J.D., Goldber, M.R., Prendergast, K., Smith, R.D., and Timmermans, P.B.M.W.M. (1996) *J. Med. Chem.*, **39**, 625–656.
21. Naik, P., Murumkar, P., Giridhar, R., and Yadav, M.R. (2010) *Bioorg. Med. Chem.*, **18**, 8148–8156.
22. Wong, P.C. and Timmermans, P.B.M.W.M. (1996) *Blood Press.*, **5**, 11–14.
23. Bladin, J.A. (1885) *Ber . Dtsch. Chem. Ges.*, **18**, 1544–1551.
24. Bladin, J.A. (1886) *Ber. Dtsch. Chem. Ges.*, **19**, 2598–2604.

25. Bamberger, E. and De Grutter, P. (1893) *Ber. Dtsch. Chem. Ges.*, **26**, 2385–2397.
26. Widman, O. (1893) *Ber. Dtsch. Chem. Ges.* **26**, 2612–2617.
27. Sealy, J.E. and Laragh, J.H. (1990) in *Hypertension: Pathophysiology, Diagnosis and Management* (eds J.H. Laragh and B.M. Brenner), Raven Press, New York, pp. 1287–1317.
28. McAreavey, D. and Robertson, J.I.S. (1990) *Drugs*, **40**, 326–345.
29. Vallotten, M.B. (1987) *Trends Pharmacol. Sci.*, **8**, 69–74.
30. Waeber, B., Nussberger, J., and Brunner, H.B. (1990) in *Hypertension: Pathophysiology, Diagnosis and Management* (eds J.H. Laragh and B.M. Brenner), Raven Press, New York, pp. 2209–2232.
31. Erodes, E.G. and Skidgel, R.A. (1986) *Hypertension*, **8**, 40–48.
32. Skidgel, R.A. and Erodes, E.G. (1987) *Clin. Exp. Hypertens.*, **A9**, 243–259.
33. Lindgren, B.R. and Andersson, R.G.G. (1989) *Med. Toxicol. Adverse Drug Exp.*, **4**, 369–380.
34. Chin, H.L. and Buchan, D.A. (1990) *Ann. Intern. Med.*, **112**, 312–313.
35. Levens, N.R., de Gasparo, M., Wood, J.M., and Bottari, S.P. (1992) *Pharmacol. Toxicol.*, **71**, 241–249.
36. Greenlee, W.J. (1990) *Med. Res. Rev.*, **10**, 173–236.
37. Ocain, T.D. and Abou-Garbia, M. (1991) *Drugs Future*, **16**, 37–51.
38. Cody, R.J. (1994) *Drugs*, **47**, 586–598.
39. Kleinert, H.D. (1994) *Expert Opin. Investig. Drugs*, **3**, 1087–1104.
40. Timmermans, P.B.M.W.M., Wong, P.C., Chiu, A.T., and Herblin, W.F. (1991) *Trends Pharmacol. Sci.*, **12**, 55–62.
41. Johnson, A.L., Carini, D.J., Chiu, A.T., Duncia, A.V., Price, W.A., Wells, G.J., Wexler, R.R., Wong, P.C., and Timmermans, P.B.M.W.M. Jr. (1990) *Drugs News Perspect.*, **3**, 337–351.
42. Furukawa, Y., Kishimoto, S., and Nishikawa, K. (1982) US Patent 4, 340,598.
43. Furukawa, Y., Kishimoto, S., and Nishikawa, K. (1982) US Patent 4, 355,040.
44. Chiu, A.T., Carini, D.J., Johnson, A.L., McCall, D.E., Price, W.A., Thoolen, M.J.M.C., Wong, P.C., Taber, R.I., and Timmermans, P.B.M.W.M. (1988) *Eur. J. Pharmacol.*, **157**, 13–21.
45. Wong, P.C., Chiu, A.T., Price, W.A., Thoolen, M.J.M.C., Carini, D.J., Johnson, A.I., Taber, R.I., and Timmermans, P.B.M.W.M. (1988) *J. Pharmacol. Exp. Ther.*, **247**, 1–7.
46. Wong, P.C., Price, W.A., Chiu, A.T., Carini, D.J., Duncia, J.V., Johnson, A.L., Wexler, R.R., and Timmermans, P.B.M.W.M. (1990) *Hypertension*, **15**, 823–834.
47. Carini, D.J., Duncia, J.V., Aldrich, P.E., Chiu, A.T., Johnson, A.L., Pierce, M.E., Price, W.A., Santella, J.B., Wells, G.J., Wexler, R.R., Wong, P.C., Yoo, S.-E., and Timmermans, P.B.M.W.M. (1991) *J. Med. Chem.*, **34**, 2525–2547.
48. Brenner, B.M., Cooper, M.E., De Zeeuw, D., Keane, W.F., Mitch, W.E., Parving, H.H., Remuzzi, G., Snapinn, S.M., Zhang, Z., and Shahinfar, S. (2001) *N. Engl. J. Med.*, **345**, 861–869.
49. Mizuno, M., Sada, T., Ikeda, M., Fukuda, N., Miyamoto, M., Yanagisawa, H., and Koike, H. (1995) *Eur. J. Pharmacol.*, **285**, 181–188.
50. Kubo, K., Kohara, Y., Yoshimura, Y., Inada, Y., Shibouta, Y., Furukawa, Y., Kato, T., Nishikawa, K., and Naka, T. (1993) *J. Med. Chem.*, **36**, 2343–2349.
51. Wilson, T.W., Lacourciere, Y., and Barnes, C.C. (1988) *Can. Med. Assoc. J*, **159**, 469–476.
52. Mihina, J.S. and Herbst, R.M. (1950) *J. Org. Chem.*, **18**, 1082–1092.
53. Ogilvie, W. and Rank, W. (1987) *Can. J. Chem.*, **65**, 166–169.
54. Finnegan, W.G., Henry, R.A., and Lofquist, R. (1958) *J. Am. Chem. Soc.*, **80**, 3908–3911.
55. Demko, Z.P. and Sharpless, B.A. (2001) *J. Org. Chem.*, **66**, 7945–7950.
56. Schmidt, B., Meid, D., and Kieser, D. (2007) *Tetrahedron*, **63**, 492–496.
57. Sisido, K., Nabika, K., Isida, T., and Kozima, S. (1971) *J. Organomet. Chem.*, **33**, 337–346.

58. Duncia, J.V., Pierce, M.E., and Santella, J.B. III (1991) *J. Org. Chem.*, **56**, 2395–2400.

59. Curran, D.P., Hadida, S., and Kim, S.-Y. (1999) *Tetrahedron*, **55**, 8997–9006.

60. Wittenberger, S.J. and Donner, B.G. (1993) *J. Org. Chem.*, **58**, 4139–4141.

61. Jin, T., Kitahara, F., Kamijo, S., and Yamamoto, Y. (2008) *Tetrahedron Lett.*, **49**, 2824–2827.

62. Amantini, D., Beleggia, R., Fringuelli, F., Pizzo, F., and Vaccaro, L. (2004) *J. Org. Chem.*, **69**, 2896–2898.

63. Buhlmayer, P., Furet, P., Criscione, L., de Gasparo, M., Whitebread, S., Schmidlin, T., Lattman, R., and Wood, J. (2010) *Bioorg. Med. Chem. Lett.*, **18**, 8148–8156.

64. Sedelmeier, G., Rampf, F.A., and Grimler, D. (2011) WO 2011,051,213.

65. Boivin, J., Husinec, S., and Zard, S.Z. (1995) *Tetrahedron*, **51**, 11737–11742.

66. Yi, K.Y. and Yoo, S. (1995) *Tetrahedron Lett.*, **36**, 1679–1682.

67. Carey, R.M. (2009) in *Renin Angiotensin System and Cardiovascular Disease Contemporary Cardiology* (eds W.C. Demello and E.D. Frohlich), Humana Press, pp. 59–79.

68. Kearney, P.M., Whelton, M., Reynolds, K., Muntner, P., Whelton, P.K., and He, J. (2005) *Lancet*, **365**, 217–223.

69. Cohuet, G. and Struijker-Boudier, H. (2006) *Pharmacol. Ther.*, **111**, 81–98.

70. Carey, R.M. (2008) in *Handbook of Physiology: Microcirculation*, vol. **2** (eds R.F. Tuma, W.N. Duran, and K. Ley), Elsevier, Amsterdam, pp. 794–895.

71. Maggioni, A.P. (2006) *Cardiovasc. Drugs Ther.*, **20**, 295–308.

72. Schmieder, R.E. (2005) *Am. J. Hypertens.*, **18**, 720–730.

73. Kaschina, E. and Unger, T. (2003) *Blood Press.*, **12**, 70–88.

74. Carey, R.M. (2005) *Hypertension*, **45**, 840–844.

75. Ashton, W.T., Hutchins, S.M., Greenlee, W.J., Doss, G.S., Chang, R.S.L., Lotti, V.J., Faust, K.A., Chen, T.-B., Zingaro, G.J., Kivlighn, S.D., and Siegl, P.K.S. (1993) *J. Med. Chem.*, **36**, 3595–3605.

76. Salimbeni, A., Canevotti, R., Paleari, F., Bonaccorsi, F., Renzetti, A.R., Belvisi, L., Bravi, G., and Scolastic, C. (1994) *J. Med. Chem.*, **37**, 3928–3938.

77. Estenne, G., Dodey, P., Renaut, P., and Leclerc, G. (1995) *Bioorg. Med. Chem. Lett.*, **5**, 15–18.

78. Lin, H.-S., Rampersaud, A.A., Zimmerman, K., Steinberg, M.I., and Boyd, D.B. (1992) *J. Med. Chem.*, **35**, 2658–2667.

79. Middlemiss, D., Drew, G.M., Ross, B.C., Robertson, M.J., Scopes, D.I.C., Dowle, M.C., Akers, J., Cardwell, K., Clark, K.L., Coote, S., Eldred, C.D., Hamblett, J., Hilditch, A., Hirst, G.C., Jack, T., Montana, J., Panchal, T.A., Paton, J.M.S., Shah, P., Stuart, G., and Travers, A. (1991) *Bioorg. Med. Chem. Lett.*, **1**, 711–716.

80. Poss, M.A., Gu, Z., Ryono, D.E., Reid, J.A., Sieber-McMaster, E., Spitzmiller, E.R., Dejneka, T., Dickinson, K.E.J., Williams, S.B., Moreland, S., Delaney, C.L., Bird, J.E., Waldron, T.L., Schaeffer, T.R., Hedberg, S.A., and Petrillo, E.W. (1994) *Bioorg. Med. Chem. Lett.*, **4**, 145–150.

81. Quan, M.L., Olson, R.E., Carini, D.J., Ellis, C.D., Hillyer, G.L., Lalka, G.K., Liu, J., VanAtten, M.K., Chiu, A.T., Wong, P.C., Wexler, R.R., and Timmermans, P.B.M.W.M. (1994) *Bioorg. Med. Chem. Lett.*, **4**, 2011–2016.

82. Norman, M.H., Smith, D.H., Andrews, W.C., Tang, F.L.M., Cowan, C.L., and Steffen, R.P. (1995) *J. Med. Chem.*, **38**, 4670–4678.

83. Kawano, K.I., Fujishima, K., Nagura, J., Yasuda, S., Hachisu, M., and Konno, F. (1998) *Eur. J. Pharmacol.*, **357**, 33–39.

84. De, B., Winn, M., Zydowsky, T.M., Kerkman, D.J., Debernardis, J.F., Lee, F., Buckner, S., Warner, R., Brune, M., Hancock, A., Opgenorth, T., and Marsh, K. (1992) *J. Med. Chem.*, **35**, 3714–3717.

85. Middlemiss, D., Ross, B.C., Eldred, C., Montana, J.G., Shah, P., Hirst, G.C., Watson, S.P., Panchal, T.A., Paton, J.M.S., Hubbard, T., Drew, G.M., Robertson, M.J., Hilditch, A., and

Clark, K.L. (1992) *Bioorg. Med. Chem. Lett.*, **2**, 1243–1246.

86. Attanasi, O.A., Colombani, S.A., Crescentini, L.D., Giorgi, R., Monti, S., Perrone, A., Perrulli, F.R., Renzetti, A.R., and Santeusanio, S. (1999) *IL Farmaco*, **54**, 64–76.
87. Huang, H.-C., Reitz, D.B., Chamberlain, T.S., Olins, G.M., Corpus, V.M., McMahon, E.G., Palomo, M.A., Koepke, J.P., Smits, G.J., McGraw, D.E., Blaine, E.H., and Manning, R.E. (1993) *J. Med. Chem.*, **36**, 2172–2181.
88. Bernhart, C.A., Perreaut, P.M., Ferrari, B.P., Muneaux, Y.A., Assens, J.-L.A., Clement, J., Haudricourt, F., Muneaux, C.F., Taillades, J.E., Vignal, M.-A., Gougat, J., Guiraudou, P.R., Lacour, C.A., Roccon, A., Cazaubon, C.F., Breliere, J.-C., Le Fur, G., and Nisato, D. (1993) *J. Med. Chem.*, **36**, 3371–3380.
89. Le Bourdonnec, B., Meulon, E., Yous, S., Goossens, J.-F., Houssin, R., and Henichart, J.-P. (2000) *J. Med. Chem.*, **43**, 2685–2697.
90. Murray, W.V., Lalan, P., Gill, A., Addo, M.F., Lewis, J.M., Lee, D.K.H., Wachter, M.P., Rampulla, R., and Underwood, D.C. (1993) *Bioorg. Med. Chem. Lett.*, **3**, 369–374.
91. Winn, M., De, B., Zydowsky, T.M., Altenbach, R.J., Basha, F.Z., Boyd, S.A., Brune, M.E., Buckner, S.A., Crowell, D., Drizin, I., Hancock, A.A., Jae, H.S., Kester, J.A., Lee, J.Y., Mantei, R.A., Marsh, K.C., Novosad, E.I., Oheim, K.W., Rosenberg, S.H., Shiosaki, K., Sorensen, B.K., Spina, K., Sullivan, G.M., Tasher, A.S., VonGeldern, T.W., Warner, R.B., Opgenorth, T.J., Kerkman, D.J., and Debernardis, J.F. (1993) *J. Med. Chem.*, **36**, 2676–2688.
92. Bantick, J.R., Beaton, H.G., Cooper, S.L., Hill, S., Hirst, S.C., McInally, T., Spencer, J., Tinker, A.C., and Willis, P.A. (1994) *Bioorg. Med. Chem. Lett.*, **4**, 127–132.
93. Nicolaie, E., Cure, G., Goyard, J., Kirchner, M., Teulon, J.M., Versigny, A., Cazes, M., Virone-Oddos, A., Caussade, F., and Cloarec, A. (1995) *Eur. J. Med. Chem.*, **30**, 365–375.
94. Murray, W.V., Lalan, P., Gill, A., Addo, M.F., Lewis, J.M., Lee, D.K.H., Rampulla, R., Wachter, M.P., Hsi, J.D., and Underwood, D.C. (1992) *Bioorg. Med. Chem. Lett.*, **2**, 1775–1779.
95. Hsi, J.D., Murray, W.V., and Gill, A. (1993) *Bioorg. Med. Chem. Lett.*, **3**, 1523–1526.
96. Kubo, K., Inada, Y., Kohara, Y., Suguira, Y., Ojima, M., Itoh, K., Furukawa, Y., Mishikawa, K., and Nakat, T. (1993) *J. Med. Chem.*, **36**, 1772–1784.
97. Mantlo, N.B., Chakravarty, P.K., Ondeyka, D.L., Siegl, P.K.S., Chang, R.S., Lotti, V.J., Faust, K.A., Chen, T.-B., Schorn, T.W., Sweet, C.S., Emmert, S.E., Parchett, A.A., and Greenlee, W.J. (1991) *J. Med. Chem.*, **34**, 2919–2922.
98. de Laszlo, S.E., Allen, E.E., Quagliato, C.S., Greenlee, W.J., Patchett, A.A., Nachbar, R.B., Siegl, P.K.S., Chang, R.S., Kivlighn, S.D., Schorn, T.S., Faust, K.A., Chen, T.-B., Zingaro, G.J., and Lotti, V.J. (1993) *Bioorg. Med. Chem. Lett.*, **3**, 1299–1304.
99. Bradbury, R.H., Allott, C.P., Dennis, M., Fisher, E., Luke, R.W.A., Major, J.S., Oldham, A.A., Pearce, R.J., Reid, A.C., Roberts, D.A., Rudge, D.A., and Russell, S.T. (1993) *Bioorg. Med. Chem. Lett.*, **3**, 899–904.
100. Kim, K.S., Qian, L., Dickinson, K.E.J., Moreland, S., Schaeffer, T.R., Waldron, T.L., Delaney, C.L., Weller, H.N., and Miller, A.V. (1993) *J. Med. Chem.*, **36**, 2335–2342.
101. Reitz, D.B. and Manning, R.E. (1992) US Patent 5, 164,403.
102. Kubo, K., Kohara, Y., Imamiya, E., Suguira, Y., Inada, Y., Furukawa, Y., Nishikawa, K., and Naka, T. (1993) *J. Med. Chem.*, **36**, 2182–2195.
103. Kohara, Y., Imamiya, E., Kubo, K., Wada, T., Inada, Y., and Naka, T. (1995) *Bioorg. Med. Chem. Lett.*, **5**, 1903–1908.
104. Bali, A., Bansal, Y., Sugumaran, Y., Saggu, J.G., Balakumar, P., Kaur, G., Bansal, G., Sharma, A., and Singh, M. (2005) *Bioorg. Med. Chem. Lett.*, **15**, 3962–3965.

105. Ries, U.J., Mihm, G., Narr, B., Hasbach, K.M., Wittneben, H., Entzeroth, M., VanMeel, J.C.A., Wienen, W., and Hauel, M.H. (1993) *J. Med. Chem.*, **36**, 4040–4051.

106. Xu, J.Y., Ran, Q., Hua, W.Y., Wu, X.M., Wang, Q.J., and Zhang, J. (2007) *Chin. Chem. Lett.*, **18**, 251–254.

107. Rivero, R.A., Kevin, N.J., and Allen, E.E. (1993) *Bioorg. Med. Chem. Lett.*, **3**, 1119–1124.

108. Mantlo, N.B., Chang, R.S.L., and Siegl, P.K.S. (1993) *Bioorg. Med. Chem. Lett.*, **3**, 1693–1696.

109. Cappelli, A., Mohr, G.P., Gallelli, A., Rizzo, M., Anzini, M., Vomero, S., Mennuni, L., Ferrari, F., Makovec, F., Menziani, C., De Benedetti, P., and Giorgi, G. (2004) *J. Med. Chem.*, **47**, 2574–2586.

110. Carpino, P.A., Sneddon, S.F., Jardine, P.S., Magnus-Ayritey, G.T., Rauch, A.L., and Burkard, M.R. (1994) *Bioorg. Med. Chem. Lett.*, **4**, 93–98.

111. Dhanoa, D.S., Bagley, S.W., Chang, R.S.L., Lotti, V.J., Chen, T.-B., Kivlighn, S.D., Zingaro, G.J., Siegl, P.K.S., Patchett, A.A., and Greenlett, W.J. (1993) *J. Med. Chem.*, **36**, 4230–2438.

11
Thiazide-Based Diuretics for the Treatment of Hypertension and Genitourinary Disorders

Jürgen Dinges

11.1
Introduction

Thiazide is the generic name for compounds with a 2*H*-benzo[*e*]-1,2,4-thiadiazine 1,1-dioxide ring system that carries a sulfonamide group in 7-position (Figure 11.1) [1]. Their corresponding 3,4-dihydro analogs are referred to as *hydrothiazides*. These names originated in the late 1950s and were initially used to differentiate direct analogs of the diuretic drug chlorothiazide (**1a**, Diuril®) [2] from nondiuretic benzothiadiazine 1,1-dioxides, which lack the 7-sulfamoyl group. At present, the term thiazide is applied more generally to diuretics that are believed to have their primary site of action in the distal tubules of nephrons, the basic secretory units of the kidneys [3]. For example, based on the latter definition, chlorthalidone (**3**) [4] is considered to be a member of the thiazide family, although it does not contain the characteristic benzothiadiazine structure (Figure 11.2) [5]. This chapter focuses on compounds with the benzothiadiazine dioxide heterocycle. Additional characteristic examples of diuretic drugs in this structural class are hydrochlorothiazide (**2a**, HCT) [6] and hydroflumethiazide (**2b**) [7].

11.2
History

The earliest report of benzothiadiazine dioxides was by Alfred Ekbom in 1902 [8]. While investigating analogs of 2-aminobenzenesulfonamide (**4**), he noticed that

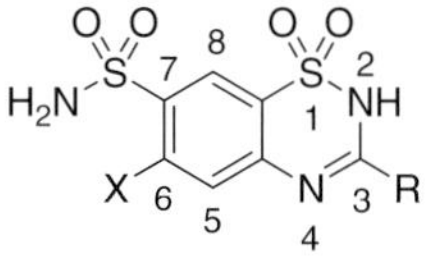

1

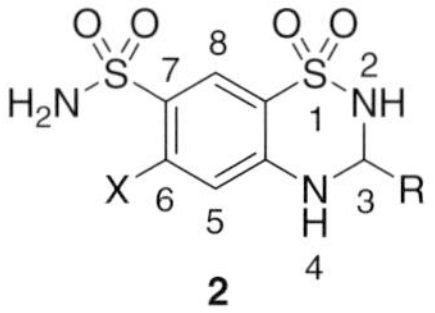

2

Figure 11.1 Basic structures and numbering conventions for thiazides (**1**) and hydrothiazides (**2**).

Bioactive Heterocyclic Compound Classes: Pharmaceuticals,
First Edition. Edited by Jürgen Dinges and Clemens Lamberth.

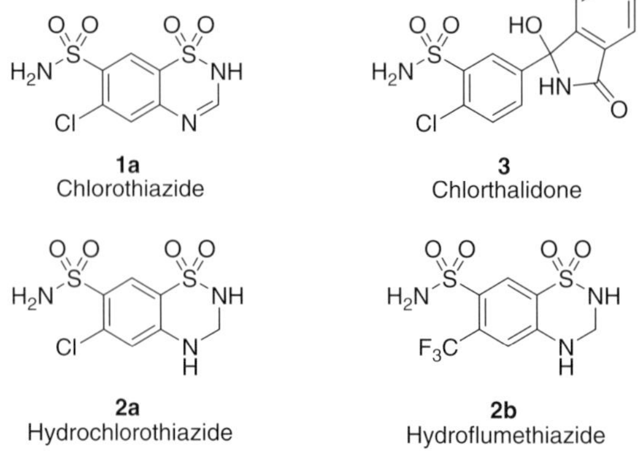

Figure 11.2 Diuretics of the thiazide family: chlorothiazide (**1a**) [2], chlorthalidone (**3**) [4], hydrochlorothiazide (**2a**) [6], and hydroflumethazide (**2b**) [7].

the formamide apparently only has transitory existence and cyclizes spontaneously to the benzothiadiazine dioxide **5a** (Scheme 11.1). The acetamide on the other hand was isolated, but then could also be cyclized to **5b** on heating above its melting point (>148 °C). Almost 50 years later, Freeman and Wagner found that the 3,4-double bond in benzothiadiazine 1,1-dioxides was stable to hydrogenation in the presence of a palladium on charcoal catalyst and only ring cleavage products were observed when the compounds were treated with sodium in alcohol [9]. However, the authors were able to synthesize 3,4-dihydro-2*H*-1,2,4-benzothiadiazine 1,1-dioxides by treating 2-aminobenzenesulfonamides with aldehydes in the presence of a base (Scheme 11.2). A series of clinical observations together with a certain degree of serendipity ultimately led to the discovery of Diuril (**1a**) by Novello and Sprague in 1957 [2]. Southworth described in 1937 that administration of large doses of sulfanilamide (**8**, Prontalbin®) (Figure 11.3) gave rise to clinical acidosis [10], which was later attributed to a renal electrolyte imbalance [11]. In 1940, Mann and Keilin [12] at the Molteno Institute at Cambridge University identified the unsubstituted sulfonamide group in sulfanilamide (**8**) as the key functionality for the inhibition of carbonic anhydrase, and at present, it is known that this group is a preferred ligand for the zinc ion in the active site of the enzyme [13]. Nine years later, William Schwartz [14] administered sulfanilamide (**8**) to three patients with severe

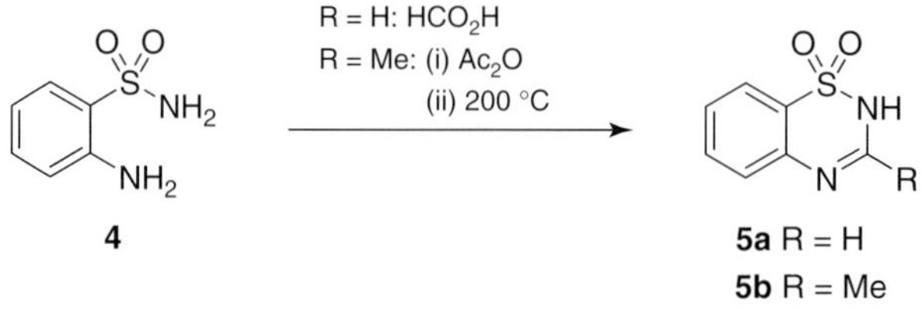

Scheme 11.1 Acylation of 2-aminobenzenesulfonamide (**4**) and subsequent intramolecular cyclization to give the benzothiadiazine dioxides **5a** and **5b**.

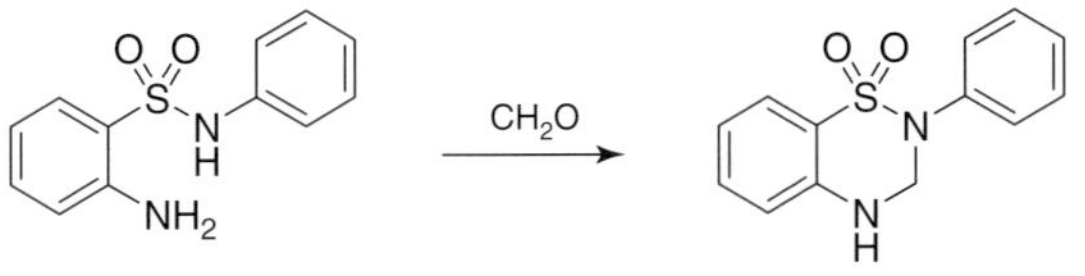

Scheme 11.2 Ring closure of 2-amino-*N*-phenylbenzenesulfonamide (**6**) with formaline to give 3,4-dihydro-2-*N*-phenyl-2*H*-1,2,4-benzothiadiazine 1,1-dioxide (**7**).

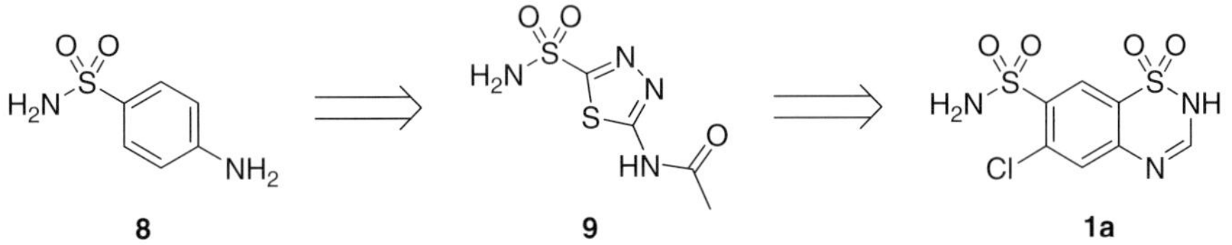

Figure 11.3 Historic development of sulfonamide-based diuretics from sulfanilamide (**8**) over acetazolamide (**9**) to chlorothiazide (**1a**, Diuril®).

congestive heart failure and attributed the observed diuretic effect to the inhibition of carbonic anhydrase. Carbonic anhydrases are metalloenzymes that are highly abundant in the kidney. These enzymes catalyze the conversion of carbon dioxide and water to bicarbonate and protons. Carbonic anhydrase inhibitors were found to increase diuresis by inhibiting the excretion of protons in the proximal tubule of the nephron. As a result, transport of sodium and potassium bicarbonate from the tubular lumen to the interstitial space is increased and is accompanied by water to maintain isotonicity [13]. The following search for improved carbonic anhydrase inhibitors initially yielded acetazolamide (**9**, Diamox®) [15]. This compound was about 300-fold more potent *in vitro* than sulfanilamide (**8**), but its failure to significantly increase the excretion of chloride, and a tendency to cause hypokalemia, limited its use as a diuretic [16]. Hypokalemia, a low concentration of potassium in the blood, has to be monitored carefully since it is associated with arrhythmias and sudden cardiac death. At Merck Sharp & Dohme, Novello and Sprague worked on analogs of 4-aminobenzene-1,3-disulfonamides when they came across the intramolecular cyclization of the formamide. One particular benzothiadiazine dioxide, Diuril (**1a**), showed outstanding carbonic anhydrase inhibition. Animal studies by Karl Beyer showed that Diuril (**1a**) was well absorbed when administered orally and displayed favorable saluretic activity (excretion of sodium and chloride) compared to existing organomercurial diuretics or other carbonic anhydrase inhibitors [17]. A key discovery was made by Freis and Wilson in 1958 when they found that oral administration of Diuril (**1a**) could effectively lower blood pressure [18]. Until this time, the treatment options for hypertension were very limited and often more harmful than beneficial (e.g., ganglionic blockers and *Rauwolfia* alkaloids). The introduction of thiazide diuretics therefore can be considered as the start of the modern era of antihypertensive therapy. In the same year, De Stevens *et al.* at Ciba Pharmaceutical Products in New Jersey synthesized HCT (**2a**), the

3,4-dihydro analog of Diuril (**1a**) [6]. With a maximal response at 75–100 mg day^{-1} in clinical trials, HCT (**2a**, Esidrix®: Ciba Ltd., Hydrodiuril®: Merck Sharp & Dohme, Oretic®: Abbott Laboratories) was about 20 times more active than Diuril (**1a**) [19]. However, its weak potency against carbonic anhydrase and the pronounced increase in chloride excretion suggested that inhibition of carbonic anhydrase was not the primary mechanism of action for the observed efficacy. Thiazide diuretics very quickly became so successful that in 1961 HCT (**2a**) was sold by 66 different companies, followed by chlorothiazide (**1a**) by 40 companies [20]. In Italy alone HCT was available under 27 different names. The explosive development of novel thiazides resulted in at least 14 additional marketed drugs, for example, hydroflumethiazide (**2b**, Saluron®: Bristol Laboratories, Diucardin®: Ayerst Research Laboratories, Rontyl®: Leo Pharmaceutical Products) [7]; methyclothiazide (**10a**, Enduron®: Abbott Laboratories) [21]; bendroflumethiazide (**2c**, Naturetin®: Bristol Myers Company) [7]; cyclopenthiazide (**2d**, Navidrix®: Ciba Pharmaceutical Products) [22]; trichlormethiazide (**2e**, Naqua®: Schering; Esmarin®: E. Merck) [23]; buthiazide (**2f**, Saltucin®: Boehringer Mannheim) [24]; and polythiazide (**10b**, Renese®: Pfizer & Co.) [25] (Figure 11.4). In comparison to Diuril (**1a**) and Hydrodiuril (**2a**) those drugs were not significantly more efficacious but had a much longer duration of action and a weaker inhibitory effect on carbonic anhydrase [5]. A combination of HCT (**2a**) with reserpine and the smooth muscle relaxant Hydralazine® was used in the Veterans Administration Cooperative study in 1967 [26]. The superiority of this combination in decreasing cardiovascular events in hypertensive patients compared to placebo clearly demonstrated the potential to reduce hypertension-related morbidity and mortality. The use of diuretics as first-line agents for hypertension started to fade in the 1970s when their patents expired [27]. As a consequence, the pharmaceutical industry shifted their marketing efforts toward newer generations of antihypertensive drugs such as angiotensin-converting enzyme (ACE) inhibitors, angiotensin II receptor blockers (ARBs), and calcium

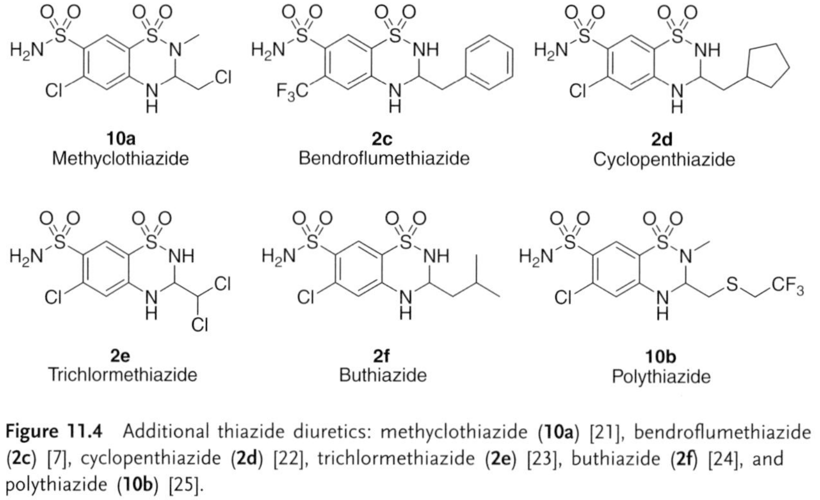

Figure 11.4 Additional thiazide diuretics: methyclothiazide (**10a**) [21], bendroflumethiazide (**2c**) [7], cyclopenthiazide (**2d**) [22], trichlormethiazide (**2e**) [23], buthiazide (**2f**) [24], and polythiazide (**10b**) [25].

channel blockers (CCBs). In this process, adverse metabolic side effects such as hypokalemia and hyperglycemia, caused by high doses of thiazides [28], were emphasized and clinical trials were sponsored to demonstrate improved safety profiles for the newer agents. With the results from the "The Antihypertensive and Lipid-Lowering Treatment to Prevent Heart Attack Trial" (ALLHAT), the so far most comprehensive comparison between a diuretic (chlorthalidone (**3**)), an ACE inhibitor (lisinopril), and a CCB (amlodipine) was published in 2002 [29]. With respect to the primary end point of fatal coronary heart disease and nonfatal myocardial infarction, no significant differences were observed for all three treatment groups. However, chlorthalidone (**3**) was found to be superior in secondary end points such as heart failure and stroke. The results pointed the Seventh Joint National Committee on Prevention, Detection, Evaluation, and Treatment of High Blood Pressure to recommend diuretics as the first-step treatment for patients with hypertension [30]. In 2008, HCT (**2a**) was the second most prescribed antihypertensive medication in the United States [31]. Hypertensive patients often require multiple drugs, either as monotherapies in a stepped care approach or as combination drugs to achieve the desired blood pressure range. Thiazide diuretics were highly recommended for combination regimens since lower doses (12.5–25 mg day^{-1}) are safe and achieve therapeutic synergy with all other classes of antihypertensive drugs [3]. Combinations of HCT (**2a**) with the ARBs valsartan (Diovan HCT®: Novartis), olmesartan (Benicar HCT®: Daiichi-Sankyo), losartan (Hyzaar®: Merck&Co.), irbesartan (Avalide®: Bristol-Myers Squibb Company; CoAprovel®: Sanofi-Aventis), and telmisartan (Micardis HCT®: Boehringer Ingelheim Pharmaceuticals, Inc., also marketed by Abbott Laboratories) were some of the best selling drugs in 2009 [32]. More recently, hydrothiazides gained interest as positive allosteric modulators of the α-amino-3-hydroxy-5-methyl-4-isoxazole-propionic acid (AMPA) receptor [33]. The AMPA receptor is one of the ionotropic (ligand-gated ion channels) glutamate receptors in the central nervous system, which mediate the fast synaptic response to extracellular glutamate. Potentiation of the AMPA receptor is being evaluated as a therapeutic strategy to treat cognitive disorders, schizophrenia, depression, and Parkinson's disease. Cyclothiazide (**2g**) (Figure 11.5) was found to modulate the desensitization of AMPA receptors and thus was able to increase the amplitude and duration of synaptic responses to glutamate.

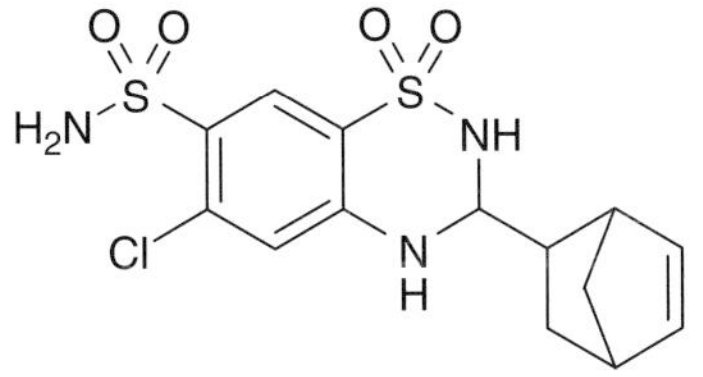

2g
Cyclothiazide

Figure 11.5 The AMPA-receptor-positive allosteric modulator cyclothiazide (**2g**).

The compound was also reported as a noncompetitive inhibitor of recombinant metabotropic (G-protein-coupled) glutamate receptors [34] and as a reversible inhibitor of γ-aminobutyric acid (GABA)-induced membrane currents [35].

11.3 Synthesis

The general routes for the synthesis of thiazides and hydrothiazides is shown in Scheme 11.3. The aniline disulfonyl chlorides (**12**) are readily prepared by heating a solution of the appropriately substituted anilines **11** in chlorosulfonic acid to 150 °C for 3–4 h [36]. X = Cl for the synthesis of chlorothiazide (**1a**) and HCT (**2a**), other typical substituents are CF_3, Me, MeO, F, Br, and NO_2. Sodium chloride is commonly added to the reaction mixtures for improved yields. The crude products can then be directly converted to the disulfonamides (**13**) with liquid ammonia [21] or ammonium hydroxide [7] at room temperature.

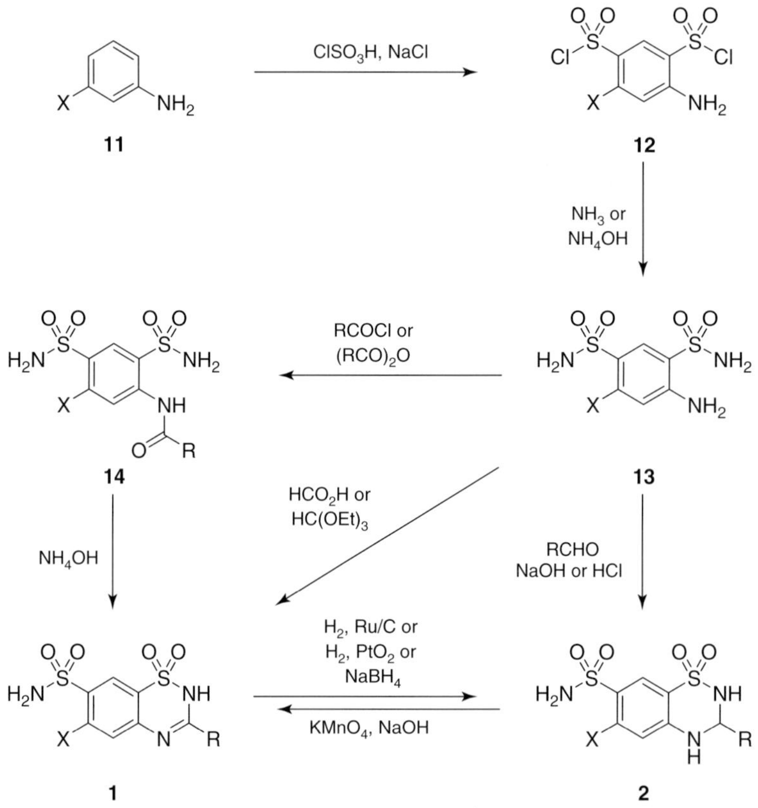

Scheme 11.3 General routes for the synthesis of thiazides and hydrothiazides.

The formation of thiazides (**1**) with a substituent in 3-position follows a two-step protocol. Acylation of **13** with acyl chlorides [36] or acetic anhydrides [37] leads to the amides **14**, which can be cyclized to the thiazide in the presence of ammonium hydroxide. For compounds without a 3-substituent, the thiazides are directly generated in high yields by refluxing **13** in a large excess of formic acid for 24 h [38]. This excess of formic acid is undesirable for the manufacturing of thiazides, for example, Diuril (**1a**), on an industrial scale. The reactions are therefore carried out with only about 4.5 equivalents of formic acid in dimethylformamide [39]. If the sulfonamide ortho to the amino group in **13** is alkylated, the more reactive triethyl orthoformate is often the reagent of choice for the ring-closure reaction [38]. However, triethyl orthoformate should not generally be used because it tends to form an *N*-ethoxymethylene adduct with the 7-sulfonamide, which subsequently has to be cleaved with ammonia. The synthesis of the hydrothiazides **2** can be accomplished by condensation of **13** with aldehydes (or to a lesser extent acetals or enol ethers), preferentially in the presence of an acid (i.e., HCl, H_2SO_4) or a base (i.e., NaOH, NH_3) [38]. The amount of aldehyde used in this reaction is critical since excess results in the formation of resinous side products that are difficult to separate from the hydrothiazides [7]. The formation of these polymeric side products presents a major problem for the industrial manufacturing of hydrothiazide drugs, such as, Hydrodiuril (**2a**), and several approaches have been described to address it. A 2007 patent from Unichem Laboratories Ltd. favors the use of sulfuric acid in methanol to catalyze the formation of the cyclic aminals and the stepwise addition of formaldehyde over the reaction time [40]. The crude material is then dissolved in hot acetone, and the side products are adsorbed onto activated carbon. The authors claim that the product obtained via this process meets ICH and USFDA guidelines of 99.9% purity and single impurity level less than 0.1%. Ketones can also be used for the cyclization reaction, either directly or in form of their ketals, to give hydrothiazides with two substituents in 3-position. However, the reaction conditions have to be adapted more carefully to the reactivity of each ketone [1]. The thiazides **1** and the hydrothiazides **2** are interconvertible. For example, chlorothiazide (**1a**) can be hydrogenated in the presence of 5% ruthenium on charcoal [38] or a platinum oxide catalyst [37] at about 2 bars. Analogs with a 3-substituent are more resistant to hydrogenation but can be reduced with sodium borohydride [6, 41]. Conversely, HCT (**2a**) can be dehydrogenated to **1a** through permanganate oxidation [38]. Substituents in the 5- and 8-positions of the thiazides and hydrothiazides are introduced via the above-mentioned reaction sequence by using anilines with the respective substitution patterns [42, 43]. The same route is also used when the benzene nucleus is replaced by a heterocycle, for example, the 2*H*-pyrido[2,3-e]-1,2,4-thiadiazine-7-sulfonamide 1,1-dioxide (pyridine nitrogen in 5-position) and its 3,4-dihydro analog [44, 45], as well as 2*H*-thieno[2,3-e]-1,2,4-thiadiazine-6-sulfonamide 1,1-dioxide [46]. The synthesis of N-alkylated analogs of **1** and **2** has been extensively studied. Direct alkylation of chlorothiazide (**1a**) with either methyl iodide [8] or dimethyl sulfate [38] predominantly leads to the 4-methylated product. A mixture of the tautomeric 2- and 4-alkylation products of **1** ($X = CF_3$, $R = H$) can be obtained

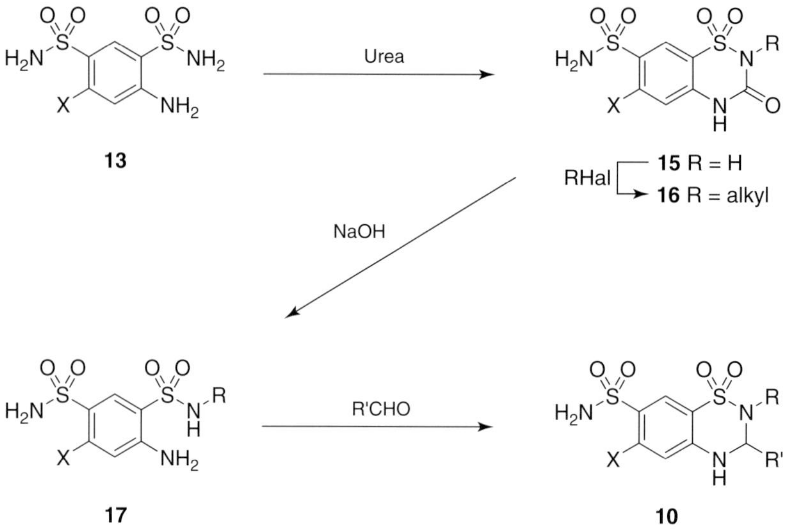

Scheme 11.4 Regioselective introduction of an alkyl substituent into the 2-position of hydrothiazides.

if triethyl orthoformate is used as the alkylating agent [47]. In the latter case, the 7-sulfonamide group does not compete as a potential alkylation site but instead forms the previously mentioned *N*-ethoxymethylene adduct. In contrast to **1a**, methyl iodide [21] or dimethyl sulfate [24] alkylate HCT (**2a**) primarily in 2-position. In addition, a small amount of a dimethylation product, in which the 7-sulfonamide group is also alkylated, is observed. While these methods usually work well for methylations, larger alkylating agents produce unsatisfactory results. The introduction of larger substituents into the 2-position of hydrothiazides therefore usually follows the protocol of Close *et al.* (Scheme 11.4) [21]. Cyclization of the disulfonamides **13** with urea at 180 °C furnishes the 3-keto thiazides **15**. These intermediates can be alkylated regioselectively under standard conditions. Basic hydrolysis (NaOH, reflux over night) and recyclization with an aldehyde then delivers the 2-alkylated hydrothiazides **10**. Alkylation in 2- or 4-position generally renders thiazides more labile to hydrolysis, particularly in the presence of a base [36]. Alkylated hydrothiazides on the other hand were found to be stable under basic as well as acidic conditions [48]. Hydrothiazides **2** with substituents in 3-position are chiral. The cyclic aminal substructures are surprisingly stable to enantiomerization [49] and it has been demonstrated that both enantiomers can have different biological activities [50]. A catalytic asymmetric synthesis of hydrothiazides has recently been established using the chiral phosphoric acid catalyst **18** (Scheme 11.5) [51].

Ring closure of **13** (X = Cl) with various aldehydes produced the corresponding hydrothiazides (*R*)-**2**(R = $-CH(CH_3)_2$, $-CH_2$-*c*-C_5H_9, $-CH_2C_6H_5$, and -*n*-C_5H_{11}) with yields of about 80% and enantiomeric excess of 90–92%.

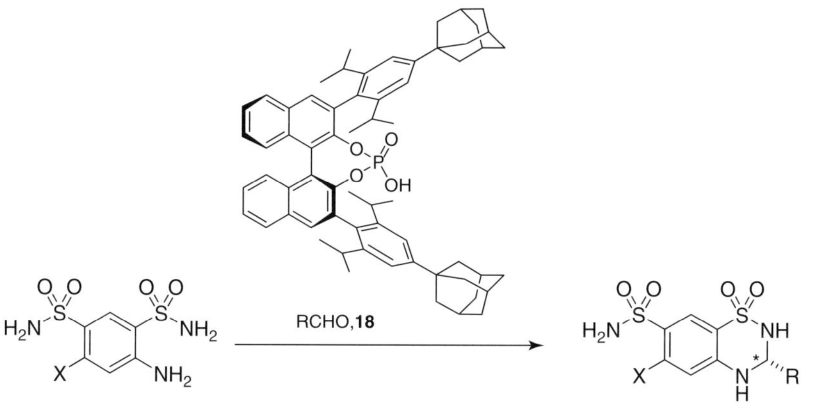

Scheme 11.5 Catalytic asymmetric synthesis of hydrothiazides.

11.4 Mode of Action

About 1700 l of blood flow through the human kidneys each day, where it is filtered at a rate of about 100 ml/min to remove water, electrolytes (Na^+, K^+, Ca^{2+}, Mg^{2+}, Cl^-, HCO_3^-, HPO_4^{2-}, etc.), and low-molecular-weight organic substances (glucose, urea, uric acid, creatinine, etc.) [20]. To accomplish this enormous task, the kidneys consist of thousands of basic secretory units called *nephrons* (Figure 11.6). Within a nephron, the blood is filtered in the glomerulus and the filtrate is then passed through the renal tubular system, where about 99% of the water and most of the electrolytes are reabsorbed back into the blood stream. The reabsorption process serves to control the electrolyte balance and to maintain a constant hydrostatic pressure. The remaining filtrate, which contains waste products such as urea, uric acid, and creatinine, reaches the collecting duct and is eventually excreted as urine. Imbalances in the glomerulus or the tubules due to reduced cardiac output or renal disease can lead to decreased excretion of sodium, chloride, and water, which may give rise to high blood pressure, congestive heart failure, and edema (retention of excess fluids in the body). Diuretic agents try to reestablish homeostasis by inhibiting the reabsorption of water and electrolytes in different regions of the tubular system. Based on the large volume of water that is reabsorbed, a slight reduction in tubular reabsorption will result in a large increase in urine volume. Thiazides inhibit the electroneutral Na^+/Cl^- cotransporter (symport) in the distal convoluted tubule, where ~7% of total sodium is reabsorbed [3]. They are able to reach their site of action because they are tightly bound to plasma proteins, which prevents them from being filtered by the glomeruli. The effect of thiazides on blood pressure can be divided into two chronological phases. In the acute response phase (a few weeks), the inhibition of sodium reabsorption induces increase of excreted urine, which in turn leads to reduction of extracellular fluid and plasma

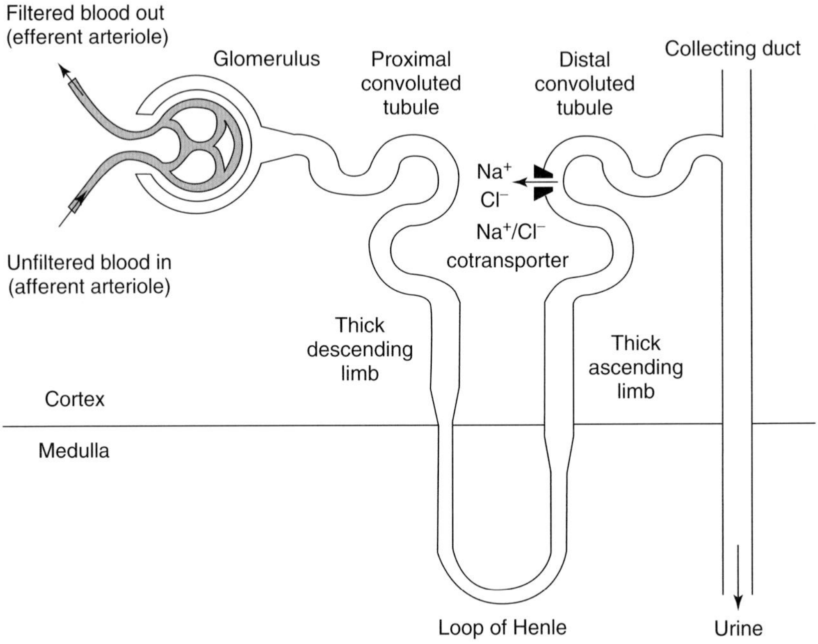

Figure 11.6 Basic anatomy of a nephron and site of action for thiazide diuretics.

volume [52]. As a consequence, venous return is diminished and cardiac output is reduced, leading to a decrease in blood pressure. However, the reduction in cardiac output triggers the activation of the sympathetic nervous system and the renin-angiotensin-aldosterone system. These systems effectuate an increase in total peripheral resistance of the vasculature followed by returning the plasma volume to almost pretreatment levels. These changes occur without an increase in blood pressure and represent the transition into the chronic response phase. At this stage, a combination of a thiazide with ARBs or ACE inhibitors can increase the antihypertensive response by counteracting the increase in vascular resistance [3]. The chronic response phase has to be based on a mechanism other than the decrease in plasma volume since the blood pressure reduction is maintained after this volume is almost fully restored [53]. This effect is most likely based on a reduction in total peripheral resistance, but the exact details remain unclear. Current hypotheses include a direct vasodilatory effect, perhaps separate from the diuretic effect; a downward shift in vascular resistance based on prolonged diuresis; direct action on the endothelium by opening the large conductance calcium channel or through inhibition of carbonic anhydrase; calcium antagonist-like activities; calcium desensitization; a process of autoregulation, which adjusts for the low cardiac output by vasodilation; and altered ion gradients across smooth muscle cells or changes in membrane-bound adenosine triphosphate activity [3, 52, 53].

11.5 Structure–Activity Relationship

The vast majority of the structure–activity relationships (SARs) for thiazides and hydrothiazides were established at a time when these compounds were believed to be inhibitors of carbonic anhydrase. Later, the Na^+/Cl^- symport was identified as their primary target. This section therefore does not list any enzyme potencies but instead focuses on the SAR that was developed for the overall diuretic and saluretic (excretion of Na^+ and Cl^-) activity after oral or intravenous dosing in rats or dogs.

Saturation of the 3,4-double bond in thiazides (Figure 11.7, structure I) to give hydrothiazides (structure III) generally increases the diuretic effect in dogs and man by about 5- to 10-fold [54]. The key functionality of all thiazides and hydrothiazides is the sulfonamide in 7-position. Without this group, the compounds are devoid of any diuretic activity [16]. Attempts to replace the sulfonamide with other electron-withdrawing groups were not successful [54]. In addition, the nitrogen atom of the sulfonamide has to be unsubstituted. A few active compounds with lower N-alkyl or N-acyl substituents were described in the literature, but on closer examination, those residues could be removed metabolically. Therefore, the free sulfonamides were assumed to be responsible for the observed efficacy. Also critical for the diuretic effect of thiazides and hydrothiazides is an appropriate substituent in 6-position (X in structures I–III), the so called activating group [54]. Compounds without such an activating group show very little activity [55]. The most potent compounds carry a chloride, a bromide, or a trifluoromethyl substituent in 6-position. Methyl and nitro derivatives also display functional activity, but their methoxy and amino counterparts are inactive. Larger groups in 6-position appear to be detrimental [56]. Additional substituents in positions 5 or 8 on the phenyl ring only reduce activity [16]. Substitutions on the two nitrogens in the heterocyclic portion of the core produce very different effects between thiazides and hydrothiazides. Thiazides with substituents in either 2- (R_2 in structure I) [16] or 4-position (R_4 in structure II) [48] are completely inactive. Initially it was suspected that this lack of activity could be due to the chemical instability of those compounds, but the more stable hydrothiazides with a 4-substituent were also found to be inactive [48]. In contrast, 2-substituents have the potential to significantly improve the diuretic activities of hydrothiazides (R_2 in structure III).

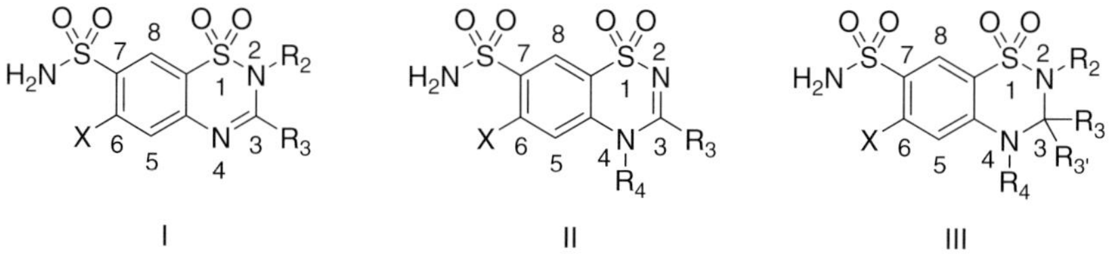

Figure 11.7 General structures for thiazides (I), 4-substituted thiazides (II), and hydrothiazides (III).

Close *et al.* [57] at Abbott Laboratories prepared 12 pairs of hydrothiazides with and without a 2-methyl group, and in each case, the *N*-methyl homolog was more active than its parent compound. Benzylation in 2-position improved the potency of HCT (**2a**) approximately fivefold [48]. However, this larger group did not allow an additional substituent in 3-position. The most extensively investigated SAR for thiazides and hydrothiazides is based on substitutions in 3-position. Introduction of a carbonyl group into hydrothiazides (structure III: $R_3{-}C{-}R_{3'} = C{=}O$) decreased activity [55]. In the thiazide series, small alkyl groups ($C_1{-}C_5$) in 3-position (R_3 in structure I) are tolerated but do not lead to a marked improvement in the animal diuretic potency [54]. Compounds with a phenyl substituent in this position are inactive. A sharp increase in activity (5- to 10-fold) was found for a chlorothiazide (**1a**) with a 3-benzylthiomethyl side chain, but this compound represents an exception. Hydrothiazides with aliphatic side chains in 3-position (R_3 or $R_{3'}$ in structure III) display improved saluretic activities, and the *n*-butyl or *i*-butyl groups appear to be optimum [58]. Cycloalkyl-containing side chains also have a beneficial effect. Mono, and more significantly, dihalomethyl analogs (Cl, Br, and I but not F) are more potent than their unsubstituted parent compounds [59]. For example, trichlormethiazide (**2e**) displayed a 4- to 14-fold improvement in its diuretic activity over HCT (**2a**), depending on the species and route of administration [58]. Benzyl- and phenethyl-substituted hydrothiazides are among the strongest saluretic agents that have been identified, but some of these compounds suffer from insufficient oral bioavailabilities [58]. Based on the established SAR, the α-carbon of the 3-substituent has to carry at least one hydrogen atom [58, 59]. Compounds with a trihalomethyl substituent, a directly attached phenyl or aryl group or with a C–C-double bond in α-position, failed to show efficacy in the *in vivo* models. Polar functionalities (alcohols, ethers, esters, and amines) in the side chain render the hydrothiazides ineffective. Surprisingly, on intravenous administration into dogs, the (2-pyridyl)-ethyl analog of HCT (**2a**) was almost 17 times more active than **2a**. However, the compound was characterized by poor oral absorption in dogs and rats [58]. Reduced diuretic activity was observed for compounds that carry two substituents in 3-position, except if one of those substituents is a chloromethyl group [16]. Spiro compounds effected no change compared to their parent thiazides.

References

1. Cragoe, E.J. Jr., Woltersdorf, O.W. Jr., Baer, J.E., and Sprague, J.M. (1962) *J. Med. Pharm. Chem.*, **5**, 896–912.
2. Novello, F.C. and Sprague, J.M. (1957) *J. Am. Chem. Soc.*, **79**, 2028–2029.
3. Ernst, M.E. and Moser, M. (2009) *N. Engl. J. Med.*, **361**, 2153–2164.
4. Graf, W., Girod, E., Schmid, E., and Stoll, W.G. (1959) *Helv. Chim. Acta*, **42**, 1085–1101.
5. Early, L.E. and Orloff, J. (1964) *Annu. Rev. Med.*, **15**, 149–165.
6. De Stevens, G., Werner, L.H., Halamandaris, A., and Ricca, S. Jr. (1958) *Experientia*, **14**, 463.
7. Holdrege, C.T., Babel, R.B., and Cheney, L.C. (1959) *J. Am. Chem. Soc.*, **81**, 4807–4810.
8. Ekbom, A. and Bihang, K. (1902) *Svenska Vet.-Akad. Handl.*, **27**, 3–24.

9. Freeman, J.H. and Wagner, E.C. (1951) *J. Org. Chem.*, **16**, 815–837.
10. Southworth, H. (1937) *Proc. Soc. Exp. Biol. Med.*, **36**, 58–61.
11. Berliner, R.W. and Orloff, J. (1956) *Pharmacol. Rev.*, **8**, 137–174.
12. Mann, T. and Keilin, D. (1940) *Nature*, **146**, 164–165.
13. Supuran, C.T. (2008) *Nat. Rev. Drug Discov.*, **7**, 168–181.
14. Schwartz, W.B. (1949) *N. Engl. J. Med.*, **240**, 173–177.
15. Roblin, R.O. and Clapp, J.W. Jr. (1950) *J. Am. Chem. Soc.*, **72**, 4890–4892.
16. Lund, F.J. and Kobinger, W. (1960) *Acta Pharmacol. Toxicol.*, **16**, 297–324.
17. Beyer, K.H. (1958) *Ann. N. Y. Acad. Sci.*, **71**, 363–379.
18. Freis, E.D., Wanko, A., Wilson, I.M., and Parrish, A.E. (1958) *J. Am. Med. Assoc.*, **166**, 137–140.
19. Richterich, R. (1958) *Experientia*, **14**, 458.
20. Schlitter, E., DeStevens, G., and Werner, L. (1962) *Angew. Chem., Int. Ed. Engl.*, **1**, 235–245.
21. Close, W.J., Swett, L.R., Brady, L.E., Short, J.H., and Vernsten, M. (1960) *J. Am. Chem. Soc.*, **82**, 1132–1135.
22. Barrett, W.E., Chart, J.J., and Renzi, A.A. (1961) *Arch. Int. Pharmacodynam. Ther.*, **131**, 325–338.
23. Taylor, R.M. and Winbury, M.M. (1960) *Nature (London)*, **187**, 603–604.
24. Werner, H.L., Halamandaris, A., Ricca, S., Dorfman, L., and DeStevens, G. Jr. (1960) *J. Am. Chem. Soc.*, **82**, 1161–1166.
25. Scriabine, A., Korol, B., Kondratas, B., Yu, M., P'an, S.Y., and Schneider, J.A. (1961) *Proc. Soc. Exp. Biol. Med.*, **107**, 864–872.
26. Veterans Administration Cooperative Study Group on Hypertensive Agents (1967) *J. Am. Med. Assoc.*, **202**, 1028–1034.
27. Ernst, M.E. and Grimm, R.H. (2008) *Curr. Hypertens. Rev.*, **4**, 256–265.
28. Greenberg, A. (2000) *Am. J. Med. Sci.*, **319**, 10–24.
29. The ALLHAT Officers and Coordinators for the ALLHAT Collaborative Research Group (2002) *J. Am. Med. Assoc.*, **288**, 2981–2997.
30. Chobanian, A.V., Bakris, G.L., Black, H.R., Cushman, W.C., Green, L.A., Izzo, J.L. Jr., Jones, D.W., Materson, B.J., Oparil, S., Wright, J.T. Jr., and Roccella, E.J. (2003) *J. Am. Med. Assoc.*, **289**, 2560–2572.
31. SDI/Verispan, VONA, 2008 Top 200 generic drugs by total prescriptions http://www.modernmedicine.com/modernmedicine/data/articlestandard/drugtopics/192009/597084/article.pdf (accessed 2011).
32. Mack, D.J., Brichacek, M., Plichta, A., and Njarðarson, J.T. Top 200 Pharmaceutical Products by Worldwide Sales in 2009 http://cbc.arizona.edu/njardarson/group/sites/default/files/Top200PharmaceuticalProductsByWorldwideSalesin2009.pdf (accessed 2011); Top 200 Drugs for 2009 by Sales http://www.drugs.com/top200_2009.html (accessed 2011).
33. Grove, S.J.A., Jamieson, C., Maclean, J.F.K., Morrow, J.A., and Rankovic, Z. (2010) *J. Med. Chem.*, **53**, 7271–7279.
34. Surin, A., Pshenichkin, S., Grajkowska, E., Surina, E., and Wroblewski, J.T. (2007) *Neuropharmacology*, **52**, 744–754.
35. Deng, L. and Chen, G. (2003) *Proc. Natl. Acad. Sci.*, **100**, 13025–13029.
36. Novello, F.C., Bell, S.C., Abrams, E.L.A., Ziegler, C., and Sprague, J.M. (1960) *J. Org. Chem.*, **25**, 965–970.
37. Whitehead, C.W., Traverso, J.J., Marshall, F.J., and Morrison, D.E. (1961) *J. Org. Chem.*, **26**, 2809–2813.
38. Novello, F.C., Bell, S.C., Abrams, E.L.A., Ziegler, C., and Sprague, J.M. (1960) *J. Org. Chem.*, **25**, 970–981.
39. Hinkley, D.F. (1958) (Merck & Co.), US2937169 Sept. 25, 1958.
40. Ramakrishnan, A., Kapkoti, G.S., Singare, D.T., and Kumar, N. (2005) (Unichem Laboratories, Ltd.) WO 2007026376, Aug. 29, 2005.
41. Whitehead, C.W., Traverso, J.J., Sullivan, H.R., and Marshall, F.J. (1961) *J. Org. Chem.*, **26**, 2814–2818.
42. Topliss, J.G., Sherlock, M.H., Clarke, F.H., Daly, M.C., Pettersen, B.W., Lipski, J., and Sperber, N. (1961) *J. Org. Chem.*, **26**, 3842–3850.
43. Short, J.H. and Biermacher, U. (1960) *J. Am. Chem. Soc.*, **82**, 1135–1138.

44. Yale, H.L., Losee, K., and Bernstein, J. (1960) *J. Am. Chem. Soc.*, **82**, 2042–2046.
45. Cragoe, E.J., Nicholson, J.A., and Sprague, J.M. Jr. (1961) *J. Med. Pharm. Chem.*, **4**, 369–383.
46. Buzas, A. and Teste, J. (1960) *Bull. Soc. Chim. Fr.*, 793–803.
47. Yale, H.L. and Sheehan, J.T. (1961) *J. Org. Chem.*, **26**, 4315–4325.
48. Whitehead, C.W. and Traverso, J.J. (1962) *J. Org. Chem.*, **27**, 951–956.
49. Cannazza, G., Braghiroli, D., Iuliani, P., and Parenti, C. (2006) *Tetrahedron Asymm.*, **17**, 3158–3162.
50. Uzunov, D.P., Zivkovich, I., Pirkle, W.H., Costa, E., and Guidotti, A. (1995) *J. Pharm. Sci.*, **84**, 937–942.
51. Cheng, X., Vellalath, S., Goddard, R., and List, B. (2008) *J. Am. Chem. Soc.*, **130**, 15786–15787.
52. Sica, D.A. (2004) *J. Clin. Hypertens.*, **6**, 661–664.
53. Duarte, J.D. and Cooper-DeHoff, R.M. (2010) *Expert Rev. Cardiovasc. Ther.*, **8**, 793–802.
54. Novello, F.C. and Sprague, J.M. (1967) *Ind. Chim. Belge*, **32**, 222–225.
55. Sprague, J.M. (1958) *Ann. N. Y. Acad. Sci.*, **71**, 328–343.
56. Feit, P.W., Tvaermose Nielsen, O.B., and Bruun, H. (1972) *J. Med. Chem.*, **15**, 437–440.
57. Close, W.J., Swett, L.R., and Nordeen, C.W. (1961) *J. Org. Chem.*, **26**, 3423–3428.
58. Taylor, R.M. and Topliss, J.G. (1963) *J. Med. Chem.*, **6**, 312–320.
59. Klosa, J. and Voigt, H. (1962) *J. Prakt. Chem.*, **16**, 264–276.

12
Tetrahydropyranone-Based HMG-CoA Reductase Inhibitors for the Treatment of Arterial Hypercholesterolemia

Hongyu Zhao

12.1
Introduction

The basic pharmacophore of the first discovered 3-hydroxy-3-methylglutaryl-coenzyme A (HMGCoA) reductase inhibitors contains a (4*R*,6*R*)-4-hydroxytetrahydro-2*H*-pyran-2-one-6-yl ring connected through a two-carbon linker to a chiral decalin moiety. Under basic conditions, the tetrahydropyranone ring is hydrolyzed to its "(3*R*,5*R*)-5-yl-3,5-dihydroxypentanoic acid," or statin form, and the reverse reaction occurs under thermal or acidic conditions. As an example, Scheme 12.1 shows the interconversion of mevastatin (formerly called *compactin* or *ML-236B*) between its tetrahydropyranone, or δ-lactone form, **1**, and its "dihydroxypentanoic acid" form, **2** [1]. Both the forward and reverse reactions can occur in the liver [2].

The name "statin" originated from Latin "stat," meaning "halt," and was first coined by Brazeau in 1973 to describe the growth hormone secretion inhibitor somatostatin [3]. As of 2011, two tetrahydropyranone-based HMGCoA reductase inhibitors remain in the major markets. They are the natural product lovastatin (**3**, Mevacor®, Merck & Co. Inc.) [4] and the semisynthetic simvastatin (**4**, Zocor®, Merck & Co. Inc.) [5] (Figure 12.1). Mevastatin (**1**) has never been approved for marketing, while its synthetically oxidized analog, pravastatin (**5**, Pravachol®, Sankyo Pharma Inc. and Bristol-Myers Squibb Company), was launched in 1989 in the open acid form [6].

In the scope of this book, this chapter mainly focuses on the history, synthesis, mode of action, and the structure–activity relationship (SAR) of the tetrahydropyranone-based HMGCoA reductase inhibitors, but also mentions marketed open acid analogs since the lactone and hydroxyl acid forms interconvert *in vivo*.

12.2
History

(4*R*,6*R*)-4-Hydroxy-6-methyltetrahydro-2*H*-pyran-2-one (**6**), the smallest possible δ-lactone with a stereochemistry identical to that of HMGCoA reductase inhibitors,

Bioactive Heterocyclic Compound Classes: Pharmaceuticals,
First Edition. Edited by Jürgen Dinges and Clemens Lamberth.

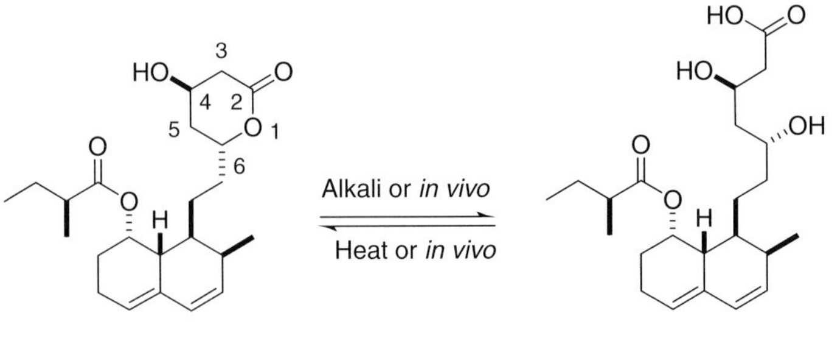

1 2

Scheme 12.1 Structure of mevastatin in its (4*R*,6*R*)-4-hydroxytetrahydro-2*H*-pyran-2-one-6-yl form, **1** and its "(3*R*,5*R*)-5-yl-3,5-dihydroxypentanoic acid" form, **2**.

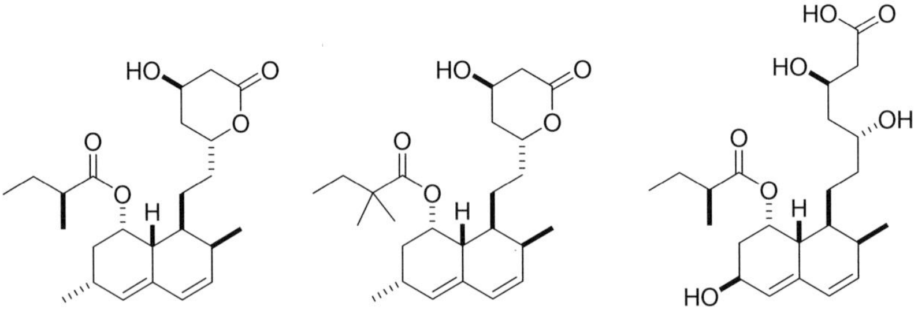

3 Lovastatin

4 Simvastatin

5 Pravastatin

Figure 12.1 Natural and semisynthetic HMGCoA reductase inhibitors.

was independently identified in 1971 by two research groups as part of the bitter tasting compound in the berries of the Finnish mountain ash (*Sorbus aucuparia*). Tschesche *et al.* [7] were able to enzymatically hydrolyze the bitter tasting compound into the aglycon **6** (Scheme 12.2) and glucose. The team then used proton nuclear magnetic resonance (^{1}H-NMR), mass spectroscopy, as well as the Cotton effect to determine the absolute stereochemistry of **6**. Pyysalo *et al.* [8] at the University of Helsinki in Finland established the first synthesis of **6** in racemic form (Scheme 12.2). Thus, deacetylation of dehydracetic acid (**7**) in concentrated sulfuric acid followed by a hydrogenation reaction gave a mixture of *rac*-**6** and *rac*-**8** at an 18% overall yield. A separation was not attempted since the products polymerized quickly on storage.

In the same year, inspired by Alexander Fleming's discovery of penicillin from blue–green mold (*Penicillium notatum*), Akira Endo [9, 10] at Sankyo, Japan, started to look for HMGCoA reductase inhibitors in mold culture broths.

In the 1970s, coronary heart disease was responsible for 40% of deaths in America. A basic manifestation of coronary heart disease is atherosclerosis caused by lipid plaque on the artery walls. Cholesterol was first found to relate to atherosclerosis in 1910 by the Russian chemist Nikolai N. Anitschlov, who concluded "without

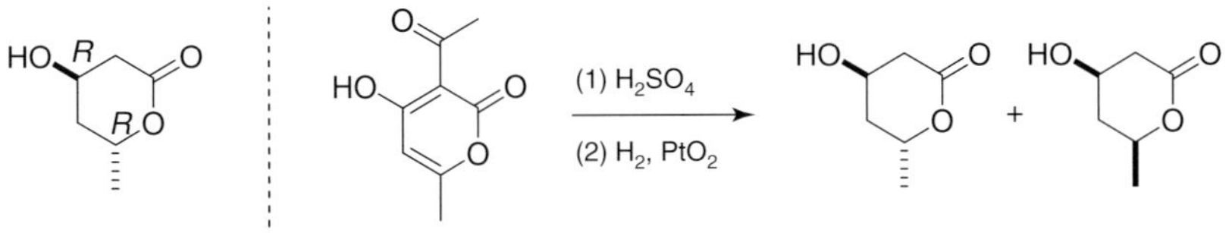

Scheme 12.2 Isolated aglycon **6** and first synthesis of racemic 4-hydroxy-6-methyltetrahydro-2*H*-pyran-2-one (*rac*-**6**).

cholesterol there can be no atherosclerosis'' [11]. Cholesterol is the main component of the atherosclerotic plaque, and elevated plasma cholesterol levels have been linked to higher incidence of heart disease by the Framingham study [12] and the Seven Countries Study [13]. Cholesterol is insoluble in plasma and needs to be transported by lipoproteins. An elevated low-density lipoprotein (LDL, the major form of total cholesterol) level is considered a main risk factor for heart disease, while high-density lipoprotein (HDL) is believed to be beneficial. In the past two decades, pharmacological reduction of LDL has been a major strategy for heart disease prevention.

The *de novo* synthesis of cholesterol is mainly regulated by the activity of HMGCoA reductase, which catalyzes the rate-limiting step in this complex process involving more than 30 enzymes. Dr. Endo hypothesized that some microbes might produce HMGCoA reductase inhibitors as a defensive mechanism against microorganisms whose growth requires mevalonate-derived isoprenoids. In 1973, after testing thousands of samples, Endo isolated mevastatin (**1**) from a culture broth of blue–green mold (*Penicillium citrinum*), a mold from the same genus from which penicillin was discovered. The compound effectively inhibited cholesterol synthesis *in vitro* but failed to show efficacy in rats. Two years of additional studies into the mode of action of mevastatin (**1**) revealed that longer term dosing of **1** into rats induced a 10-fold increase of hepatic HMGCoA reductase, which overwrote the compound's inhibitory effect. Around the same time, mevastatin (**1**) was also discovered at Beecham Pharmaceuticals from another blue–green mold (*Penicillium brevicompactum*) during a search for antifungal agents [14]. However, its development as a cholesterol-lowering agent was halted because of the lack of efficacy in rats. In contrast, Endo demonstrated that mevastatin (**1**) lowered plasma cholesterol by 50% in laying hens after one month of dosing and also showed efficacy in dogs and monkeys. In a clinical study that was not sanctioned by Sankyo, mevastatin (**1**) was given to an 18-year-old woman with severe familial hypercholesterolemia in 1978. At a daily dose of 500 mg, her plasma cholesterol levels were only marginally lowered but elevated blood transaminase levels and muscular dystrophy were observed. These side effects were reversible when the treatment was discontinued. The treatment of five heterozygous patients with familial hypercholesterolemia and three patients with combined hyperlipidemia was more successful and lowered plasma cholesterol levels by 30% without showing any severe side effects. On the basis of these results, Sankyo started phase I clinical trials in late 1978, followed by a phase II trial in 1979. However, the development of mevastatin (**1**) was discontinued in 1980, probably because of lymphomas found in dogs.

In 1976, Merck initiated a collaboration on the statin project with Sankyo. After confirming the potency of mevastatin (**1**) in-house, Merck set out to find its own statin. Three years later, Alfred Albert isolated lovastatin (**3**, formerly called *mevinolin* or *monacolin K*) from the fungus *Aspergillus terreus*. However, the development of lovastatin (**3**) (Figure 12.1) was halted soon after the clinical trials with mevastatin (**1**) were discontinued. In the next three years, Merck spent a significant amount of resources to understand the nature of the effects of lovastatin (**3**) in dogs and concluded that the lymphomas observed were not neoplastic. This cleared the safety concerns for lovastatin (**3**). The FDA was convinced by Merck's analysis, the clinical trials resumed, and lovastatin (**3**) was approved in 1987. In 1989, Sankyo launched pravastatin (**5**) (Figure 12.1).

Like all breakthrough drugs, the discovery and development of statins have been difficult [15]. In the late 1970s, target-based rational drug discovery was a relatively novel and unproven approach. The clinical effects of lowering LDL with diet or drugs available at the time were less than impressive and the side effects (such as flushing, gastrointestinal discomfort, etc.) were unpleasant [16]. As recently as 1989, the cholesterol hypothesis was seriously questioned [17], and rightfully so, given the fact that convincing clinical evidence would not emerge until five years later. The possible negative consequences of inhibiting the biosynthesis of cholesterol, an essential component of cell membrane and a precursor of several hormones, represented a major risk.

The initial labeling for lovastatin (**3**) was limited to LDL reduction. The landmark Scandinavian Simvastatin Survival Group (4S) study in 1994 with 4444 enrolled patients confirmed the connection between LDL lowering by statins and secondary prevention of heart attacks [18]. Since then, several large-scale clinical trials have been conducted and the beneficial effects of statins and their benign safety profile have been unequivocally established. Lovastatin (**3**) was originally positioned as a niche product for high-risk patients. Much of the beneficial effects beyond LDL reduction and the safety profile of the statins in chronic settings were established years after lovastatin (**3**) was approved. A significant amount of resource was required to determine clinical results and to inform the medical community and the public about the benefits of LDL reduction by statins. It is no easy feat for the statins to sail through this complex maze because, more often than not, these expensive efforts only prove that the medicine of interest is unsuitable for the patients. The current efforts to delineate the complex effects of inhibiting the cholesterol ester transfer protein to raise HDL levels can be considered as a sequel of the statin thriller [19].

For the follow-on statins, different risks and challenges were involved, and it took no less determination to successfully bring such products to the market. The complex structure of natural statins made them difficult to chemically optimize. In the 1980s, scientists at Merck discovered that a 2-(4′-fluorophenyl)-benzene ring can replace the decalin ring in natural statins without losing potency (see Section 12.5 for details) [20]. This key discovery allowed the design and synthesis

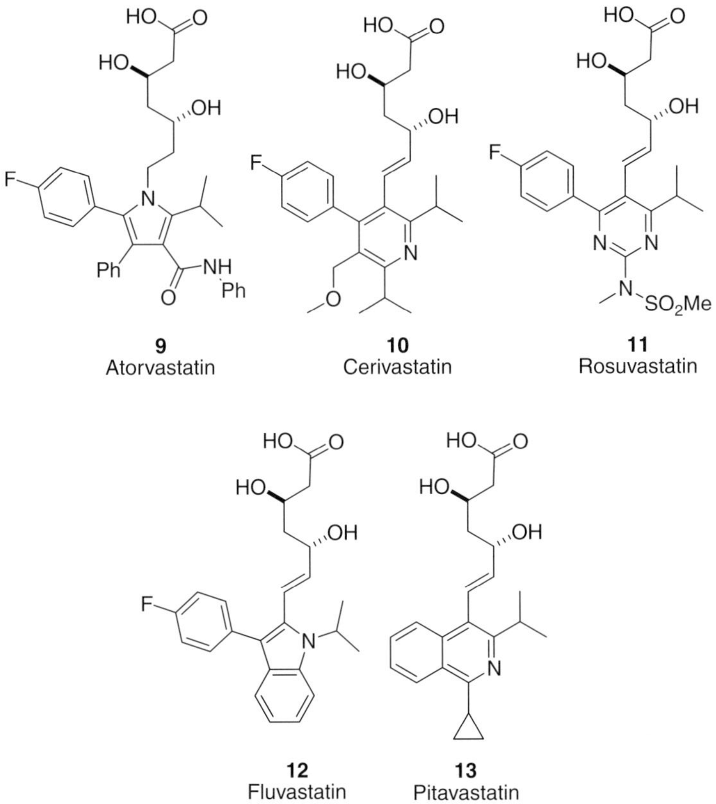

Figure 12.2 Synthetic statins.

of a large number of synthetic statins, which included atorvastatin (**9**, Lipitor®, Pfizer), cerivastatin (**10**), rosuvastatin (**11**), fluvastatin (**12**), and pitavastatin (**13**) (Figure 12.2). Atorvastatin (**9**), being the fifth statin to be launched, was not considered favorably as a clinical candidate by Parke-Davis. The project champion, Dr. Roger Newton, got on one knee in a meeting to beg the management to sponsor a phase I study on atorvastatin (**9**) [21]. Cerivastatin (**10**) was withdrawn form the market in 2001 due to rhabdomyolysis (the breakdown of muscle fibers). Even for validated targets, it is not trivial to locate the narrow safety window with a proper compound. This is also true for other classes of drugs such as proton pump inhibitors and H_2 receptor antagonists where early candidates were found to be unsafe.

The success of statins and several other classes of drugs helped to create a booming era for the pharmaceutical industry, and in the end, the patients benefited the most. By 2002, the death rate due to heart disease dropped by 52% [22]. Medicines such as statins played a critical role in this achievement [23]. At present, the pharmaceutical industry is under pressure again and the sustainability of the current feast-or-famine R&D model is being questioned [24, 25]. I hope the difficult but successful journey of statins will serve as a reminder for the improvements of quality of life and life expectancy that drugs can provide.

12.3 Synthesis

Mevastatin (**1**) and lovastatin (**3**) are natural products, whereas pravastatin (**5**) and simvastatin (**4**) are semisynthetic analogs. Although natural statins are produced by fermentation on a commercial scale, their total syntheses can be achieved in the laboratory. Scheme 12.3 shows the final steps of the total synthesis of mevastatin (**1**) to illustrate that the open hydroxyl acid form can be readily converted into the lactone form [26]. Thus, the *tert*-butyldimethylsilyl (TBS) and acetylide groups in **14** were removed by HF and the resulting hydroxy ester cyclized under the reaction conditions to give lactone **15**. The C4 hydroxy group was selectively eliminated via a three-step sequence to provide mevastatin (**1**).

Atorvastatin (**9**), cerivastatin (**10**), rosuvastatin (**11**), fluvastatin (**12**), and pitavastatin (**13**) are open acid forms and are produced by chemical synthesis. The convergent synthesis of atorvastatin (**9**) is described here as a representative example.

One efficient way of synthesizing the intermediate **21** (Scheme 12.4) begins with isoascorbic acid (**16**). Oxidation of **16** with hydrogen peroxide under basic conditions followed by neutralization of the reaction mixture via ion exchange gave the trihydroxyacid salt (**17**), with a 92% yield. One report claimed that the process was "so satisfactory that we recommend it as a laboratory preparation for the instruction of students" [27]. Selective bromination of **17** afforded the dibromo hydroxyester **18**, which was selectively monodebrominated to produce the bromohydrin (**19**) with 60% overall yield [28]. TBS protection of the alcohol in **19** gave the silyl ether (**20**), which was converted to the versatile intermediate **21** by displacing the bromo substituent with a cyano group.

Intermediate **21** can be converted to the amino building block **24** using different reaction sequences [29]. One sequence is shown in Scheme 12.5. The methyl ester in **21** was hydrolyzed and activated using carbonyldiimidazole (CDI). Reaction with

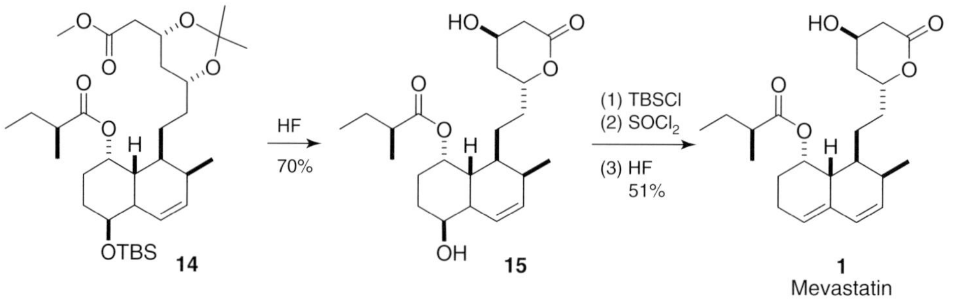

Scheme 12.3 Finishing steps in a total synthesis of mevastatin (**1**).

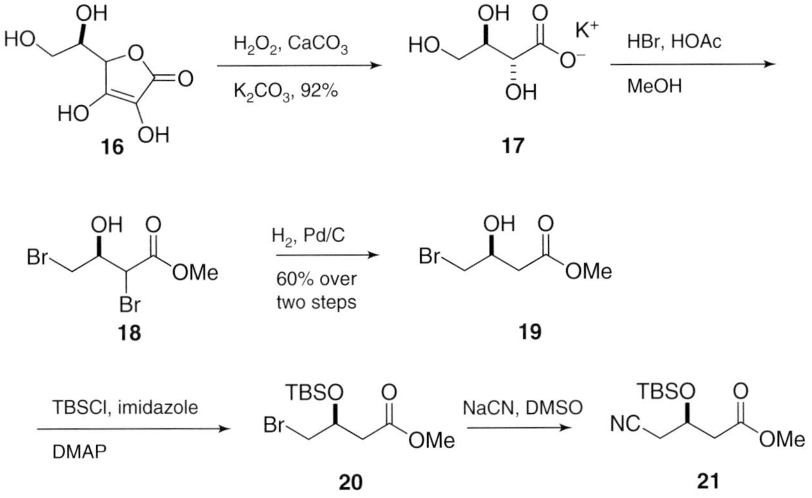

Scheme 12.4 Synthesis of the atorvastatin (**9**) intermediate **21**.

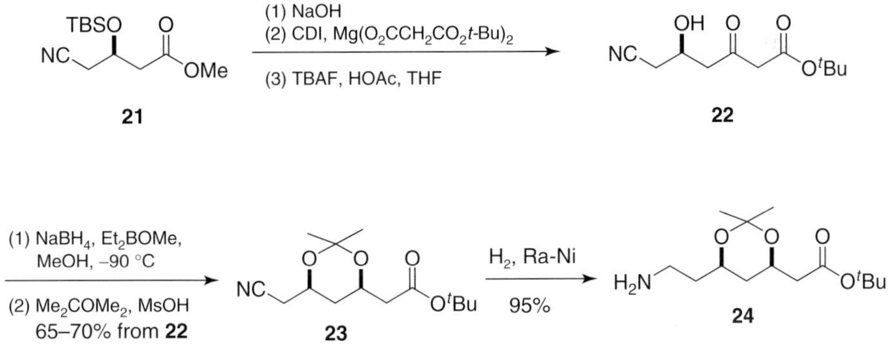

Scheme 12.5 Synthesis of the amino building block **24**.

the magnesium salt of *tert*-butyl malonate and subsequent cleavage of the TBS group using tetrabutylammonium fluoride (TBAF) produced the ketoester (**22**).

The keto functionality in **22** could be reduced stereoselectively using sodium borohydride in the presence of diethyl(methoxy)borane, and acetylide protection of the resulting diol gave compound **23** in good yields (65–70% from **22**). Raney-nickel-catalyzed hydrogenation of the cyano group in **23** provided **24**, with 95% yield.

The synthesis of the hydrophobic portion of atorvastatin (**9**) started with the ketoamide (**25**) (Scheme 12.6). Knoevenagel-type condensation between **25** and benzaldehyde provided the enoate (**26**). The following Stetter-type addition of *para*-fluorobenzaldehyde, catalyzed by the thiazolium salt (**27**), produced the diketone (**28**).

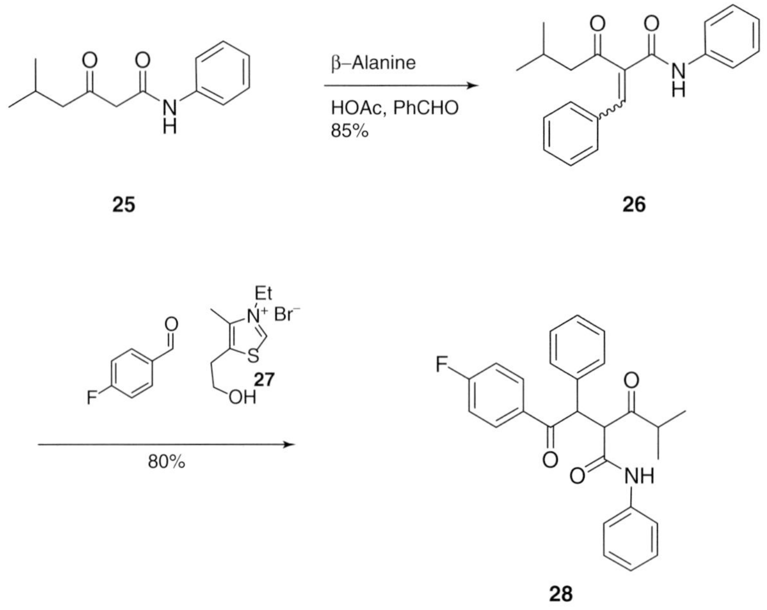

Scheme 12.6 Synthesis of the diketone building block **28**.

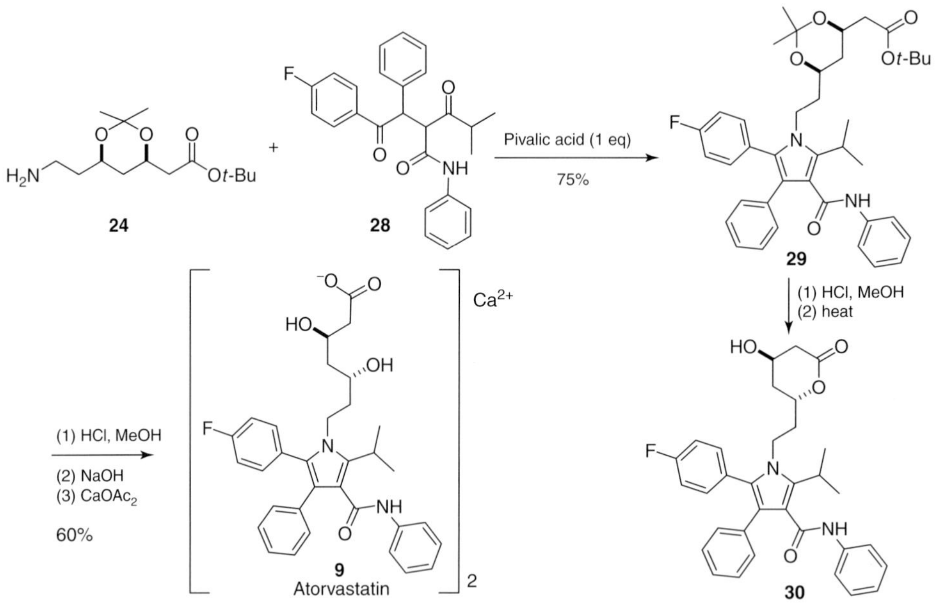

Scheme 12.7 Synthesis of the atorvastatin (**9**).

To complete the synthesis, the amine (**24**) and the diketone (**28**) were heated in the presence of 1 equivalent of pivalic acid in a mixture of toluene-heptane-tetrahydrofuran (THF) (1 : 4 : 1) to give the Paal–Knorr product **29**. Removal of the acetylide and hydrolysis of the *tert*-butyl ester followed by ion exchange with calcium acetate produced the calcium salt of atorvastatin (**9**), with 60% yield from **29** with 99.5% enantiomeric purity (Scheme 12.7) [30]. The lactone form (**30**) can also readily be obtained from **29** via a two-step sequence [31].

12.4 Mode of Action

Statins inhibit HMGCoA reductase, the enzyme that catalyzes the rate-limiting step in the cholesterol biosynthesis, that is, the reduction of HMGCoA (**31**) to mevalonate (**32**) (Scheme 12.8). It was found in the 1960s that dietary cholesterol can suppress the hepatic synthesis of mevalonate (**32**) but not β-3-hydroxy-β-methylglutarate, which indicated that HMGCoA reductase is the main feedback point in the regulation of cholesterol [32]. Also, the activity of HMGCoA reductase was shown to be regulated under several other conditions and was related to cholesterol synthesis [33]. Reduction of the mevalonate levels leads to slower cholesterol biosynthesis, but inhibiting HMGCoA reductase also results in the upregulation of HMGCoA reductase [34] and other cholesterol synthesis enzymes [35, 36], which compromises the effects of statins. Fortunately, inhibiting HMG-CoA reductase also leads to the upregulation of LDL receptors, which enhances the plasma cholesterol clearance [37, 38]. In patients with homozygous familial hypercholesterolemia with impaired LDL receptor synthesis, atorvastatin (**9**) was less effective in lowering LDL levels than the heterozygous counterparts [39]. In summary, decreased cholesterol synthesis and upregulation of LDL receptor through HMGCoA reductase inhibition appear to be the key modes of action of statins, whereas upregulation of HMGCoA reductase as a result of its inhibition is insufficient to override the positive effects of statins in most species including humans.

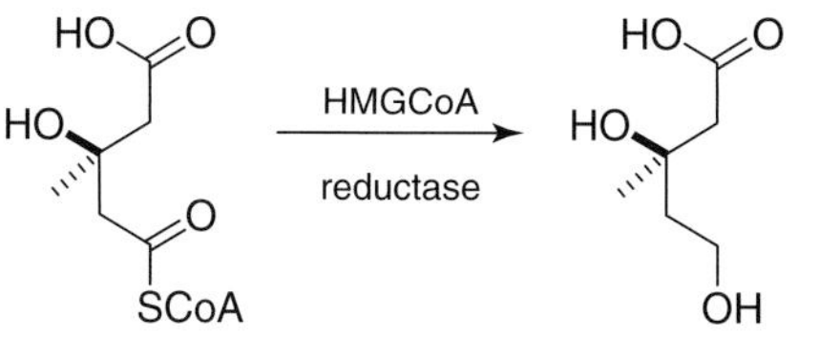

31
HMGCoA

32
Mevalonate

Scheme 12.8 HMGCoA reductase catalyzes the conversion of HMGCoA (**31**) to mevalonate (**32**).

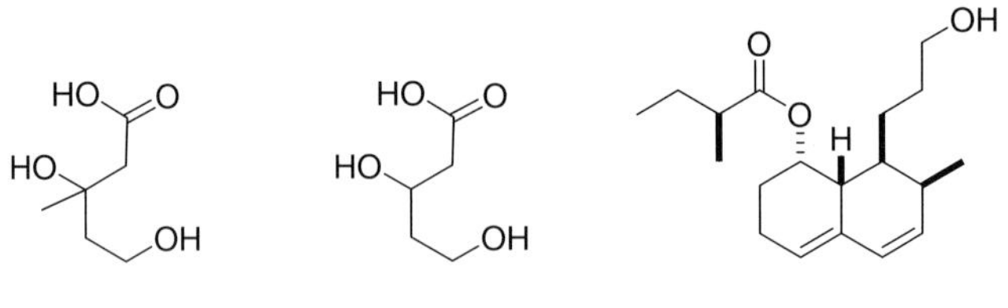

33 34 35

Figure 12.3 Fragments of statins.

12.5 Structure–Activity Relationship

12.5.1 The Hydrophilic Dihydroxypentanoic Acid Portion

The initial SAR study revealed that the dihydroxypentanoic acid portion is the key binding motif. The open acid form is more active than the lactone form under enzymatic assay conditions. The inhibition constants of *DL*-mevalonate (**33**) (Figure 12.3) and *DL*-3,5-dihydroxyvalerate (**34**) were found to be 11 and 33 mm, respectively, while compound **35** was totally inactive [40]. The K_m value for HMGCoA (**31**) is ~10 μM. The relative and absolute stereochemistry at the 3- and 5-hydroxy stereogenic centers in statins are important for the activity with the (3*R*,5*S*)-isomer being the most active [41]. Conversion of the hydroxyl groups in the "dihydroxypentanoic acid" portion of mevastatin (**1**) to a methyl ester abolished its activity as does the conversion of the acid group to an amide [40]. Oxidation of the 3-OH to a ketone group also greatly reduced potency [41]. The "dihydroxypentanoic acid" portion thus needs to be maintained, while the hydrophobic portion can be modified in a variety of different ways without losing potency or other key properties. The distance between the two regions is important for activity, with a two-atom linker being optimal [41].

12.5.2 The SAR of Natural Statins

A number of related natural statins have been isolated from fungal cultures, and some SAR can be derived from these compounds. For example, removal of the acyl group at the 1-position of the decalin moiety in mevastatin (**36**) (Figure 12.4) resulted in a 25-fold loss of HMGCoA reductase inhibitory activity, while having no substituent at all in this position (**37**) only led to a fivefold decrease in potency. Saturating the 4a,4-olefin (**38**) had no effect on the overall activity [10]. Introduction of a hydrophobic methyl group at the 3-position slightly increased the potency by twofold (lovastatin (**3**) Figure 12.1), whereas the additional introduction of a polar keto group into the ester side chain (**39**) (Figure 12.4) decreased the potency by fivefold.

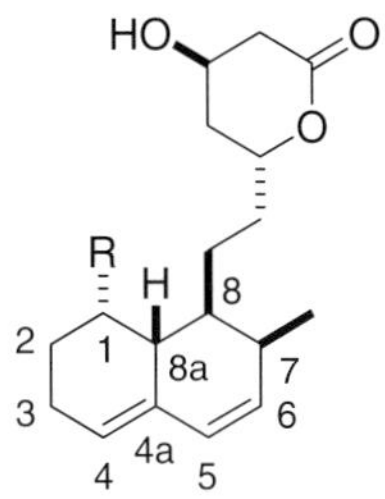

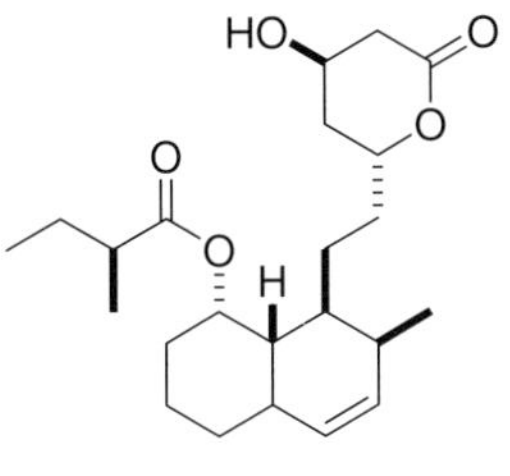

38

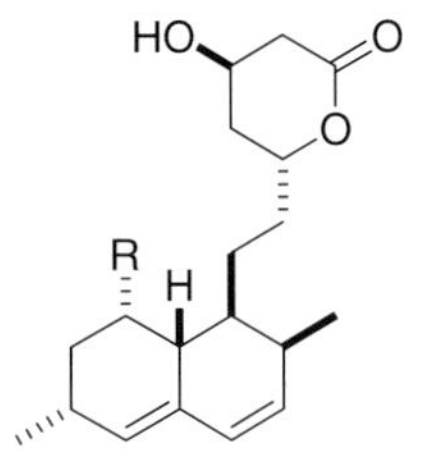

1 R = (*S*)-EtCH(Me)CO_2

36 R = OH

37 R = H

39 R= (*S*)-$CH_3C(O)CH(Me)CO_2$

40 R = $MeCO_2$

41 R = (*R*)-EtCH(Me)CO_2

42 R = 4-F-$C_6H_4CO_2$

43 R = CH_2=CHCH(Me)O

44 R = 4-Cl-$C_6H_4CH_2O$

45 R = 4-F-$C_6H_4CH_2O$

Figure 12.4 Analogs of mevastatin (**1**).

Consistent with these observations, another SAR study of the acyl side chain on the decalin ring revealed that hydrophobic esters are desirable for potency [5]. So the acetate (**40**) is 100-fold less active than lovastatin (**3**). When the chain was systematically elongated, the potency was gradually recovered, but beyond a 4-carbon length, the potency plateaued. The stereochemistry at the α-carbon of the ester group in lovastatin (**3**) was found unimportant. Epimer **41** was equipotent to lovastatin (**3**). Further addition of a methyl group at this position improved the potency by 2.5-fold (simvastatin (**4**)) (Figure 12.1) compared to lovastatin (**3**). Aromatic esters were found less active. For example, ester **42** is 17-fold less active than lovastatin (**3**).

Aliphatic ether analogs at the 1-position of the decalin ring are generally less active than their ester counterparts. For example, allyl ether (**43**) and 4-chlorobenzyl ether (**44**) were 60- and 80-fold less active than lovastatin (**3**), respectively. However, the 4-fluorobenzyl ether (**45**) was about equipotent to mevastatin (**1**), which could be due to a fluorine hydrogen bonding effect (see Section 12.5.5 for details) [42].

Simvastatin (**4**) was derivatized chemically to produce several analogs (Figure 12.5) [43]. Limited by synthetic accessibility, the SAR study was not systematic. The 6′*S* and 6′*R* oxidized analogs **46** and **47** with rearranged double bonds are 33- and 13-fold less active than simvastatin (**4**), respectively, whereas the analogs **48** and **49** are inactive.

The 4-oxo (**50**), 4-keto (**51**), and 4-hydroxy (**52**) analogs of simvastatin (**4**) (Figure 12.6) are all about equipotent to mevastatin (**1**). Further addition of hydrophobic groups to **52** moderately increased the potency. For example, the carbamate **53** and the ether **54** (Figure 12.6) are two- and threefold more potent, respectively [44]. From this SAR study, the ether **54** was the most efficacious compound in *in vivo* models.

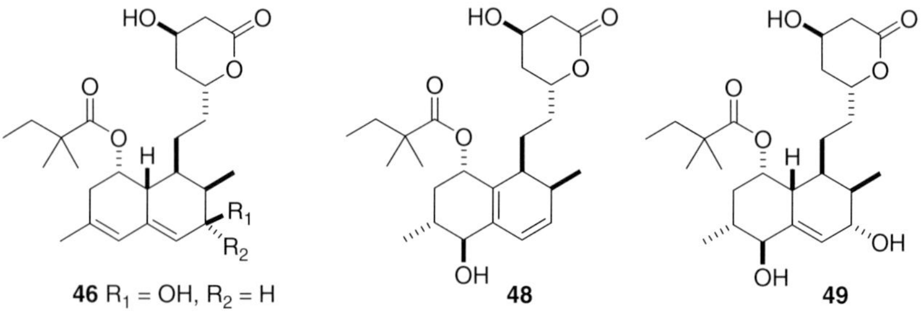

Figure 12.5 Modifications of the decalin ring of simvastatin (**4**).

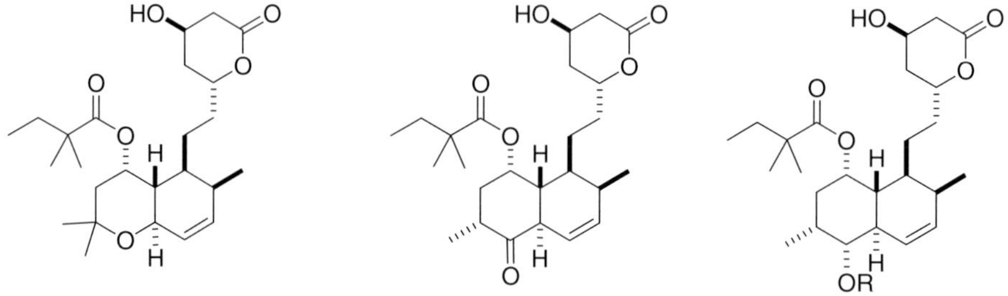

Figure 12.6 Modifications at the 3- and 4-positions of the decalin ring of simvastatin (**4**).

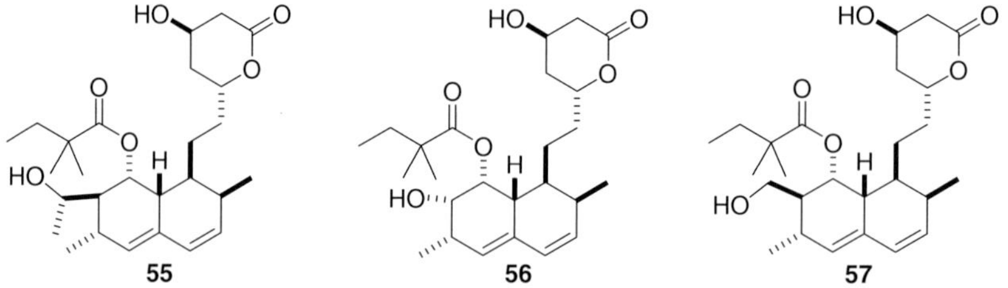

Figure 12.7 Modifications at the 2-position of simvastatin (**4**).

The 2-position of simvastatin (**4**) is fairly SAR neutral. Polarity, size, or stereochemistry of substituents at this position had had no impact on activity based on a limited number of produced analogs. So, for example, compounds **55–57** in Figure 12.7 are practically equipotent to the parent simvastatin (**4**) [45].

Overall, hydrophobic groups on the tetralin ring are favorable for potency, which is consistent with the hydrophobic nature of the enzyme pocket that this portion of the molecule binds to. Limited by synthetic accessibility of relevant analogs, the published SAR studies were not always systematic and only inhibitors with modestly improved activities compared to the parent mevastatin (**1**) were identified.

12.5.3
SAR of Synthetic Statins

The narrow SAR of natural statin analogs prompted a search for new scaffolds. A variety of simpler aliphatic and aromatic carbocyclic systems have been investigated to replace the complex decalin ring system in the natural statins. Not surprisingly, the earliest attempt was made by the scientists from Sankyo Co., where mevastatin (**1**) was first isolated [46]. Probably due to the misassignment of the relative stereochemistry in the lactone portion of mevastatin (**1**) by the initial X-ray crystallography study [14], a less-active diastereomer was used in their effort. Nonetheless, a simple benzene replacement for the decalin moiety in mevastatin (**1**) was identified. For example, compound **58** (Figure 12.8) showed modest activity in HMGCoA reductase inhibition with an IC_{50} value (reported as ID_{50}) of 30 μM. Soon after, scientists at Merck reported several carbocyclic replacements for the decalin system in mevastatin (**1**) with the correct stereochemistry in the lactone [41]. The more hydrophobic dichlorobenzene analog **59** showed an IC_{50} value of 15 μM, compared to an IC_{50} of 10 nM for mevastatin (**1**) in the same assay. Interestingly, around the same time, a similar compound, **60**, was reported to be equipotent to mevastatin (**1**) with a K_i value of ~0.25 nM for the inhibition of yeast HMGCoA reductase [40].

A major breakthrough was the utilization of a 4-fluorophenyl group at the 6-position of the benzene ring in **59** to fill a hydrophobic pocket in the enzyme. In 1986, scientists from Merck reported the discovery of compound **61** (IC_{50} = 5 nM), which was far more potent than **59** [47]. It was later revealed that the fluoro group in **61** forms a hydrogen bond with Arg^{590} in HMGCoA reductase and provides a critical interaction with the enzyme [48]. This compound became the prototype synthetic statin. The design of the so-called superstatins (atorvastatin (**9**), cerivastatin (**10**), rosuvastatin (**11**), and fluvastatin (**12**)) took advantage of this pharmacophore, and these compounds proved to be more effective in reducing LDL cholesterol. The

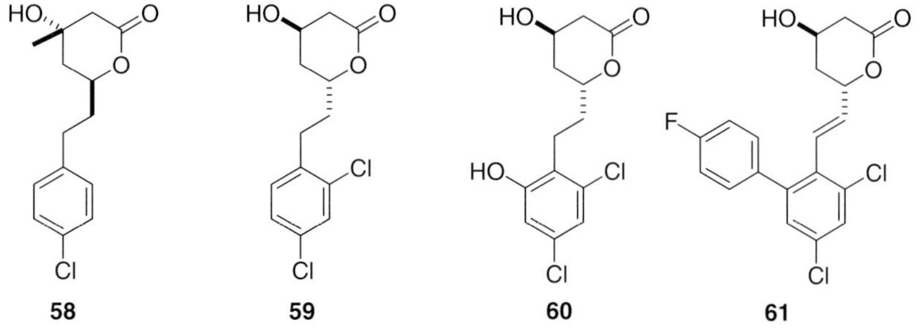

Figure 12.8 Early attempts to find synthetic statins.

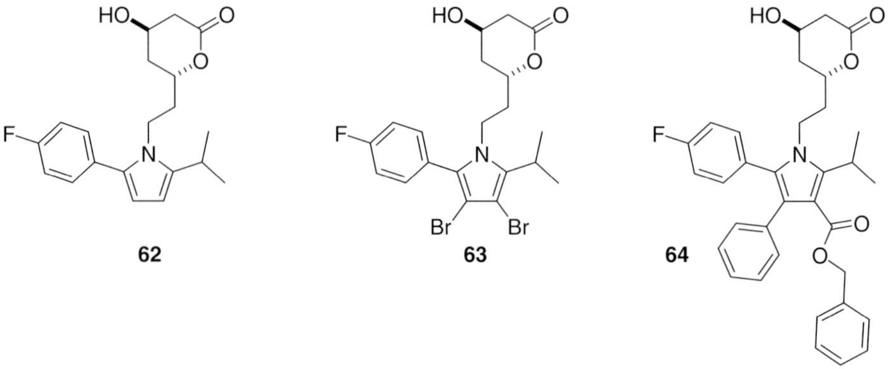

Figure 12.9 Atorvastatin (**9**) precursors.

ethenyl linker in **61** provided ~10-fold boost of potency compared to the saturated ethyl linker.

Compound **61** provided a road map for the later synthetic statins. It captured the essential hydrophobic interactions between HMGCoA reductase and the statins and a key polar interaction between the fluoro atom and Arg^{590}. This scaffold is much easier to chemically modify, and a large number of systematic analogs were made in the following decade. Thanks to the promiscuity of the hydrophobic pocket in HMGCoA reductase, a variety of heterocyclic rings can be used to replace the central phenyl ring without losing potency, and several second-generation statins were successfully discovered and marketed. For example, cerivastatin (**10**) (Figure 12.2) contains a pyridine ring, whereas rosuvastatin (**11**) (Figure 12.2) [49] and fluvastatin (**12**) (Figure 12.2) [50] have a pyrimidine and an indole ring, respectively, at the same position [51].

In the late 1980s, scientists from Warner-Lambert Company designed a series of pyrrole-based HMGCoA reductase inhibitors, such as **62** (Figure 12.9) [52]. This modestly looking molecule with moderate potency ($IC_{50} = 0.4\ \mu M$) served as the starting point for further lead optimization efforts. A dibromo analog, **63**, was soon found to be ~10-fold more potent, but this compound displayed significant toxicity in rodents. This was likely due to the high plasma drug levels resulting from the high lipophilicity of **63** compared to other inhibitors [53]. Introduction of a more polar ester group at the 4-position of the pyrrole ring (**64**) retained potency. Changing the ester to a phenylamide further increased polarity and boosted the potency into the low nanomolar range [31]. This compound would later be named atorvastatin (**9**) (Lipitor) and become the most successful drug in history [54].

The only synthetic statin that does not contain a 4-fluorobenzene fragment is pitavastatin (**13**) (Figure 12.2), which contains a quinoline ring [55] but was claimed to be more effective than simvastatin (**4**), pravastatin (**5**), and atorvastatin (**9**) with a longer duration of action [56]. The naphthalene ring proved less effective in building potency [57]. The parent naphthalene analog **65** (Figure 12.10) was 2300-fold less active than mevastatin (**1**), but adding small hydrophobic substituents to the ring

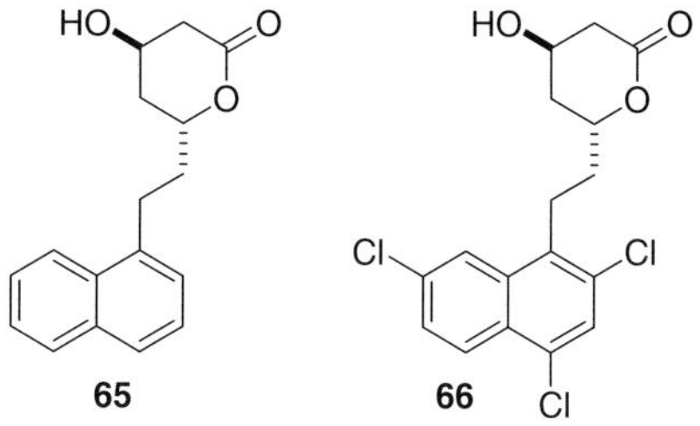

65

66

Figure 12.10 Naphthalene scaffolds.

gradually enhanced potency. A wide range of SAR was observed in this series, resulting in compound **66** which is only twofold less active than mevastatin (**1**).

12.5.4 Liver-Selective HMGCoA Inhibitors

The liver is the major organ that generates and uses cholesterol, and so liver-selective statins have the potential to be less toxic if their systemic exposure can be minimized. It was found that more hydrophilic compounds tend to have higher liver HMGCoA reductase inhibitory activity, whereas more hydrophobic compounds show more activity in peripheral tissues. Specific structural features appeared to be less important than the $c\log P$ value [58]. It was hypothesized that more lipophilic ligands reach the targeted tissue by passive diffusion, and thus, lower selectivity was observed among different tissues. More hydrophilic ligands, on the other hand, rely on active hepatic transporters to penetrate cell membranes and higher exposure in the liver can be achieved [59]. Using this strategy to improve tissue selectivity, however, is limited by the already low permeabilities and oral bioavailabilities of the statins. A further increase of polarity often results in even less permeable ligands with lower oral bioavailabilities.

12.5.5 X-Ray Crystal Structures

The X-ray crystal structures of HMGCoA reductase [60] and its complexes with statins have been determined. For example, the structure of atorvastatin (**9**) in complex with HMGCoA reductase is shown in Figure 12.11 [48]. The dihydroxypentanoic acid portion forms several polar interactions with Ser^{684}, Asp^{690}, Lys^{691}, Lys^{692}, Glu^{559}, and Asp^{767}. The terminal carboxylate forms a salt bridge with Lys^{735}. A hydrophobic pocket formed by Leu^{562}, Val^{683}, Leu^{853}, Ala^{856}, and Leu^{857} provides key interactions between the statins and HMGCoA reductase. The pocket is flexible to allow statins to adopt different conformations to maximize contacts. In addition to these van der Waals forces, several polar residues in HMGCoA reductase such as Arg^{568}, Ser^{565}, and Arg^{590} can form dipole–dipole or hydrogen bonding interactions with the polar groups on some statins. Of note, a polar interaction between Arg^{590} and the fluorine atom on synthetic statins was observed. The sulfone group in

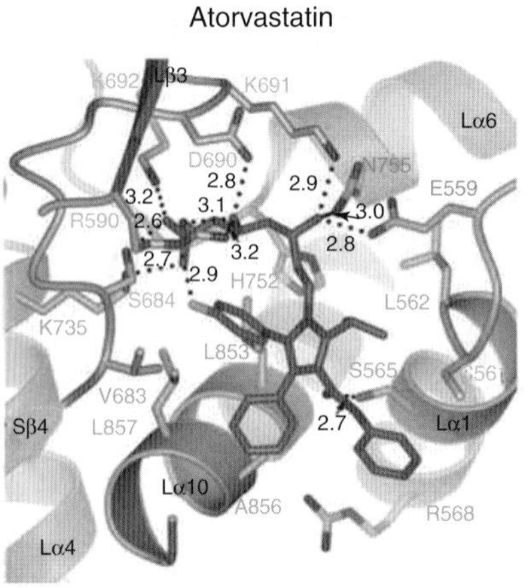

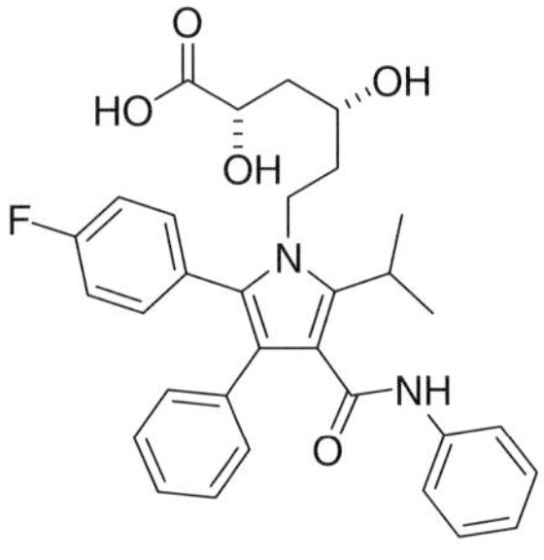

9

Figure 12.11 Binding mode of atorvastatin (**9**) to human HMGCoA reductase. (Source: Permitted by the *Science* magazine.)

rosuvastatin (**11**), the statin that has the most interactions with the enzyme, forms a unique hydrogen bonding interaction with Arg^{568}. The abundance in stabilizing residues and the nonspecific nature of hydrophobic interactions explain why a diverse set of statins can tightly bind to the enzyme. These structural features are consistent with the SAR observed.

References

1. Endo, A., Kuroda, M., and Tsujita, Y. (1976) *J. Antibiot.*, **29**, 1346–1348.
2. Vickers, S., Duncan, C.A., Chen, I.W., Rosegay, A., and Duggan, D.E. (1990) *Drug Metab. Dispos.*, **18**, 138–145.
3. Brazeau, P., Vale, W., Burgus, R., Ling, N., Butcher, M., Rivier, J., and Guillemin, R. (1973) *Science*, **179**, 77–79.
4. Alberts, A.W., Chen, J., Kuron, G., Hunt, V., Huff, J., Hoffman, C., Rothrock, J., Lopez, M., Joshua, H., Harris, E., Patchett, A., Monaghan, R., Currie, S., Stapley, E., Albers-Schonberg, G., Hensens, O., Hirshfield, J., Hoogsteen, K., Liesch, J., and Springer, J. (1980) *Proc. Natl. Acad. Sci. U.S.A.*, **77**, 3957–3961.
5. Hoffman, W.F., Alberts, A.W., Anderson, P.S., Chen, J.S., Smith, R.L., and Willard, A.K. (1986) *J. Med. Chem.*, **29**, 849–852.
6. Tsujita, Y., Kuroda, M., Shimada, Y., Tanzawa, K., Arai, M., Kaneko, I., Tanaka, M., Masuda, H., Tarumi, C., Watanabe, Y., and Fujii, S. (1986) *Biochim. Biophys. Acta*, **877**, 50–60.
7. Tschesche, T., Hoppe, H.J., Snatzke, G., Wulff, G., and Fehlhaber, H.W. (1971) *Chem. Ber.*, **104**, 1420–1428.
8. Pyysalo, H. and Kuusi, T. (1971) *Suom. Kemistilehti B*, **44**, 393–396.
9. Endo, A. (1992) *J. Lipid Res.*, **33**, 1569–1582.
10. Endo, A. (1985) *J. Med. Chem.*, **28**, 401–405.
11. Klimov, A.N. and Nagornev, V.A. (2002) *Pediatr. Pathol. Mol. Med.*, **21**, 307–320.
12. Dawber, T.R., Moore, F.E., and Mann, G.V. (1957) *Am. J. Public Health*, **47**, 4–24.
13. Keys, A., Menotti, A., Karvonen, M.J., Aravanis, C., Blackburn, H., Buzina, R., Djordjevic, B.S., Dontas, A.S., Fidanza, F., Keys, M.H., Kromhout, D.,

Nedeljkovic, S., Punsar, S., Seccareccia, F., and Toshima, H. (1986) *Am. J. Epidemiol.*, **124**, 903–915.
14. Brown, A.G., Smale, T.C., King, T.J., Hansenkamp, R., and Thompson, R.H. (1976) *J. Chem. Soc., Perkin Trans. 1*, **11**, 1165–1170.
15. Endo, A. (2008) *Nat. Med.*, **14**, 1050–1052.
16. Stossel, T.P. (2008) *Cell*, **134**, 903–904.
17. Moore, T.J. (1989) *Atl. Mon.*, **264**, 37.
18. Scandinavian Simvastatin Survival Study Group (1994) *Lancet*, **344**, 1383–1389.
19. Katsiki, N., Athyros, V.G., and Mikhailidis, D.P. (2011) *Expert Opin. Investig. Drugs*, **20**, 311–314.
20. Stokker, G.E., Alberts, A.W., Anderson, P.S., Cragoe, E.J., Deana, A.A., Gilfillan, J.L., Hirshfield, J., Holtz, W.J., and Hoffman, W.F. Jr. (1986) *J. Med. Chem.*, **29**, 170–181.
21. LaMattina, J.L. (2009) *Drug Truths: Dispelling the Myths about Pharma R&D*, John Wiley & Sons, Inc.
22. Jemal, A., Ward, E., Hao, Y., and Thun, M. (2005) *J. Am. Med. Assoc.*, **294**, 1255–1259.
23. Yusuf, S. (2002) *Lancet*, **360**, 2–3.
24. Cuatrecasas, P. (2006) *J. Clin. Invest.*, **116**, 2837–2842.
25. Cutler, D.M. (2007) *N. Engl. J. Med.*, **356**, 1292–1293.
26. Hirama, M. and Uei, M. (1982) *J. Am. Chem. Soc.*, **104**, 4251–4253.
27. Isbell, H.S. and Frush, H.L. (1979) *Carbohydr. Res.*, **72**, 301–304.
28. Sletzinger, M., Verhoeven, T.R., Volante, R.P., McNamara, J.M., Corley, E.G., and Liu, T.M.H. (1985) *Tetrahedron Lett.*, **26**, 2951–2954.
29. Brower, P.L., Butler, D.E., Deering, C.F., Le, T.V., Millar, A., Nanninga, T.N., and Roth, B.D. (1992) *Tetrahedron Lett.*, **33**, 2279–2282.
30. Baumann, K.L., Butler, D.E., Deering, C.F., Mennen, K.E., Millar, A., Nanninga, T.N., Palmer, C.W., and Roth, B.D. (1992) *Tetrahedron Lett.*, **33**, 2283–2284.
31. Roth, B.D., Blankley, C.J., Chucholowski, A.W., Ferguson, E., Hoefle, M.L., Ortwine, D.F., Newton, R.S., Sekerke, C.S., Sliskovic, D.R., and Wilson, M. (1991) *J. Med. Chem.*, **34**, 357–366.
32. Siperstein, M.D. and Fagan, V.M. (1966) *J. Biol. Chem.*, **241**, 602–609.
33. Siperstein, M.D. (1970) *Curr. Top. Cell. Regul.*, **2**, 65–100.
34. Brown, M.S. and Goldstein, J.L. (1980) *J. Lipid Res.*, **21**, 505–517.
35. Bergstrom, J.D., Wong, G.A., Edwards, P.A., and Edmond, J. (1984) *J. Biol. Chem.*, **259**, 14548–14553.
36. Balasubramaniam, S., Goldstein, J.L., and Brown, M.S. (1977) *Proc. Natl. Acad. Sci. U.S.A.*, **74**, 1421–1425.
37. Kovanen, P.T., Bilheimer, D.W., Goldstein, J.L., Jaramillo, J.J., and Brown, M.S. (1981) *Proc. Natl. Acad. Sci. U.S.A.*, **78**, 1194–1198.
38. Ma, P.T., Ma, P.T., Gil, G., Südhof, T.C., Bilheimer, D.W., Goldstein, J.L., and Brown, M.S. (1986) *Proc. Natl. Acad. Sci. U.S.A.*, **83**, 8370–8374.
39. Marais, A.D., Naoumova, R.P., Firth, J.C., Penny, C., Neuwirth, C.K., and Thompson, G.R. (1997) *J. Lipid Res.*, **38**, 2071–2078.
40. Nakamura, C.E. and Abeles, R.H. (1985) *Biochemistry*, **24**, 1364–1376.
41. Stokker, G.E., Hoffman, W.F., Alberts, A.W., Cragoe, E.J., Deana, A.A., Gilfillan, J.L., Huff, J.W., Novello, F.C., Prugh, J.D., Smith, R.L., and Willard, A.K. (1985) *J. Med. Chem.*, **28**, 347–358.
42. Lee, T.J., Holtz, W.J., Smith, R.L., Alberts, A.W., and Gilfillan, J.L. (1991) *J. Med. Chem.*, **34**, 2474–2477.
43. Stokker, G.E. (1994) *Bioorg. Med. Chem. Lett.*, **4**, 1767–1770.
44. Duggan, M.E., Alberts, A.W., Bostedor, R., Chao, Y.S., Germershausen, J.I., Gilfillan, J.I., Halczenko, W., Hartman, G.D., Hunt, V., Imagire, J.S., Schwartz, M.S., Smith, R.L., and Stubbs, R.J. (1991) *J. Med. Chem.*, **34**, 2489–2495.
45. Hartman, G.D., Halczenko, W., Duggan, M.E., Imagire, J.S., Smith, R.L., Pitzenberger, S.M., Fitzpatrick, S.L., Alberts, A.W., Bostedor, R., Chao, Y.S., Germershausen, J.I., Gilfillan, J.L., and Hunt, V. (1992) *J. Med. Chem.*, **35**, 3813–3821.
46. Sato, A., Ogiso, A., Noguchi, H., Mitsui, S., Kaneko, I., and Shimada, Y. (1980) *Chem. Pharm. Bull.*, **28**, 1509–1525.

47. Stokker, G.E., Alberts, A.W., Anderson, P.S., Cragoe, E.J., Deana, A.A., Gilfillan, J.L., Hirshfield, J., Holtz, W.J., Hoffman, W.F., Huff, J.W., Lee, T.J., Novello, F.C., Prugh, J.D., Rooney, C.S., Smith, R.L., and Willard, A.K. (1986) *J. Med. Chem.*, **29**, 170–181.
48. Istvan, E.S. and Deisenhofer, J. (2001) *Science*, **292**, 1160–1164.
49. Hirai, K., Ishiba, T., Koike, H., and Watanabe, M. (1993) Eur. Pat. Appl., EP 521,471.
50. Kathawala, F.G. (1984) PCT Int. Appl., WO 8, 402,131.
51. Angerbauer, R., Fey, P., Huebsch, W., Philipps, T., Bischoff, H., Petzinna, D., and Schmidt, D. (1992) Ger. Offen., DE 4, 040,026.
52. Roth, B.D., Ortwine, D.F., Hoefle, M.L., Stratton, C.D., Sliskovic, D.R., Wilson, M.W., and Newton, R.S. (1990) *J. Med. Chem.*, **33**, 21–31.
53. Sigler, R.E., Dominick, M.A., and Mcguire, E.J. (1992) *Toxicol. Pathol.*, **20**, 595–602.
54. Roth, B.D. (2002) *Prog. Med. Chem.*, **40**, 1–22.
55. Mutsukado, M. and Suzuki, M. (2000) *Joho Kagaku Toronkai, Kozo Kassei Sokan Shinpojiumu Koen Yoshishu*, **23-28**, 192–193.
56. Mukhtar, R.Y., Reid, J., and Reckless, J.P. (2005) *Int. J. Clin. Pract.*, **59**, 239–252.
57. Prugh, J.D., Alberts, A.W., Deana, A.A., Gilfillian, J.L., Huff, J.W., Smith, R.L., and Wiggins, J.M. (1990) *J. Med. Chem.*, **33**, 758–765.
58. Roth, B.D., Bocan, T.M., Blankey, C.J., Chucholowski, A.W., Creger, P.L., Creswell, M.W., Ferguson, E., Newton, R.S., O'Brien, P., Picard, J.A., Roark, W.H., Sekerke, C.S., Sliskovic, D.R., and Wilson, M.W. (1991) *J. Med. Chem.*, **34**, 463–466.
59. Hamelin, B.A. and Turgeon, J. (1998) *Trends Pharmacol. Sci.*, **19**, 26–37.
60. Istvan, E.S., Palnitkar, M., Buchanan, S.K., and Deisenhofer, J. (2000) *EMBO J.*, **19**, 819–830.

Part III
Infectious Diseases

Bioactive Heterocyclic Compound Classes: Pharmaceuticals,
First Edition. Edited by Jürgen Dinges and Clemens Lamberth.

13
Adenine-Based Reverse Transcriptase Inhibitors as Anti-HIV Agents

Alastair Donald

13.1
Introduction

Adenine (**1**) is the common name given to 6-amino-imidazo[4,5-d]pyrimidine (Figure 13.1). It is the aglycone of adenosine (**2**), one of the five nucleotides that form the backbone of RNA and DNA, and the congeners of the adenine ring system are ubiquitously utilized as substrates and cofactors in countless biochemical processes across every kingdom of life on earth (e.g., cyclic AMP, ATP, NAD^+/NADH). It is therefore not surprising that chemists have been intensely interested in studying this important heterocycle from near the very beginnings of organic chemistry itself.

The adenine core must surely be one of the most "privileged" structures in all the arsenal of medicinal chemistry [1]: compounds based on an adenine or adenosine core have found use in a broad spectrum of therapeutic classes [2–7]. However, this chapter focuses on adenine-based antiretroviral compounds for the treatment of human immunodeficiency virus (HIV), exemplified by adefovir (**3**), tenofovir (**4**), and their phosphonate ester prodrugs (Figure 13.2) [8, 9]. While adefovir dipivoxil (**5**) was found to be effective for the treatment of herpes and hepatitis viruses, it failed to demonstrate clinical benefit in the treatment of HIV.

13.2
History

Adenine (**1**) was first isolated from the pancreas of the ox by Kossel [10] in 1885. It was not until 1898 that the seminal synthetic work of Fisher [11, 12] definitively established the structure of adenine and the three alternate nucleobases of DNA (Scheme 13.1).

Chlorination of uric acid (**7**) by phosphorus pentachloride yielded 2,6,8-trichloropurine (**8**), which was further elaborated into adenine (**1**) via introduction of an amino group at the most reactive 6-position (yielding **9**), followed by reduction of the remaining two chloro groups with tetraphenylphosphonium

Bioactive Heterocyclic Compound Classes: Pharmaceuticals,
First Edition. Edited by Jürgen Dinges and Clemens Lamberth.

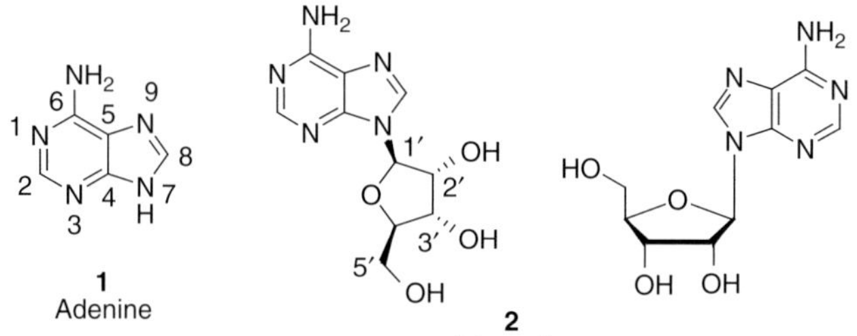

1 Adenine

2 Adenosine

Figure 13.1 Adenine (**1**) and adenosine (**2**) – structure, numbering, and drawing conventions.

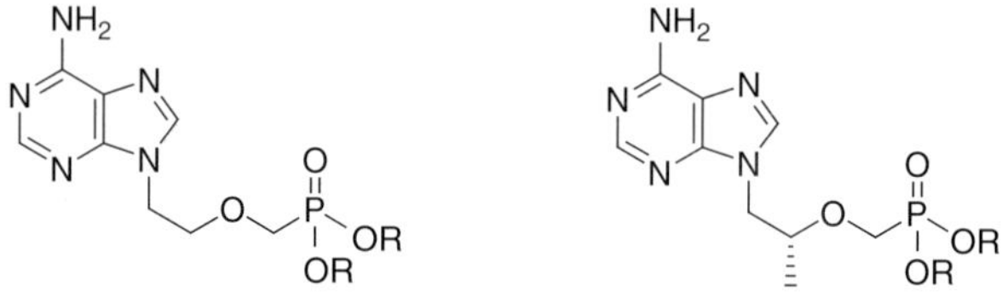

3 R = H : adefovir
5 R = CH_2OPiv: adefovir dipivoxil

4 R = H : tenofovir
6 R = $CH_2OC(O)O$*i*-Pr tenofovir disoproxil

Figure 13.2 Adenine-based antiretroviral drugs.

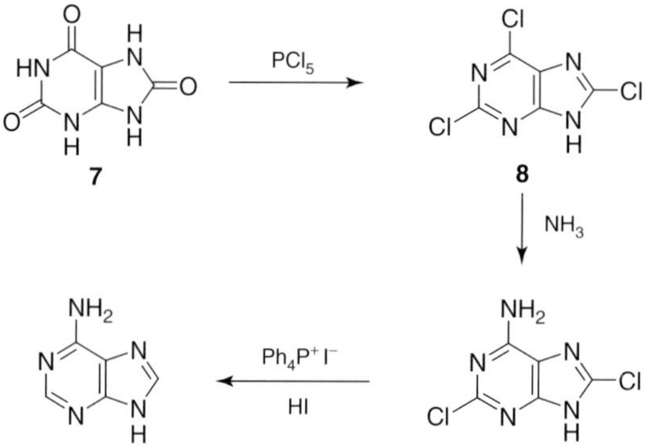

Scheme 13.1 Fischer's synthesis of adenine (**1**).

iodide in hydroiodic acid. Analogs of these nucleobases and nucleosides have been of special interest to medicinal chemists, a fact that remains as true today as it was in 1902, when Fischer noted in his Nobel prize speech [13]

> Such is the privilege of organic synthesis; where it is successful in penetrating such a field it far surpasses narrowly restricted Nature, a fact strikingly exemplified by the purine group.

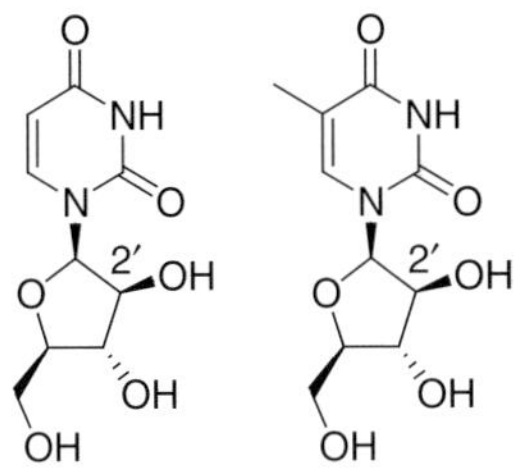

10 11

Figure 13.3 Spongouridine (**10**) and spongothymidine (**11**).

The first significant efforts in examining the medicinal properties of adenine-based compounds started in 1926 when I.G. Farbenindustrie AG investigated the use of animal organ extracts (principally beef heart extracts consisting mostly of AMP and adenosine (**2**)) in the treatment of heart disease. Although these efforts were ultimately unsuccessful, adenosine (**2**) was subsequently licensed for medical use (albeit over 50 years later, in 1989) for the termination of *paroxysmal supraventricular tachycardia* under the trade name Adenocard® (Astellas Pharma Inc.) [14]. In 1949, the first report of the antiviral effects of simple nucleoside analogs was disclosed by Thomson (Wellcome Research Laboratories) [15]. Given the discovery of such effects, attention soon turned to the discovery and development of superior analogs. In the period 1951–1955, Bergmann [16, 17] identified the nucleoside analogs spongouridine (3-β-D-arabofuranosyluracil, **10**) and spongothymidine (3-β-D-arabofuranosylthymine, **11**) as components of the acetone extracts of the Bahamian sponge *Cryptotethia crypta* (Figure 13.3).

These natural products differ from the naturally occurring nucleosides only by the stereochemistry at the 2′ position of the arabinose. Further investigation and exploration into analogs of Bergmann's unprecedented nucleosides revealed significant anticancer and antiviral activities [18].

In 1962, two research teams (Upjohn, Underwood) independently demonstrated that in addition to its anticancer effects, cytosine arabinoside (Ara-C, cytarabine, **12**) possessed antiviral effects against vaccinia and herpes simplex (both DNA viruses) [19, 20]. In 1964, adenine arabinoside (Ara-A, vidaribine, **13**) was also reported to have interesting antiviral effects against DNA viruses and was subsequently the first antiviral agent to be approved for use in humans: in 1976, it was licensed for the treatment of herpes zoster virus (Figure 13.4) [21].

Early synthetic efforts around the chemistry of adenosine analogs proved to be challenging because of the stereochemical complexity, instability of various building blocks, and the difficulties of controlling and assigning the ratio of anomers[1)] formed on glycosidation of N9 of the purine ring. The first synthesis of adenosine (**2**), reported in 1948 by Lord Todd (awarded the 1953 Nobel prize for his work in nucleotide and nucleoside chemistry), is illustrative of the difficulties encountered (Scheme 13.2) [22, 23]. Reaction of peracetylated D-ribose (**14**) with anhydrous hydrochloric acid at 0 °C for three days resulted in the formation of the

1) An anomer is one of two stereoisomers of a cyclic saccharide – they differ only by their configuration at the anomeric (hemiacetal or hemiketal) carbon, designated alpha (α) or beta (β).

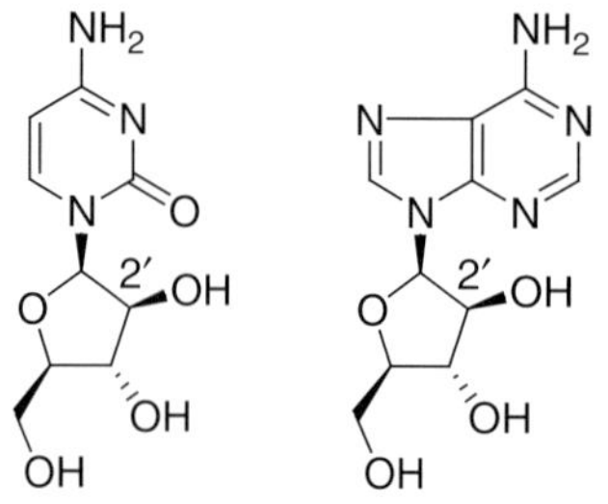

12 13

Figure 13.4 Cytarabine (**12**) and vidaribine (**13**).

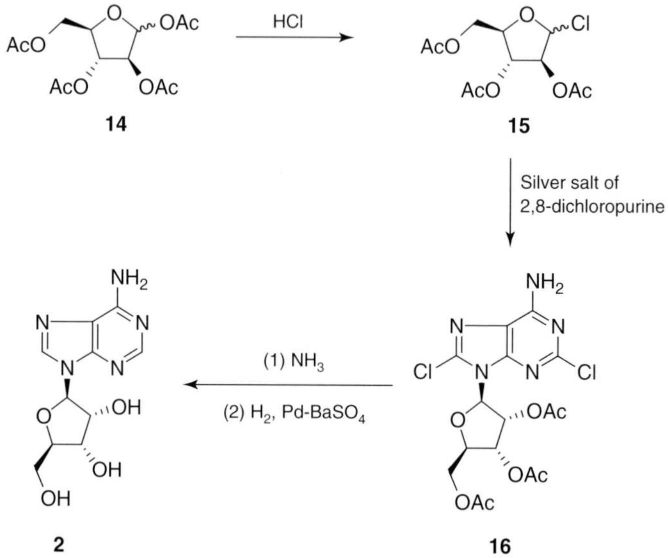

Scheme 13.2 Todd's synthesis of adenosine (**2**).

1-halo adduct (**15**), which was not isolated but reacted directly with the silver salt of 2,8-dichloropurine to give **16**. Subsequent deacylation with methanolic ammonia followed by reductive removal of the two purine chlorine atoms yielded a sample of adenosine (**2**) identical in all regards to that previously described by Bredereck *et al.* [24].

Early synthetic work in the area of 2′-deoxy nucleosides (such as the preparation of cladribine (**17**) by Christensen and Broom [25] is also illustrative of the challenges that had to be overcome (Scheme 13.3). Interest in these 2′-deoxy analogs was at that time due to their antiproliferative and antibacterial activities, principally their activity against cultured leukemia cell lines.

Glycosidation of 2,6-dichloropurine (**18**) with 1,3,5-tri-*O*-acetyl-2-deoxy-D-*erythro*-pentofuranone resulted in an α/β anomeric mixture with a ratio of 6 : 5, which

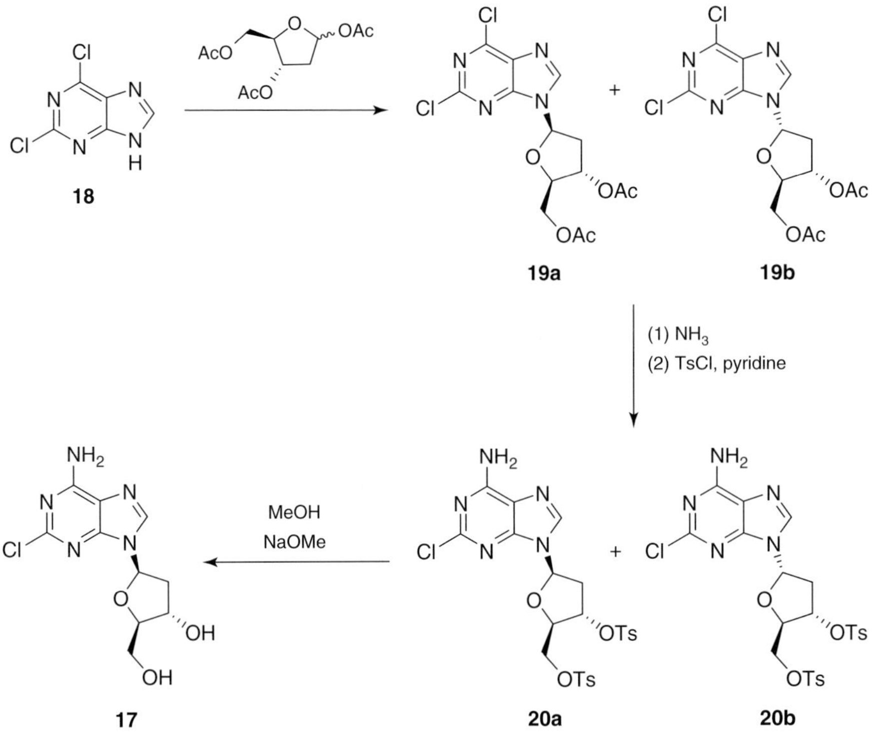

Scheme 13.3 Preparation of cladribine (**17**).

could not be separated (**19a** and **19b**). Deacylation-amination with liquid ammonia gave an inseparable mixture of the 6-aminonucleosides, which was subsequently tosylated to yield a separable mixture of the 3′, 5′-ditosyl congeners **20a** and **20b**. Removal of the tosyl groups with methanolic sodium methoxide yielded the desired 2-chloro-2′-deoxyadenosine (**17**). In March 1993, some 22 years after its preparation, this compound was approved for use in the treatment of various leukemias (Leustatin®, Ortho Biotech). Cladribine (**17**) requires intracellular phosphorylation by deoxycytidine and deoxyguanosine kinases followed by further phosphorylation by nucleoside phosphate kinases to exert its effects as an inhibitor of DNA polymerase, DNA ligase, and ribonucleotide reductase [26]. Such basic discoveries on the use of antimetabolites for the treatment of viral infections spurred a great amount of interest in the area and led to the discovery and development of the acyclic nucleoside analogs such as acylovir (**21**), a guanosine analog approved for topical use in 1982 for the treatment of varicella zoster and herpes simplex viruses (Figure 13.5). For their work and discoveries in this area, Gertrude Elion and George Hitchings were awarded the Nobel prize for medicine in 1988 [27]. Subsequently, valacyclovir (**22**), the valine ester prodrug of acyclovir (**21**), was also licensed for use in 1995 (Valtrex®, GlaxoSmithKline). Valacyclovir (**22**) demonstrates significantly improved oral bioavailability compared with acyclovir (**21**).

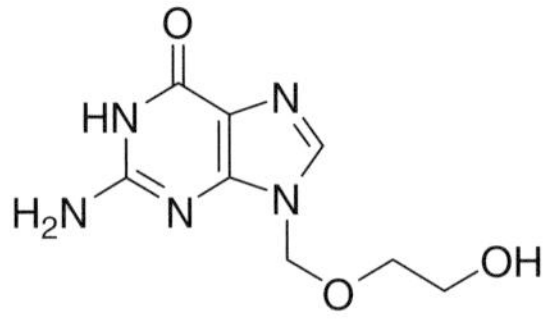

21

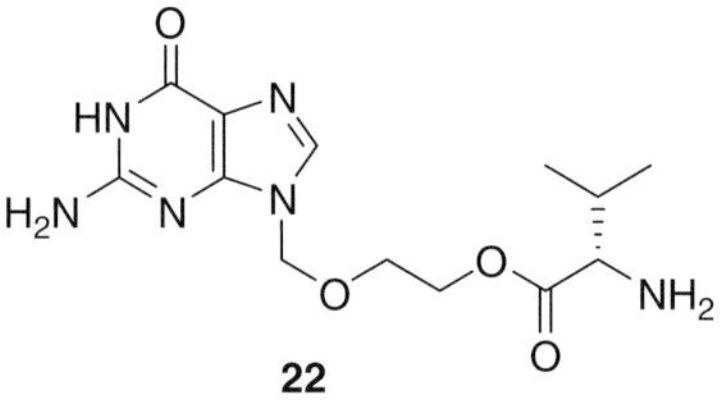

22

Figure 13.5 Acyclovir (**21**) and valacyclovir (**22**).

At the time of the discovery of acyclovir (**21**), the threat of HIV and its ultimately fatal consequence, acquired immune deficiency syndrome (AIDS) was just starting to emerge: the first AIDS patients were admitted to the NIH hospital in June 1981 and January 1982. Many more soon followed. By June 1983, just two years after the first case emerged, 1641 cases of AIDS had been reported (with 644 deaths) in the United States alone. The cause of their then mysterious illness was not identified until the autumn of 1984; three groups independently identified a retrovirus as the cause, which they named HTLV-III, LAV, and ARV, the virus we now know as *HIV* [28]. Given their leading role in the discovery and development of acyclovir (**21**) and their extensive expertise in antiviral drug discovery, Burroughs Wellcome met with the NIH in September of 1984 to develop plans to screen compounds as potential inhibitors of retroviruses to tackle this rapidly growing epidemic [29]. By November of that same year, they had identified the compound BW A509U (azidothymidine, AZT, **23**) from their collection as a promising agent with activity against feline leukemia retroviruses (Figure 13.6). AZT (**23**) had originally been synthesized as a viral reverse transcriptase inhibitor and potential anticancer agent (a prevailing theory at that time being that a number of cancers were caused by retroviruses) [30].

After sending samples to the NIH, AZT (**23**) was soon identified as being highly active against HIV – a flurry of activity (scientific, legal, and regulatory) saw AZT (**23**) rapidly brought into clinical trials. The investigational new drug application (IND) for AZT (**23**) was filed on 15 June 1985, and the first patient enrolled on 3 July. By March 1987, enough clinical evidence of efficacy and safety had been gathered for the FDA to approve AZT (**23**), the first nucleoside reverse transcriptase

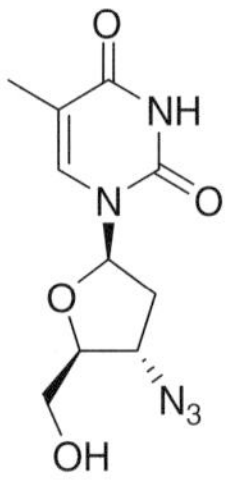

23

Figure 13.6 Azidothymidine (**23**).

inhibitor (NRTI) for use in the treatment of HIV, only 25 months after the laboratory demonstration of an anti-HIV effect. Despite this unprecedented success, there were many significant drawbacks for the use of AZT (**23**) in the treatment of HIV. Side effects were numerous and debilitating, and the dosing regimen (400 mg every 4 h) was highly inconvenient and difficult to adhere to. Owing to the absence of a proofreading function of HIV reverse transcriptase enzymes, a high rate of mutation was observed and drug resistance was seen to emerge [31]. Selectivity issues also emerged as being important for the field – toxicity as a result of the inhibition of human polymerases was thought to be an area of further investigation. As a result of these numerous issues, the race was on to develop superior therapeutics. As the understanding of the mechanism by which NRTIs exerted their effects in cells improved, the discovery process became more refined. Similar to cladribine (**20**) (Scheme 13.2), it was found that NRTIs require activation via phosphorylation by host kinases to exert their effects and that these processes could vary quite substantially between cell lines and organisms. In particular, in humans, it was found that efficient phosphorylation of AZT monophosphate by thymidylate kinase was particularly rate limiting [32, 33]. Attempts to circumvent such bottlenecks saw efforts made to switch the thymidine nucleus of AZT (**23**) for an alternate base that would be more efficiently phosphorylated *in vivo* by a different kinase. Retention of the 2′, 3′-dideoxy ribose moiety (a key pharmacophore element in NRTIs) led to the identification of didanosine – developed by Broder, Mitsuya, and Yarchoan at the National Cancer Institute (NCI) [34]. Didanosine, an inosine nucleoside analog, is metabolized *in vivo* to dideoxyadenosine triphosphate (in part by phosphorylation of the monophosphate by the adenylate kinase – Section 13.4). Although dideoxyadenosine was first prepared some years earlier, the nonnaturally occurring inosine analog (didanosine) was found to be superior due to lesser nephrotoxicity [35]. The discovery that acyclic nucleoside analogs (and in particular acyclic nucleoside phosphonates) were highly active against the HIV virus led to a significant amount of effort in this area and ultimately to the discovery of adefovir (**3**) and tenofovir (**4**) by De Clercq and Holý [36].

13.3 Synthesis

As outlined previously, many of the major complications in the synthesis of adenine-based NRTIs are a result of the formation of mixtures of anomers on alkylation of the purine N9 when using acylated ribofuranosyl halides as electrophiles, as well as the formation of N6 and N7 nucleoside by-products [22]. Original methods for the formation of the glycosidic linkage such as mercurate and silver salt additions (Fischer–Helferich methods) [37] and the Hilbert–Johnson methods [38] have been superseded by the Vorbrüggen, fusion, and sodium salt methods [39–41]. The Vorbrüggen glycosylation (also known as the *silyl Hilbert–Johnson method*) utilizes a persilylated nucleophilic component that reacts under acid catalysis with a glycosyl donor (**24**) (Scheme 13.4). Excellent selectivity

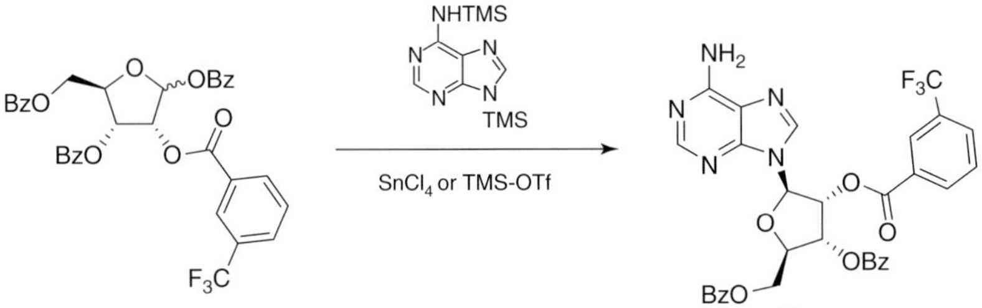

24

25

Scheme 13.4 Vorbrüggen method for the synthesis of adenine nucleosides.

for the desired β-anomer is achieved due to anchimeric assistance from the neighboring benzoyl group. Although introduced for the preparation of pyrimidine bases, the Vorbrüggen reaction has been extended to encompass the preparation of adenines such as **25** (Scheme 13.4) and both 2′-deoxy and 2′-hydroxy nucleosides [42].

2′-Deoxy nucleosides can be accessed from these adducts via photoinduced electron transfer of the 2′ protecting group. Efforts to improve stereocontrol in the synthesis of 2′-deoxynucleosides led to the introduction of additional methods utilizing neighboring group participation, particularly the benzylthio, phenylthio, and phenylselenyl groups [43–45]. A comprehensive study by Hildebrand and Wright [46] on the product isomer distribution (Table 13.1) from the sodium salt method in the alkylation of purines is highly illustrative of this approach (Scheme 13.5).

Brik *et al.* [47] found that the use of tetrabutylammonium fluoride greatly accelerated the rate of alkylation of the purine scaffold and often resulted in excellent N9/N7 regiochemical control, particularly when a large group was present on N6 of the purine ring.

Mitsunobu reactions using **28** have also been found to be effective in alkylating the purine N9 (**29**, 22% yield), while adenines (**1**) with substituted amines at the

Table 13.1 Product distribution from the sodium salt alkylation of purines.

		% yield				
		26a	**26b**	**27a**	**27b**	
Compound	**Substituents**	**9-α**	**9-β**	**7-α**	**7-β**	**Yield nucleosides**
1	2,6-diCl	1.5	50.0	–	15.0	66.5
2	2,6-diBr	12.0	50.0	3.8	6.6	72.4
3	2,6-diSMe	0.3	62.0	0.2	19.0	81.5
4	6-Cl	1.4	58.0	1.2	3.0	63.6
5	6-Br	2.6	45.0	1.4	13.0	62.0
6	2-Br-6-SMe	1.9	68.3	0.8	14.9	85.9

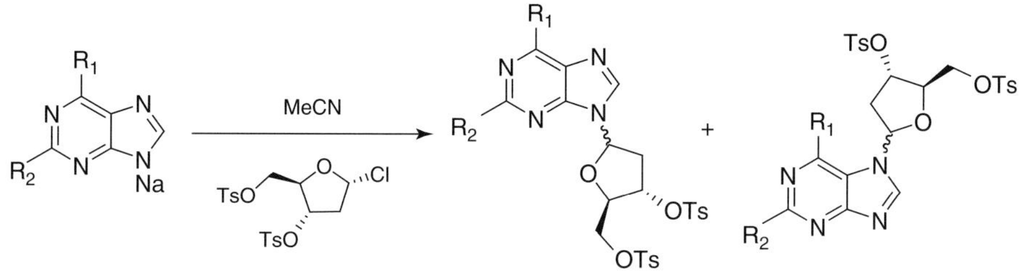

26a = α

26b = β

27a = α

27b = β

Scheme 13.5 Sodium salt method product distribution.

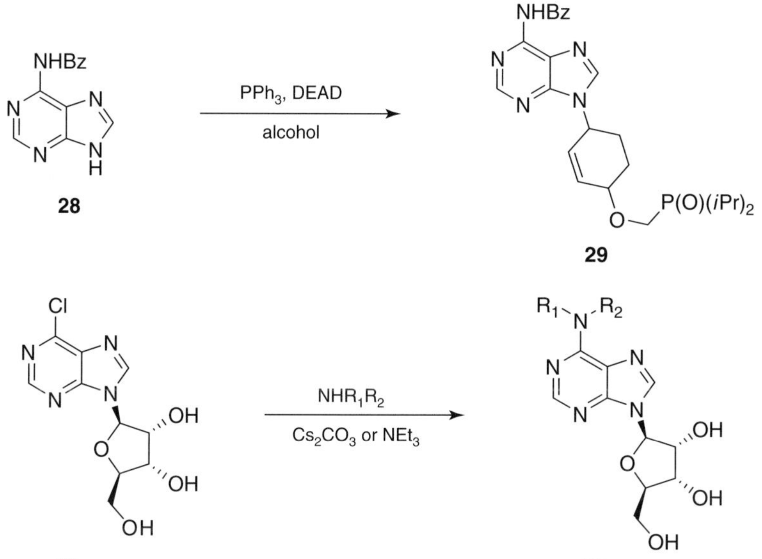

28

29

31

30

Scheme 13.6 Mitsunobu alkylation of adenine (**28**) and the preparation of cytokinin analogs by 6-halo substitution of **31**.

6-position (**30**) can be conveniently prepared from 6-chloropurine riboside (**31**), useful for the preparation of cytokinin analogs (Scheme 13.6) [48–50].

A recent focus of synthetic efforts in the preparation of adenine-based NRTIs has been in the development of more efficient and lower cost large-scale routes of these important drugs, so as to facilitate access to these medicines in the developing world. Tenofovir disoproxil (**6**) is prepared on an industrial scale from adenine (**1**) [51]. A critical modification to the process was the addition of tetrabutylammonium bromide to the reaction mixture for the alkylation of the phosphonic acid, which gave higher yields, increased product stability, and led to a significant improvement in the overall efficiency of the process [51, 52].

13.4 Mode of Action

Reverse transcriptase is an essential enzyme for retroviral replication: as a DNA polymerase, it transcribes the RNA-based genetic information of the HIV retrovirus into a DNA form usable by the host cell machinery. Absence or pharmacological inhibition of the reverse transcriptase enzyme disables the incorporation of the viral genome into the host cell [53]. Before phosphorylation, phosphonate esters (such as **6**, Viread®, Gilead Sciences, Inc.) are converted to phosphonic acids in the intestinal lumen and plasma by diester hydrolysis. These intracellular phosphonates are then phosphorylated in sequential steps to monophosphates and to the active metabolites, the diphosphates. The antiviral activity of acyclic nucleoside analogs is the result of the higher affinity of the diphosphorylated metabolite for viral DNA polymerases in comparison to the cellular DNA polymerases. The diphosphate metabolites compete with their natural nucleotide counterpart, deoxyadenosine-5′-triphosphate, for incorporation into the newly forming viral DNA strand: incorporation of the NRTI diphosphates causes termination of the elongating DNA chain. Phosphorylation of adefovir (**3**) and tenofovir (**4**) to their monophosphates is mediated by kinases such as adenylate kinase. Mutant cells deficient in such kinase activities appear to be resistant to these drugs, indicating that these enzymes are key factors in the phosphorylation events. Phosphorylation of NRTIs lacking a phosphonate (such as didanosine) is thought to be a limiting factor in the activation of such NRTIs. Thus, NRTIs such as tenofovir (**4**) that already incorporate this first phosphorus generally demonstrate superior activity. In addition to this, the prolonged intracellular half-life of tenofovir diphosphate is thought to give rise to improved antiviral activity.

13.5 Structure–Activity Relationship

Analysis of structure–activity relationship (SAR) trends in NRTI series in which multiple cellular activation events are required is challenging as the activity of inhibitors not only depends on the affinity for the ultimate target (reverse transcriptase) but also relies on the successful processing by the activating enzymes. In such situations, phenotypic and pharmacology-based approaches (as opposed to pure structure-based approaches) are commonly utilized.

2′-Deoxy and 2′,3′-dideoxy nucleoside substructures are common pharmacophore elements in NRTIs, often with replacement of the hydroxyls with azido and/or fluoro groups [54]. However, such nucleoside substructures have metabolic liabilities, and thus, significant efforts have been undertaken to improve their stability by replacing the oxygen of the glycosyl component with carbon (carbasugars) – the lability of glycosyl bonds following oral dosing was one obvious limiting factor to oral bioavailability. A number of such compounds (e.g., carbovir (**32**)) have been shown

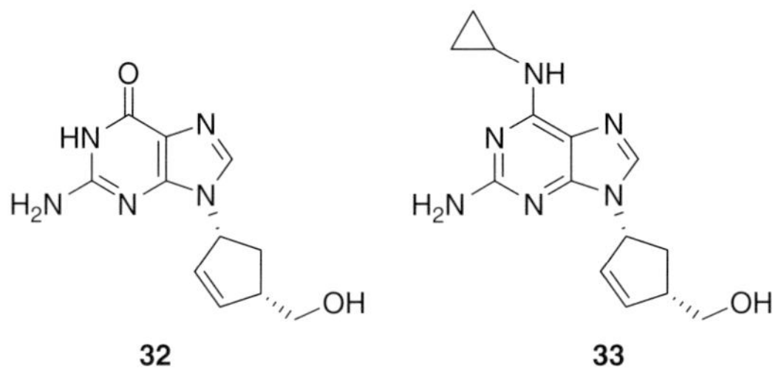

Figure 13.7 Carbovir (**32**) and abacavir (**33**).

to possess good antiviral activity, and some have reached the market (abacavir (**33**), Ziagen®, ViiV Healthcare) (Figure 13.7) [55, 56].

A variety of saturated and unsaturated adenine-based analogs incorporating ribose replacements of various ring sizes have been prepared – a number of these retain antiviral activity [57–61]. Acyclic nucleoside analogs such as tenofovir (**4**) obviously benefit from the absence of a glycosidic stereocenter, both in terms of synthetic and stereochemical complexity. The discovery that they were active inhibitors of HIV was one of the major breakthroughs in the area, as was the incorporation of the phosphonate group, which overcame problems associated with phosphate hydrolysis. Alternative approaches to phosphorus prodrugs have also been described such as the phosphoramidates [62]. GS-9131 is a phosphoramidate prodrug of GS-9148 that is selectively activated in lymphatic tissues by hydrolysis of the prodrug moiety by cathepsin A (Figure 13.8) [63]. A number of acetal and thioacetal moieties have been found to be effective NRTI ribose isosteres – emtricitabine (Emtriva®, Gilead Sciences Inc.), amdoxovir, racivir, and lamivudine (Epivir®, ViiV Healthcare) all contain such groups [9]. Adenine-containing analogs of such acetals and thioacetals have also been prepared, although these have been studied in less detail than cytidine analogs [64–66]. Resistance to NRTIs by way of a K65R mutation in highly active antiretroviral therapy is observed in regimens containing tenofovir (**4**) and lamivudine [67]. A number of adenine-based analogs have been prepared in efforts to overcome such limitations and have resulted in the identification of compounds that retain activity against clinically relevant

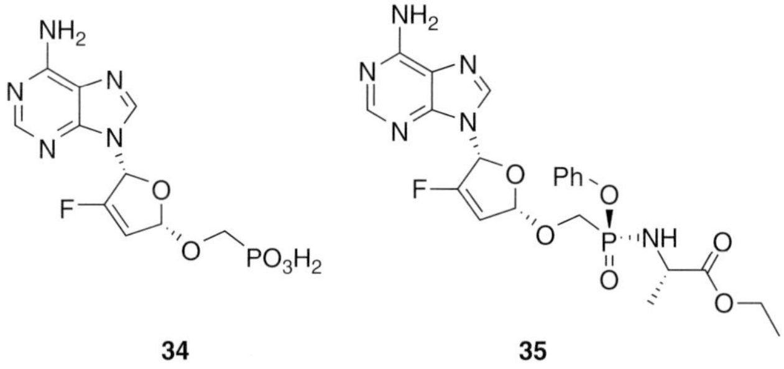

Figure 13.8 GS-9148 (**34**) and GS-9131 (**35**).

mutants, exemplified by GS-9148 (**34**) and its phosphoramidate prodrug GS-9131 (**35**) (Figure 13.8) [68, 69].

These and other members of the next generation of NRTIs offer significant promise in further reducing the medical burden of HIV while reducing the side effect profiles of this important drug class.

References

1. Muller, G. (2003) *Drug Discov. Today*, **8**, 681–691.
2. Cristalli, G., Lambertucci, C., Marucci, G., Volpini, R., and Dal Ben, D. (2008) *Curr. Pharm. Des.*, **14**, 1525–1552.
3. Chiosis, G. (2006) *Curr. Top. Med. Chem.*, **6**, 1183–1191.
4. Gessi, S., Merighi, S., Varani, K., and Borea, P.A. (2011) *Adv. Pharmacol.*, **61**, 41–75.
5. Jacobsen, K.A., Zablocki, J., and Bhagwat, S. (2009) *Purinergic Signal.*, **5**, 1.
6. Carroll, A.R., Duffy, S., and Avery, V.M. (2010) *J. Org. Chem.*, **75**, 8291–8294.
7. Forsman, J.J. and Leino, R. (2011) *Chem. Rev.*, **111**, 3334–3357.
8. De Clercq, E. (2010) *Curr. Opin. Pharmacol.*, **10**, 507–515.
9. De Clercq, E. (2011) *Annu. Rev. Pharmacol. Toxicol.*, **51**, 1–24.
10. Kossel, A. (1885) *Chem. Ber.*, **18**, 1928–1930.
11. Fischer, E. (1897) *Chem. Ber.*, **30**, 2226–2254.
12. Fischer, E. (1899) *Chem. Ber.*, **32**, 435–504.
13. Nobel Prize Webpage. *http://nobelprize.org/nobel_prizes/chemistry/laureates/1902/fischer-lecture.pdf* (accessed June 6 2011).
14. Nagelhout, J.J. (1992) *AANA J.*, **60**, 287–292.
15. Thomson, R.L., Wilkin, M.L., Hitchings, C.H., Elion, G.B., Falco, E.A., and Russell, P.B. (1949) *Science*, **110**, 454.
16. Bergmann, W. and Feeney, R.J. (1951) *J. Org. Chem.*, **16**, 981–987.
17. Bergmann, W. and Burke, D.C. (1955) *J. Org. Chem.*, **20**, 1501–1507.
18. Suckling, C.J. (1991) *Sci. Prog.*, **75**, 323–329.
19. Underwood, G.E. (1962) *Virology*, **21**, 482–498.
20. Newman, D.J. and Cragg, G.M. (2004) *J. Nat. Prod.*, **67**, 1216–1238.
21. Johnson, A.W. and Jervey, E.D. (1964) *Arch. Ophthalmol.*, **72**, 826–828.
22. Davoll, J., Lythgoe, B., and Todd, A.R. (1948) *J. Chem. Soc.*, 967–969.
23. Brown, D.M. and Kornberg, H. (2000) *Biogr. Mem. Fellows R. Soc.*, **46**, 517–532.
24. Bredereck, H., Martini, A., and Richter, F. (1941) *Chem. Ber.*, **74**, 694–697.
25. Christensen, J.F. and Broom, A.D. (1972) *J. Med. Chem.*, **15**, 735–739.
26. Grever, M.R. and Lozanski, G. (2011) *J. Clin. Oncol.*, **29**, 583–590.
27. Nobel Prize Webpage. *http://nobelprize.org/nobel_prizes/medicine/laureates/1988* (accessed 6 June 2011).
28. National Institutes of Health Webpage. *http://history.nih.gov/NIHInOwnWords/docs/page_04.html* (accessed 6 June 2011).
29. National Institutes of Health Webpage. *http://history.nih.gov/NIHInOwnWords/docs/page_29.html* (accessed 6 June 2011).
30. De Clercq, E., Balzarini, J., Descamps, J., and Eckstein, F. (1980) *Biochem. Pharmacol.*, **29**, 1849–1851.
31. Roberts, J.D., Bebenek, K., and Kunkel, T.A. (1988) *Science*, **242**, 1171–1173.
32. King, D.H. (1988) *J. Am. Acad. Dermatol.*, **18** (1, Part 2), 176–179.
33. Lavie, A. and Konrad, M. (2004) *Mini Rev. Med. Chem.*, **4**, 351–359.
34. McGowan, J.J., Tomaszewski, J.E., Cradok, J., Hoth, D., Grieshaber, C.K., Broder, S., and Mitsuya, H. (1990) *Clin. Infect. Dis.*, **12**, S5-S513–S5-S520.
35. Plunkett, W. and Cohen, S.S. (1975) *Cancer Res.*, **35**, 1547–1554.

36. De Clercq, E., Holý, A., Rosenberg, T., Sakuma, T., Balzarini, J., and Maudgal, P.C. (1986) *Nature*, **323**, 464–467.
37. Fischer, E. and Helferich, B. (1914) *Chem. Ber.*, **47**, 210–235.
38. Hilbert, G.E. and Johnson, T.B. (1930) *J. Am. Chem. Soc.*, **52**, 4489–4494.
39. Vorbrüggen, H. and Ruh-Pohlenz, C. (2000) *Org. React.*, **55**, 1–630.
40. Sato, T., Shimadate, T., and Ishido, Y. (1960) *Nippon Kagaku Zasshi*, **81**, 1440–1442.
41. Kazimierczuk, Z., Cottam, H.B., Revankar, G.R., and Robins, R.K. (1984) *J. Am. Chem. Soc.*, **106**, 6379–6382.
42. Wang, Z., Prudhomme, D.R., Buck, J.R., Park, M., and Rizzo, C.J. (2000) *J. Org. Chem.*, **65**, 5969–5985.
43. Ryan, K.J., Acton, E.M., and Goodman, L. (1971) *J. Org. Chem.*, **36**, 2646–2657.
44. Kawakami, H., Ebata, T., Koseki, K., Matsumoto, K., Matsushita, H., Naoi, Y., and Itoh, K. (1991) *Heterocycles*, **32**, 2451–2470.
45. Chu, C.K., Babu, J.R., Beach, J.W., Ahn, S.K., Huang, H., Jeony, L.S., and Lee, S.J. (1990) *J. Org. Chem.*, **55**, 1418–1420.
46. Hildebrand, W. and Wright, G.E. (1992) *J. Org. Chem.*, **57**, 1808–1813.
47. Brik, A., Wu, C.-Y., Best, M.D., and Wong, C.-H. (2005) *Bioorg. Med. Chem.*, **13**, 4622–4626.
48. Toyota, A., Katagiri, N., and Kaneko, C. (1993) *Heterocycles*, **36**, 1625–1630.
49. Pérez-Pérez, M.-J., Rozenski, J., Busson, R., and Herdewijn, P. (1995) *J. Org. Chem.*, **60**, 1531–1537.
50. Fleysher, H.M. (1972) *J. Med. Chem.*, **15**, 187–191.
51. Brown Ripin, D.H., Teager, D.S., Fortunak, J., Mahaboob Basha, S., Bivins, N., Boddy, C.N., Byrn, S., Catlin, K.K., Houghton, S.R., Jagadeesh, S.T., Anesh Kumar, K., Melton, J., Muneer, S., Rao, L.N., Roa, R.V., Ray, P.C., Reddy, N.G., Reddy, R.M., Shekar, K.C., Silverton, T., Smith, D.T., Stringham, R.W., Subbaraju, G.V., Talley, F., and Williams, A. (2011) *Org. Process Res. Dev.*, **14**, 1194–1201.
52. Schoofs, M. (2011) Researchers Manipulate Drug's Chemistry in Bid to Lower Treatment Cost. The Wall Street Journal (May 13), p. 12.
53. De Clercq, E. and Holý, A. (2005) *Nat. Rev. Drug Discov.*, **4**, 928–940.
54. Mikhailopulo, I.A., Poopeiko, N.E., Pricota, T.I., Sivets, G.G., Kvasyuk, E.I., Balzarini, J., and De Clercq, E. (1991) *J. Med. Chem.*, **34**, 2195–2202.
55. Jeong, L.S. and Lee, J.A. (2004) *Antivir. Chem. Chemother.*, **15**, 235–250.
56. De Clercq, E. (2009) *Int. J. Antimicrob. Agents*, **33**, 307–320.
57. Gourdel-Martin, M.-E. and Huet, F. (1997) *J. Org. Chem.*, **62**, 2166–2172.
58. Csuk, R. and von Scholz, Y. (1995) *Tetrahedron*, **51**, 7193–7206.
59. Rosenquist, Å., Kvarnström, I., Classon, B., and Samuelsson, B. (1996) *J. Org. Chem.*, **61**, 6282–6288.
60. Roy, A., Chakrabarty, K., Dutta, P.R., Bar, N.C., Basu, N., Achari, B., and Mandal, S.B. (1999) *J. Org. Chem.*, **64**, 2304–2309.
61. Wachmeister, J., Classon, B., Samuelsson, B., and Kvarnström, I. (1995) *Tetrahedron*, **51**, 2029–2038.
62. Ray, A.S., Vela, J.E., Boojamra, C.G., Zhang, L., Hui, H., Callebaut, C., Stray, K., Lin, K.-Y., Gao, Y., Mackman, R.L., and Cihlar, T. (2008) *Antimicrob. Agents Chemother.*, **52**, 648–654.
63. Birkus, G., Wang, R., Liu, X., Kutty, N., MacArthur, H., Cihlar, T., Gibbs, C., Swaminathan, S., Lee, W., and McDermott, M. (2007) *Antimicrob. Agents Chemother.*, **51**, 543–550.
64. Jeong, L.S., Schinazi, R.F., Beach, J.W., Kim, H.O., Shanmuganathan, K., Nampalli, S., Chun, M.W., Chung, K.-W., Choi, B.G., and Chu, K.C. (1993) *J. Med. Chem.*, **36**, 2627–2638.
65. Jeong, L.S., Schinazi, R.F., Beach, J.W., Kim, H.O., Nampalli, S., Shanmuganathan, K., Alves, A.J., McMillan, A., Chu, K.C., and Mathis, R. (1993) *J. Med. Chem.*, **36**, 181–195.
66. Wang, W., Jin, H., and Mansour, T.S. (1994) *Tetrahedron Lett.*, **35**, 4739–4742.
67. Parikh, U.M., Koontz, D.L., Chu, K.C., Schinazi, R.F., and Mellors, J.W. (2005) *Antimicrob. Agents Chemother.*, **49**, 1139–1144.

68. Mackman, R.L., Boojamra, C.G., Prasad, V., Zhang, L., Lin, K.-Y., Petrakovsky, O., Babusis, D., Chen, J., Douglas, J., Grant, D., Hui, H.C., Kim, C.U., Markevitch, D.Y., Vela, J., Ray, A., and Cihlar, T. (2007) *Bioorg. Med. Chem. Lett.*, **17**, 6785–6789.

69. Boojamra, C.G., Mackman, R.L., Markevitch, D.Y., Prasad, V., Ray, A.S., Douglas, J., Grant, D., Kim, C.U., and Cihlar, T. (2008) *Bioorg. Med. Chem. Lett.*, **18**, 1120–1123.

14
Guanine-Based Nucleoside Analogs as Antiviral Agents

Maurizio Franzini

14.1
Introduction

This chapter focuses on guanosine analogs with the general structure of 2-amino-1,9-dihydro-9-alkyl-6*H*-purin-6-ones (Figure 14.1). Analogs with NH_2 at C-6 are not covered here, except in a few cases where they undergo *in vivo* metabolic transformation into the corresponding guanine derivative.

Aciclovir (**3**, ACV) was the first representative of this family to be introduced as pharmacological agent against viruses of the herpes family (Figure 14.2). Approved in 1982, it has been the gold standard for the treatment of cutaneous and systemic herpes simplex virus type 1 (HSV-1) and HSV-2 infections (including herpes encephalitis and genital herpes infections) and varicella zoster virus (VZV) infections (shingles, chickenpox) [1]. It can also be used in the prophylaxis of genital and orofacial HSV infections and is currently widely distributed under a variety of commercial names, Zovirax® likely being the best known. For the treatment of immunocompromised patients, ACV (**3**) is typically administered intravenously in order to obtain the desired efficacy. Valaciclovir (**4**, VACV, Valtrex®, Zelitrex®), the oral L-valyl ester prodrug of ACV (**3**), discovered in 1988, achieves much higher plasma concentrations than **3** and has become particularly useful in the prevention of human cytomegalovirus (HCMV) disease after organ transplantation [2].

Of all the ACV analogs currently available in the market, the most relevant are ganciclovir (**5**, GCV, Cymevene®, Cytovene®), preferred for the treatment of HCMV infections, and its L-valyl ester oral prodrug valganciclovir (**6**, Valcyte®), approved in 2001. Penciclovir (**7**, PCV, Denavir®) has a longer intracellular half-life than ACV (**3**). The compound is also sold as Vectavir® cream for recurrent herpes labialis in immunocompromised patients. Its prodrug famciclovir (**8**, FCV, Famvir®) was introduced for the treatment of recurrent genital herpes, acute VZV infections, and postherpetic neuralgia. Abacavir (**9**, Ziagen®), an anti-HIV agent, is the prodrug of carbovir (**10**, CBV) (Figure 14.3). Amdoxovir (**11**, DAPD), in itself a prodrug, recently exited phase II clinical trials in combination with already approved HIV drugs [3]. Among other antivirals that have reached the development stage (Figure 14.3) [4] buciclovir (**12**, BCV) and the GCV isomer H2G

Bioactive Heterocyclic Compound Classes: Pharmaceuticals,
First Edition. Edited by Jürgen Dinges and Clemens Lamberth.

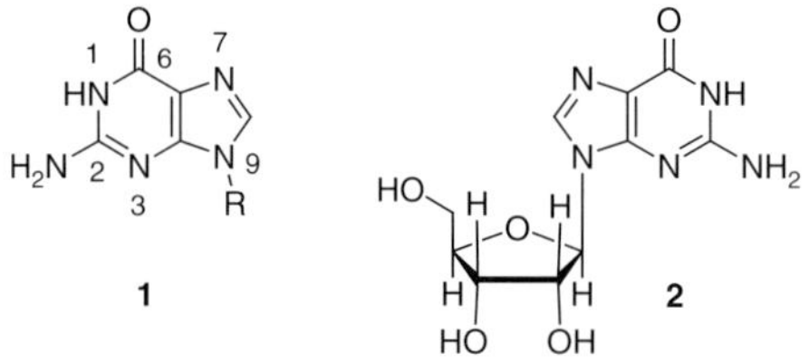

Figure 14.1 Basic structures and numbering for guanine derivatives (**1**) and guanosine (**2**).

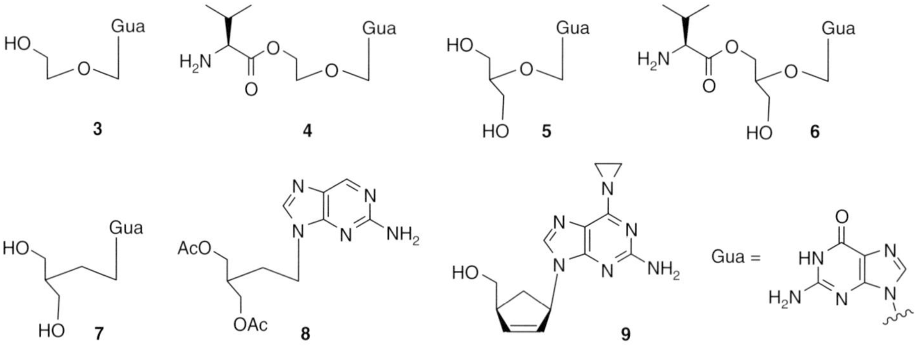

Figure 14.2 Approved antiviral analogs of guanosine (**2**): aciclovir (**3**), valaciclovir (**4**), ganciclovir (**5**), valganciclovir (**6**), penciclovir (**7**), famciclovir (**8**), and abacavir (**9**).

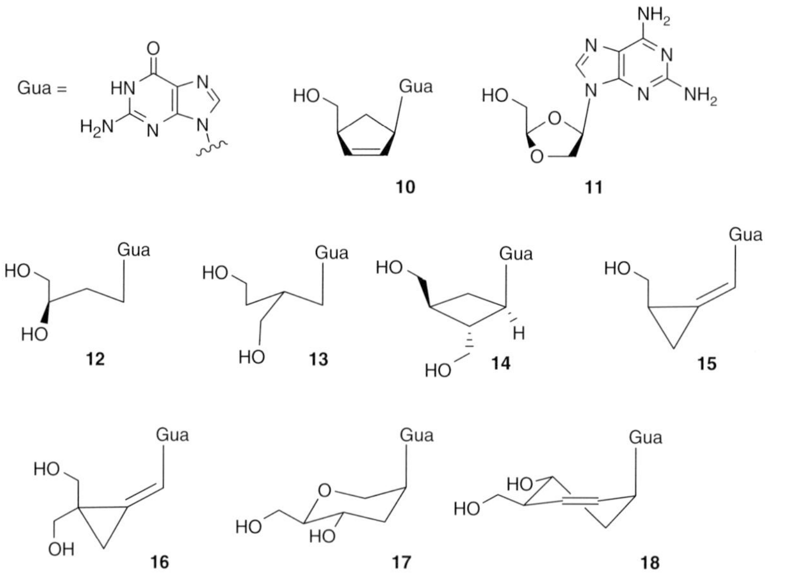

Figure 14.3 Other antiviral guanosine analogs: carbovir (**10**), amdoxovir (**11**), buciclovir (**12**), H2G (**13**), lobucavir (**14**), synguanol (**15**), cyclopropavir (**16**), anhydrohexitol guanine (**17**), and cyclohexenyl guanine (**18**).

(**13**, (-)2HM-HBG), which is very active against VZV; lobucavir (**14**, (*R*)-BHCG), effective against human herpes viruses HHV-6 and HHV-7; synguanol (**15**), a much weaker HSV inhibitor but more potent against HCMV and Epstein-Barr virus (EBV), the cause of mononucleosis; cyclopropavir (**16**, CPV); anhydrohexitol guanine (**17**); and D/L-cyclohexenyl guanine (**18**) are of interest.

14.2 History

ACV (**3**) was discovered in 1974 as an inhibitor of herpes simples viruses by Gertrude B. Elion and coworkers at the Wellcome Research Laboratories in Research Triangle Park, NC, during the search for adenosine deaminase inhibitors that could enhance the potency of vidarabine (ara-A), an antiviral drug no longer in use [5]. The synthesis of **3** was inspired by spongosine [6], an arabinosyl nucleoside isolated from the Caribbean sponge *Cryptotethia crypta* with analgesic properties [7], whose structure was first elucidated in 1956 [8]. Despite the lack of an intact cyclic carbohydrate moiety, acyclo-Guo, as **3** was initially named, can be phosphorylated by the virus thymidine kinase (TK) to the corresponding triphosphate (ACV-TP), which in turn inhibits the viral DNA polymerase [9]. ACV (**3**) was thus found to be at least 10 times more potent *in vitro* than idoxuridine, which at that time was still used in clinical practice. The compound was also found to be effective against HSV-2.

Soon after these first disclosures, ACV (**3**) became the subject of several double-blind placebo-controlled clinical trials, which demonstrated its efficacy in the topical treatment of dendritic corneal ulcers [10], of primary genital herpes [11], and in the systemic treatment or prophylaxis of mucocutaneous HSV-1 infections in immunocompromised patients [12]. ACV (**3**) also entered full-fledged development for VZV infections [13]. The limited bioavailability of **3** due to poor water solubility (0.2% w/w at 25 °C) spurred researchers at the Rega Institute in Leuven, Belgium, to synthesize a variety of aminoacyl esters of **3** in order to increase solubility and therefore to allow for an easier topical or parenteral delivery, in lieu of the intravenous administration as a bolus infusion. All the aminoacyl esters prepared turned out almost as potent as the parent ACV (**3**) in rabbit kidney cell cultures, an indication that they were readily hydrolyzed to the active principle, without showing any significant cytotoxicity [14]. In the end, the valyl ester of ACV, VACV (**4**), emerged as the preferred oral prodrug for **3** to be administered for the suppression of recurrent genital herpes in HIV-infected individuals [15] and for the treatment of herpes zoster infections in immunocompetent adults [16].

GCV (**5**), initially dubbed as DHPG, was designed with a second hydroxymethyl group to mimic the ribosylic 3′-hydroxy functionality. The compound was independently synthesized by Biologics, Toronto, ON [17], and by Syntex, Palo Alto, CA, as a potent and broad-acting antiherpetic agent [18]. GCV (**5**) was more water soluble and slightly more bioavailable than **3** in monkeys, and it was claimed to be safe for lack of toxicity in mice and absence of mutagenic activity in the Ames test. GCV (**5**) was also independently reported (as 2′-NDG) by researchers at Merck

Sharp & Dohme as at least 10-fold more potent than ACV (**3**) in cell cultures in the inhibition of HCMV replication and EBV-induced lymphocyte transformation into lymphoblasts [19]. By statistical analyses on survival times, orally administered **5** was found to be up to 50-fold more efficacious than **3** in treating murine systemic or local (orofacial) HSV-1 infections and intravaginal HSV-2 infections.

A later finding about adverse toxic effects in animals and humans caused by GCV (**5**) spurred more research efforts into alternative acyclonucleosides [20]. Although first reported in 1972 [21], it was not until 1985 that the synthesis of PCV (**7**) would be made practical by researchers at Beecham in the United Kingdom [22]. When tested in virus reduction assays in human lung fibroblasts, **7** was found equipotent to ACV (**3**) and not cytotoxic at the concentrations at which it was active. A couple of years later, the same group reported on potential prodrugs of PCV (**7**); amongst those, a simple 6-deoxy variant of **7** and some of its diesters showed good gastrointestinal absorption on oral administration. In particular, its diacetate, FCV (**8**), provided a plasma concentration of **7** that was 15-fold higher than that obtained by administration of **7** alone [23]. This diester was identified as the preferred prodrug for PCV (**7**) and found to be clinically equivalent to VACV (**4**) [24].

BCV (**12**), an isomer of **7**, was introduced at around the same time by a research group in Sweden. The (*R*)-enantiomer was determined as the most active against HSV-1 and HSV-2 [25]. However, as it is the case with other acyclic guanosine derivatives, **12** only had a limited effect against herpes virus infections once the virus had crossed the blood–brain barrier and entered the central nervous system (CNS) [26]. The same group later presented 2HM-HBG (**13**), a homolog of **12**, slightly more potent than the latter against HSV-1 but equipotent against HSV-2 [27] and very effective in inhibiting VZV in cell cultures [28].

The synthesis and antiviral properties of cyclobut-G, the racemic form of lobucavir (**14**), and of related compounds were reported by scientists at Bristol-Myers Squibb in 1989 [29]. The proposal of attaching a cyclobutyl group to guanine (**1**) was inspired by the natural product oxetanocin (**19**) (Figure 14.4), which had been recently isolated and was found to have anti-HIV activity [30].

Racemic cyclobut-G (*rac*-**14**) was determined to be equipotent to ACV *in vitro* (**3**) against HSV-1 and 2, but more potent against VZV and EBV, with a therapeutic index greater than 50. It was also effective in a murine model of cytomegalovirus (CMV). The *in vitro* anti-HIV activity was comparable to that of zidovudine (AZT). After achieving stereospecific synthesis of **14**, it was found that the active enantiomer displayed the geometrical relationship of the hydroxyl groups on the cyclobutyl moiety found in the purine base (the "natural configuration") [31].

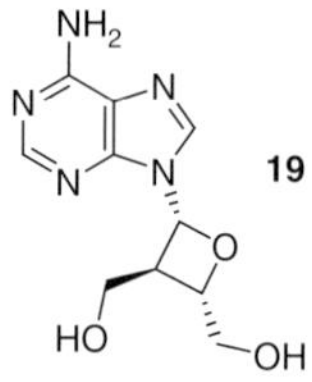

19

Figure 14.4 Structure of oxetanocin (**19**).

HSV remains highly prevalent worldwide in provoking genital ulcerations and neonatal infectious encephalitis. Transmission of HSV from mother to newborn can be the cause of fatal disease in the child, especially when the woman experiences infection during pregnancy. Prophylaxis with oral ACV (**3**) or VACV (**4**) during late pregnancy can limit viral shedding at the time of delivery, thus reducing the risk of neonatal HSV disease [32]. A very large clinical study on live births recently conducted in Denmark did not find any statistically significant association of birth defects risk with ACV (**3**), VACV (**4**), and FCV (**8**) exposure, a fairly strong reassurance on the safety of these antiviral drugs [33].

HSV infections are also deemed a cofactor in fueling the HIV epidemic [34]. Recent randomized controlled trials have shown that although ACV (**3**) does not prevent HIV-1 transmission, it decreases its viral RNA load levels in HSV-2-infected patients, possibly through an indirect mechanism of activation by HSV-2 of **3** against HIV-1 reverse transcriptase (RT) [35]. There is also more recent evidence that ACV-TP is a direct inhibitor of HIV-RT in HSV-negative $CD4^{(+)}$ T cells [36].

14.3 Synthesis

The syntheses of cyclic, carbocyclic, and acyclic guanosine derivatives have been summarized in excellent reviews up to the year 2000 [37]. Possibly the most difficult hurdle in preparing N^9-substituted guanines, the unwanted competing formation of the respective N^7-regioisomers and their laborious separation, has spurred a variety of different endeavors to overcome this pitfall since the early 1970s.

Attempts to perform a direct alkylation of the poorly soluble guanine (**1**) have all afforded mediocre results, both in terms of overall yield and regioselectivity. Guanine has a propensity to undergo electrophilic attack at several possible sites of substitution (N^1, 2-NH_2, N^3, and 6-O, in addition to the already mentioned N^7 and N^9). A large number of early methods utilize the trimethylsilyl group to transiently mask active sites on the purine scaffold, for example, the first patented synthesis by Wellcome, which delivered ACV (**3**) with an overall yield of 24% [38]. According to the modified silyl-Hilbert-Johnson protocol, a moisture-sensitive bis-silylated 2-NH_2-acetylated guanine must be reacted *in situ* with alkylating agents such as peracetylated sugars or chloromethyl ether derivatives in the presence of tetrabutylammonium fluoride (TBAF) or a Lewis acid catalyst ($SnCl_4$ or TMSOTf). Expensive hexamethyldisilazide (HMDS) or bis(trimethylsilyl)amide (BSA) is typically used as the silylating reagent. Phase-transfer conditions are sometimes preferred to increase the regioselectivity of the substitution, and higher reaction temperatures supposedly facilitate the conversion of the kinetically favored N^7-isomer into the desired N^9-isomer. For most of these methods, a chromatographical purification of the final product is needed. Although high yields and selectivities were reported for some glycosylations, which could be carried out in a one-pot modality, the preparation of guanosines containing acyclic sugar mimetics lacking the activating β-oxygen atom (e.g., in **7**) was vexed by poor yields and/or low N^9 versus N^7 selectivity. An improved procedure that includes a trisilylated guanine precursor was more recently described by Mitra (Scheme 14.1). The alkylation product **20** was obtained with complete regioselectivity in 93% yield and without the need of a chromatographic purification [39].

In many cases an "enol" form of the purine base, such as a 6-alkoxy purine or a 6-halopurine, is preferred as the starting material, mainly because of its better solubility compared to the parent guanine. The following examples of alkylations are reported in good yields (after column chromatography) and with minimal amounts of the undesired N^7 isomer. For instance, 6-chloropurine was the starting material for an efficient and highly regioselective synthesis of **7**, performed on a gram scale [40]. In another example, CBV (**10**) was prepared with exquisite N^9 selectivity by researchers at Glaxo via a key [$Pd(PPh_3)_4$]-catalyzed alkylation of 2-amino-6-chloropurine (**23**) with cyclopentadiene monoepoxide [41]. The Pd(0)-catalyst approach to nucleoside analogs, initially pioneered by Trost *et al.* [42], was further advanced by researchers at Tohoku University for the preparation of **10**. They utilized 2-azabicyclo[2.2.1]hept-5-en-3-one (ABH), which is produced on an industrial scale, as the electrophile for a proficient alkylation of the tetrabutylammonium salt of 2-formamido-6-chloropurine [43]. Another example is provided by Crimmins *et al.*, who utilized Pd(0)-catalyzed N^9-allylic alkylations of purines with carbocyclic electrophiles, yielding abacavir (**9**, ABV) with optimized $N^9:N^7$ selectivity of 95 : 5. Trost also reported a method of synthesis of **10** that takes advantage of π-allyl palladium chemistry [44], which was later expediently applied at SmithKline Beecham to the preparation of an acyclic nucleoside (**8**) with an overall remarkable 53% yield, a testament to the power of palladium-catalyzed allylation substitutions (Scheme 14.2) [45].

Palladium(0)-catalysis did not, however, deliver the alkylation products for the reactions of 2-amino-6-chloropurine (**23**) with cyclohexenol-based electrophiles. In this case, researchers at the Rega Institute resorted to a Mitsunobu-type condensation of an allylic alcohol with the purine base, which delivered the target cyclohexenyl guanosine product **18** (Figure 14.3) in 46% yield after removal of all protecting groups [46]. In this analog, the ring oxygen of the natural furanose has been replaced by a double bond.

Geen *et al.* [47] noted that the $N^9:N^7$ alkylation ratio can be improved from 5.5 : 1 to 9 : 1 on switching from 2-amino-6-chloro-purine to 2-amino-6-iodo-purine. Scientists at Bristol-Myers Squibb successfully coupled a tetralkylammonium salt of 2-amino-6-iodo-purine with an equimolar amount of cyclobutyl triflate in a highly regioselective synthesis to obtain lobucavir (**14**) [48]. The same procedure was later adopted by Poisson *et al.* [49], who developed a novel asymmetric synthesis of cyclobutanones, including the precursor required to install the carbocyclic unit of **14**.

2-Amino-6-halopurines, however, are not ideal starting materials because of their difficult preparation and their proven mutagenicity [50]. Therefore Kobe attempted to find conditions for a selective N^7-alkylation of guanosine (**2**), cheaply manufactured on large scale by fermentation. The obtained 7-benzylguanine (**24**)

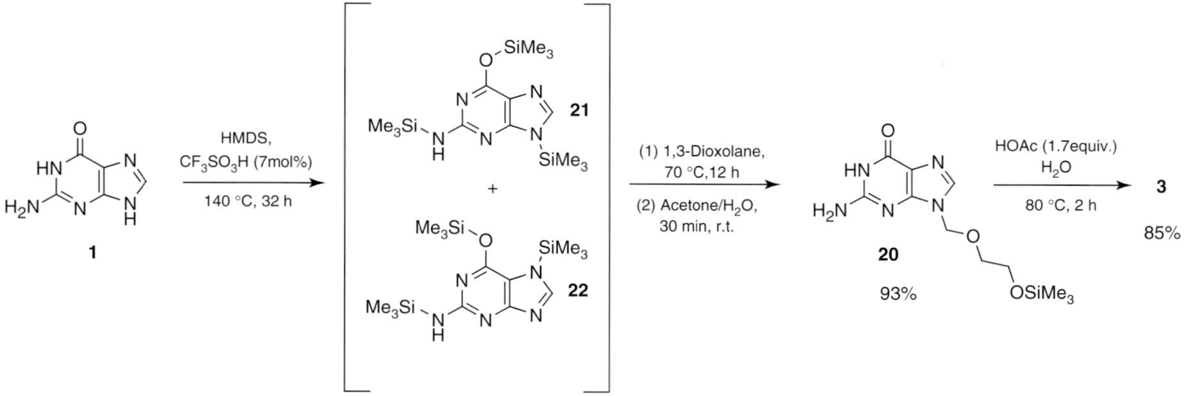

Scheme 14.1 Synthesis of aciclovir (**3**) via persilylated guanine (**21**/**22**).

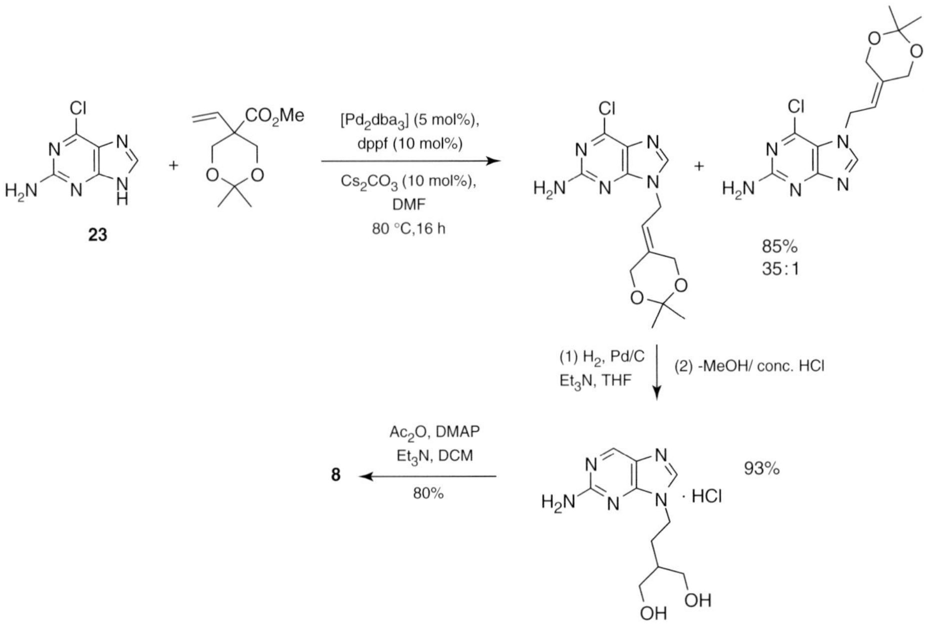

Scheme 14.2 Synthesis of famciclovir (**8**) via a key Pd(0)-catalyzed allylic alkylation.

was subsequently converted into N^9 derivatives, an example of which is given in Scheme 14.3 [51].

A team of industrial scientists in Japan independently reported on a similar protocol for improved synthesis of PCV (**7**) and its prodrug, FCV (**8**). They opted for 7-benzyl-N^2-acetyl guanine (**25**) as a substrate in the alkylation with 6,6-dimethyl-5,7-dioxaspiro[2.5]octane-4,8-dione (**26**) over four steps to provide PCV (**7**) in an excellent 57% yield (Scheme 14.4) [52], a significant improvement over the 23% yield of a prior procedure starting from 2-amino-6-chloropurine (**23**) [53].

A couple of recent methods to prepare prodrugs of guanosines containing acyclic sugar mimetics are worth considering here. Scientists at Bristol-Myers Squibb devised a regioselective aminoacylation with valine of one of the two hydroxyl groups of lobucavir (**14**), obtaining the corresponding prodrug, BMS 233866, which was submitted to preclinical development [54]. The process required the selective enzymatic hydrolysis of a diacetylated lobucavir intermediate, performed with Chiro-CLEC™ PC, a cross-linked crystalline version of *Pseudomonas cepacia* lipase. A high-yielding production of VACV (**4**) via a solid-to-solid biotransformation of ACV (**3**) was recently described using ChiroCLEC-BL, a cross-linked form of subtilisin Carlsberg [55]. The enzymatic reverse hydrolysis of an intermediate diacyl-ACV has potential as an alternative to the current commercial process that relies on dicyclohexylcarbodiimide-mediated coupling, followed by a Pd-catalyzed hydrogenolysis of the protecting group.

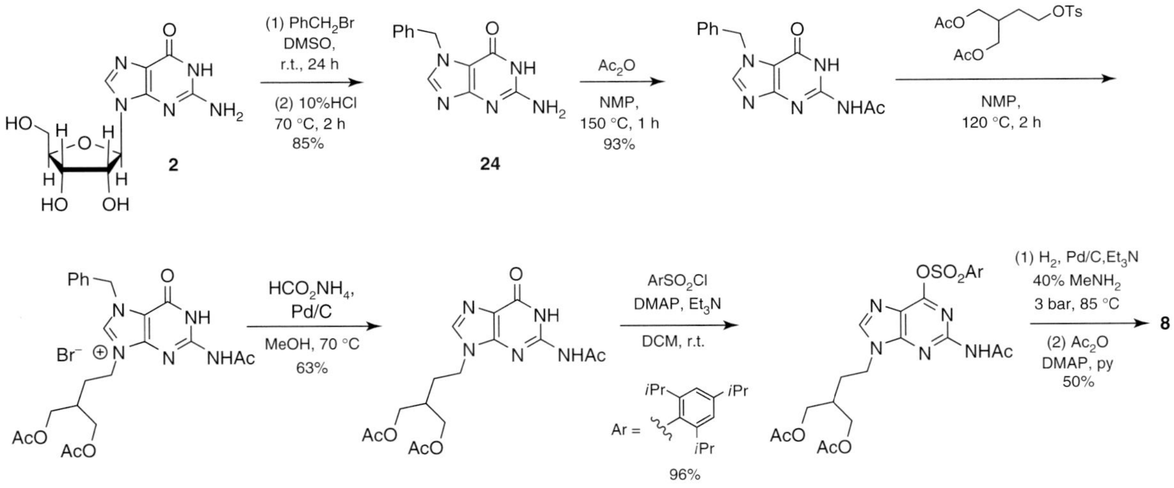

Scheme 14.3 Alternate synthesis of famciclovir (**8**) via 7-benzylguanine (**24**).

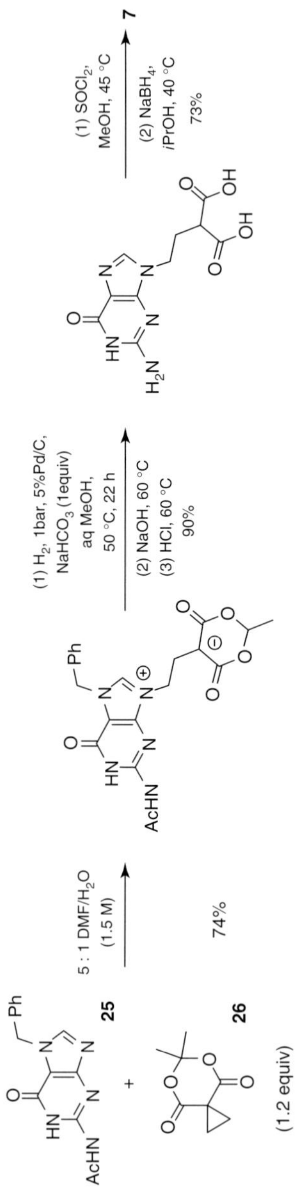

Scheme 14.4 Improved synthesis of penciclovir (**7**).

14.4
Mode of Action

Only eight distinct viruses of the greater than 80-member Herpesviridae family are known to cause infections in humans [56]. By virtue of the availability of more competent and selective antiviral compounds, a better understanding has been achieved of the replication and disease-inducing state of α-herpes viruses in the past two decades [57]. Those include agents causing neurotropic infections such as HSV-1 (most commonly manifested in gingivostomatitis and orolabial infections), HSV-2 (in genital warts and ulcers), and VZV (in shingles and varicella), as well as lymphotropic infections agents HCMV and EBV. Still, relatively little is known about HHV-6, HHV-7, and HHV-8, as they were isolated and identified only in the early 1990s [58]. HHV-6 and HHV-7, together with CMV, belong to the β-herpes virus family, whereas HHV-8 and EBV are γ-herpes viruses. After primary infection, all HHVs enter long-term latency, only to reactivate intermittently, especially during periods of immunosuppression. Infections in most cases are self-limited and usually do not require antiviral therapy; however, they can be the source of unbearable pain (VZV) or can be potentially life-threatening (HCMV) in AIDS patients and organ transplant recipients [59]. The emergence among immunocompromised patients of resistance against most of the approved drugs has reached levels of concern [60], in particular, because of their shared mechanism of action.

The Wellcome team led by Trudy Eliot, who discovered ACV (**3**), also gave the first fundamental contribution to the understanding of its mechanism of action. The team found that **3** was phosphorylated up to 120-fold faster in HSV-infected cells than in uninfected Vero cells [61]. They identified the responsible enzyme as the virus-coded thymidine kinase, TK (EC 2.7.1.75). ACV (**3**) was inactive in mutant strains of virus deficient in TK, and therefore, the first phosphorylation was the necessary metabolic activation step that would also trap the now negatively charged molecule inside the HSV-infected cell. The host cell GMP kinase (EC 2.7.4.8) catalyzes the transformation of aciclovir monophosphate (ACV-MP) into the triphosphate ACV-TP, eventually causing the inhibition of viral DNA polymerase (Pol-UL42) [62]. The polymerase-associated 3′,5′-exonuclease cannot excise ACV-MP that incorporates into the 3′-end of the primer [63]; consequently, inhibition of HSV-1 DNA polymerase occurs on binding of the next coded 5′-TP, forming a dead-end complex [64]. Conversely, ACV-TP is only a weak inhibitor of the host DNA polymerase α [65]. The crystal structure of HSV-1 DNA polymerase (resolved at 2.7 Å) enabled the proposal of a detailed model for the ACV-TP · polymerase dead-end complex, thus providing an explanation at the molecular level of drug resistance ensuing from mutant forms of DNA polymerase [66]. A similar mechanism of action was found for GCV (**5**, GCV), although GCV-MP is a more efficient substrate than ACV-MP for the GMP kinase, which resulted in a much higher rate of complete phosphorylation to GCV-TP [19]. Only one of the two prochiral hydroxyl groups gets phosphorylated by the viral TK to give the (*S*)-monophosphate, which is preferred by GMP kinase over the (*R*)-enantiomer [67]. The crystal structures (refined to 2.8 Å) of HSV-1 TK complexed with natural substrate deoxythymidine and with GCV (2.2 Å) reveals that one of the hydroxyl groups of the ligand is in the same pocket environment as the O-5′ of deoxythymidine [68]. A more recent kinetics analysis has shed further light on the mechanism of DNA polymerase inhibition. It was determined that the natural substrate deoxy-GTP is incorporated by the Pol-UL42 enzyme with a favorable discrimination factor of about 50 over ACV-TP. Once incorporated, ACV (**3**) is excised by the proofreading exonuclease with a half-life of ~1 h, which results in a meager therapeutic index of 42 in favor of viral replication suppression over human mitochondrial DNA polymerase poisoning [69]. However, mitochondrial toxicity is relatively low because **3** is activated only in infected cells by the promiscuous viral TK.

Although acting through the same mechanism, the antiviral activity of PCV (**7**) in cell cultures persists longer than with ACV (**3**), because PCV-TP is more stable than ACV-TP (half-life of 10–20 h vs. 0.7–1 h) within the virus-infected cells [70]. Similar considerations are valid for BCV (**12**), although BCV-TP was deemed to inhibit the HSV-DNA polymerase without being incorporated into DNA [71].

A different mechanism of action was purported for cyclohexenyl guanosine analogs (such as **18**). Surprisingly, both the D- and the "unnatural" L-cyclohexenyl nucleosides were selectively potent against HSV, CMV, and VZV, which raises the possibility that these analogs directly inhibit the viral TK, known to be promiscuous toward D- and L-β-nucleoside substrates [72]. Although the TK enzyme encoded by HSV-1 is not required for efficient viral replication, it is supposed to be important in its reactivation from latency. The viral TK is a potential target to curb the recurrence of infection via inhibition by an L-nucleoside that is not expected to show any activity against the host enzymes, thus potentially avoiding toxicity [73].

Different enzymes are involved with the metabolic activation of (−)-CBV (**10**, CBV), which is phosphorylated by the mammalian 5′-nucleotidase to give (−)-CBV-MP. The latter, having the "natural" configuration of β-D-nucleoside monophosphates, is an efficient substrate of GMP kinase for complete phosphorylation to (−)-CBV-TP, which inhibits HIV-RT by termination of transcription through competitive incorporation of (−)-CBV 5′-MP into newly synthesized DNA [74]. Despite its potency, *in vivo* preclinical studies revealed poor oral bioavailability, limited CNS penetration (which could be overcome to a great extent with the prodrug ABV (**9**) [75]), and dose-limiting kidney and cardiac toxicities, problems that sealed the fate of CBV (**10**).

The first proposed prodrug of ACV (**3**) was the more soluble 6-deoxyaciclovir that can be readily oxidized to **3** by xanthine oxidase (EC 1.2.3.2). A study of 6-deoxyaciclovir in humans confirmed a fivefold increase of mean urinary recovery of **3** [76].

Since only L-amino acid esters of **3** behaved as prodrugs, but not the corresponding D-antipodes, an active transporter was invoked to warrant for the increased systemic bioavailability of VACV (**4**). A carrier-mediated intestinal absorption, via the hPEPT1 peptide transporter, was found responsible for the transport of VACV (**4**) and valganciclovir (**6**) [77]. VACV (**4**) is hydrolyzed *in vivo* to ACV (**3**) by the action of a serine hydrolase, human biphenyl hydrolase-like protein (hBph-rp) [78].

The "third-generation" nucleoside prodrug FCV (**8**) is orally well absorbed in mice, rats, and humans, and then converted to PCV (**7**) by enzymatic removal of two *O*-acetyl groups, followed by oxidation at C-6 on the purine moiety by xanthine oxidase [79].

Other prodrugs of guanosines containing acyclic sugar mimetics also contain a phosphate or phosphonate group. In an application of the so-called "ProTide" technology, debuted by an industry–academia joint effort, the phenylmethoxyalaninyl phosphoramidate of ABV (**9**) achieved an up to 60-fold increase in *in vivo* anti-HIV potency [80]. The intracellular metabolic activation of these phosphoramidates proceeds through an initial methyl ester hydrolysis, followed by carboxylate-assisted phenol expulsion to form a transient anhydride, which in turn is hydrolyzed to give a stable phosphoramidate acid (Scheme 14.5).

Finally, a phosphoramidase is supposed to cleave the amino acid and generate ABV-MP. The detected intermediate ABV-MP is converted to CBV-MP by the action of a specific cytosolic deaminase [81]. DAPD (**11**), another HIV-RT inhibitor, is analogously deaminated *in vivo* to the active metabolite, β-D-dioxolane guanosine [82]. The nucleoside "kinase bypass" also revealed its potential with ProTide ACV, active also against TK-deficient HSV-1 [83].

14.5 Structure–Activity Relationship

Among the modifications of the guanine core, substitution on C-8 (as in methyl-, amino-, bromo-, and iodoaciclovir) gave rise to derivatives with only moderately lower *in vitro* antiviral potency than the parent drug, but with higher specificity for the inhibition of viral DNA synthesis [84]. Replacement of the 6-oxo functionality with a thiol or amino group greatly reduced potency in GCV (**5**) [85]. Deletion of N^9 (to give an oxypyrimidopyrrole system) or addition of N^8 (oxypyrimidotriazole) all but erased antiviral activity, confirming the necessity to maintain the purine nucleus intact [86]. In another approach, only monomethylation on N^1 was tolerated, albeit the N^1-methyl-ACV analog was 10-fold less active than the parent compound. Methylation of ACV (**3**) on 2-NH_2 or on N^7 vastly reduced potency [87]. The mono- and diesters of PCV (**7**) with the 6-oxo functionality replaced by a 6-fluoro substituent were readily absorbed in the gastrointestinal tract and hydrolyzed by esterases. Conversion of these compounds to **7** by adenosine deaminase was less efficient than that of FCV (**8**) [88]. However, 6-fluoro analogs of ACV (**3**) and GCV (**5**) are 12 and 8 times, respectively, more efficiently metabolized to their parent compounds (in terms of V_{max}/K_m) than the corresponding 6-amino derivatives [89].

Regarding the acyclic side chain, one of first published reports was concerned with analogs containing amine functionalities. Of those, only one compound, the analog of ACV (**3**) with an amino functionality in place of the hydroxyl group, showed some activity, although 80-fold less potent than **3** [90]. Analogs of GCV (**5**) with S, SO, or SO_2 in place of the hemiaminal ether oxygen exhibited slightly

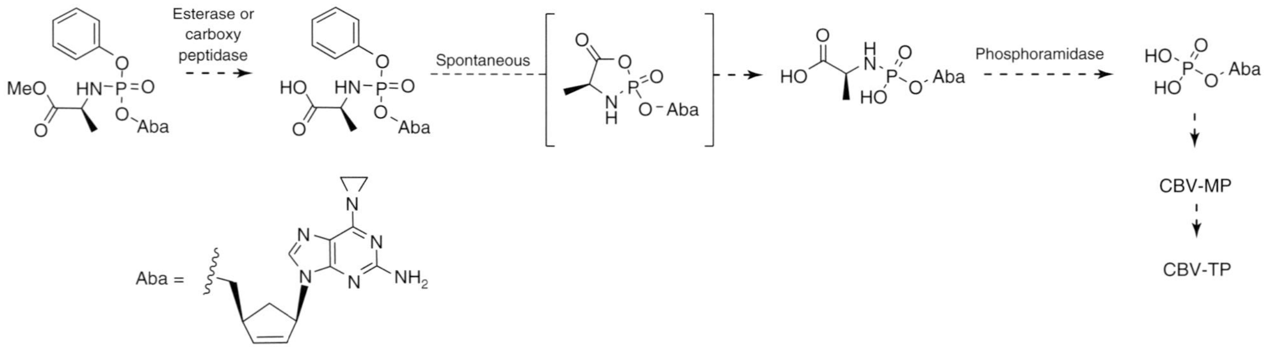

Scheme 14.5 Metabolic activation of Abacavir ProTide.

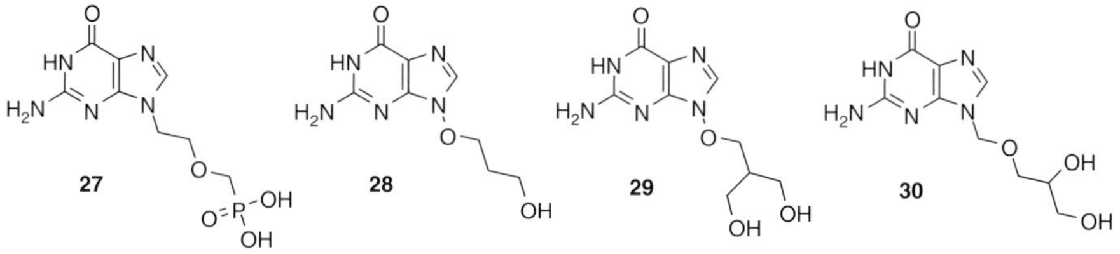

Figure 14.5 Structures of PMEG (**27**) and positional isomers of aciclovir (**3**): **28** and ganciclovir (**5**): **29** and **30**.

reduced activity against HSV-1, but were much weaker against HSV-2 [91]. The cyclic phosphate of **5** was ascertained to penetrate the cell membrane, bypass the TK-promoted phosphorylation step, and to maintain activity [92]. Of a series of phosphonate ACV analogs, optimal activity was achieved when two carbon atoms separated the phosphonomethoxy functionality from the purine base. The obtained lead compound, PMEG (**27**) (Figure 14.5) was up to 13-fold more potent than ACV (**3**) against HSV-1 and HSV-2 in tissue cultures and 1000-fold more active against HCMV [93]. The β-oxygen atom in the phosphonomethyl ether functionality enhances the acidity of the phosphonate, thus lowering the pK_a value closer to that of a phosphate group.

The ACV positional isomer **28** and the corresponding GCV isomer **29** (Figure 14.5) showed potent and selective activity against HSV-1, HSV-2, and VZV in cell culture, comparable to or better than **3** [94].

Elongation of the ACV side chain with a hydroxymethyl group, as in iNDG (**30**), and replacement of one of the OH groups with an F substituent, gave active derivatives, although weaker than ACV (**3**) [95]. A significant difference in HSV-1 DNA polymerase inhibition was noted for the two enantiomers of the iNDG-TP. Specifically, (*S*)-iNDG was up to 25-fold more active than the antipode and had an ED_{50} comparable to that of GCV (**5**); both enantiomers were, however, inactive against HCMV [96]. Similarly, GCV analogs where one of the OH groups was replaced by a variety of functionalities (amino, azido, chloro, fluoro, and methoxy) resulted in compounds that were far less active *in vitro* than the parent compound, although they were still functional TK substrates [97].

All nucleoside analogs in which the furanose was replaced by a pyranose were found to be devoid of antiviral activity. Six-membered rings are conformationally less flexible than their five-membered counterparts, thus compromising the steric and electronic fit with the viral TK enzyme [98]. By relocating the 5′-oxygen atom in the six-membered ring one position away from the guanine, such as in anhydrohexitol guanine **17**, the geometry and conformation of the nucleoside changes, placing the purine base now in an axial position with respect to the pseudosugar [99]. The new series of 1,5-anhydro-2,3-dideoxy-D-ribohexitol nucleosides exhibits anti-HSV-1, HSV-2, and VZV activity, without evidence of cytotoxicity [100]. Indeed, when one of the proposed anhydrohexitol nucleosides was cocrystallized with HSV-1 TK, the ligand adopted a different conformation than the one that was calculated to be the most thermodynamically stable. This is an indication that the imparted flexibility in the nucleoside allows the enzyme to induce a conformational change in the ligand on binding [101].

"Carbocyclic" nucleoside analogs are those in which a cyclopentane ring replaces the purine-bound pentofuranose, conferring metabolic stability to phosphorylase enzymes that cleave the glycosidic linkage of "regular" nucleosides [102]. Among the first examples to be proposed, 1-guaninyl-3-hydroxy-4-hydroxymethyl cyclopentane (2′-CDG) was a potent anti-HSV agent [103]. Monofluorinated variants of this compound (on C2′ or C5′) were later prepared, either as racemic mixtures or enantiomerically enriched variants. Some of these analogs were significantly more potent than ACV (**3**) against HSV-1 and HSV-2 in *in vivo* murine survival efficacy studies [104].

In addition to cyclobut-G (**14**), another pair of acyclic nucleosides with two hydroxymethyl groups mimicking the 3′- and the 5′-OH groups of the 2′-deoxyribose moiety was prepared, where a *cis*-bishydroxymethyl cyclopropyl group is linked to the guanine N^9 via a methylene linker. Both enantiomers of this compound, considered 2HM-HBG analogs, manifested strong inhibitory activity against HSV-1 and were 10-fold more potent than ACV (**3**) against VZV [105]. The cyclopropane ring is supposed to impose geometrical and conformational restrictions on the two hydroxyl groups, oriented to positions close to those of the corresponding hydroxyl groups in 2′-deoxyguanosine. Not surprisingly, the *trans*-isomers were inactive.

As a sterically constrained bioisostere of ACV (**3**), synguanol (**15**) showed anti-HCMV activity comparable to GCV (**5**) in the plaque reduction assay [106], was orally effective *in vivo* in murine models, but lacked potency against HSV-1 and HSV-2 [107]. The *Z*-configuration of the double bond is essential for these methylenecyclopropyl analogs to undergo phosphorylation in infected cells. Analogs of **15** featuring one fluorine atom on the cyclopropyl ring also proved effective against HCMV [108]. CPV (**16**) was presented as an extension of the concept of bioisosterism to GCV (**5**) [109]; as an antiviral agent it was particularly effective against HCMV and EBV [110] and was one of the few compounds active against HHV-6 and HHV-8 [111]. CPV (**16**) is currently being developed as a new investigational drug for the treatment of HCMV infections [112]. However, recent findings showed that among the most common mutations described in GCV-resistant clinical cytomegalovirus isolates, some conferred CPV (**16**) cross-resistance with up to 20-fold increases in EC_{50} values [113]. The phosphonate and the corresponding cyclized phosphonate prodrugs of **16** were potent and noncytotoxic inhibitors of HCMV and EBV. Most interestingly, both compounds were active against two HCMV Towne strains with a mutation in *UL97*, the protein kinase encoding gene in HCMV, thus circumventing resistance [114].

References

1. Field, H.J. and De Clerq, E. (2008) *Antiviral Chem. Chemother.*, **19**, 51–62.
2. Vigil, K.J. and Chemaly, R.F. (2010) *Exp. Opin. Pharmacother.*, **11**, 1901–1913.

3. The Body(R), The Complete HIV/AIDS Resource; Nucleoside Analog Reverse Transcriptase Inhibitors in Development. *http://www.thebody.com/content/art6097.html#rti* (accessed 24 April 2012).
4. De Clerq, E. (2010) *J. Med. Chem.*, **53**, 1438–1450.
5. Elion, G.B., Furman, P.A., Fyfe, J.A., De Miranda, P., Bauchamp, L., and Schaeffer, H.J. (1977) *Proc. Natl. Acad. Sci. U.S.A.*, **74**, 5716–5720.
6. Ebada, S.S. and Proksch, P. (2011) *Mini-Rev. Med. Chem.*, **11**, 225–246.
7. Richardson, P., Lee, K., and Lione, L. (2004) PCT Int. Appl. WO 2004078183 A1 20040916.
8. Bergmann, W. and Burke, D.C. (1956) *J. Org. Chem.*, **21**, 226–228.
9. Schaeffer, H.J., Beauchamp, L., de Miranda, P., Elion, G.B., Bauer, D.J., and Collins, P. (1978) *Nature*, **272**, 583–585.
10. Jones, B.R., Coster, D.J., Fison, P.N., Thompson, G.M., Cobo, L.M., and Falcon, M.G. (1979) *Lancet*, **1**, 243–244.
11. Mindel, A., Adler, M.W., Sutherland, S., and Fiddian, A.P. (1982) *Lancet*, **1**, 697–700.
12. Saral, R., Burns, W.H., Laskin, O.L., Santos, G.W., and Lietman, P.S. (1981) *New Engl. J. Med.*, **305**, 63–67.
13. Peterslund, N.A., Seyer-Hansen, K., Ipsen, J., Esmann, V., Schnheyder, H., and Juhl, H. (1981) *Lancet*, **2**, 827–830.
14. Colla, L., De Clerq, E., Busson, R., and Vanderhaeghe, H. (1983) *J. Med. Chem.*, **26**, 602–604.
15. De Jesus, E., Wald, A., Warren, T., Schacker, T.W., Trottier, S., Shahmanesh, M., Hill, J.L., and Brennan, C.A. (2003) *J. Infect. Dis.*, **188**, 1009–1016.
16. Beutner, K.R., Friedman, D.J., Forszpaniak, C., Andersen, P.L., and Wood, M.J. (1995) *Antimicrob. Agents Chemother.*, **39**, 1546–1553.
17. Smith, K.O., Galloway, K.S., Dennell, W.L., Ogilvie, K.K., and Radatus, B.K. (1980) *Antimicrob. Agents Chemother.*, **22**, 55–61.
18. Martin, J.C., Dvorak, C.A., Smee, D.F., Matthews, T.R., and Verheyden, J.P.H. (1983) *J. Med. Chem.*, **26**, 758–761.
19. Field, A.K., Davies, M.E., DeWitt, C., Perry, H.C., Liou, R., Germershausen, J., Karkas, J.D., Ashton, W.T., Johnston, D.B.R., and Tolman, R.L. (1983) *Proc. Natl. Acad. Sci. U.S.A.*, **80**, 4139–4143.
20. Shepp, D.H., Dandliker, P.S., de Miranda, P., Burnette, T.C., Cedeberg, D.M., Kirk, L.E., and Meyers, J.D. (1985) *Ann. Intern. Med.*, **103**, 368–373.
21. Pandit, U.K., Grose, W.F.A., and Eggelte, T.A. (1972) *Synth. Commun.*, **2**, 345–351.
22. Harnden, M.R. and Jarvest, R.L. (1985) *Tetrahedron Lett.*, **26**, 4265–4268.
23. Harnden, M.R., Jarvest, R.L., Boyd, M.R., Sutton, D., and Vere Hodge, R.A. (1989) *J. Med. Chem.*, **32**, 1738–1743.
24. Chakrabarty, A., Tyrubg, S.K., Beutner, K., and Rauser, M. (2004) *Antiviral Chem. Chemother.*, **15**, 251–253.
25. Larsson, A., Oeberg, B., Alenius, S., Hagberg, C.E., Johansson, N.G., Lindborg, B., and Stening, G. (1983) *Antimicrob. Agents Chemother.*, **23**, 664–670.
26. Lundgren, B., Ericson, A.C., Berg, M., and Datema, R. (1986) *Antimicrob. Agents Chemother.*, **29**, 294–297.
27. Larrson, A., Stenberg, K., Ericson, A.C., Haglund, U., Yisak, W.A., Johansson, N.G., Oeberg, B., and Datema, R. (1986) *Antimicrob. Agents Chemother.*, **30**, 598–605.
28. Abele, G., Karlstroem, A., Harmenberg, J., Shigeta, S., Larsson, A., Lindborg, B., and Wahren, B. (1987) *Antimicrob. Agents Chemother.*, **31**, 76–80.
29. Slusarchyk, W.A., Young, M.G., Bisacchi, G.S., Hockstein, D.R., and Zahler, R. (1989) *Tetrahedron Lett.*, **30**, 6453–6456.
30. Hoshino, H., Shimuzu, N., Shimada, N., Takita, T., and Takeuchi, T. (1987) *J. Antibiot.*, **40**, 1077–1078.
31. Bisacchi, G.S., Braitman, A., Cianci, C.W., Clark, J.M., Field, A.K., Hagen, M.E., Hockstein, D.R., Malley, M.F.,

Mitt, T., Slusarchyk, W.A., Sundeen, J.E., Terry, B.J., Tuomari, A.V., Weaver, E.R., Young, M.G., and Zahler, R. (1991) *J. Med. Chem.*, **34**, 1415–1421.
32. Jones, C.A. (2009) *Drugs*, **69**, 421–434.
33. Mills, J.L. and Carter, T.C. (2010) *J. Am. Med. Assoc.*, **304**, 905–906.
34. Wilson, S.S., Fakioglu, E., and Herold, B.C. (2009) *Exp. Rev. Anti-Infect. Ther.*, **7**, 559–568.
35. Vanpouille, C., Lisco, A., and Margolis, L. (2009) *Curr. Opin. Infect. Dis.*, **22**, 583–587.
36. McMahon, M.A., Parson, T.L., Shen, L., Siciliano, J.D., and Siciliano, R.F. (2011) *J. Virol.*, **85**, 4618–4622.
37. Gao, H. and Mitra, A.K. (2000) *Synthesis*, **3**, 329–351.
38. Schaeffer, H.J. (1974) Ger. Patent No. 2, 539, 963
39. Gao, H. and Mitra, A.K. (2001) *Synth. Commun.*, **31**, 1399–1419.
40. Toyokuni, T., Walsh, J.C., Namavari, M., Shinde, S.S., Moore, J.R., Barrio, J.R., and Satyamurthy, N. (2003) *Synth. Commun.*, **33**, 3897–3905.
41. Peel, M.R., Sternbacj, D.D., and Johnson, M.R. (1991) *J. Org. Chem.*, **56**, 4990–4993.
42. Trost, B.M., Kuo, G.-H., and Benneche, T. (1988) *J. Am. Chem. Soc.*, **110**, 621–622.
43. Katagiri, N., Takebayashi, M., Kokufuda, H., Kaneko, C., Kanehira, K., and Torihara, M. (1997) *J. Org. Chem.*, **62**, 1580–1581.
44. Trost, B.M., Madsen, R., Guile, S.G., and Elia, A.E.H. (1996) *Angew. Chem., Int. Ed. Engl.*, **35**, 1569–1572.
45. Freer, R., Geen, G.R., Ramsay, T.W., Share, A.C., Slater, G.R., and Smith, N.M. (2000) *Tetrahedron*, **56**, 4589–4595.
46. Wang, J., Froeyen, M., Hendrix, C., Andrei, G., Snoeck, R., De Clercq, E., and Herdewijn, P. (2000) *J. Med. Chem.*, **43**, 736–745.
47. Geen, G.R., Grinter, T.J., Kincey, P.M., and Choudary, B.M. (1990) *Tetrahedron*, **46**, 6903–6914.
48. Bisacchi, G.S., Singh, J., Godfrey, J.D. Jr., Kissick, T.P., Mitt, T., Malley, M.F., Di Marco, J.D., Gougoutas, J.Z., Mueller, R.H., and Zahler, R. (1995) *J. Org. Chem.*, **60**, 2902–2905.
49. Darses, B., Greene, A.E., Coote, S.C., and Poisson, J.-F. (2008) *Org. Lett.*, **10**, 821–824.
50. Kaina, B. and Aurich, O. (1984) *Mutat. Res.*, **139**, 149–154.
51. Kalayanvov, G., Jaška, S., Scarcia, T., and Kobe, J. (2004) *Synthesis*, **12**, 2026–2034.
52. Torii, T., Yamashita, K., Kojima, M., Suzuki, Y., Hijiya, T., and Izawa, K. (2006) *Nucleosides Nucleotides Nucleic Acids*, **25**, 625–634.
53. Choudary, B.M., Geen, G.R., Kincey, P.M., Parratt, M.J., Dales, J.R.M., Johnson, G.P., O'Donnell, S., Tudor, D.W., and Woods, N.A. (1996) *Nucleosides Nucleotides*, **15**, 981–994.
54. Hanson, R.L., Shi, Z., Brzozowski, D.B., Banerjee, A.B., Kissick, T.P., Singh, J., Pullockaran, A.J., North, J.T., Fan, J., Howell, J., Durand, S.C., Montana, M.A., Kronenthal, D.R., Mueller, R.H., and Patel, R.N. (2000) *Bioorg. Med. Chem.*, **8**, 2681–2687.
55. McClean, K., Preston, C., Spence, D., Sutton, P.W., and Whittall, J. (2011) *Tetrahedron Lett.*, **52**, 215–218.
56. Coen, D.M. and Schaffer, P.A. (2003) *Nat. Rev.*, **2**, 278–288.
57. Villareal, E. (2003) *Prog. Drug Res.*, **60**, 263–307.
58. De Clercq, E., Naesens, L., De Bolle, L., Schols, D., Zhang, Y., and Neyts, J. (2001) *Rev. Med. Virol.*, **11**, 381–395.
59. Eid, A.J. and Razonable, R.R. (2010) *Drugs*, **70**, 965–981.
60. Piret, J. and Boivin, G. (2011) *Antimicrob. Agents Chemother.*, **55**, 459–472.
61. Fyfe, J.A., Keller, P.M., Furman, P.A., Miller, R.L., and Elion, G.B. (1978) *J. Biol. Chem.*, **253**, 8721–8727.
62. Miller, W.H. and Miller, R.L. (1980) *J. Biol. Chem.*, **255**, 7204–7207.
63. Derse, D., Cheng, Y.-C., Furman, P.A., St. Clair, M.H., and Elion, G.B. (1981) *J. Biol. Chem.*, **256**, 11447–11451.
64. Reardon, J.E. and Spector, T. (1989) *J. Biol. Chem.*, **264**, 7405–7411.
65. Reardon, J.E. (1989) *J. Biol. Chem.*, **264**, 19039–19044.
66. Liu, S., Knafels, J.D., Chang, J.S., Waszak, G.A., Baldwin, E.T.,

Deibel, M.R. Jr., Thomsen, D.R., Homa, F.L., Wells, P.A., Tory, M.C., Poorman, R.A., Gao, H., Qiu, X., and Seddon, A.P. (2006) *J. Biol. Chem.*, **281**, 18193–18200.
67. Karkas, J.D., Germerhausen, J., Tolman, R.L., MacCoss, M., Wagner, A.F., Liou, R., and Bostedor, R. (1987) *Biochim. Biophys. Acta*, **911**, 127–135.
68. Brown, D.G., Visse, R., Sandhu, G., Davies, A., Rizkallah, P.J., Melitz, C., Summers, W.C., and Sanderson, M.R. (1995) *Nat. Struct. Biol.*, **2**, 876–881.
69. Hanes, J.W., Zhu, Y., Parris, D.S., and Johnson, K.A. (2007) *J. Biol. Chem.*, **282**, 25159–25167.
70. Boyd, M.R., Bacon, T.H., Sutton, D., and Cole, M. (1987) *Antimicrob. Agents Chemother.*, **31**, 1238–1242.
71. Stenberg, K., Larsson, A., and Datema, R. (1986) *J. Biol. Chem.*, **261**, 2134–2139.
72. Spadari, S., Maga, G., Focher, F., Ciarrocchi, G., Manservigi, R., Aracmone, F., Capobianco, M., Carcuro, A., Colonna, F., Iotti, S., and Garbesi, A. (1992) *J. Med. Chem.*, **35**, 4214–4220.
73. Tenser, R.B. (1991) *Intervirology*, **32**, 76–92.
74. Miller, W.H., Daluge, S.M., Garvey, E.P., Hopkins, S., Reardon, J.E., Boyd, F.L., and Miller, R.L. (1992) *J. Biol. Chem.*, **29**, 21220–21224.
75. Daluge, S.M., Good, S.S., Faletto, M.B., Miller, W.H., St. Clair, M.H., Boone, L.R., Tisdale, M., Parry, N.R., Reardon, J.E., Dornsife, R.E., Averett, D.R., and Krenitsky, J.A. (1997) *Antimicrob. Agents Chemother.*, **41**, 1082–1093.
76. Krenitsky, T.A., Hall, W.W., de Miranda, P., Beauchamp, L.M., Schaeffer, H.J., and Whiteman, P.D. (1984) *Proc. Natl. Acad. Sci. U.S.A.*, **81**, 3209–3212.
77. Kim, I., Chu, X.-Y., Kim, S., Provoda, C.J., Lee, K.-D., and Amidon, G.L. (2003) *J. Biol. Chem.*, **278**, 25348–25356.
78. Puente, X.S. and López-Otin, C. (1995) *J. Biol. Chem.*, **270**, 12926–12932.
79. Pue, M.A. and Benet, L.Z. (1993) *Antivir. Chem. Chemother.*, **4** (Suppl. 1), 47–55.
80. McGuigan, C., Harris, S.A., Daluge, S.M., Gudmundsson, K.S., McLean, E.W., Burnette, T., Marr, H., Hazen, R., Condreay, L.D., Johnson, L., De Clercq, E., and Balzarini, J. (2005) *J. Med. Chem.*, **48**, 3504–3515.
81. Faletto, M.B., Miller, W.H., Garvey, E.P., St. Clair, M.H., Daluge, S.M., and Good, S.S. (1997) *Antimicrob. Agents Chemother.*, **41**, 1099–1107.
82. Mehellou, Y. and De Clercq, E. (2010) *J. Med. Chem.*, **53**, 521–538.
83. Derudas, M., Carta, D., Brancale, A., Vanpouille, C., Lisco, A., Margolis, L., Balzarini, J., and McGuigan, C. (2009) *J. Med. Chem.*, **52**, 5520–5530.
84. Robins, M.J., Hatfield, P.W., Balzarini, J., and De Clercq, E. (1984) *J. Med. Chem.*, **27**, 1486–1492.
85. Martin, J.C., Jeffrey, G.A., McGee, D.P.C., Tippie, M.A., Smee, D.F., Matthews, T.R., and Verheyden, J.P.H. (1985) *J. Med. Chem.*, **28**, 358–362.
86. Beauchamp, L.M., Dolmatch, B.L., Schaeffer, H.J., Collins, P., Bauer, D.J., Keller, P.M., and Fyfe, J.A. (1985) *J. Med. Chem.*, **28**, 982–987.
87. Boryski, J., Golankiewicz, B., and De Clercq, E. (1988) *J. Med. Chem.*, **31**, 1351–1355.
88. Kim, D.-K., Lee, N., Kim, H.-T., Im, G.-J., and Kim, K.H. (1999) *Bioorg. Med. Chem.*, **7**, 565–570.
89. Kim, D.-K., Kim, H.-K., and Chae, Y.-B. (1994) *Bioorg. Med. Chem. Lett.*, **4**, 1309–1312.
90. Kelley, J.L., Krochmal, M.P., and Schaeffer, H.J. (1981) *J. Med. Chem.*, **24**, 1528–1531.
91. McGee, D.P.C., Martin, J.C., Smee, D.F., Matthews, T.R., and Verheyden, J.P.H. (1985) *J. Med. Chem.*, **28**, 1242–1245.
92. Prisbe, E.J., Martin, J.C., McGee, D.P.C., Barker, M.F., Smee, D.F., Duke, E., Matthews, R., and Verheyden, J.P.H. (1986) *J. Med. Chem.*, **29**, 671–675.
93. Kim, C.U., Luh, B.Y., Misco, P.F., Bronson, J.J., Hitchock, M.J.M., Ghazzouli, I., and Martin, J.C. (1990) *J. Med. Chem.*, **33**, 1207–1213.

94. Harnden, M.R., Wyatt, P.G., Boyd, M.R., and Sutton, D. (1990) *J. Med. Chem.*, **33**, 187–196.
95. Karkas, J.D., Ashton, W.T., Canning, L.F., Liou, R., Germershausen, J., Bostedor, R., Arison, B., Field, A.K., and Tolman, R.L. (1986) *J. Med. Chem.*, **29**, 842–848.
96. Ashton, W.T., Canning, L.F., Reynolds, G.F., Tolman, R.L., Karkas, J.D., Liou, R., Davies, M.-E.M., DeWitt, C.M., Perry, H.C., and Filed, A.K. (1985) *J. Med. Chem.*, **28**, 926–933.
97. Martin, J.C., McGee, D.P.C., Jeffrey, G.A., Hobbs, D.W., Smee, D.F., Matthews, T.R., and Verheyden, J.P.H. (1986) *J. Med. Chem.*, **29**, 1384–1389.
98. Hossain, N., Rozenski, J., De Clercq, E., and Herdewijn, P. (1997) *J. Org. Chem.*, **62**, 2442–2447.
99. Verheggen, I., Van Aerschot, A., Van Meervelt, L., Rozenski, J., Wiebe, L., Snoeck, R., Andrei, G., Balzarini, J., Claes, P., De Clercq, E., and Herdewijn, P. (1995) *J. Med. Chem.*, **38**, 826–835.
100. Verheggen, I., Van Aerschot, A., Toppet, S., Snoeck, R., Janssen, G., Balzarini, J., De Clercq, E., and Herdewijn, P. (1993) *J. Med. Chem.*, **36**, 2033–2040.
101. Ostrowski, T., Wroblowski, B., Busson, R., Rozenski, J., De CLercq, E., Bennett, M.S., Champness, J.N., Summers, W.C., Sanderson, M.R., and Herdewjin, P. (1998) *J. Med. Chem.*, **41**, 4343–4353.
102. Crimmins, M.T., King, B.W., Zuercher, W.J., and Choy, A.L. (2000) *J. Org. Chem.*, **65**, 8499–8509.
103. Shealy, Y.F., O'Dell, C.A., Shannon, W.M., and Arnett, G. (1984) *J. Med. Chem.*, **27**, 1416–1421.
104. Borthwick, A.D., Kirk, B.E., Biggadike, K., Exall, A.M., Butt, S., Roberts, S.M., Knight, D.J., Coates, J.A.V., and Ryan, D.M. (1991) *J. Med. Chem.*, **34**, 907–914.
105. Sekiyama, T., Hatsuya, S., Tanaka, Y., Uchiyama, M., Ono, N., Iwayama, S., Oikawa, M., Suzuki, K., Okunishi, M., and Tsuji, T. (1998) *J. Med. Chem.*, **41**, 1284–1298.
106. Qiu, Y.-L., Ksebati, M.B., Ptak, R.G., Fan, B.Y., Breitenbach, J.M., Lin, J.-S., Cheng, Y.-C., Kern, E.R., Drach, J.C., and Zemlicka, J. (1998) *J. Med. Chem.*, **41**, 10–23.
107. Rybak, R.J., Zemlicka, J., Qiu, Y.-L., Hartline, C.B., and Kern, E.R. (1999) *Antivir. Res.*, **43**, 175–188.
108. Zhou, S., Kern, E.R., Gullen, E., Cheng, Y.-C., Drach, J.C., Tamiya, S., Mitsuya, H., and Zemlicka, J. (2006) *J. Med. Chem.*, **49**, 6120–6128.
109. Zhou, S., Breitenbach, J.M., Borysko, K.Z., Drach, J.C., Kern, E.R., Gullen, E., Cheng, Y.-C., and Zemlicka, J. (2004) *J. Med. Chem.*, **47**, 566–575.
110. Zemlicka, J. and Chen, X. (2004) in *Frontiers in Nucleosides and Nucleic Acids* (eds R.F. Schinazi and D.C. Liotta), IHL Press, Tucker, GA, pp. 267–307.
111. Zemlicka, J. (2007) in *Advances in Antiviral Drug Design* (ed. E. De Clercq), Elsevier, Amsterdam, pp. 113–165.
112. Gentry, B.G., Kamill, J.P., Coen, D.M., Zemlicka, J., and Drach, J.C. (2010) *Antimicrob. Agents Chemother.*, **54**, 3093–3098.
113. Chou, S. and Bowlin, T.L. (2011) *Antimicrob. Agents Chemother.*, **55**, 382–384.
114. Mhaske, S.B., Ksebati, B., Prochard, M.N., Drach, J.C., and Semlicka, J. (2009) *Bioorg. Med. Chem.*, **17**, 3892–3899.

15
Penicillin and Cephalosporin Antibiotics

Michael Z. Hoemann

15.1
Introduction

The discovery of penicillin by Nobel laureate Alexander Flemming in 1928 [1] through serendipitous observation was heralded as one of the great discoveries in medicine, and he was recognized as one of Time Magazine's most important 100 people of the twentieth century [2]. His discovery led to the development of the β-lactam antibiotic family of which two of the main members are the penicillins (**1**) and the cephalosporins (**2**). These two family members are structurally related based on their β-lactam-containing bicyclic ring systems (Figure 15.1).

A number of semisynthetic analogs as well as nonnatural analogs of medical importance were prepared in the decades to come to complement the natural products (i.e., penicillin G (**1a**)) isolated by Flemming's group. A few examples are illustrated in Figure 15.2.

This chapter focuses on the history, chemistry discoveries, mode of action, and structure–activity relationship (SAR) of both the penicillin (**1**) and cephalosporin (**2**) antibiotics. A key part of the chemistry discoveries is a review of methods to prepare core systems along with the functionalization at a number of key positions that were important for medical reasons.

15.2
History

Flemming's [1] initial report that a contamination of *Staphylococcus* colonies, a *Penicillium* mold, displayed antibacterial activity was a significant breakthrough in what was to become an ever-growing area of antibiotic research. After Flemming's safety assessment in animals, it was widely accepted that "penicillin" was useful as a topical antiseptic. The real implication of Flemming's discovery for the medical community was not realized until Chain and Florey demonstrated that animals infected with *streptococci* and *staphylococci* could be treated with subcutaneous injections of "penicillin" [6]. Flemming used the term "*penicillin*" for the mold

Bioactive Heterocyclic Compound Classes: Pharmaceuticals,
First Edition. Edited by Jürgen Dinges and Clemens Lamberth.

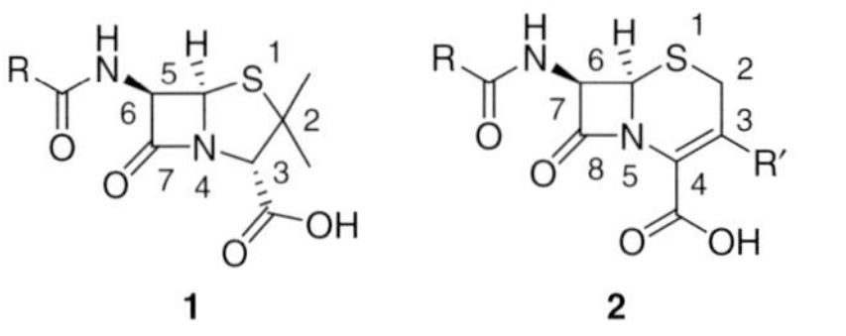

Figure 15.1 Basic structures and numbering conventions for penicillins (**1**) and cephalosporins (**2**).

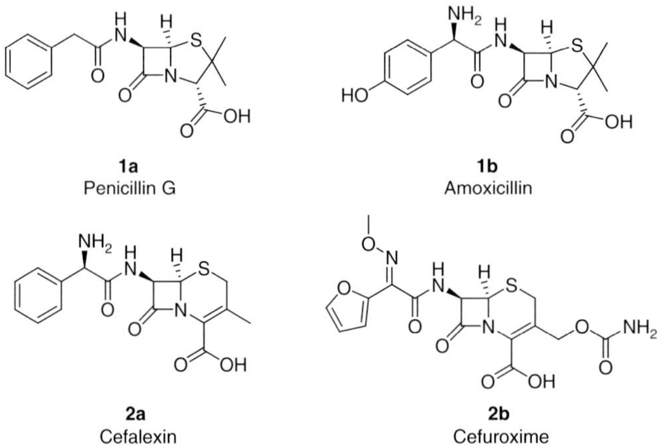

Figure 15.2 Representatives of the penicillin (**1**) and cephalosporin (**2**) antibiotics: penicillin G (**1a**) [1], amoxicillin (**1b**) [3], cefalexin (**2a**) [4], and cerfuroxime (**2b**) [5].

broth filtrates that he isolated, and after his discovery, significant effort was put forth to identify the chemical substances responsible. This opened the door for much of the early synthetic work around penicillins (**1**), focused on the biosynthesis from various *Penicillium* and *Cephalosporium* species. Initially, the starting materials for the ring system were identified as L-cysteine (**4**) and L-valine (**5**) through labeling experiments (Scheme 15.1) [7]. The first step in the biosynthesis of penicillin G (**1a**, generically often just called *penicillin*) is the condensation of three amino acids, L-aminoadipate (**3**), L-cysteine (**4**), and L-valine (**5**), into the tripeptide **6**. During the condensation process, the stereocenter of L-valine (**5**) epimerizes to the D-isomer. The tripeptide then undergoes an oxidative cyclization to form the ring-fused β-lactam known as *isopenicillin* (**7**). Subsequent hydrolysis followed by acylation of the nitrogen yields penicillin G (**1a**). In fact, this biosynthetic pathway was the basis for Baldwin to claim the first stereocontrolled total synthesis of the penicillin core structure starting from a dipeptide as well as for Kishi to complete a formal synthesis of 7-methoxycephalosporin C [8].

One of the major challenges for penicillin (**1a**) was obtaining enough material to run the required clinical trials during world war II. It was through the combined efforts of Florey and Heatley with US Pharmaceutical companies (Merck, Squibb, Pfizer, Abbott, and Winthrop) that a consortium was formed to provide for

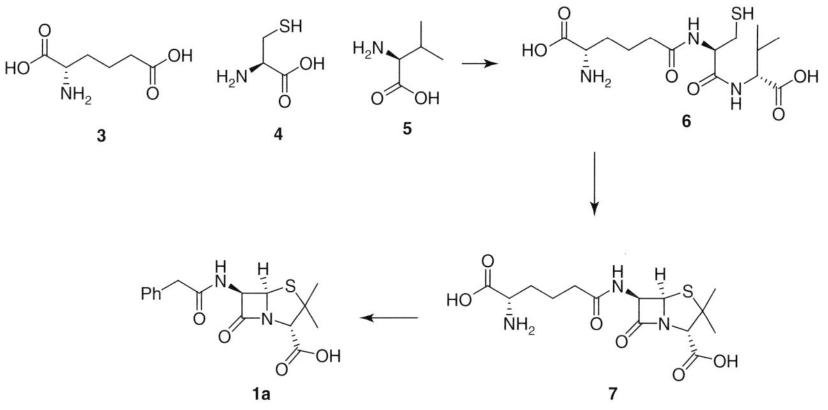

Scheme 15.1 The biosynthetic pathway for penicillin antibiotics (**1**).

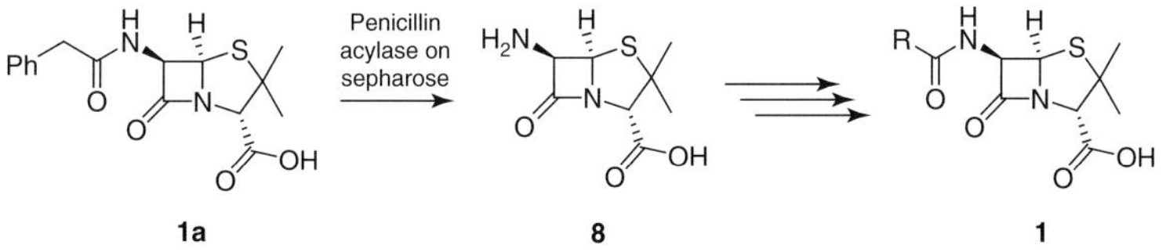

Figure 15.3 Preparation of 6-APA (**8**) and its conversion to semisynthetic penicillin analogs (**1**).

the production of material for clinical trials. The ultimate breakthrough came from the Northern Regional Research Laboratory where a microbiologist, Moyer [9], discovered that the addition of corn steeping liquor to the growth medium increased yields of penicillin (**1a**) by a factor of 10, which enabled the production of the required amounts. This commercialization paved the way for a process to supply not only the much needed antibiotics but also a key intermediate, 6-aminopenicillanic acid (**8**, 6-APA), for the preparation of a large number and variety of semisynthetic penicillins (**1**). Three routes have been identified to obtain 6-APA (**8**): (i) direct fermentation, (ii) total synthesis, and (iii) hydrolysis of naturally occurring penicillin (**1a**). Of the three, the last option, hydrolysis of naturally occurring penicillin (**1a**), proved to be the most commercially viable one and in theory, could deliver tons of material a year (Figure 15.3) [10].

15.3 Synthesis

15.3.1 Total Syntheses

The first total synthesis of natural penicillin G (**1a**), albeit in very low yield, was reported in 1946 by Vigneaud and coworkers [11] by condensing

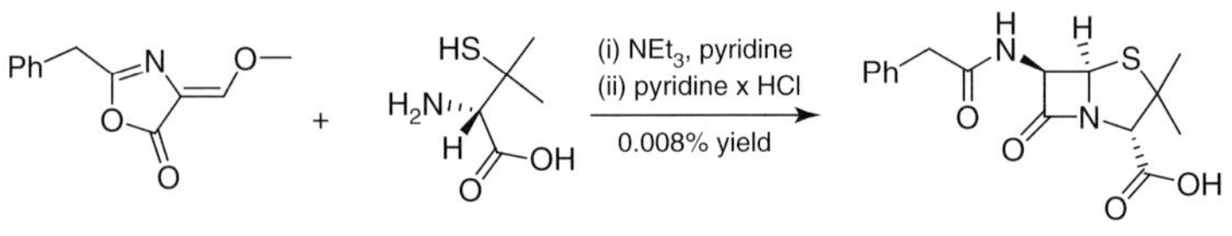

Scheme 15.2 First total synthesis of penicillin G (**1a**).

2-benzyl-4-methoxymethylene-5(4)-oxazazolone (**9**) with D-penicillamine (**10**) (Scheme 15.2). The reaction conditions for the condensation were triethyl amine in pyridine followed by treatment with pyridinium chloride to give penicillin G (**1a**) in 0.008% yield based on the biological activity of the crude product. While this demonstrated that penicillin and active analogs could be made through synthetic means, it was at a time when the exact chemical structure was under debate; therefore, the synthesis was most useful in helping to establish the identity of the β-lactam core .

A second, more scalable approach was developed over the course of the next decade by Sheehan (Scheme 15.3). The syntheses of both penicillin V and penicillin G (**1a**) were reported along with subsequent nonnatural penicillin analogs [12]. This approach started with the condensation of *t*-butylphthalimidemalonaldeydate (**11**) with D-penicillamine (**12**) to give the penicilloic ester **13** [13]. Removal of both the phthalimide and the *t*-butyl groups afforded penicilloic acid (**14**). The amine was protected with a trityl group, and **15** was cyclized with CDI to afford the β-lactam core (**16**) in moderate yield. Removal of the protecting groups and acylation afforded the first practical synthesis of penicillin G (**1a**).

A second Nobel prize, this time for chemistry, was connected with the β-lactam antibiotics when Woodward [14] presented on the first total synthesis of cephalosporin C (**2c**) at his acceptance lecture. As in nature, Woodward began his synthesis with

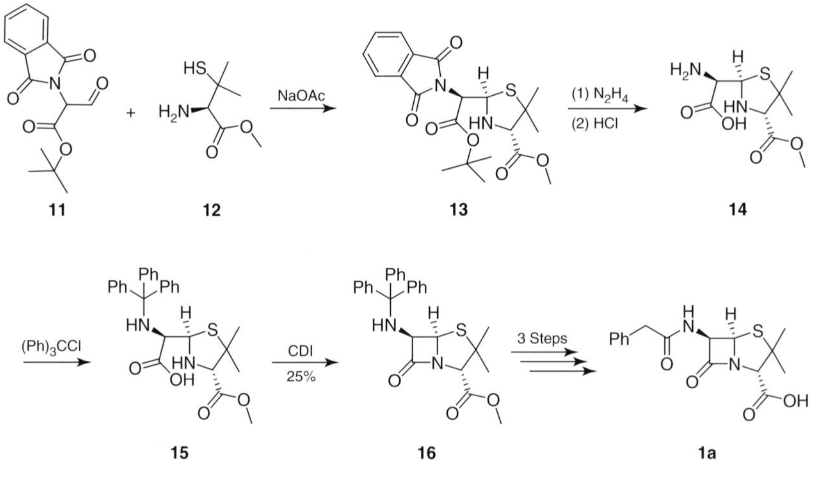

Scheme 15.3 More practical total synthesis of penicillin G (**1a**).

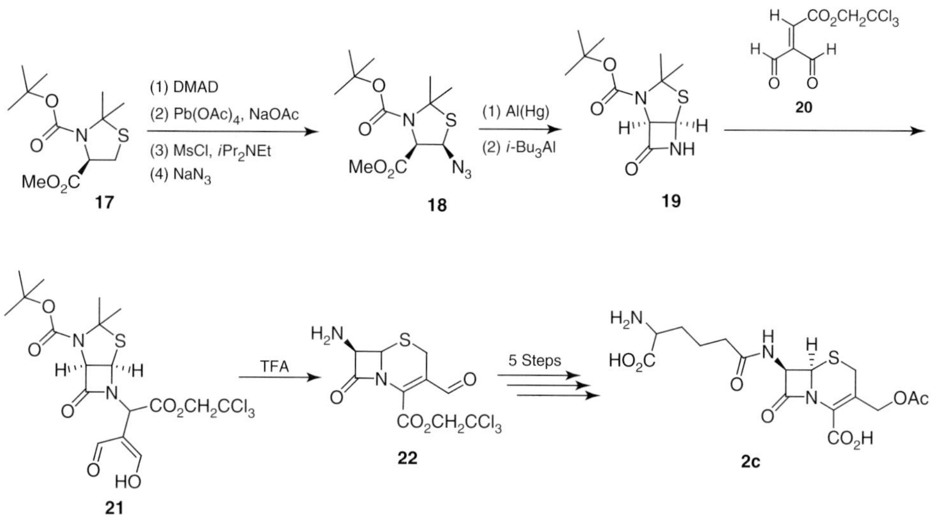

Scheme 15.4 First total synthesis of cephalosporin C (**2c**).

L-cysteine (**4**), which was suitably protected as the 2,2-dimethylthiazolidine (**17**) (Scheme 15.4). This allowed having one of the two chiral centers of the core already in place. The first key challenge was the introduction of what ultimately would be the β-lactam nitrogen at the methylene group adjacent to the sulfur. This had to be done in a stereospecific manner to provide the second chiral center of the core. In a four-step process, an azide was introduced stereospecifically in cis orientation to the ester. Reaction of **17** with dimethylazodicarboxylate (DMAD) formed a hydrazodiester, which was oxidized with $Pb(OAc)_4$ to the trans hydroxyl ester. Subsequent mesylation and S_N2 displacement yielded the cis azido ester (**18**). Reduction of the azide and treatment with a Lewis acid provided the β-lactam (**19**). The next stage of the synthesis was to set up the annulation of the second ring via Michael addition of **19** to the bisaldehyde (**20**) to give the intermediate **21**. Deprotection of the thiazolidine allowed the annulation to occur via a 1,4-addition, followed by elimination of water. The cephalosporin core **22** could then be transformed to cephalosporin C (**2c**) in five steps.

Woodward seemed to discount an approach to the cephalosporin core through β-lactam formation as the last step, based on the lack of a naturally occurring equivalent to penicilloic acid (**14**) for the cephalosporins (**2**). Another approach subsequently showed that the cephalosporin core can be prepared in racemic form from the thiazine intermediate **25** (Scheme 15.5) [15]. In the key step, **25** was N-acylated with α-azidoacetyl chloride (**26**) and the resulting acyliminium intermediate closed to the β-lactam (**27**). One drawback was that the azide group was introduced with the wrong stereochemistry. Therefore, it was necessary to reduce the azide (**27**), convert the resulting amine to the Schiff base (**28**), racemize the 7-position, and then separate the stereoisomers. The cephalosporin (**29**), with the

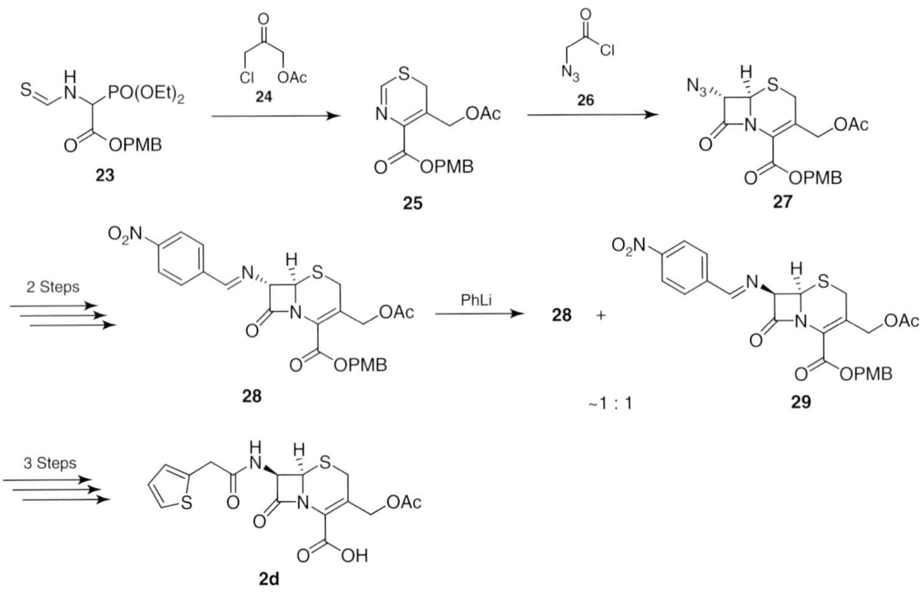

Scheme 15.5 First total synthesis of cephalothin (**2d**).

appropriate stereochemistry, was then used to finish the synthesis of cephalothin (**2d**).

Finally, as with the penicillins (**1**), it was highly desirable to obtain useful intermediates for the synthesis of semisynthetic cephalosporin analogs. Biochemical methods were not as successful as they were for the cleavage of the side chains of cephalosporins (**2**), so chemical means were investigated to obtain larger quantities of 7-aminocephalosporonic acid (**31**, 7-ACA). Morin *et al.* [16] demonstrated that the reactivity of the aminoadipyl side chain of **2c** was amenable to oxidation of the amino group with nitrosyl chloride and subsequent intramolecular displacement to yield intermediate **30** (Scheme 15.6). Hydrolysis with formic acid then gave 7-ACA (**31**) in 40% yield.

An alternative strategy was also reported by Morin. He discovered that the penicillin scaffold, on oxidation of the sulfur to a sulfoxide followed by treatment with acid, underwent a rearrangement that directly yielded cephalosporins (**2**) with the side chains already in place (Scheme 15.7) [17]. Treatment of phenoxymethyl penicillin (**1d**) with sodium periodate and esterification with diazomethane yielded the sulfoxide (**32**). Refluxing **32** with *p*-toluene sulfonic acid (TSA) in xylene affected the core rearrangement to give the cephalosporin ester (**33**). Hydrolysis of the ester under basic conditions was complicated by the isomerization of the double bond to give a mixture of the (**2e**) and (**34**) in a ratio of 3 : 7. The problem could be avoided by using ester groups that could be cleaved under reductive conditions such as a benzyl group. This approach and variants thereof have been used to prepare a number of interesting semisynthetic cephalosporins (**2**), such as cephalexin [18], as well as compounds substituted at the 3′-position [19].

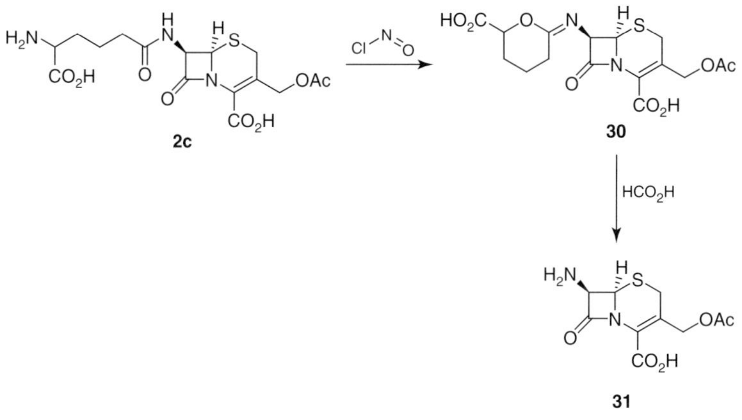

Scheme 15.6 Conversion of cephalosporin C (**2c**) to 7-ACA (**31**).

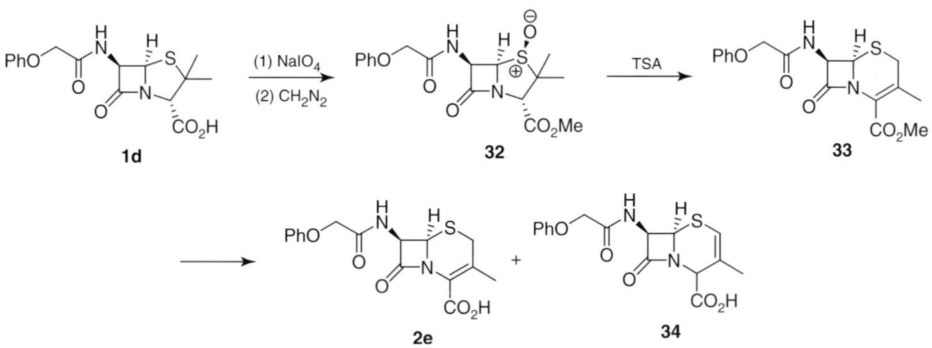

Scheme 15.7 Conversion of penicillin (**1a**) to cephalosporin (**2e**).

While there were a number of subsequent total syntheses methods developed for the cephalosporin scaffold, none of them had the eloquence of the Woodward route or the efficiency of the two synthesis strategies proposed by Morin [20].

15.3.2
Core Modifications

A significant amount of effort was placed on the functionalization of both the 6α-position of the penicillin core and the 7α-position of the cephalosporin core with a variety of substituents. This was due in large part to findings that natural cephalosporins (**2**) containing a 7α-methoxy group had improved stability to β-lactamases, known to confer resistance to microorganisms. It is important to note that, in many cases, the chemistries and subsequent stereochemical outcomes for penicillins (**1**) translate quite well to the cephalosporins (**2**) and vice versa. There were two methods widely used for the substitution at these

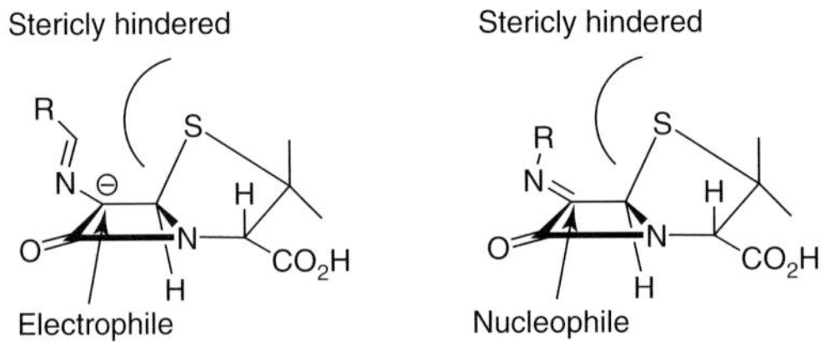

Figure 15.4 Stereochemistry of nucleophilic and electrophilic reactions at C6 of the penicillin core.

positions, (i) electrophilic substitution of a stabilized carbanion and (ii) nucleophilic substitution of an activated imine or imine equivalent. In terms of stereochemistry, both electrophiles and nucleophiles prefer an approach from the α-face of the core due to the planar nature of the β-lactam combined with the steric bulk of the sulfur atom and the ring attached to it (Figure 15.4). The stereochemical outcomes of many of the early substitutions were established through small molecule X-ray crystal structures and subsequently through comparison of chemical shifts and coupling constants in ^{1}H-NMR spectra [21].

For the first method, the electrophilic substitution of a stabilized carbanion, initial attempts to deprotonate the C6 position of penicillins (**1**) directly, using the amide as a stabilizing group, only led to the inversion of the stereocenter to give the inactive 6α-epimer [22]. Therefore, the formation of an imine was necessary to stabilize the C6 anion without epimerization. These anions were conveniently prepared by formation of a Schiff base from a suitably protected 6-APA, or 7-ACA core, followed by treatment with a base (Scheme 15.8). One of the earliest examples (method A) was based on the Schiff base of the 6-APA methyl ester (**36a**) with salicylaldehyde (**35**). Formation of a cuprate complex and subsequent addition of a benzaldehyde gave a hydroxymethylene compound **38** [23]. This methodology was expanded on to include alkylations with alkyl halides (method B) [21a, 24], formation of thioethers with reaction of disulfides (method C) [21b, 25], reaction with formaldehyde and acetaldehyde to introduce functionalized alkyl substituents [26], halogenation (method D) [21b, 27], Michael additions with α,β-unsaturated nitriles (method E) [26b], and acylations with acid chlorides or chloroformates [26b, 28].

It is important to note that the β-lactam has been found to be reasonably stable to basic reagents as shown in a study by Koppel [29] where treatment of penicillin G with LiOMe in tetrahydrofuran (THF) at −80°C for 1 min afforded a mixture of penicillin G and the lactam ring-opening product in a 9 : 1 ratio. Reactive electrophiles can therefore successfully compete with this side reaction. Other stabilizing groups have been successfully used, such as an isocyanide (Scheme 15.9) [30]. The isocyanide (**49**) was prepared from the formamide (**48**) by treatment with phosgene. Deprotonation with a tertiary amine yielded an anion that behaved in a similar manner to the imine-stabilized anions and reacted with a variety of electrophiles.

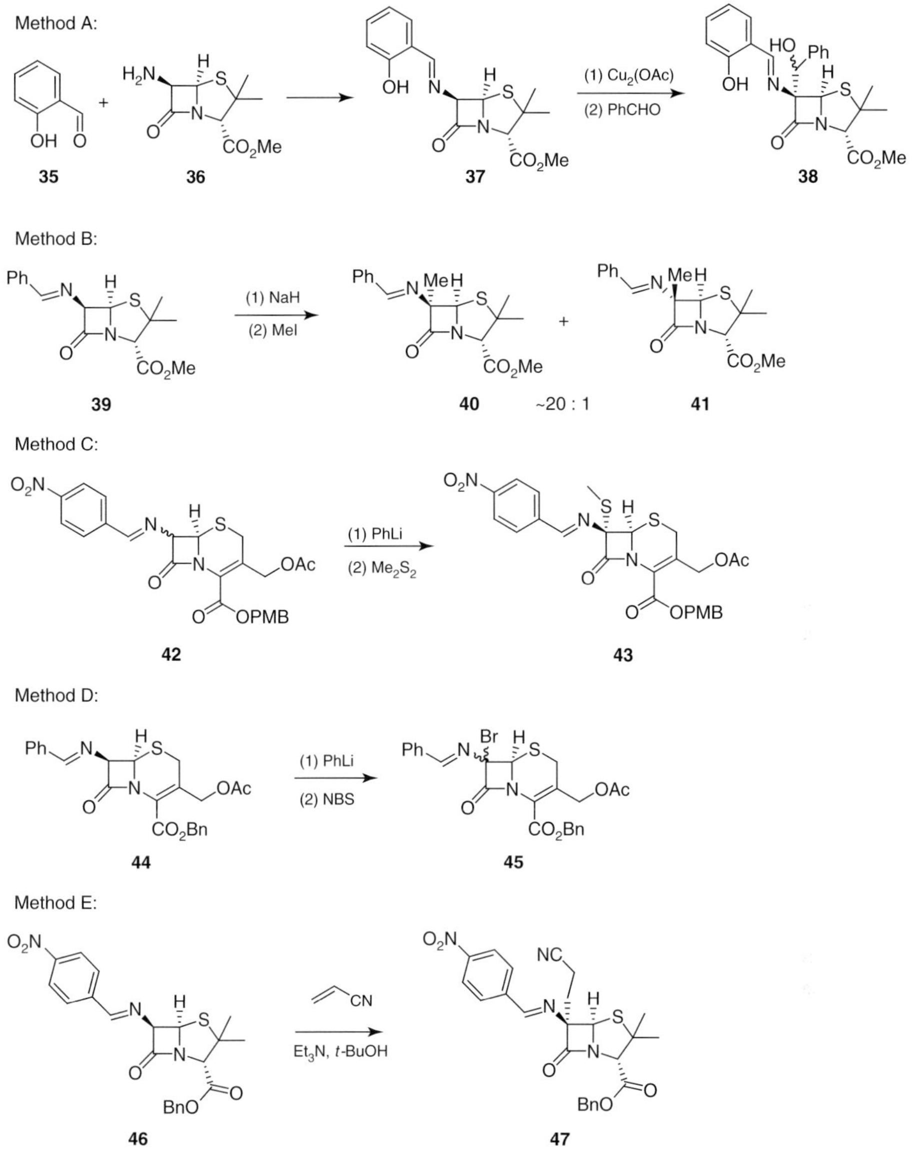

Scheme 15.8 Some examples of electrophilic substitutions at C6 of penicillins (**1**) and C7 of cephalosporins (**2**).

In another approach, a nitrogen ylide was used to promote an intramolecular delivery of an electrophile equivalent (Scheme 15.10) [31]. Treatment of the quaternary allyl ammonium species **51** with NaH in a mixture of DMF and benzene afforded the 6-allyl-substituted derivative **52**.

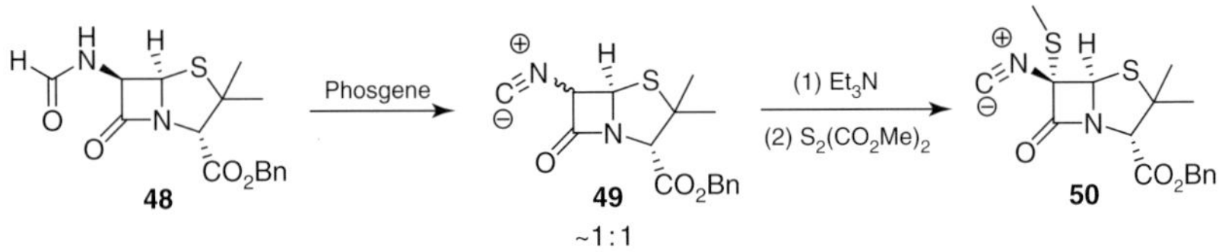

Scheme 15.9 Electrophilic substitution at C6 on a penicillin with an isocyanide stabilizing group.

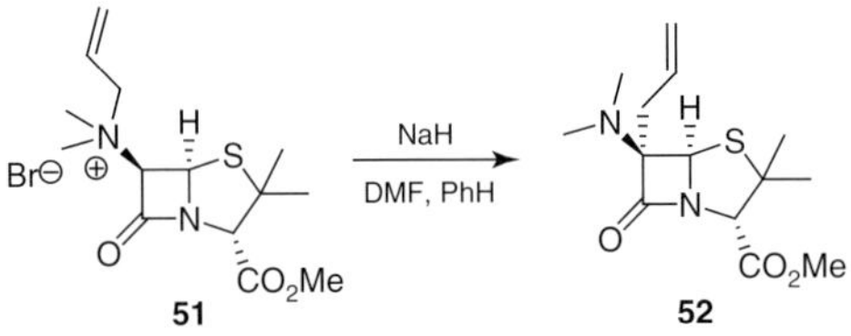

Scheme 15.10 Intermolecular rearrangement of an ylide.

For the functionalization of the 6α-position of the penicillin core and the 7α-position of the cephalosporin core through a nucleophilic substitution, there are two possible routes. Either the C–N bond can be oxidized directly or a leaving group is temporarily introduced and is then eliminated under formation of an imine.

Some examples of the oxidative approach are shown in Scheme 15.11. The earliest report demonstrated that 6-APA benzyl ester (**36b**) could be converted to a diazopenicillinate **53** [32] with $NaNO_2$. Reaction with bromine azide then afforded a mixture of the 6-bromo-6-azidopenicillinates **54** and **55** [33]. In another approach, penicillin benzyl ester (**56**) was oxidized to the *N*-chloroamide (**57**). In the presence of a base, **57** yielded the activated acyl imine (**58**), which was susceptible to nucleophilic attack [29, 34]. It further could be demonstrated that treating the 6-APA benzyl ester bis-trifluoromethanesulfonamide (**60**) with triethyl amine in methanol gave rise to the 6β-methoxy-substituted penicillantate (**62**) through transient formation of the activated sulfonylimine (**61**) [35].

A group in Japan, engineered an imine, **63**, which could undergo an oxidative rearrangement to form the quinone methide (**64**) (Scheme 15.12, method A) [36]. Conjugate addition of nucleophiles to **64** generated the Schiff base (**65**), which could be cleaved using standard procedures. This method allowed for a variety of nucleophiles to be used including cyanides and azides [37], Grignard reagents [38], as well as phosphites [39]. Another method used the treatment of the C6 amino group in **66** with a sulfenyl chloride in the presence of propylene oxide and 4 Å molecular sieve as acid scavengers to generate the thiooxime (**67**). This intermediate could then rearrange to introduce the sulfide into the 6-position [40].

Alternatively to the direct oxidation, a suitable leaving group can be introduced at C6 through any of the electrophilic methods described above. A reactive imine can then be generated through elimination of this leaving group (Scheme 15.13). While halogens could be used as leaving groups [21b, 27, 33],

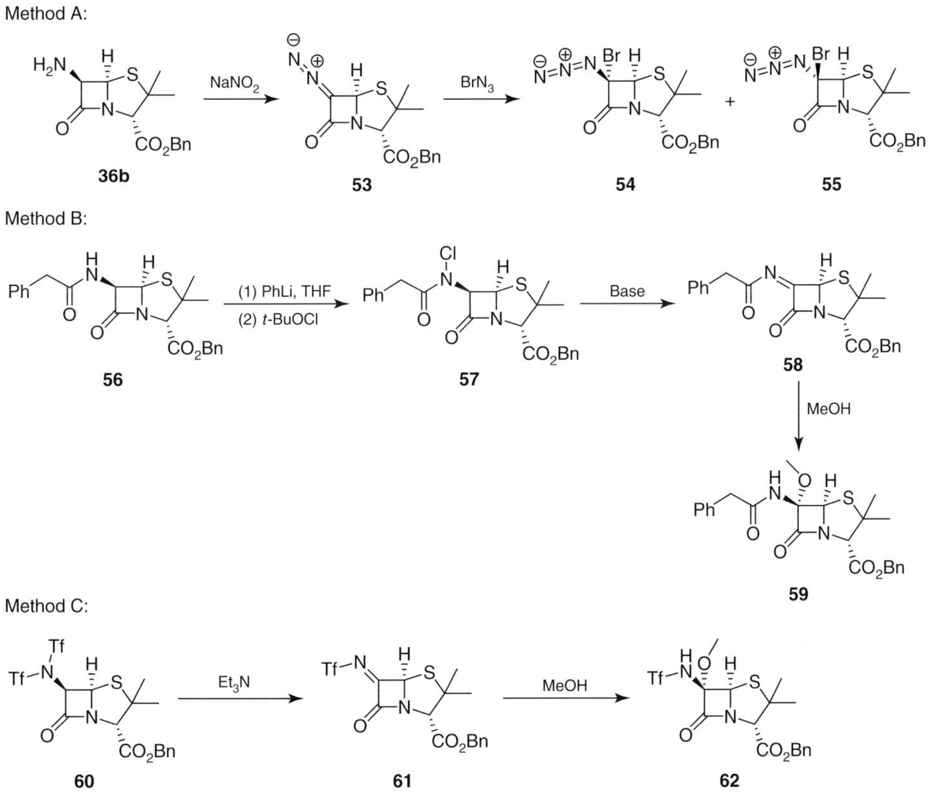

Scheme 15.11 Examples of nucleophilic substitutions at C6 of penicillins (**1**) through oxidation of the 6-amino group.

the preferred method to access the much sought after 6α-methoxypenicillins as well as the 7α-methoxycephalosporins was through a 6α-methylthio substituent [21b,c, 25, 41]. Typically, these reactions are carried out using a mercury salt to sequester the sulfide and facilitate the formation of the imine (**69** → **70**, method A). This methodology was expanded to 6α-aminopenicillins (**71** → **72**, method B) and 6α-formamidopenicillins (**73** → **74**, method C) using ammonia and *N,N*-bistrimethylsilylformamide as nucleophiles, respectively [42]. Alternatively, the sulfide could be converted to a better leaving group through oxidation to a sulfoxide, which both oxygen (**75** → **76**, method D) and nitrogen nucleophiles could displace under thermal conditions [43]. Another leaving group that was utilized for the incorporation of nitrogen and oxygen nucleophiles was the succinimido-oxy group [44]. The advantage of this group was that it also allowed to add carbon nucleophiles such as cyanide and Grignard reagents (**77** → **78**, method E).

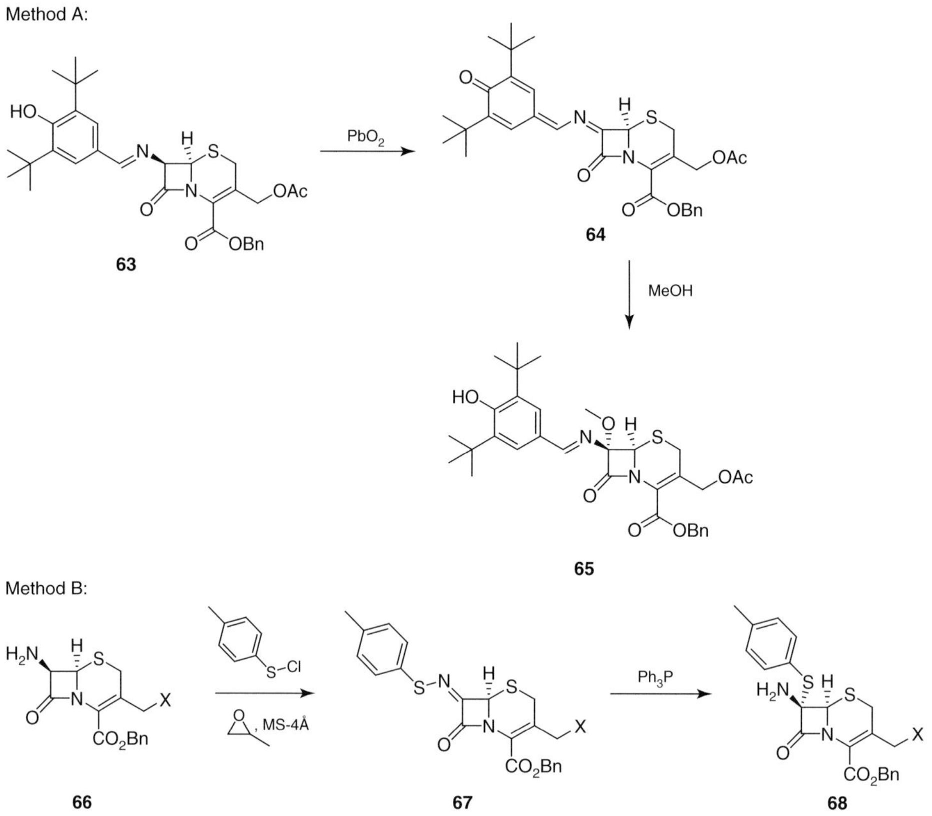

Scheme 15.12 Examples of nucleophilic substitution at C7 for cephalosporins (**2**) through oxidation of the 6-amino group.

15.4 Mode of Action

The β-lactam antibiotics work through inhibition of the bacterial cell wall synthesis. Specifically, they disrupt the formation of the peptidoglycan cross-links by binding to an enzyme, DD-transpeptidase, that links the peptidoglycan molecules in bacteria. The failure to form new cross-links while others are degraded leads to a weakening of the cell wall and ultimately to lysis due to osmotic pressure. This is particularly effective for gram-positive bacteria whose cell wall predominantly composed of peptidoglycan but may not be as effective for gram-negative bacteria that have an additional outer membrane composed of lipopolysaccharide (LPS).

Specific mechanisms of bacterial resistance to the β-lactam antibiotics have been identified. These include enzymatic hydrolysis of the β-lactam ring by β-lactamases and generation of a modified penicillin-binding protein with increased affinity for β-lactam antibiotics. Bacteria may inherently possess genes that code for β-lactamases or acquire them from other microorganisms through plasmid transfer.

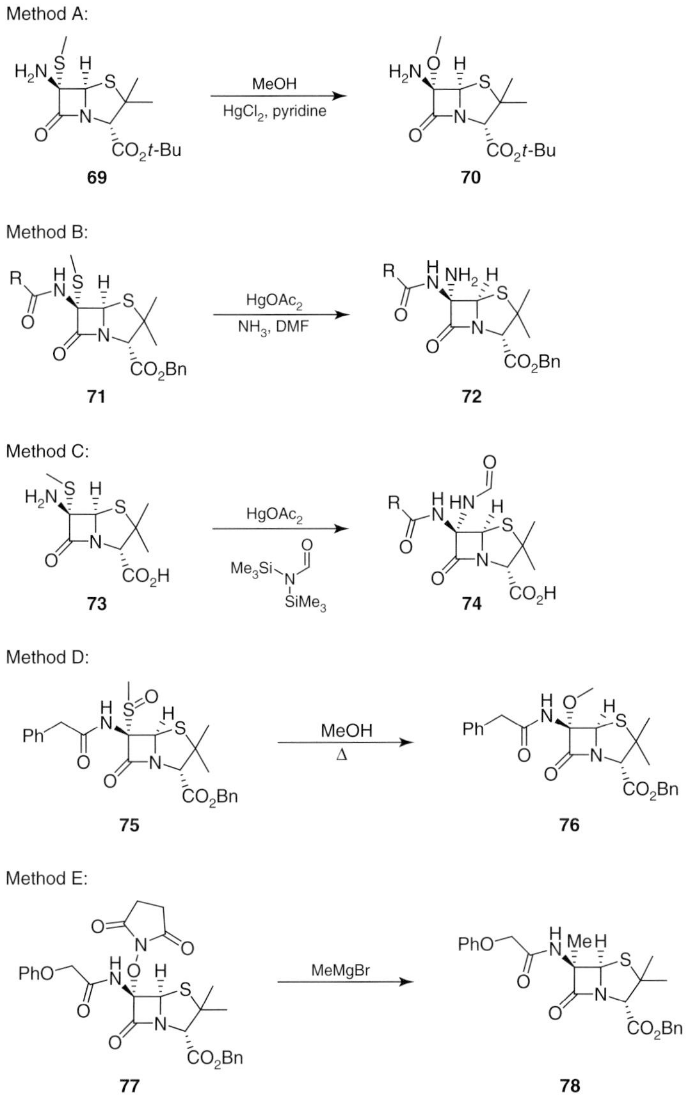

Scheme 15.13 Examples of nucleophilic substitution at C6 for penicillins (**1**) after generating the respective Schiff base through transient introduction of a leaving group.

In addition, the genes may be induced on exposure to β-lactam antibiotics. Significant efforts have been devoted to the identification of β-lactams that are resistant to β-lactamases and to the coadministration of β-lactamase inhibitors, for example, amoxicillin and clavulanic acid (Augmentin). The modified penicillin-binding proteins can result in serious challenges in the treatment of bacterial infections with β-lactams in the future and usually result in the necessity to choose other classes of antibiotics. Some notable examples are methicillin-resistant *Staphylococcus aureus* (MRSA) and pencillin-resistant *Streptococcus pneumoniae.*

15.5 Structure-Activity Relationships

The core structures of penicillins (**1**) and cephalosporins (**2**) tolerate very little change. Only carbapenems (**79**) had some commercial success (Figure 15.5). All other changes have led to compounds with substantially diminished activities. In addition, changes in the stereochemistry of the core stereocenters led to significant losses in activities. For the penicillins (**1**) and cephalosporins (**2**), there are only two main areas in which SARs can be developed, the amide side chain represented by R_1 and the substituent at the α-position of the exocyclic nitrogen represented by R_2. Cephalosporins (**2**) can also be modified in the 3′-position represented by R_3.

Owing to the ease of synthesis, N-acylations of either 6-APA (**8**) or 7-ACA (**31**) have been primarily investigated for SAR development. From penicillin G (**1a**), numerous analogs were made containing amides that can be either lipophilic or polar in nature (Figure 15.6). The polar side chains sometimes serve in a prodrug capacity. The main intent of these changes was to increase resistance to β-lactamases and in some cases, broaden the spectrum of activity from predominantly gram-positive to both gram-positive and gram-negative bacteria. Representative examples are amoxicillin (**1b**), carbenicillin (**1e**), piperacillin (**1f**), meticillin (**1g**), oxacillin (**1h**), cefdinir (**2f**), and cefalotin (**2g**).

A fair amount of synthetic effort was also dedicated to the identification of groups at the α-position to the exocyclic nitrogen that would not interfere with the binding to DD-transpeptidase but would protect the β-lactams from degradation by β-lactamases. Unfortunately, most of the evaluated substituents led to a loss in potency. Only the methoxy group maintained sufficient activity to produce marketed antibiotics such as temocillin (**1e**), cefoxitin (**2h**), and cefmetazole (**2i**) (Figure 15.7).

Finally, the 3′-position of the cephalosporins (**2**) has been targeted to modulate the physicochemical properties of these compounds in order to improve their pharmacokinetic profiles. Examples here are cefoxitin (**2h**), cefmetazole (**2i**), cefaloridine (**2j**), cefaclor (**2k**), and cefipim (**2l**) (Figure 15.8).

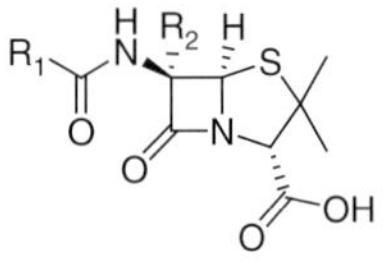

1

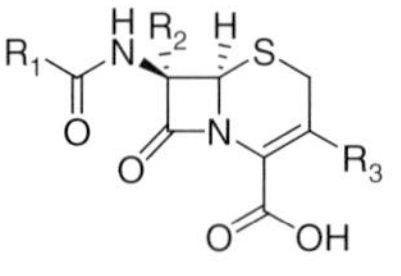

2

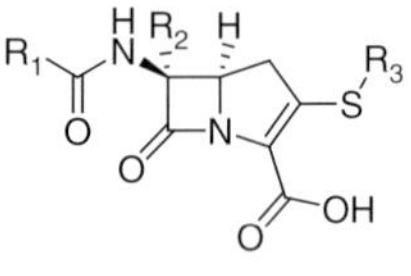

79

Figure 15.5 Sites in penicillins (**1**) and cephalosporins (**2**) where modifications are tolerated. General structure of carbapenems (**79**).

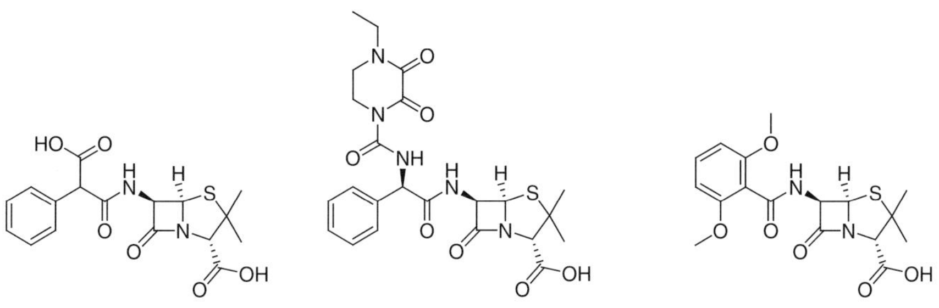

1e 1f 1g

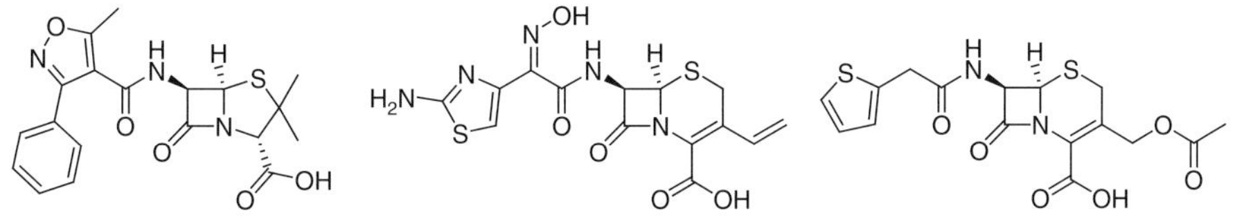

1h 2f 2g

Figure 15.6 Examples of N-acyl analogs of penicillins (**1**) and cephalosporins (**2**).

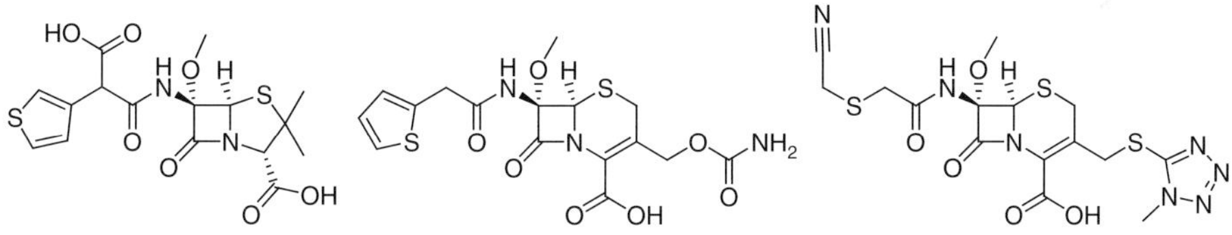

1i 2h 2i

Figure 15.7 Examples of penicillins (**1**) and cephalosporins (**2**) with a methoxy group at the α-position to the exocyclic nitrogen.

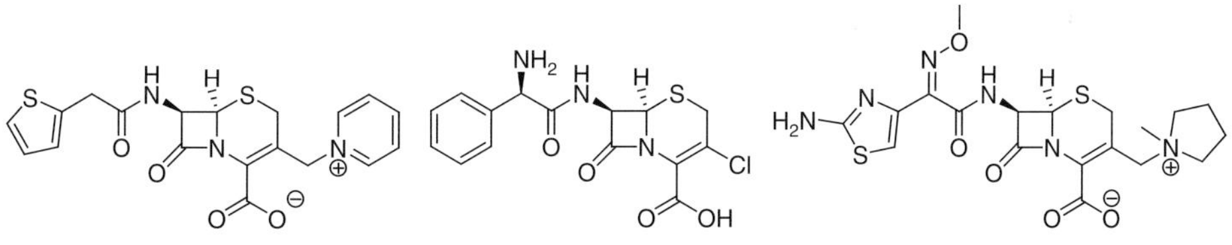

2j 2k 2l

Figure 15.8 Examples of 3′-substituted cephalosporins (**2**) with improved pharmacokinetic profiles.

References

1. Flemming, A. (1929) *Br. J. Exp. Pathol.*, **10**, 226.
2. Ho, D. (1999) *Time*, **29**, 117.
3. Handsfield, H.H., Clark, H., Wallace, J.F., Holmes, K.K., and Turck, M. (1973) *Antimicrob. Agents Chemother.*, **3**, 262.

4. Wick, W.E. (1967) *Appl. Microbiol.*, **15**, 765.
5. O'Callaghan, C.H., Sykes, R.B., Ryan, D.M., Foord, R.D., and Murrleton, P.W. (1976) *J. Antibiot.*, **29**, 29.
6. Chain, E., Florey, H.W., Gardner, A.D., Heatley, N.G., Jennings, M.A., Orr-Ewing, J., and Sanders, A.G. (1940) *Lancet*, **2**, 226.
7. (a) Arnstein, H.R.V., and Morris, D. (1960) *Biochem. J.*, **73**, 323; (b) Baldwin, J.E., and Abraham, E. (1988) *Nat. Prod. Rep.*, **5**, 129.
8. (a) Baldwin, J.E., Christie, M.A., Haber, S.B., and Kruse, L.I. (1976) *J. Am. Chem. Soc.*, **98**, 3045; (b) Nakatsuka, S., Tanino, H., and Kishi, Y. (1975) *J. Am. Chem. Soc.*, **97**, 5008.
9. Moyer, A.J. (1949) GB 624411A.
10. Tulin, E.E., Thomas, C.R., and Fish, N.M. (1986) *NSTA Technol. J.*, **11**, 18.
11. du Vigneaud, V., Carpenter, F.H., Holley, R.W., Livermore, A.H., and Rachele, J.R. (1946) *Science*, **104**, 431.
12. (a) Sheehan, J.C. and Henery-Logan, K.R. (1975) *J. Am. Chem. Soc.*, **79**, 1262; (b) Sheehan, J.C. and Henery-Logan, K.R. (1959) *J. Am. Chem. Soc.*, **81**, 5838; (c) Bolhofer, W.A., Sheehan, J.C., and Abrams, E.L.A. (1960) *J. Am. Chem. Soc.*, **82**, 3437.
13. Sheehan, J.C. and Henery-Logan, K.R. (1959) *J. Am. Chem. Soc.*, **81**, 3089.
14. Woodward, R.B. (1966) *Science*, **153**, 487.
15. Ratcliffe, R.W. and Christensen, B.G. (1973) *Tetrahedron Lett.*, **14**, 4649.
16. (a) Morin, R.B., Jackson, B.G., Flynn, E.H., and Roeske, R.W. (1962) *J. Am. Chem. Soc.*, **84**, 3400; (b) Morin, R.B., Jackson, B.G., Flynn, E.H., Roeske, R.W., and Andrews, S.L. (1969) *J. Am. Chem. Soc.*, **91**, 1396.
17. (a) Morin, R.B., Jackson, B.G., Mueller, R.A., Lavagnino, E.R., Scanlon, W.B., and Andrews, S.L. (1963) *J. Am. Chem. Soc.*, **85**, 1896; (b) Morin, R.B., Jackson, B.G., Mueller, R.A., Lavagnino, E.R., Scanlong, W.B., and Andrews, S.L. (1969) *J. Am. Chem. Soc.*, **91**, 1401.
18. (a) Chauvette, R.R., Pennington, P.A., Ryan, C.W., Cooper, D.G., José, F.L., Wright, I.G., Van Heyningen, E.M., and Huffman, G.W. (1971) *J. Org. Chem.*, **36**, 1259; (b) Gottstein, W.J., Misco, P.F., and Cheney, L.C. (1972) *J. Org. Chem.*, **37**, 2765.
19. (a) Brain, E.G., Eglington, A.J., Nayler, J.H.C., Osborne, N.F., Pearson, M.J., Smale, T.C., Southgate, R., Tolliday, R., Basker, M.J., and Sutherland, R. (1977) *J. Med. Chem.,*, **20**, 1082; (b) Gordon, E.M. and Cimarusti, E.M. (1977) *Tetrahedron Lett.*, **18**, 3425.
20. (a) Heymes, R., Amiard, G., and Nominé, C.R. (1966) *Acad. Sci.*, **263**, 170; (b) Dolfini, J.E., Schwartz, J., and Weisenborn, F. (1969) *J. Org. Chem.*, **34**, 1582; (c) Edwards, J.A., Guzman, A., Johnson, R., Beeby, P.J., and Fried, J.H. (1974) *Tetrahedron Lett.*, **15**, 2031.
21. (a) Bohme, E.H.W., Applegate, H.E., Toeplitz, B., and Dolfini, J.E. (1971) *J. Am. Chem. Soc.*, **93**, 4324; (b) Slusarchyk, W.A., Applegate, H.E., Funke, P., Koster, W., Puar, M.S., Young, M., and Dolfini, J.E. (1973) *J. Org. Chem.*, **38**, 943; (c) Applegate, H.E., Dolfini, J.E., Puar, M.S., Slusarchyk, W.A., and Toeplitz, B. (1974) *J. Org. Chem.*, **39**, 2794.
22. (a) Johnson, D.A., Mania, D., Panetta, C.A., and Silverstri, H.H. (1968) *Tetrahedron Lett.*, **9**, 1903; (b) Wolfe, S. and Lee, W.S. (1968) *Chem. Commun.*, 242; (c) Clayton, J.P., Naylor, J.H.C., Southgate, R., and Stove, E.R. (1969) *Chem. Commun.*, 129.
23. Rainer, R. and Zeller, P. (1968) *Helv. Chim. Acta*, **51**, 1905.
24. (a) Johnston, D.B.R., Schmitt, S.M., Firestone, R.A., and Christensen, B.G. (1972) *Tetrahedron Lett.*, **13**, 375; (b) Bohme, E.H.W., Applegate, H.E., Ewing, J.B., Funke, P.T., Puar, M.S., and Dolfini, J.E. (1973) *J. Org. Chem.*, **38**, 230.
25. (a) Ratcliffe, R.W. and Christensen, B.G. (1973) *Tetrahedron Lett.*, **14**, 4653; (b) Spitzer, W.A. and Goodson, T. (1973) *Tetrahedron Lett.*, **14**, 273.
26. (a) Firestone, R.A., Schelechow, N., Johnston, D.B.R., and Christensen, B.G. (1972) *Tetrahedron Lett.*, **13**, 4917; (b) Rasmusson, G.H., Reynolds, G.F., and Arth, G.E. (1973) *Tetrahedron Lett.*, **14**, 145.

27. Cama, L.D. and Christensen, B.G. (1973) *Tetrahedron Lett.*, **14**, 3505.

28. Spitzer, W.A., Goodson, T., Smithey, R.J., and Wright, I.G. (1972) *J. Chem. Soc., Chem. Commun.*, 1138.

29. Koppel, G.A. and Koehler, R.E. (1973) *J. Am. Chem. Soc.*, **95**, 2403.

30. Bentley, P.H. and Clayton, J.P. (1974) *J. Chem. Soc., Chem. Commun.*, 278.

31. Kaiser, G.V., Ashbrook, C.W., and Baldwin, J.E. (1971) *J. Am. Chem. Soc.*, **93**, 2342.

32. von Hauser, D. and Sigg, H.P. (1968) *Helv. Chim. Acta*, **50**, 1327.

33. Cama, L.D., Leanza, W.J., Beattie, T.R., and Christensen, B.G. (1972) *J. Am. Chem. Soc.*, **94**, 1408.

34. (a) Baldwin, J.E., Urban, F.J., Cooper, R.D.G., and Jose, F.L. (1973) *J. Am. Chem. Soc.*, **95**, 2401; (b) Firestone, R.A. and Christensen, B.G. (1973) *J. Org. Chem.*, **38**, 1436.

35. (a) Mézes, P.S.F., Freisen, R.W., Viswanatha, T., and Dmitrienko, G.I. (1982) *Heterocycles*, **19**, 1207; (b) Branch, C.L., Pearson, M.J., and Smale, T.C. (1988) *J. Chem. Soc., Chem. Commun.*, 2865.

36. Yanagisawa, H., Fukushima, M., Ando, A., and Nakao, H. (1975) *Tetrahedron Lett.*, **16**, 2705.

37. Yanagisawa, H., Fukushima, M., Ando, A., and Nakao, H. (1976) *Tetreahedron Lett.*, **17**, 259.

38. Yanagisawa, H. and Nakao, H. (1976) *Tetreahedron Lett.*, **18**, 1815.

39. Yanagisawa, H. and Nakao, H. (1976) *Tetreahedron Lett.*, **18**, 1811.

40. (a) Gordon, E.M., Chang, H.W., and Cimarusti, C.M. (1977) *J. Am. Chem. Soc.*, **99**, 5504; (b) Kobayashi, T., Iino, K., and Hiraoka, T. (1977) *J. Am. Chem. Soc.*, **99**, 5505; (c) Gordon, E.M., Chang, H.W., Cimarusti, C.M., Toeplitz, B., and Gougoutas, J.Z. (1980) *J. Am. Chem. Soc.*, **102**, 1690.

41. Jen, T., Frazee, J., and Hoover, J.R.E. (1973) *J. Org. Chem.*, **38**, 2587.

42. (a) Milner, P.H., Guest, A.W., Harrington, F.P., Ponsford, R.J., Smale, T.C., and Stachulski, A.V. (1984) *J. Chem. Soc., Chem. Commun.*, 1335; (b) Brown, A.C., Guest, A.W., and Milner, P.H. (1989) *Tetrahedron Lett.*, **30**, 2433.

43. Kaura, A.C. and Pearson, M.J. (1985) *Tetrahedron Lett.*, **26**, 2597.

44. Stachulski, A.V. (1986) *J. Chem. Soc., Chem. Commun.*, 401.

Part IV
Oncology

Bioactive Heterocyclic Compound Classes: Pharmaceuticals,
First Edition. Edited by Jürgen Dinges and Clemens Lamberth.

16
Pyrimidine-Based Kinase Inhibitors in Cancer Chemotherapy

Robert Mah

16.1
Introduction

Pyrimidines are six-membered aromatic heterocycles containing two nitrogen atoms in 1- and 3-positions. Pyrimidines can carry a single amine substituent in the 2-, 4-, or 5-position (Figure 16.1). In the case of diaminopyrimidines, any combination of the three positions is possible.

A number of aminopyrimidine-containing compounds are either in advanced clinical development or already in the market as kinase inhibitors for the treatment of cancer. Representative examples of this broad class of compounds include the Breakpoint Cluster Region-Abelson (BCR-ABL) inhibitors imatinib (**1**, STI571, Glivec®, Gleevec®) [1] and nilotinib (**2**, AMN107, Tasigna®) [2], the Aurora kinase inhibitor ENMD-2076 (**3**, ENMD-981693) [3], the B-Raf kinase inhibitor dabrafenib (**4**, GSK-2118436) [4], the phosphatidylinositol 3-kinase (PI3K) inhibitor NVP-BKM120 (**5**) [5], the Janus kinase (JAK) inhibitor CYT387 (**6**) [6], the multiple kinase inhibitor dasatinib (**7**, BMS-354825, Sprycel®) [7], and the spleen tyrosine kinase (Syk) inhibitor fostamatinib (**8**, R-406/788) [8] (Figure 16.2).

On account of the extensive amount of structural and mechanistic information available on Bcr-Abl kinase inhibitors [9], this chapter focuses on this class of compounds, including bafetinib (**9**, INNO-406, NS-187) [10] and flumatinib (**10**, HHGV-678) [11] for the treatment of chronic myeloid leukemia (CML) [12].

16.2
History

The successful syntheses of aminopyrimidines were reported in the late nineteenth/early twentieth centuries (Scheme 16.1). The first 4-aminopyrimidine, 4-amino-2,6-diethyl-5-methylpyrimidine (**11**), was unknowingly prepared in 1848 by Frankland and Kolbe [13].

In an attempt to generate an ethyl radical, propionitrile was added dropwise to potassium metal, which, in the authors' own words, led to a "...very vigorous

Bioactive Heterocyclic Compound Classes: Pharmaceuticals,
First Edition. Edited by Jürgen Dinges and Clemens Lamberth.

Figure 16.1 Basic structures and numbering conventions for 2-, 4-, and 5-amino-pyrimidines.

reaction, often accompanied by fire and rapid gas evolution. . ." Although more interested in the gas that was produced (which turned out to be butane rather than the desired ethane), the viscous yellow residue was also analyzed. Following workup and recrystallization from boiling water, a small quantity of pearl white crystals could be isolated. The substance was named "kyanethine" (**11**) as a reminder that it originated from cyanoethane (propionitrile) and was shown to be the aforementioned pyrimidine derivative some 32 years later by von Meyer [14]. The earliest synthesis of a 2-aminopyrimidine, 2-amino-4,6-dimethylpyrimidine (**12**), was described by Combes and Combes [15] in 1892 by condensing acetylacetone with guanidine carbonate. In 1901, Gabriel and Colman [16] prepared a 5-aminopyrimidine, 5-amino-2-chloro-4-methylpyrimidine (**13**), via the zinc-mediated reduction of a nitro-containing precursor.

One of the first studies on the physiological activity of aminopyrimidines was reported by Walton [17] in 1882 based on a request from von Meyer. Walton used methylkyanethine, the N-methyl analog of kyanethine (**11**), and found it to be a powerful poison acting on the CNS in dogs, rabbits, and frogs. In subsequent years, aminopyrimidines were shown to have effects in numerous indications as exemplified by the launched products pyrimethamine [18a] (**14**, dihydrofolate reductase inhibitor; Burroughs Wellcome, 1953), piribedil [18b] (**15**, dopamine D2 agonist; Servier, 1969), buspirone [18c] (**16**, 5-HT1A agonist; Bristol Myers, 1985), moxonidine [18d] (**17**, imidazoline I1 agonist; Lilly, 1991), tirilazad [18e] (**18**, lipid peroxidation inhibitor; Upjohn, 1995), bosentan [18f] (**19**, ETA/ETB antagonist; Actelion, 2001), rosuvastatin [18g] (**20**, HMG-CoA reductase inhibitor; AstraZeneca, 2003), revaprazan [18h] (**21**, H^+/K^+-ATPase inhibitor; GlaxoSmithKline, 2007), and etravirine [18i] (**22**, reverse transcriptase inhibitor; Janssen, 2008) (Figure 16.3).

With respect to treatments for CML [12], the first recorded treatment to be used involved the use of Fowler's solution, also known as *Kali arseniatum*, which was prepared as a variable composition (containing ~1% potassium arsenite (KH_2AsO_3) as its active ingredient) by boiling arsenic trioxide in aqueous potassium bicarbonate and which was originally used in the late eighteenth century as a general tonic for people and animals. Two of the earliest reports for using arsenic to treat CML were from Lissauer [19] in 1865 and from Cowan Doyle [20] in 1882 (Note: the author is recorded as Arthur Cowan Doyle but this is probably a printer's error for Arthur Conan Doyle, the Sherlock Holmes author and also a general practitioner in England at the time). Of note is that Cell Therapeutics' proprietary formulation containing arsenic trioxide (Trisenox®) received FDA approval for acute

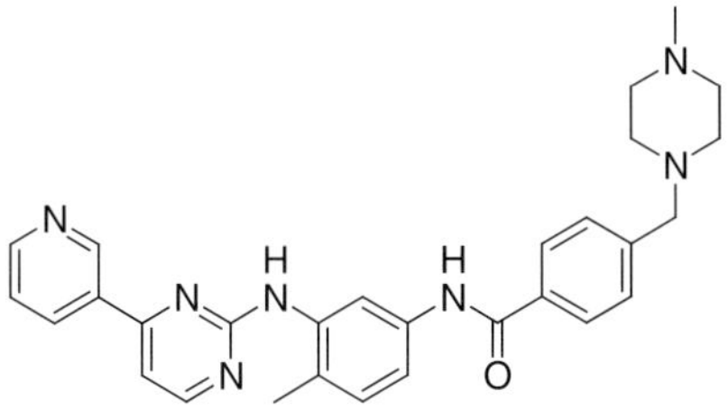

1
Imatinib

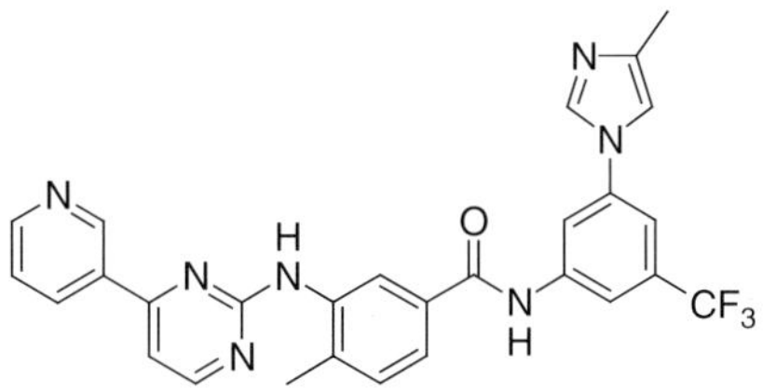

2
Nilotinib

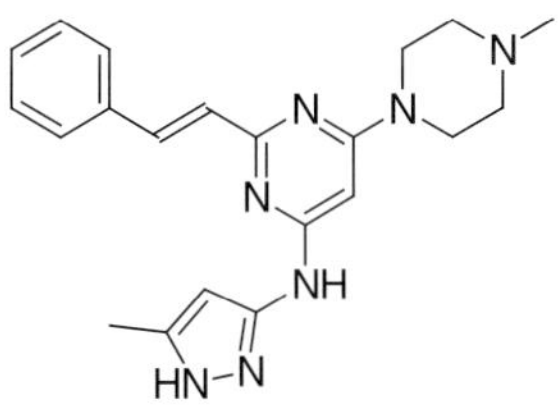

3
ENMD-2076

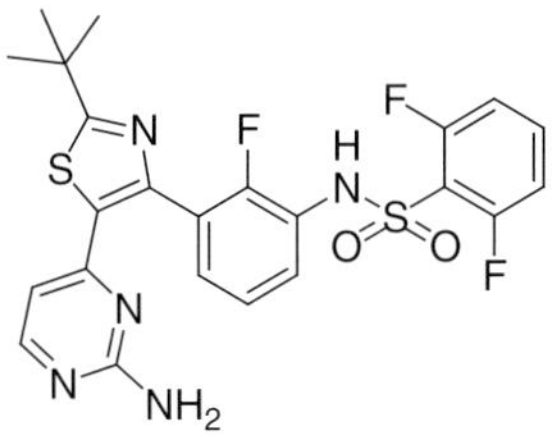

4
Dabrafenib

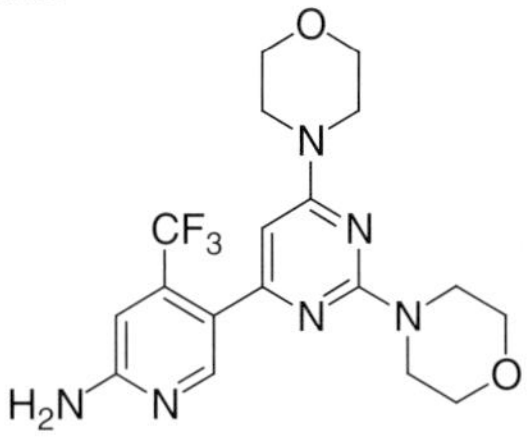

5
NVP-BKM120

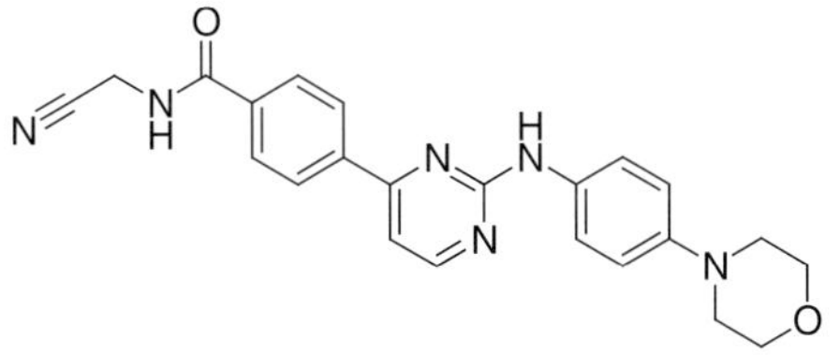

6
CYT387

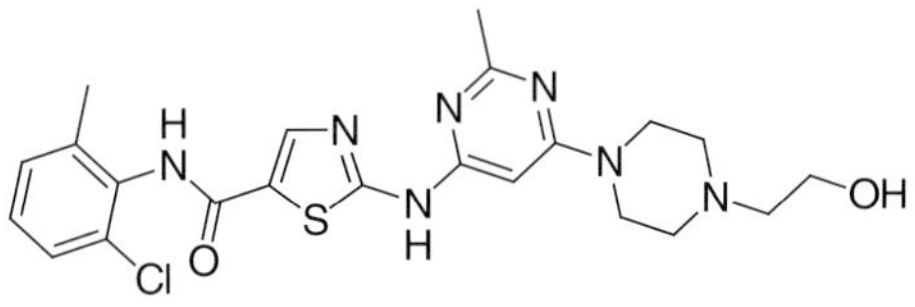

7
Dasatinib

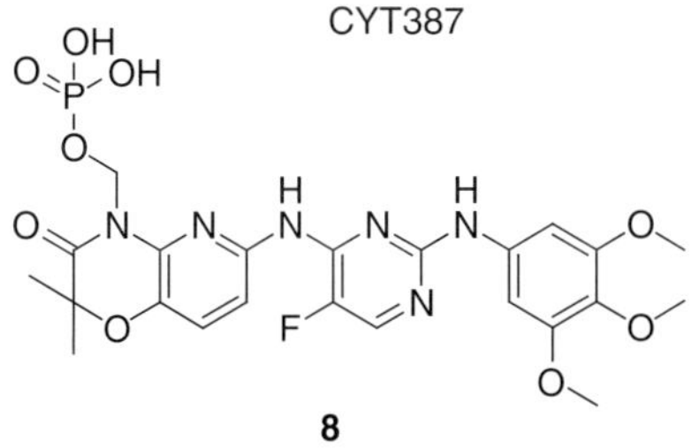

8
Fostamatinib

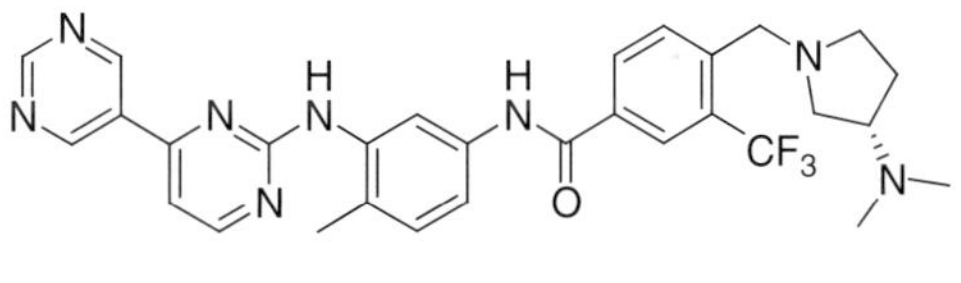

9
Bafetinib

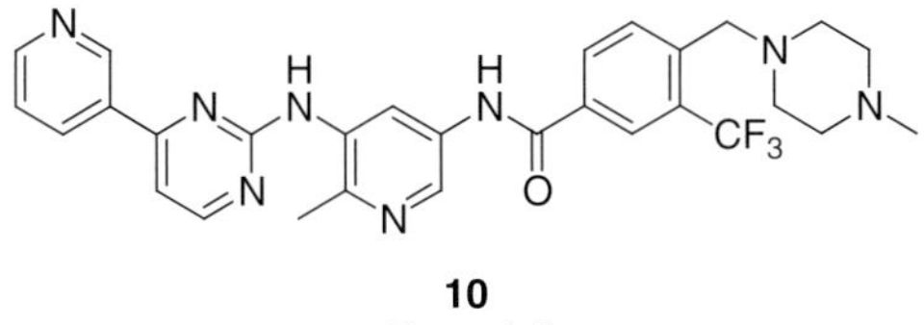

10
Flumatinib

Figure 16.2 Aminopyrimidine kinase inhibitors: imatinib (**1**), nilotinib (**2**), ENMD-2076 (**3**), dabrafenib (**4**), NVP-BKM120 (**5**), CYT387 (**6**), dasatinib (**7**), fostamatinib (**8**), bafetinib (**9**), and flumatinib (**10**).

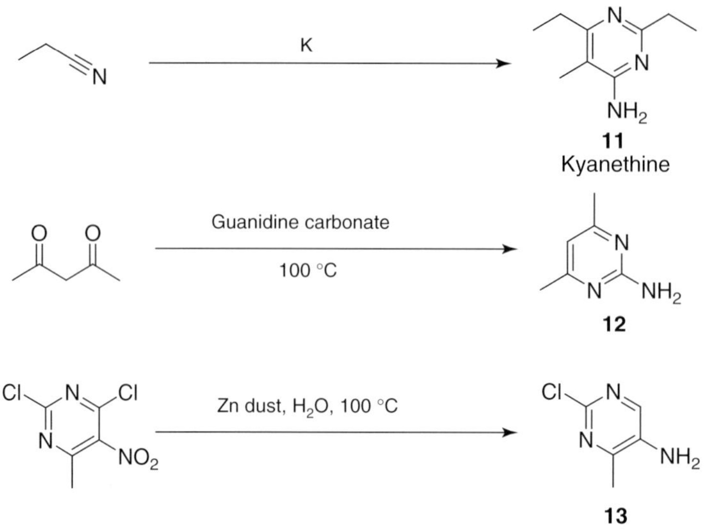

Scheme 16.1 Initial syntheses of 4-aminopyrimidines, 2-aminopyrimidines, and 5-aminopyrimidines.

promyelocytic leukemia in 2001 [21]. Spleen irradiation with X-rays was often used for CML patients in the early 1900s and, although not life prolonging, improved the patient's quality of life. Following the observation that the chemical warfare agent mustard gas caused myelosuppression, it was subsequently successfully used [22] in CML therapy and led to a search for more specific and less toxic alkylating agents, which resulted in the introduction of Burroughs Welcome's busulfan (Myleran®) in 1953 [23]. Busulfan was the first agent that consistently and significantly demonstrated a survival benefit and became the treatment of choice for CML. This was followed by hydroxyurea (Hydrea®) from Bristol-Myers Squibb (BMS) in the 1970s and interferon-α (IFNα) from Roche in the 1980s. In contrast to earlier agents, the use of IFNα, an endogenous immunomodulating agent, produced sustained cytogenic responses in almost one-third of all patients and prolonged life by one to two years in comparison with busulfan and hydroxyurea. Another promising treatment developed in the mid-1980s was hematopoietic stem cell transplantation (HCT). Provided that the patient could withstand the procedure of high-dose chemotherapy and whole body irradiation to destroy their bone marrow and there was a suitable donor of healthy bone marrow, this was the first method to provide hope that CML could be cured, as many patients enjoyed long-term disease-free lives following HCT. With the introduction of imatinib (**1**) by Novartis in 2001, a new gold standard for CML therapy was established. In over a decade of use, imatinib (**1**) has proved to be very effective, with up to 80% response rates in chronic-phase CML patients, and well tolerated with a good safety profile [24]. For patients with advanced phases of CML where imatinib (**1**) is often less effective, more potent tyrosine kinase inhibitors such as dasatinib (**7**, BMS, 2006), nilotinib (**2**, Novartis, 2007), bafetinib (**9**, Kyoto University Hospital/Nippon

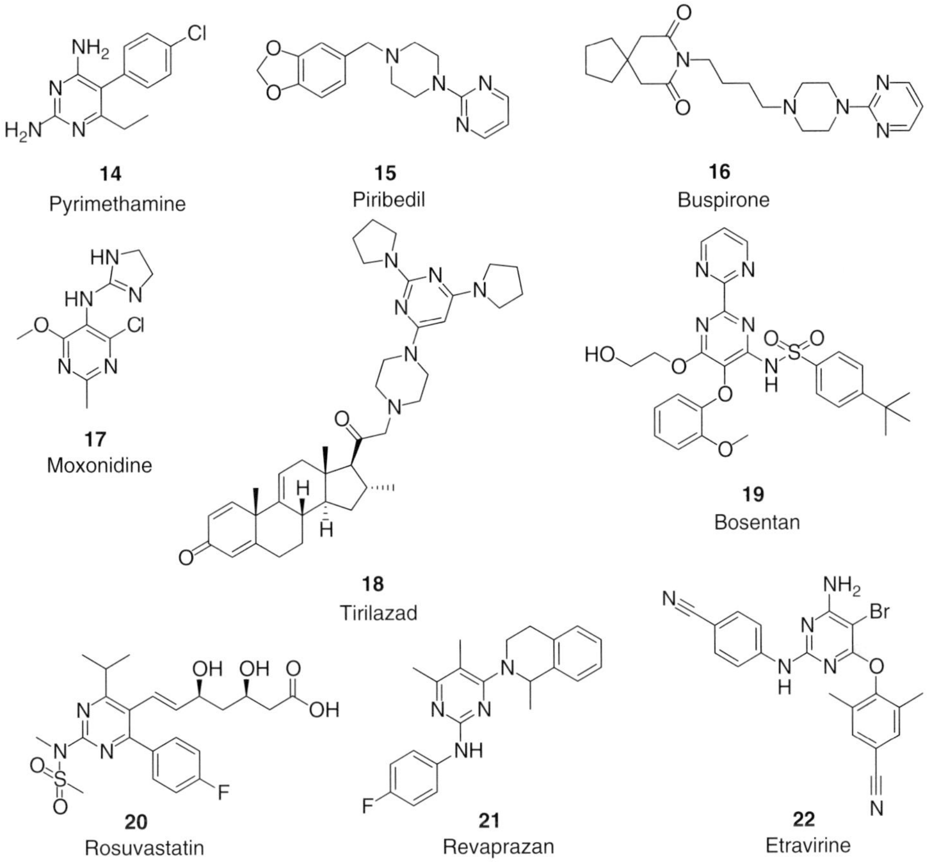

Figure 16.3 Marketed aminopyrimidines: pyrimethamine (**14**), piribedil (**15**), buspirone (**16**), moxonidine (**17**), tirilazad (**18**), bosentan (**19**), rosuvastatin (**20**), revaprazan (**21**), and etravirine (**22**).

Shinyaku/CytRx, Ph II), and flumatinib (**10**, Shanghai Institute/Jiangsu Haosen, Ph I) have been introduced/are being developed.

16.3 Synthesis

As originally reported [25], the general routes for the syntheses of imatinib (**1**), nilotinib (**2**), and bafetinib (**9**) are based on the reaction between an aryl-substituted guanidine and an enaminone to generate a substituted pyrimidine [26], as shown in Scheme 16.2.

The phenyl guanidines **24** are readily prepared by overnight refluxing of an alcohol solution of the substituted anilines **23** (R = NO_2 for imatinib (**1**) and bafetinib

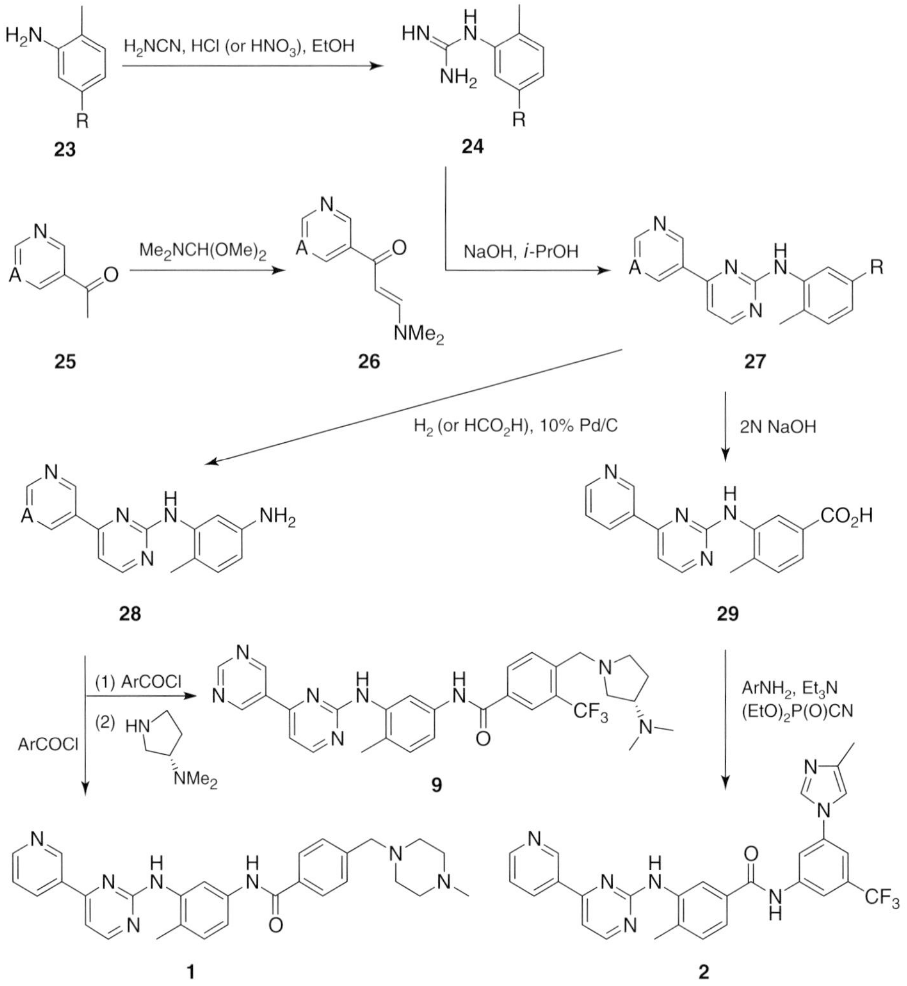

Scheme 16.2 General routes for the syntheses of imatinib (**1**), nilotinib (**2**), and bafetinib (**9**).

(**9**), R = CO_2Et for nilotinib (**2**)) with cyanamide [27]. Aniline mineral acid salts are normally used for accelerated reaction rates and for improved yields [28]. The enaminones **26** are easily obtained by reaction of *N,N*-dimethylformamide dimethyl acetal or Bredereck's reagent [29] with the heteroaromatic methyl ketones **25** (A = CH for imatinib (**1**) and nilotinib (**2**), A = N for bafetinib (**9**)). Following construction of the pyrimidine ring, reduction ($NO_2 \rightarrow NH_2$) or saponification ($CO_2Et \rightarrow CO_2H$) and subsequent amide bond formation provide the final products. In the case of bafetinib (**9**), amide formation is performed with a benzylic-bromide-containing acid chloride and the bromide is then alkylated with dimethyl-(*S*)-pyrrolidin-3-yl-amine.

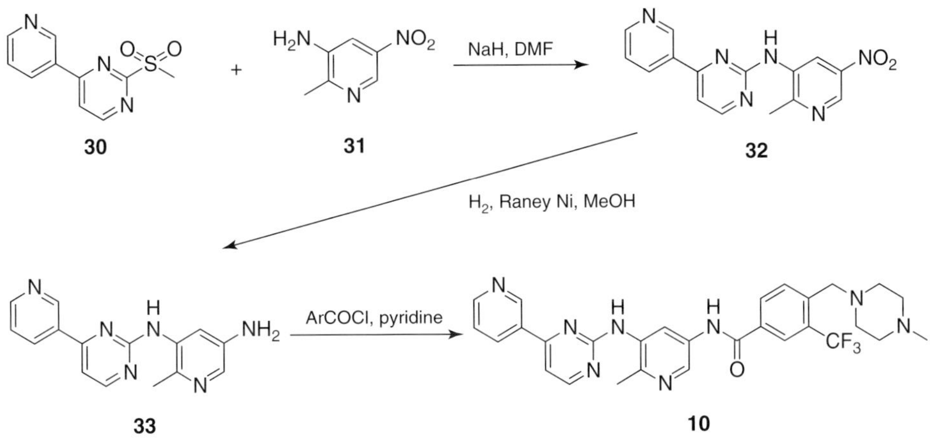

Scheme 16.3 Synthesis of flumatinib (**10**).

A number of alternative methods of syntheses of imatinib (**1**) have been disclosed, which (i) form the pyrimidine core with guanidine nitrate [30] (thus avoiding the need for cyanamide) and then use a palladium [31] or copper-catalyzed [30] coupling with an aryl bromide derivative; (ii) form the amide bond directly from the corresponding ester derivative in the presence of a Lewis acid such as $AlCl_3$, $AlMe_3$, or $Al(i\text{-}Bu)_3$ [31]; (iii) form the amide bond directly from the corresponding ester derivative in the presence of an organic base such as NaOMe, NaOEt, or KOBu [32]; or (iv) utilize an aminocarbonylation reaction to form the amide bond [33]. For nilotinib (**2**), a similar organic-base-mediated (using KO-*t*-Bu) formation of the amide bond from an ester derivative has also been reported [34].

The synthetic route for flumatinib (**10**) [35] is similar to the general routes for imatinib (**1**), nilotinib (**2**), and bafetinib (**9**) but takes advantage of the high reactivity of 2-sulfonyl-pyrimidines with amines [36] to form the aminopyrimidine core. Thus, treatment of a mixture of 2-methanesulfonyl-4-pyridin-3-yl-pyrimidine (**30**) and the substituted aniline **31** with NaH in DMF gives the aminopyrimidine **32** in 83% yield (Scheme 16.3).

A similar approach has recently been published [37] for the preparation of imatinib (**1**), in which the key 2-sulfonyl-pyrimidine **30** can be synthesized in a number of different ways (Scheme 16.4) [37, 38]. The precursor thioether **34** has been prepared either in two steps by condensation of the enaminone **26** with thiourea followed by S-methylation (87% overall) or directly via the reaction of **26** with *S*-methylisothiourea sulfate (56%). Alternatively, **34** has been prepared using a cross-coupling reaction of the 4-chloropyrimidine derivative **35** either with a pyridylboronic acid derivative under Pd-catalysis (93%) or with a pyridyl Grignard reagent catalyzed by $Fe(acac)_3$ (43%). Subsequent oxidation of **34** with either H_2O_2 (81%) or Oxone® (61%) then provides **30**.

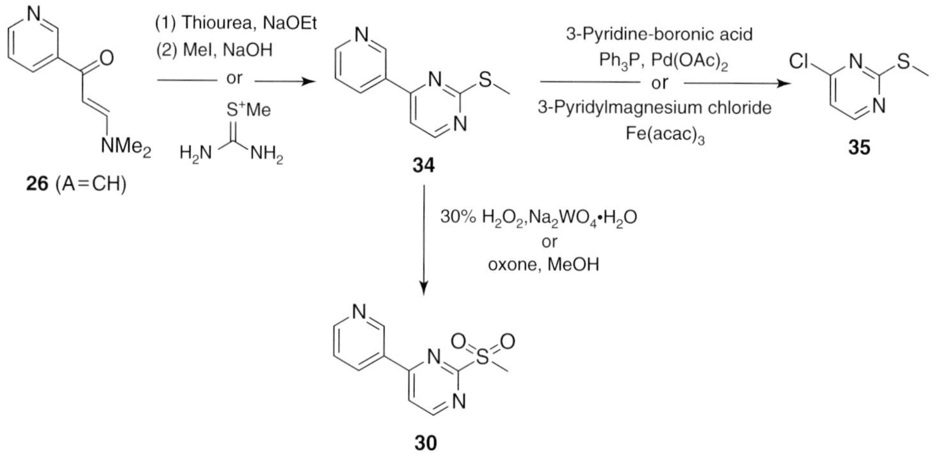

Scheme 16.4 Syntheses of 2-methanesulfonyl-4-pyridin-3-yl-pyrimidine (**30**).

16.4
Mode of Action

CML, also known as *chronic granulocytic leukemia*, was first described in 1845 by Bennett [39a] and Virchow [39b] and is a blood stem cell malignancy in which too many abnormal granulocytes develop in the blood and bone marrow. In 1960, Nowell and Hungerford [40] published several chromosome studies on normal and leukemic human leukocytes and established the presence of small chromosomal changes in CML patients. The chromosomal abnormality (termed the *Philadelphia (Ph) chromosome* after the city in which it was discovered) was characterized in 1973 by Rowley [41] and is the result of a reciprocal *t*(9;22) chromosome translocation. The result is the generation of the *BCR-ABL* gene, which encodes a fusion protein that has constitutively active Abl tyrosine kinase activity (Figure 16.4). The dysregulation of kinase activity leads to enhanced cell proliferation, impaired cell differentiation, reduced apoptosis, and reduced cell motility, resulting in the onset of pathological conditions. The Philadelphia chromosome is detected in 90–95% of CML patients and in 15–20% of Ph+ acute lymphoblastic leukemia (ALL) patients. Its causal relationship with CML was demonstrated in 1990 by several independent groups [42], which showed that either transgenic BCR-ABL p190 mice or mice infected with a retrovirus encoding Bcr-Abl developed a myeloproliferative disease that mirrored the pathogenesis of human CML.

Once identified as the protein product of an oncogene, Bcr-Abl became an attractive target for the treatment of CML with a selective Abl tyrosine kinase inhibitor and these research efforts culminated with the discovery of imatinib (**1**) [25b, 43]. On the basis of promising preclinical results, the compound entered and quickly progressed through clinical trials (phase I – June 1998; phase II – August 1999; phase III – June 2000) and was approved by the FDA on 10 May 2001 for CML patients in blast crisis, accelerated phase or in chronic phase after failure of

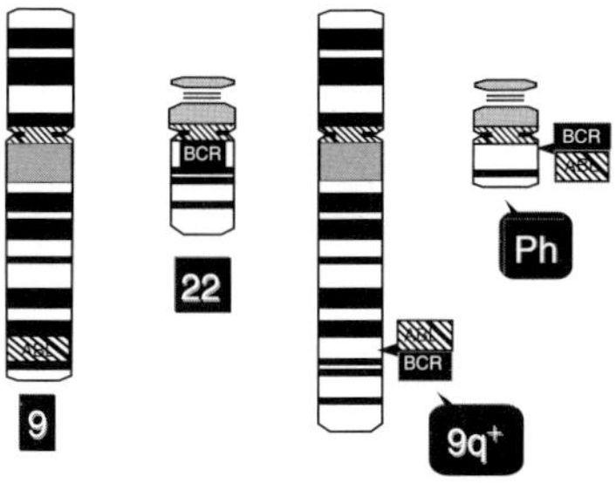

Figure 16.4 Reciprocal chromosomal translocation between chromosomes 9 and 22 to produce the Philadelphia chromosome. (Source: Reproduced with permission from Ref. [24b]. Copyright 2003 by The American Society for Pharmacology and Experimental Therapeutics.)

IFNα therapy, and as first-line treatment for newly diagnosed CML in December 2002.

Looking back, however, a prevailing notion in the early 1990s was that it was unlikely that a selective protein kinase inhibitor could ever be found given the high degree of homology of the ATP binding site of different kinases. Abl is a member of the nonreceptor tyrosine kinase (NRTK) subfamily of protein kinases that transfer the γ-phosphate of ATP to the hydroxyl group of a tyrosine in protein substrates. The catalytic domain of protein kinases consists of two subdomains (lobes), the N-lobe containing a five-stranded β-sheet and a conserved α-helix (helix C and helix αC) and a helical C-lobe, connected by a single polypeptide chain (linker region) that acts as a hinge to allow rotation of the two lobes with respect to each other on binding of ATP and/or the substrate without disrupting the secondary structure of the kinase (Figure 16.5).

A phosphate-binding loop (P-loop) connects the β1 and β2 strands and contains a glycine-rich sequence that, because of its reduced steric requirements, (i) allows close contact of the loop with the phosphates of ATP and binding via backbone interactions and (ii) provides flexibility in the absence of ATP, thus facilitating the binding of low-molecular-weight inhibitors. Additional structural features include a catalytic loop located in the C-lobe and containing aspartate and asparagine residues that are crucial for catalysis and an activation loop (A-loop and T-loop), also contained in the C-lobe and with a conserved Asp-Phe-Gly (DFG) motif located at its N-terminal and an Ala-Pro-Glu (APE) motif at its C-terminal. The catalytic activity of most protein kinases is controlled by the activation loop in a phosphorylation-dependent manner; so when the protein kinase is in a fully activated state, one or more residues in the activation loop has been phosphorylated and the loop is in an open conformation ("DFG-in") in which the DFG aspartate can interact with a magnesium ion that coordinates the ATP phosphate groups and the rest of the loop points away from the catalytic center. In the inactive state, the positions of the DFG aspartate and phenylalanine side chains are interchanged ("DFG-out") and the activation loop blocks the usual ATP/substrate binding site but exposes an additional hydrophobic binding site (sometimes termed the *allosteric pocket*) directly adjacent the ATP binding site [45].

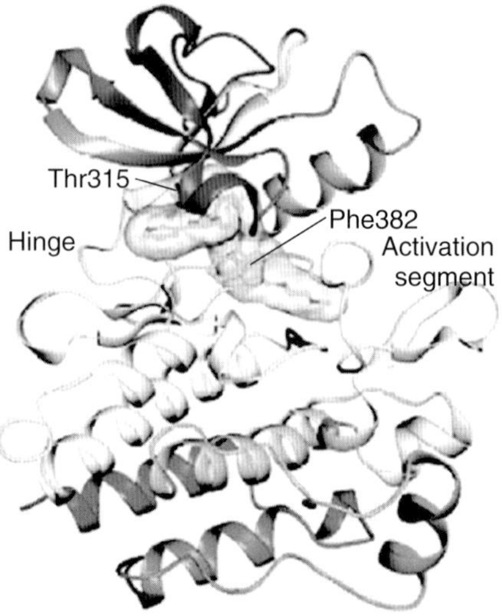

Figure 16.5 Structure of a protein kinase with a bound inhibitor. (Source: Reproduced with permission from Ref. [44]. Copyright 2005 by John Wiley & Sons.)

The majority of kinase inhibitors reported to date are ATP competitive and bind to the kinase in its activated form. The fact that mainly the so-called "type I" inhibitors have been discovered is likely a consequence of the screening conditions that have usually been used (i.e., highly active recombinant kinase domains with a phosphorylated activation loop and low ATP concentrations). Dasatinib (**7**) is an example of a type I inhibitor and, since the active conformation of Src is similar to the active conformation of Abl, it exhibits dual Src-Abl kinase inhibitory activity. Another class of kinase inhibitors (type II) is the one that binds the DFG-out conformation of the kinase and thereby prevents its activation. Although less prevalent, one thought is that it may be easier to achieve higher selectivity with type II inhibitors due to the less conserved nature of the allosteric binding pocket. The aminopyrimidines imatinib (**1**), nilotinib (**2**), bafetinib (**9**), and flumatinib (**10**) are all examples of type II inhibitors, and their high selectivity for Abl kinase has been attributed to both the binding of the inactive form of the kinase and the size of the threonine gatekeeper side chain at the back of the ATP binding pocket [46].

16.5
Structure–Activity Relationship

As might be inferred from the structures of imatinib (**1**), nilotinib (**2**), bafetinib (**9**), and flumatinib (**10**), a large number of modifications involving optimization of the C-ring substituents (Figure 16.6) have been investigated. Nevertheless, in the seminal papers on phenylaminopyrimidine (PAP) derivatives,

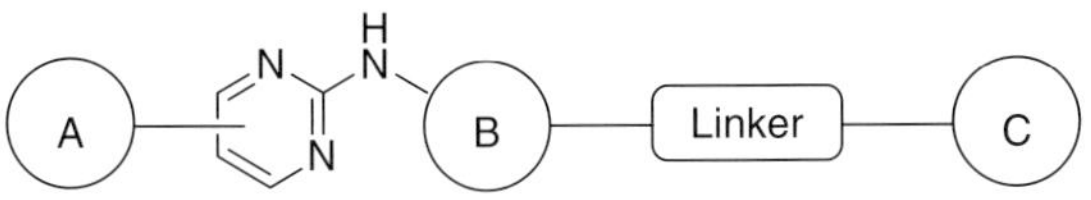

Figure 16.6 Schematic representation of aminopyrimidine-based Bcr-Abl inhibitor structures.

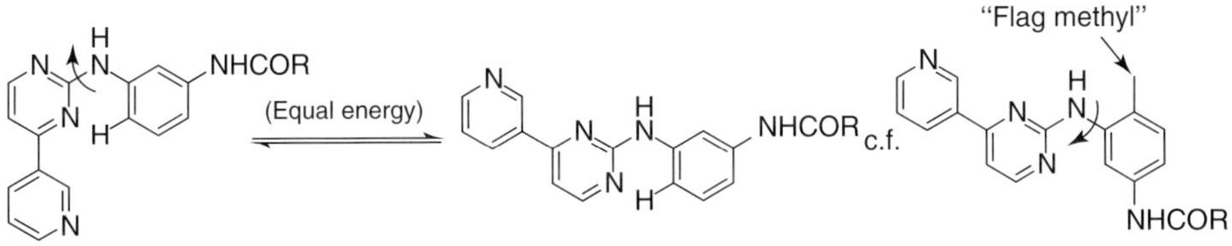

Figure 16.7 Preferred conformation induced by the "flag methyl" group.

Zimmermann *et al.* [47, 48] showed that both the potency and the selectivity can also be greatly affected by variations of the A- and/or B-rings and/or their substituents. Substituted phenyl, heterocycles, and even alkyl residues were tolerated as C-ring groups to provide moderately potent v-Abl (Abl1) kinase inhibitors (enzymatic IC_{50} values of 100–300 nM, e.g., 38 nM for imatinib (**1**)). For example, the B-ring desmethyl analog of imatinib (**1**) shows, in addition to a slight (two- to fourfold) decrease in activity against both v-Abl and platelet-derived growth factor receptor (PDGFR), an increased (>10-fold) albeit still weak ($IC_{50} = 7.8\ \mu M$) activity against c-Src. The effect on kinase selectivity due to the presence or absence of this methyl group (termed *"flag methyl"* by the authors) is particularly pronounced with respect to the inhibition of PKC-α, with several compounds lacking the methyl group being dual v-Abl/PKC-α inhibitors. This effect is attributed to the different preferred conformations of the two systems (Figure 16.7) and the resulting unfavorable interaction of the methyl group within the ATP-binding pocket of the protein kinase C (PKC) family.

Likewise, addition of a single methyl group at the 6-position of the pyrid-3-yl A-ring of imatinib (**1**) provides a compound that exhibits a slightly enhanced inhibition of the stem cell factor receptor kinase, KIT ($K_i \approx 43$ nM cf. 55 nM for imatinib (**1**)), and a drastically reduced affinity for Abl (−75%) and all other Abl variants (−95%). Also, in this case, it is postulated that the added methyl group stabilizes a loop region (C673-G676) of C-Kit while destabilizing the corresponding region (M318-G321) in Bcr-Abl [49].

Interestingly, exchanging the pyrid-3-yl A-ring with an *N*-(3-aminopropyl)-indol-4-yl group and having just 3-(1,1,2,2-tetrafluoroethoxy)-phenyl as the B-ring (and no linker/C-ring) results in a compound **36** showing little to no selectivity against v-Abl, epidermal growth factor receptor (EGFR), c-Src, PDGFR, protein kinase A (PKA), PKC-α, and PKC-δ (IC_{50}'s of 0.18–5 μM) (Figure 16.8) [48].

In another study, replacement of the amide/inverse amide linker with a urea linker led to several compounds (**37**) that, although able to effectively inhibit both recombinant c-Abl in a cell-free assay ($IC_{50} \approx 50$ nM) and PDGFR

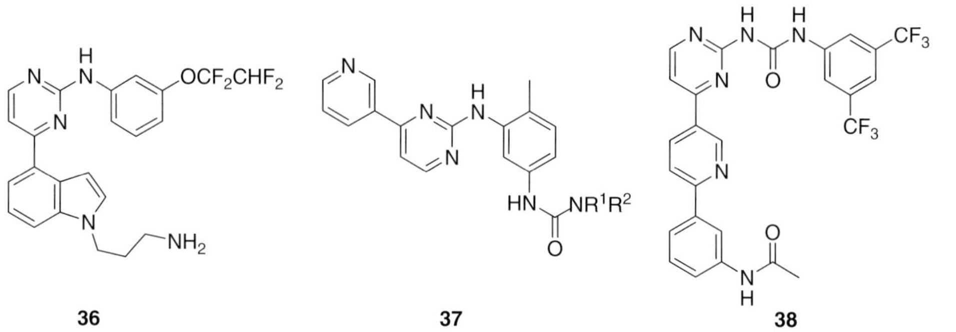

Figure 16.8 Imatinib (**1**) "analogs" **36–38**.

autophosphorylation in A31 cells ($IC_{50} \approx 20$ nM), are unexplainably $\geq 10-$ fold less active in a cellular Bcr-Abl kinase assay [50]. Incorporation of the urea linkage directly on the pyrimidine ring (with omission of the B-ring) and further substituting the pyrid-3-yl A-ring in the 6-position with an additional substituted phenyl ring results in a compound **38** with submicromolar IC_{50} values in T-47D and MDA-MB-468 breast cancer cell lines [51].

With respect to the C-ring, the *N*-methylpiperazine group of imatinib (**1**) was originally introduced to improve the aqueous solubility of the molecule and also resulted in an increased oral bioavailability. With the observation of decreased response rates and/or relapse in patients being treated with imatinib (**1**) (often a result of point mutations within the kinase domain of Bcr-Abl) [52], coupled with structural information from the imatinib-Abl complex [53], one strategy for improving the efficacy of inhibitor binding to both wild-type and mutant Bcr-Abl kinase was to replace the basic and partially solvent-exposed piperazine group with groups having more lipophilic interactions. This approach, reported by Manley *et al.* [2b, 54], ultimately led to nilotinib (**2**), which, in addition to having a better topological fit to the DFG-out pocket, also has a weak electrostatic interaction between a fluorine atom of the trifluoromethyl group and the backbone carbonyl of A381 [55]. Nilotinib (**2**) is 20- to 30-fold more potent than imatinib (**1**) toward Bcr-Abl and is able to maintain activity against 32 of 33 imatinib-resistant Bcr-Abl mutants (the lone exception being the T315I mutant). Similar to imatinib (**1**), nilotinib (**2**) is also active against discoidin domain receptor (DDR), PDGFR, and Kit but the relative order of affinity is DDR > Bcr-Abl > PDGFR > Kit (cf. DDR > PDGFR > Kit > Bcr-Abl for imatinib (**1**)) [56].

In a related approach also starting from the published structure of the imatinib-Abl complex, Asaki *et al.* [57] modified the linker/C-ring of imatinib (**1**) with a series of 3-substituted benzamides. As with nilotinib (**2**), a 3-trifluoromethyl group provided the most potent derivative, a finding that was rationalized on the basis of restricted rotation of the piperazinylmethyl group because of the adjacent substituent and a better fit in the kinase domain. In addition, motivated by considerations of the interaction of Y253 with the A-ring, the pyridine ring was replaced with a pyrimidine ring. Further fine-tuning resulted in the design of bafetinib (**9**), and evaluation of the overall profile (including animal pharmacokinetics and toxicity) led to its selection as a development candidate. Bafetinib (**9**) is 25- to 55-fold more active than imatinib (**1**) toward Bcr-Abl in K562 (Bcr-Abl-positive CML cell line) and 293T (transfected with wild-type Bcr-Abl) cells and is able to inhibit the phosphorylation of 12 of 13 known mutant kinases with again no effect on the T315I mutant. Similar to imatinib (**1**), bafetinib (**9**) is active against PDGFR ($IC_{50} = 56$ nM cf. 100 nM for imatinib (**1**) in normal human dermal fibroblasts (NHDF) cells) and, in addition, potently inhibits Lyn kinase (enzymatic $IC_{50} = 19$ nM cf. 352 nM for imatinib (**1**)) [57a]. A related 3,4-disubstituted benzamide (i.e., 3-trifluoromethyl-4-nitrophenyl with a pyrid-3-yl A-ring) has been disclosed by Skobridis *et al.* [58] and is stated to be equally effective as imatinib (**1**) in K562 cells and in inhibiting the phosphorylation of Crkl (a commonly used biomarker to evaluate *in vivo* Abl signaling).

In the case of flumatinib (**10**), only limited SAR data has been published [59]. In K562 cells, flumatinib (**10**) is reported to show about 20–40 times more potent antiproliferative effect than imatinib (**1**) and about 5 times more potent effect than nilotinib (**2**). Substitution of the trifluoromethyl group with a chloro group results in a 10-fold loss in antiproliferative activity.

References

1. (a) Capdeville, R., Buchdunger, E., Zimmermann, J., and Matter, A. (2002) *Nat. Rev. Drug Dis.*, **1**, 493–502; (b) Radford, I.R. (2002) *Curr. Opin. Invest. Drugs*, **3**, 492–499; (c) de Bree, F., Sorbera, L.A., Fernandez, R., and Castaner, J. (2001) *Drugs Future*, **26**, 545–552.
2. (a) Cortes, J.E., Hochhaus, A., le Coutre, P.D., Rosti, G., Pinilla-Ibarz, J., Jabbour, E., Gillis, K., Woodman, R.C., Blakesley, R.E., Giles, F.J., Kantarjian, H.M., and Baccarani, M. (2011) *Blood*, **117**, 5600–5606; (b) Manley, P.W., Drueckes, P., Fendrich, G., Furet, P., Liebetanz, J., Martiny-Baron, G., Mestan, J., Trappe, J., Wartmann, M., and Fabbro, D. (2010) *Biochim. Biophys. Acta*, **1804**, 445–453; (c) Davies, S.L., Bolos, J., Serradell, N., and Bayes, M. (2007) *Drugs Future*, **32**, 17–25.
3. (a) Fletcher, G.C., Brokx, R.D., Denny, T.A., Hembrough, T.A., Plum, S.M., Fogler, W.E., Sidor, C.F., and Bray, M.R. (2011) *Mol. Cancer Ther.*, **10**, 126–137; (b) Diamond, J.R., Bastos, B.R., Hansen, R.J., Gustafson, D.L., Eckhardt, S.G., Kwak, E.L., Pandya, S.S., Fletcher, G.C., Pitts, T.M., Kulikowski, G.N., Morrow, M., Arnott, J., Bray, M.R., Sidor, C., Messersmith, W., and Shapiro, G.I. (2011) *Clin. Cancer Res.*, **17**, 849–860.
4. (a) Ribas, A. and Flaherty, K.T. (2011) *Nat. Rev. Clin. Oncol.*, **8**, 426–433; (b) Nimmagadda, N. and Tawbi, H. (2011) *Drugs Future*, **36**, 63–68; (c) Adams, J.L., Dickerson, S.H., Johnson, N.W., Kuntz, K., Petrov, K., Ralph, J.M., Rheault, T.R., Schaaf, G., Stellwagen, J., Tian, X., Uehling, D.E., Waterson, A.G., and Wilson, B. (2008) (SmithKline Beecham), WO2009 137391, May 6, 2008.
5. Burger, M.T., Pecchi, S., Wagman, A., Ni, Z.J., Knapp, M., Hendrickson, T., Atallah, G., Pfister, K., Zhang, Y., Bartulis, S., Frazier, K., Ng, S., Smith, A., Verhagen, J., Haznedar, J., Huh, K., Iwanowicz, E., Xin, X., Menezes, D., Meritt, H., Lee, I., Wiesmann, M., Kaufman, S., Crawford, K., Chin, M., Bussiere, D., Shoemaker, K., Zaror, I., Maira, S.M., and Voliva, C.F. (2011) *ACS Med. Chem. Lett.*, **2**, 774–779.

6. (a) Tyner, J.W., Bumm, T.G., Deininger, J., Wood, L., Aichberger, K.J., Loriaux, M.M., Druker, B.J., Burns, C.J., Fantino, E., and Deininger, M.W. (2010) *Blood*, **115**, 5232–5240; (b) Burns, C.J., Bourke, D.G., Andrau, L., Bu, X., Charman, S.A., Donohue, A.C., Fantino, E., Farrugia, M., Feutrill, J.T., Joffe, M., Kling, M.R., Kurek, M., Nero, T.L., Nguyen, T., Palmer, J.T., Phillips, I., Shackleford, D.M., Sikanyika, H., Styles, M., Su, S., Treutlein, H., Zeng, J., and Wilks, A.F. (2009) *Bioorg. Med. Chem. Lett.*, **19**, 5887–5892.
7. (a) Das, J., Chen, P., Norris, D., Padmanabha, R., Lin, J., Moquin, R.V., Shen, Z., Cook, L.S., Doweyko, A.M., Pitt, S., Pang, S., Shen, D.R., Fang, Q., de Fex, H.F., McIntyre, K.W., Shuster, D.J., Gillooly, K.M., Behnia, K., Schieven, G.L., Wityak, J., and Barrish, J.C. (2006) *J. Med. Chem.*, **49**, 6819–6832; (b) Copland, M., Hamilton, A., Elrick, L.J., Baird, J.W., Allan, E.K., Jordanides, N., Barow, M., Mountford, J.C., and Holyoake, T.L. (2006) *Blood*, **107**, 4532–4539; (c) McIntyre, J.A., Castaner, J., and Bayes, M. (2006) *Drugs Future*, **31**, 291–303; (d) Lombardo, L.J., Lee, F.Y., Chen, P., Norris, D., Barrish, J.C., Behnia, K., Castaneda, S., Cornelius, L.A.M., Das, J., Doweyko, A.M., Fairchild, C., Hunt, J.T., Inigo, I., Johnston, K., Kamath, A., Kan, D., Klei, H., Marathe, P., Pang, S., Peterson, R., Pitt, S., Schieven, G.L., Schmidt, R.J., Tokarski, J., Wen, M.L., Wityak, J., and Borzilleri, R.M. (2004) *J. Med. Chem.*, **47**, 6658–6661; (e) Shah, N.P., Tran, C., Lee, F.Y., Chen, P., Norris, D., and Sawyers, C.L. (2004) *Science*, **305**, 399–401.
8. (a) Friedberg, J.W., Sharman, J., Sweetenham, J., Johnston, P.B., Vose, J.M., LaCasce, A., Schaefer-Cutillo, J., DeVos, S., Sinha, R., Leonard, J.P., Cripe, L.D., Gregory, S.A., Sterba, M.P., Lowe, A.M., Levy, R., and Shipp, M.A. (2010) *Blood*, **115**, 2578–2585; (b) Sweeny, D.J., Li, W., Clough, J., Bhamidipati, S., Singh, R., Park, G., Baluom, M., Grossbard, E., and Lau, D.T.W. (2010) *Drug Metab. Dispos.*, **38**, 1166–1176; (c) Bajpai, M. (2009) *IDrugs*, **12**, 174–185.
9. (a) Schenone, S., Bruno, O., Radi, M., and Botta, M. (2010) *Med. Res. Rev.*, **31**, 1–41; (b) Weisberg, E., Manley, P.W., Cowan-Jacob, S.W., Hochhaus, A., and Griffin, J.D. (2007) *Nat. Rev. Cancer*, **7**, 345–356; (c) Cowan-Jacob, S.W., Fendrich, G., Floersheimer, A., Furet, P., Liebetanz, J., Rummel, G., Rheinberger, P., Centeleghe, M., Fabbro, D., and Manley, P.W. (2007) *Acta Cryst.*, **D63**, 80–93; (d) Liu, Y. and Gray, N.S. (2006) *Nat. Chem. Biol.*, **2**, 358–364; (e) Okram, B., Nagle, A., Adrian, F.J., Lee, C., Ren, P., Wang, X., Sin, T., Xie, Y., Wang, X., Xia, G., Spraggon, G., Warmuth, M., Liu, Y., and Gray, N.S. (2006) *Chem. Biol.*, **13**, 779–786; (f) Cowan-Jacob, S.W., Guez, V., Fendrich, G., Griffin, J.D., Fabbro, D., Furet, P., Liebetanz, J., Mestan, J., and Manley, P.W. (2004) *Mini Rev. Med. Chem.*, **4**, 285–299.
10. (a) Takeuchi, M., Kimura, S., Ashihara, E., Maekawa, T., Castaner, R., and Bolos, J. (2009) *Drugs Future*, **34**, 261–269; (b) Yokota, A., Kimura, S., Masuda, S., Ashihara, E., Kuroda, J., Sato, K., Kamitsuji, Y., Kawata, E., Deguchi, Y., Urasaki, Y., Terui, Y., Ruthardt, M., Ueda, T., Hatake, K., Inue, K., and Maekawa, T. (2007) *Blood*, **109**, 306–314.
11. (a) Luo, H., Quan, H., Xie, C., Xu, Y., Fu, L., and Lou, L. (2010) *Leukemia*, **24**, 1807–1809; (b) Gong, A., Chen, X., Deng, P., and Zhong, D. (2010) *Drug Metab. Dispos.*, **38**, 1328–1340; (c) Qiu, L., Wang, X.D., Yu, B.H., Lu, R.Z., Ge, F., Wang, X.L., Chen, L.J., Han, B.H., Zhan, Z.M., Zhang, B.L., and Ma, J. (2008) *J. Exp. Hematol.*, **111**, 1039–1043.
12. (a) Signorovitch, J.E., Wu, E.Q., Betts, K.A., Parikh, K., Kantor, E., Guo, A., Bollu, V.K., Williams, D., Wei, L.J., and DeAngelo, D.J. (2011) *Curr. Med. Res. Opin.*, **27**, 1263–1271; (b) Schenone, S., Bruno, O., Radi, M., and Botta, M. (2010) *Med. Res. Rev.*, **31**, 1–41; (c) Jabbour, E., Cortes, J.E., Giles, F.J., O'Brien, S., and Kantarjian, H.M. (2007) *Cancer*, **109**, 2171–2181; (d) Goldman,

J.M. and Daley, G.Q. (2007) in *Myeloproliferative Disorders* (eds J.V. Melo and J.M. Goldman), Springer, pp. 1–13; (e) Quintas-Cardama, A. and Cortes, J.E. (2006) *Mayo Clin. Proc.*, **81**, 973–988.

13. Frankland, E. and Kolbe, H. (1848) *Liebigs Ann. Chem.*, **65**, 269–287.
14. von Meyer, E. (1880) *J. Prakt. Chem.*, **22**, 261–288.
15. Combes, A. and Combes, C. (1892) *Bull. Soc. Chim. Fr.*, **7**, 788–793.
16. Gabriel, S. and Colman, J. (1901) *Chem. Ber.*, **34**, 1234–1257.
17. Walton, G.L. (1882) *J. Physiol.*, **3**, 349–356.
18. (a) Covell, G., Shute, P.G., and Maryon, M. (1953) *Br. Med. J.*, **1**, 1081–1083; (b) Corrodi, H., Fuxe, K., and Ungerstedt, U. (1971) *J. Pharm. Pharmacol.*, **23**, 989–991; (c) Skolnick, P., Weissman, B.A., and Youdim, M.B.H. (1985) *Br. J. Pharmacol.*, **86**, 637–644; (d) Penner, S.B. and Smyth, D.D. (1994) *Br. J. Pharmacol.*, **112**, 1089–1094; (e) Mori, E., Ember, J., Copeland, B.R., Thomas, W.S., Koziol, J.A., and del Zoppo, G.J. (1995) *Cerebrovasc. Dis.*, **5**, 342–349; (f) Channick, R.N., Simonneau, G., Sitbon, O., Robbins, I.M., Frost, A., Tapson, V.F., Badesch, D.B., Roux, S., Rainisio, M., Bodin, F., and Rubin, L.J. (2001) *Lancet*, **358**, 1119–1123; (g) Jones, P.H., Davidson, M.H., Stein, E.A., Bays, H.E., McKenney, J.M., Miller, E., Cain, V.A., and Blasetto, J.W. (2003) *Am. J. Cardiol.*, **93**, 152–160; (h) Han, K.S., Kim, Y.G., Yoo, J.K., Lee, J.W., and Lee, M.G. (1998) *Biopharm. Drug. Dispos.*, **19**, 493–500; (i) Thakur, A., Thakur, M., Bharadwaj, A., and Thakur, S. (2008) *Eur. J. Med. Chem.*, **43**, 471–477.
19. Lissauer, H. (1865) *Berl. Klin. Wochenschr.*, **2**, 403–404.
20. CowanDoyle, A. (1882) *Lancet*, **119**, 490.
21. Zhu, J., Chen, Z., Lallemand-Breitenbach, V., and de Thé, H. (2002) *Nat. Rev. Cancer*, **2**, 705–714.
22. (a) Jacobson, L.O., Spurr, C.L., Barron, E.S.G., Smith, T., Lushbaugh, C., and Dick, G.F. (1946) *J. Am. Med. Assoc.*, **132**, 263–271; (b) Goodman, L.S., Wintrobe, M.M., Dameshek, W., Goodman, M.J., Gilman, A., and McLennan, M.T. (1946) *J. Am. Med. Assoc.*, **132**, 126–132.
23. (a) Galton, D.A.G. (1953) *Lancet*, **261**, 208–213; (b) Haddow, A. and Timmis, G.M. (1953) *Lancet*, **261**, 207–208.
24. (a) Druker, B.J. (2008) *Blood*, **112**, 4808–4817; (b) Deininger, M.W.N. and Druker, B.J. (2003) *Pharmacol. Rev.*, **55**, 401–423.
25. (a) Zimmermann, J. (1992) (Ciba-Geigy) EP 564409, Apr. 3, 1992; (b) Zimmermann, J. (1994) (Ciba-Geigy) US 5521184, Apr. 28, 1994; (c) Breitenstein, W., Furet, P., Jacob, S., and Manley, P.W. (2002) (Novartis) WO2004 005281, July 5, 2002; (d) Asaki, T., Sugiyama, Y., and Segawa, J. (2003) (Nippon Shinyaku) WO2005 063709, Dec. 25, 2003.
26. (a) Bredereck, H., Effenberger, F., and Botsch, H. (1964) *Chem. Ber.*, **97**, 3397–3406; (b) Benary, E. (1930) *Chem. Ber.*, **63**, 2601–2608.
27. (a) Hughes, J.L., Liu, R.C., Enkoji, T., Smith, C.M., Bastian, J.W., and Luna, P.D. (1975) *J. Med. Chem.*, **18**, 1077–1088; (b) Kaempf, A. (1904) *Chem. Ber.*, **37**, 1681–1684.
28. (a) Paul, R., Hallett, W.A., Hanifin, J.W., Reich, M.F., Johnson, B.D., Lenhard, R.H., Dusza, J.P., Kerwar, S.S., Lin, Y., Pickett, W.C., Seifert, C.M., Torley, L.W., Tarrant, M.E., and Wrenn, S. (1993) *J. Med. Chem.*, **36**, 2716–2725; (b) Bennett, G.B., Mason, R.B., Alden, L.J., and Roach, J.B. Jr. (1978) *J. Med. Chem.*, **21**, 623–628.
29. Bredereck, H., Simchen, G., Rebsdat, S., Kantlehner, W., Horn, P., Wahl, R., Hoffmann, H., and Grieshaber, P. (1968) *Chem. Ber.*, **101**, 41–50.
30. Liu, Y.F., Wang, C.L., Bai, W.J., Han, N., Jiao, J.P., and Qi, X.L. (2008) *Org. Process Res. Dev.*, **12**, 490–495.
31. Loiseleur, O., Kaufmann, D., Abel, S., Buerger, H.M., Meisenbach, M., Schmitz, B., and Sedelmeier, G. (2003) (Novartis), WO2003 066613, Feb. 6, 2003.
32. Kamath, A.A., Pai, G.G., Ujagare, A.M., He, X., Zu, S., Shen, X., Yang, J., and Zhan, H. (2009) (Arch Pharmalabs, Shanghai Parling Pharmatech), WO2011 070588, Dec. 10, 2009.

33. McCormack, P. and Cross, D. (2004) (Stylacats), WO2006 061332, Dec. 8, 2004.
34. McKenna, J. and Shieh, W.C. (2005) (Novartis), WO2006 135641, June 9, 2005.
35. Xu, G., Shen, H., Tong, T., Lu, A., and Gou, S. (2010) *Synth. Commun.*, **40**, 2564–2570.
36. Gayo, L.M. and Suto, M.J. (1997) *Tetrahedron Lett.*, **38**, 211–214.
37. Liu, H., Xia, W., Luo, Y., and Lu, W. (2010) *Monatsh. Chem.*, **141**, 907–911.
38. Boully, L., Darabantu, M., Turck, A., and Ple, N. (2005) *J. Heterocyclic Chem.*, **42**, 1423–1428.
39. (a) Bennett, J.H. (1845) *Edinb. Med. Surg. J.*, **64**, 413–423; (b) Virchow, R. (1845) *Frorieps Notizen*, **36**, 151–156.
40. (a) Nowell, P.C. and Hungerford, D.A. (1960) *Science*, **132**, 1497; (b) Nowell, P.C. and Hungerford, D.A. (1960) *J. Natl. Cancer Inst.*, **25**, 85–109.
41. Rowley, J.D. (1973) *Nature*, **243**, 290–293.
42. (a) Kelliher, M.A., McLaughlin, J., Witte, O.N., and Rosenberg, N. (1990) *Proc. Natl. Acad. Sci. U.S.A.*, **87**, 6649–6653; (b) Elefanty, A.G., Hariharan, I.K., and Cory, S. (1990) *EMBO J.*, **9**, 1069–1078; (c) Daley, G.Q., VanEtten, R.A., and Baltimore, D. (1990) *Science*, **247**, 824–830; (d) Heisterkamp, N., Jenster, G., ten Hoeve, J., Zovich, D., Pattengale, P.K., and Groffen, J. (1990) *Nature*, **344**, 251–253.
43. (a) Deininger, M.W.N., Goldman, J.M., Lydon, N., and Melo, J.V. (1997) *Blood*, **90**, 3691–3698; (b) Druker, B.J., Tamura, S., Buchdunger, E., Ohno, S., Segal, G.M., Fanning, S., Zimmermann, J., and Lydon, N.B. (1996) *Nat. Med.*, **2**, 561–566; (c) Buchdunger, E., Zimmermann, J., Mett, H., Meyer, T., Mueller, M., Druker, B.J., and Lydon, N.B. (1996) *Cancer Res.*, **56**, 100–104.
44. Nollert, P., Feese, M.D., Staker, B.L., and Kim, H. (2005) in *Drug Discovery Handbook*, Chapter 10 (ed. S.C. Gad), John Wiley & Sons, Inc., pp. 373–455.
45. (a) Nolen, B., Taylor, S., and Ghosh, G. (2004) *Mol. Cell*, **15**, 661–675; (b) Scapin, G. (2002) *Drug Discov. Today*, **7**, 601–611; (c) Huse, M. and Kuriyan, J. (2002) *Cell*, **109**, 275–282.
46. Vajpai, N., Strauss, A., Fendrich, G., Cowan-Jacob, S.W., Manley, P.W., Grzesiek, S., and Jahnke, W. (2008) *J. Biol. Chem.*, **283**, 18292–18302.
47. Zimmermann, J., Buchdunger, E., Mett, H., Meyer, T., Lydon, N.B., and Traxler, P. (1996) *Bioorg. Med. Chem. Lett.*, **6**, 1221–1226.
48. Zimmermann, J., Buchdunger, E., Mett, H., Meyer, T., and Lydon, N.B. (1997) *Bioorg. Med. Chem. Lett.*, **7**, 187–192.
49. Fernandez, A., Sanguino, A., Peng, Z., Ozturk, E., Chen, J., Crespo, A., Wulf, S., Shavrin, A., Qin, C., Ma, J., Trent, J., Lin, Y., Han, H.D., Mangala, L.S., Bankson, J.A., Gelovani, J., Samarel, A., Bornmann, W., Sood, A.K., and Lopez-Berestein, G. (2007) *J. Clin. Invest.*, **117**, 4044–4054.
50. Manley, P.W., Breitenstein, W., Brueggen, J., Cowan-Jacob, S.W., Furet, P., Mestan, J., and Meyer, T. (2004) *Bioorg. Med. Chem. Lett.*, **14**, 5793–5797.
51. El-Deeb, I.M. and Lee, S.H. (2010) *Bioorg. Med. Chem.*, **18**, 3860–3874.
52. Warmuth, M., Simon, N., Mitina, O., Mathes, R., Fabbro, D., Manley, P.W., Buchdunger, E., Forster, K., Moarefi, I., and Hallek, M. (2003) *Blood*, **101**, 664–672.
53. (a) Nagar, B., Bornmann, W.G., Pellicena, P., Schindler, T., Veach, D.R., Miller, W.T., Clarkson, B., and Kuriyan, J. (2002) *Cancer Res.*, **62**, 4236–4243; (b) Schindler, T., Bornmann, W.G., Pellicena, P., Miller, W.T., Clarkson, B., and Kuriyan, J. (2000) *Science*, **289**, 1938–1942.
54. Manley, P.W., Cowan-Jacob, S.W., and Mestan, J. (2005) *Biochim. Biophys. Acta*, **1754**, 3–13.
55. Manley, P.W., Cowan-Jacob, S.W., Fendrich, G., and Mestan, J. (2005) *Blood*, **106**, 940a–941a.
56. Manley, P.W., Stiefl, N., Cowan-Jacob, S.W., Kaufman, S., Mestan, J., Wartmann, M., Wiesmann, M., Woodman, R., and Gallagher, N. (2010) *Bioorg. Med. Chem.*, **18**, 6977–6986.

57. (a) Horio, T., Hamasaki, T., Inoue, T., Wakayama, T., Itou, S., Naito, H., Asaki, T., Hayase, H., and Niwa, T. (2007) *Bioorg. Med. Chem. Lett.*, **17**, 2712–2717; (b) Askai, T., Sugiyama, Y., Hamamoto, T., Higashioka, M., Umehara, M., Naito, H., and Niwa, T. (2006) *Bioorg. Med. Chem. Lett.*, **16**, 1421–1425.

58. Skobridis, K., Kinigopoulou, M., Theodorou, V., Giannousi, E., Russell, A., Chauhan, R., Sala, R., Brownlow, N., Kiriakidis, S., Domin, J., Tzakos, A.G., and Dibb, N.J. (2010) *ChemMedChem*, **5**, 130–139.

59. Lu, A., Yang, B., Hu, C., and Wang, C. (2005) EP 1840122, Dec. 26, 2005.

17
Benzyl Triazole-Based Aromatase Inhibitors for the Treatment of Breast Cancer

Dawn George and Stacy Van Epps

17.1
Introduction

Triazole is the generic name for compounds with a 1*H*-1,2,4-triazole ring. In the case of aromatase inhibitors, these all bear a benzyl moiety at the N1 position (Figure 17.1).

This chapter highlights compounds with the benzyl triazole moiety, specifically those targeting aromatase for the treatment of breast cancer. Examples of aromatase inhibitors in this structural class include anastrozole (**2a**, Arimidex®), first introduced by Ciba Geigy/Novartis, letrozole (**2b**, Femara®), developed by Zeneca, and vorozole (**2c**, Rivizor®) developed by Janssen (Figure 17.2).

17.2
History

Aromatase inhibitors for the treatment of hormone-dependent breast cancer were first identified in the 1970s. Aminoglutethimide, when administered in combination with hydrocortisone, was found to give clinical efficacy [1]. However, the treatment also caused dose-limiting side effects including skin rash, lethargy, and ataxia, and thus, more potent "second-generation" steroidal aromatase inhibitors superseded the development of aminoglutethimide. One of these second-generation compounds, 4-hydroxyandrostenedione (4-OHA, formestane), was found to be well tolerated and efficacious in breast cancer patients. Treatment resulted in partial to complete response in 26% of patients and disease stabilization in an additional 25% of patients [2]. 4-OHA was the first approved aromatase inhibitor in Europe. To avoid the well-known side effects associated with steroid treatment, a third generation of aromatase inhibitors was developed. These inhibitors comprise nonsteroidal compounds of the benzyl triazole class and are the subject of this chapter.

Bioactive Heterocyclic Compound Classes: Pharmaceuticals,
First Edition. Edited by Jürgen Dinges and Clemens Lamberth.

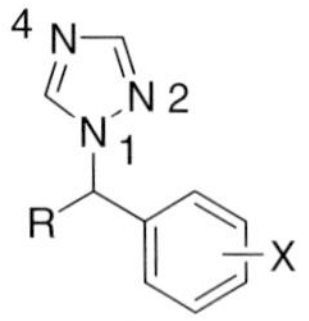

1

Figure 17.1 Basic structure and numbering convention for benzyl triazoles (**1**).

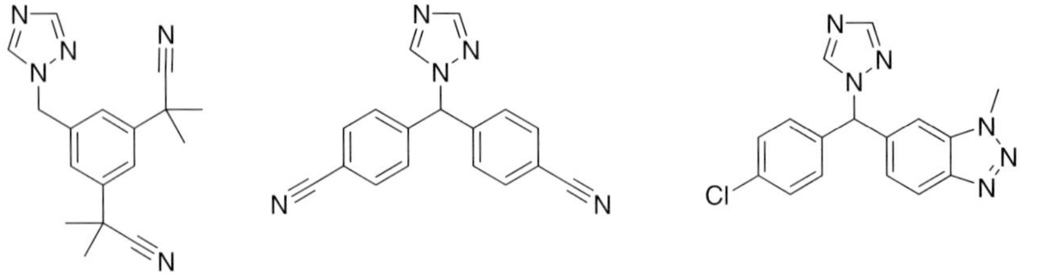

2a Anastrozole | **2b** Letrozole | **2c** Vorozole

Figure 17.2 Aromatase inhibitors **2a–c** of the benzyl triazole family.

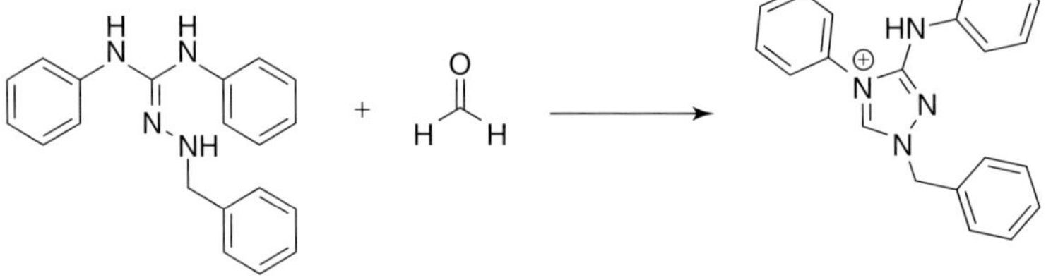

3 | **4** | **5**

Scheme 17.1 First reported synthesis of a benzyl triazole.

In 1902, Busch [3] reported the first synthesis of a benzyl triazole (**5**), prepared via the cyclization of benzyl aminoguanidine (3) with formaldehyde (Scheme 17.1).

However, it was Kaiser in the 1950s who first identified the simplistic conditions to append a benzyl group to an already constructed 1,2,4-triazole (**7**), by alkylation with a benzyl halide (**6**) (Scheme 17.2) [4].

Using similar chemistry, benzyl triazoles with otherwise unsubstituted triazole rings were later explored by Ainsworth *et al.* [5]. Subsequently, this motif was introduced into nonsteroidal aromatase inhibitors. Original analogs such as fadrazole incorporated a benzyl *imidazole* group. Dosing of fadrazole was limited by its ability to suppress aldosterone levels [6], whereas incorporation of a benzyl triazole group provided improved derivatives such as anastrozole (**2a**) and letrozole (**2b**). These compounds gave effective aromatase inhibition with improved selectivity profiles that eliminated undesired activity versus second-generation steroid targets. Benzyl triazoles (**1**) were ultimately successfully marketed for the treatment of hormone-dependent breast cancer [7]. There is continued interest in identifying improved aromatase inhibitors based on the benzyl triazole scaffold [8].

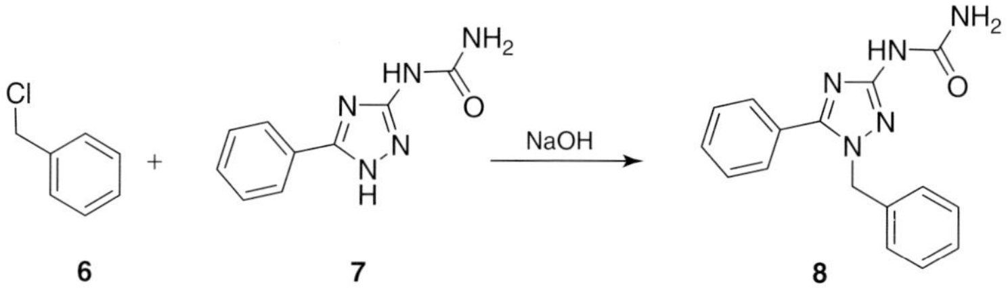

Scheme 17.2 Formation of benzyl triazole (**8**) by alkylation of **7** with a benzyl halide (**6**).

17.3 Synthesis

Although there are a multitude of approaches known in the literature for preparing triazole rings, the benzyl triazole moiety of aromatase inhibitors is typically installed by using simple alkylation chemistry through reaction of a benzyl halide with sodium or potassium 1,2,4-triazole. The general approaches used to prepare 1-benzyl-1*H*-1,2,4-triazoles (**1**) are summarized in Scheme 17.3 in which the 1,2,4-triazole (**9**) is alkylated with either a benzyl bromide **10** or a benzyl chloride **13**. There are two common methods for the preparation of the requisite aryl halides: the chlorination of a benzyl alcohol **12** by treatment with thionyl chloride (Scheme 17.3) and an N-bromosuccinimide (NBS)-mediated bromination of a tolyl

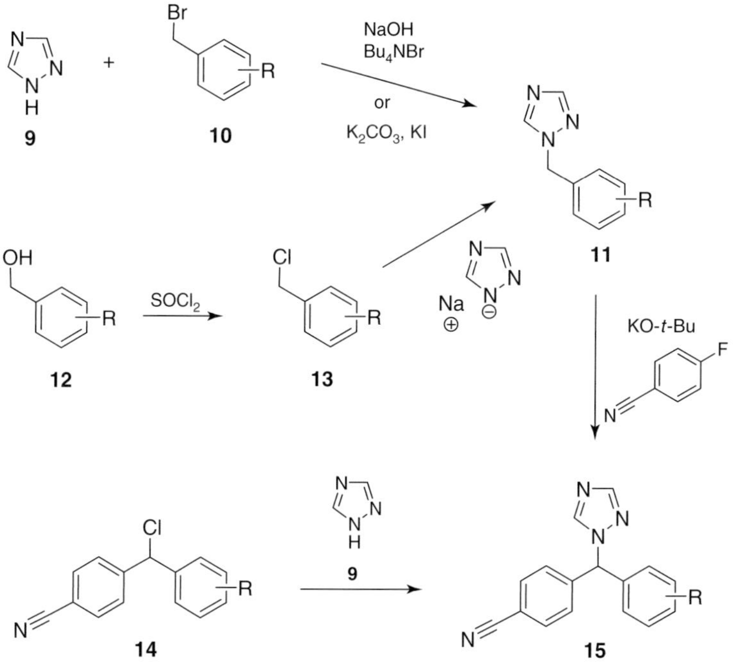

Scheme 17.3 General routes for the synthesis of benzyl triazoles (**1**).

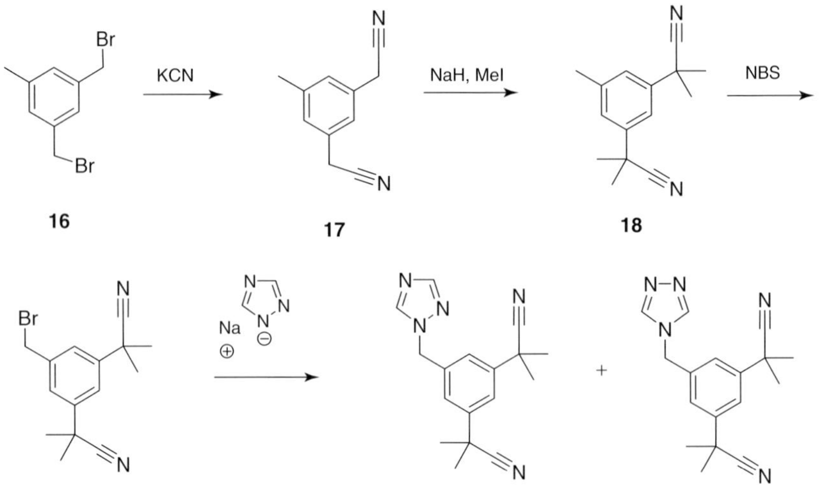

Scheme 17.4 Original synthetic route to anastrozole (**2a**).

derivative as shown by the conversion of **18** to **19** in Scheme 17.4. In the case of the more elaborate letrozole- or vorozole-type analogs **15**, an additional step is required. Thus, deprotonation of the benzylic position of **11** and subsequent reaction with an activated fluorophenyl reagent generates diaryl compounds **15** (Scheme 17.3). Alternatively, a phenyl benzyl chloride **14** may be reacted with 1,2,4-triazole (**9**) in refluxing N,N,-dimethylformamide (DMF) to give benzyl triazole (**15**) directly.

A key issue encountered in the synthesis of benzyl triazoles (**1**) is the formation of regioisomers during the alkylation step because of the ambident nature of the 1,2,4-triazole nucleophile. Two products are commonly observed from this reaction, the desired N1-substituted 1,2,4-triazole (**2a**) as well as the undesired 4-benzyl-1,2,4-triazole (**20**) as exemplified in Scheme 17.4 detailing the synthesis of anastrozole (**2a**) [9, 10].

Various methods to prevent the formation of this side product have been reported. In 2005, Natco Pharma reported that the use of DMF in addition to a nonpolar solvent such as toluene, xylenes, hexane, or heptane resulted in reduced amounts of impurity that could be removed by crystallization of an acid addition salt [11]. An additional report by Sicor claimed that using N-methylpyrrolidine (NMP) as a reaction solvent with subsequent crystallization of the product did not require a salt formation step and delivered anastrozole in 99.75% purity containing only 0.05% of regioisomer **20**. A 2006 patent from Generics Inc. addresses this issue by forming the triazole ring in the final step by reaction of a benzyl hydrazine (**24**) with s-triazine (Scheme 17.5) [12].

In another approach, reaction of the benzyl bromide intermediate **19** with a 4-amino-1,2,4-triazole minimizes the formation of the undesired regioisomer, providing **25** as the quaternary ammonium salt (Scheme 17.6) [10]. A subsequent deamination step using sodium nitrite yields anastrozole (**2a**). Although this route

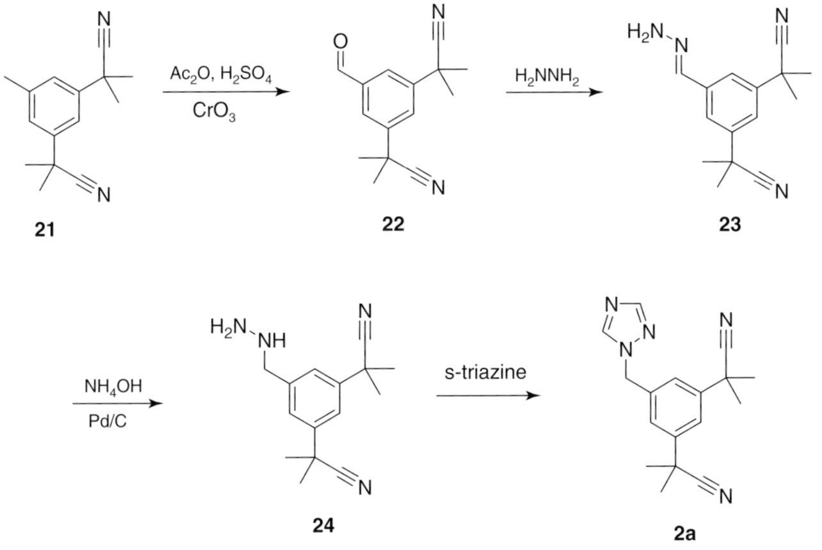

Scheme 17.5 Alternative synthetic route to anastrozole (**2a**).

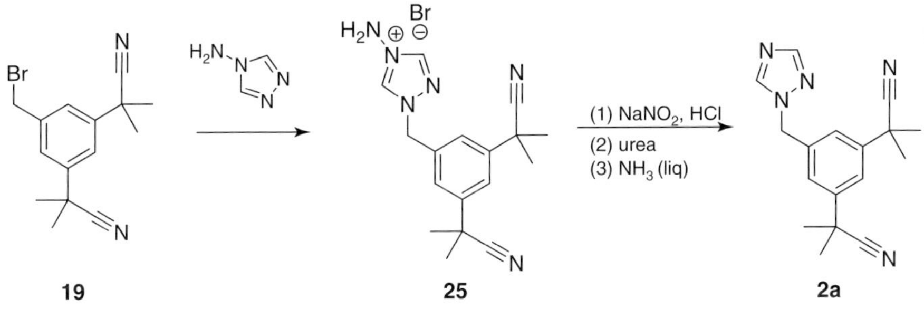

Scheme 17.6 Revised synthetic route to anastrozole (**2a**).

elegantly avoids the formation of the undesired regioisomer, it involves an extra step and introduces the hazards of working with sodium nitrite, which is not preferred on large scale. A similar method using the 4-amino-1,2,4-triazole as a selective nucleophile has also been applied to the synthesis of letrozole (**2b**) [13].

More recently, tosyl triazole (**27**) has been employed for the selective alkylation reaction with the metal salt of the requisite benzyl alcohol (**28**) (Scheme 17.7) [14]. The product is readily crystallized as the mesylate from ethyl acetate and then neutralized to give pure anastrozole (**2a**).

A reductive etherification approach to prepare letrozole (**2b**) was recently reported by Shilpa Medicare (Scheme 17.8). In this case, trimethylsilyl-1,2,4-triazole is reacted with trichlorimidate (**30**) in the presence of borontrifluoride etherate to selectively yield the desired compound.

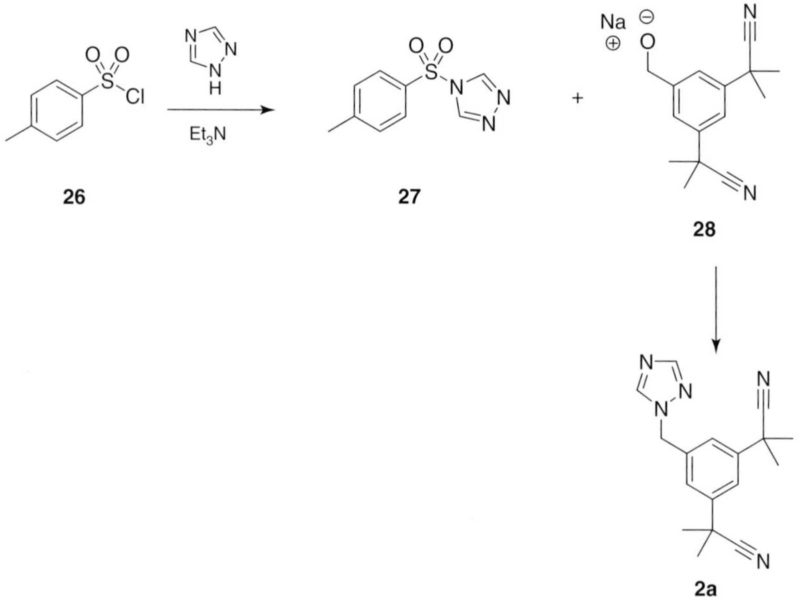

Scheme 17.7 Alternative synthetic route to anastrozole (**2a**).

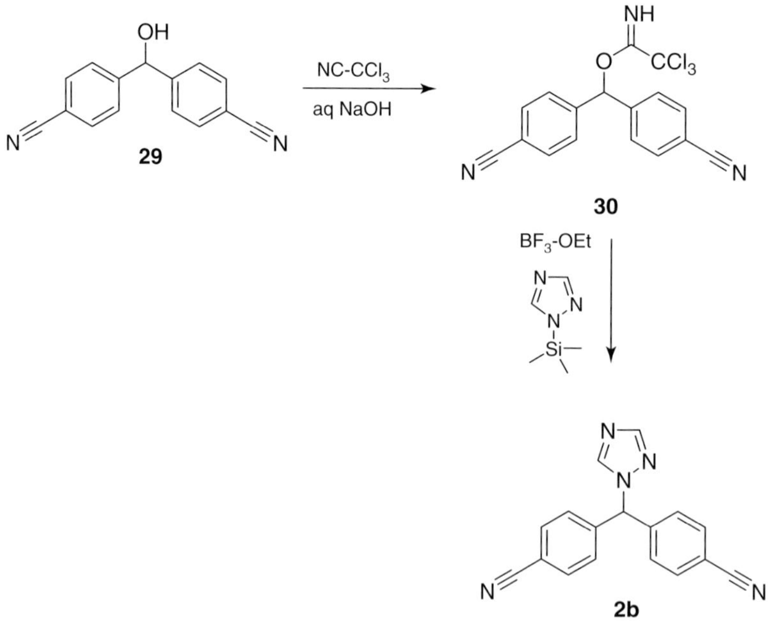

Scheme 17.8 Reductive etherification approach to letrozole (**2b**).

17.4
Mode of Action

Estrogens play a major role in several different physiological processes. In women, estrogens are involved in the development and maintenance of sexual organs, the reproductive cycle, and neuroendocrine functions [15]. In premenopausal women, the ovaries are the main source of estradiol, a biosynthetic precursor to estrogen [16]. In postmenopausal women, estrogen is produced in a number of extragonadal sites, including adipose tissue in the breast, osteoblasts and chondrocytes in the bone, vascular endothelium, aortic smooth muscle, and several sites in the brain. These secondary sites of estrogen production have essential pathways that differ vastly from those of the ovaries. Estrogen is biosynthetically produced via the conversion of androgens into estrogens by the cytochrome *P*450 enzyme complex. This enzyme – which catalyzes the aromatization of the steroid scaffold to produce estrogen – ultimately led to its name, aromatase. Estrogen synthesized within the extragonadal sites acts predominantly at the local tissue level, providing a relatively high local concentration of estrogen that can exert biological influence.

Along with normal function, these hormones also play a role in certain disease states, such as mammary and endometrial carcinomas. Breast cancer is the most common cancer among women, accounting for nearly one in four cancers diagnosed in women in the United States [17]. Previous studies of estrogen-dependent breast cancer suggest that tumors are dependent on the hormone for development and continued growth. Aromatase activity localized in breast adipose tissue supplies cancer cells with the estrogen necessary for cancerous growth [18]. Postmenopausal women diagnosed with breast cancer have a 20-fold increase of estrogen in their plasma [19]. Inhibition of the aromatase enzyme suppresses the final step of estrogen production by blocking the conversion of androgens into estrogens [20]. Although the methods for estrogen deprivation are complex, a basic understanding of the mechanism of aromatase inhibition provides a rationale for the development of aromatase inhibitors for the treatment of hormone-dependent breast cancer.

Aromatase inhibitors are not used to treat breast cancer in premenopausal women. Estrogen is continuously being produced by the ovaries; therefore, the low levels of hormone present in the plasma activate the endocrine system to increase the production of androgens. This pathway is counterproductive to the effects of an aromatase inhibitor; therefore, this therapy cannot be used as an effective treatment for premenopausal breast cancer patients. However, in postmenopausal women, aromatase inhibitors act through pure estrogen deprivation and deny cancer cells of estrogen necessary for growth. Through this mechanism, the removal of all endogenous ligands for the estrogen receptor (ER) prevents activation of the ER and all downstream ER-mediated events. An essential feature of the extragonadal biosynthesis of estrogen is that it is dependent on external C_{19} androgenic precursors. These sites are incapable of converting cholesterol to C_{19} androgen steroids; therefore, the body finds other sources of C_{19} androgens. Both testosterone and androstenedione act as prohormones that circulate in high concentrations in postmenopausal women and are subsequently converted

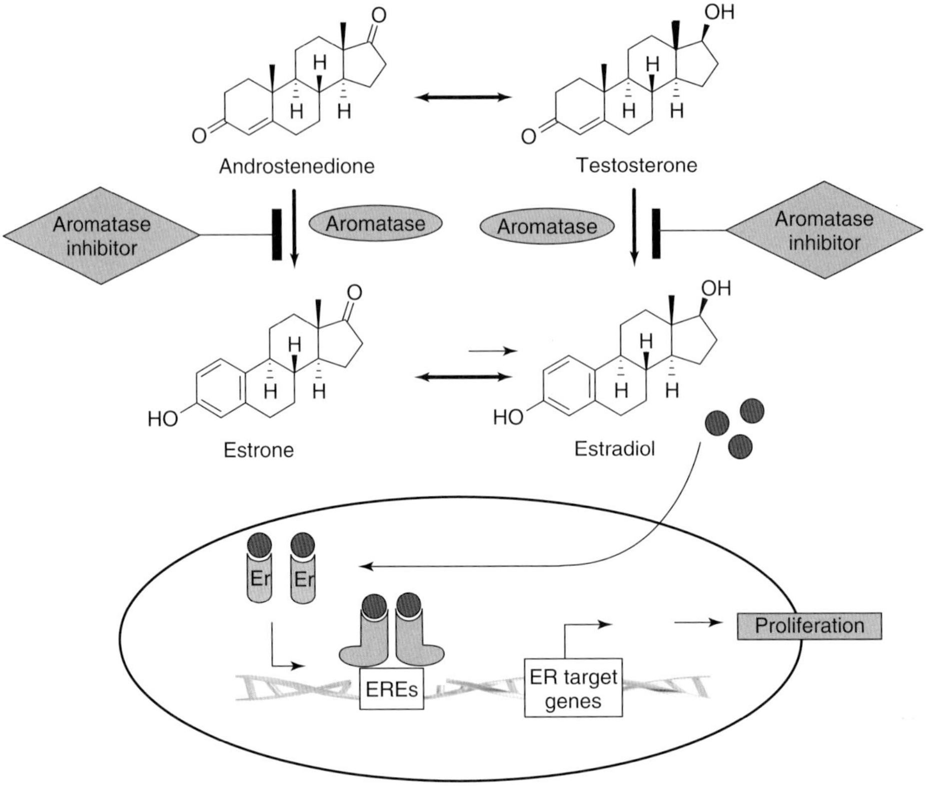

Figure 17.3 Mechanism of action of triazole aromatase inhibitors.

into estradiol and estrone (Figure 17.3). Under normal physiological conditions, estradiol diffuses into the cell and binds to the known nuclear receptor proteins, ER-α and ER-β. The receptors are activated on ligand binding, which leads to successive dimerization and conformational change. This conformational change allows for binding at the estrogen-response elements (EREs) in DNA, upstream of estrogen-regulated genes. EREs have a specific DNA sequence that is recognized by the DNA-binding domain of the ligated ER dimers. Gene transcription is then activated, which ultimately leads to proliferation. In the presence of an aromatase inhibitor, this sequence is inhibited, leading to the prevention of estrogen-dependent transcription.

17.5 Structure–Activity Relationship

Anastrozole (**2a**) and letrozole (**2b**) are two benzyl triazole aromatase inhibitors that have shown to be efficacious and safe in the clinic [8]. To date, there are few literature reports describing structure–activity relationships (SARs) for anastrozole (**2a**) and letrozole (**2b**). However, owing to recent success in the clinic, several research groups have focused their interests on developing improved benzyl triazole analogs [7, 21]. More recently, SAR of benzyl triazoles as dual aromatase sulfatase inhibitors (DASIs) has been reported. Both aromatase and sulfatase are validated targets for treating hormone-dependent breast cancer (HDBC); therefore, it has been reasoned that dual inhibition of both enzymes would provide complete estrogen deprivation.

Until recently, structure-based design of aromatase inhibitors was carried out in the absence of a crystal structure. Homology models of CYP2C9 were published and used for docking of known inhibitors to help rationalize SAR [22]. With the disclosure of a high-resolution human aromatase X-ray structure [23], research groups can now readily use structure-based design. The azaheterocycle was identified as an important pharmacophore for nonsteroidal aromatase inhibitors. It is hypothesized that the electron lone pair of the heterocycle nitrogen binds to the heme Fe^{2+} ion located in the active site of cytochrome *P*450 aromatase, thus interfering with steroid hydroxylation. While coordination to the heme is an essential interaction for triazole inhibitors to bind to the active site, improved potency is achieved with the addition of a hydrogen-bond acceptor proximal to the heterocycle. The nitrile groups incorporated into the structures of both anastrozole (**2a**) and letrozole (**2b**) are effective as hydrogen-bond acceptors, providing a favorable interaction with Ser 478 in the aromatase active site [8].

Before the discovery of anastrozole (**2a**) and letrozole (**2b**), benzyl triazoles suffered from both poor aromatase potency as well as a lack of selectivity against desmolase, a cholesterol side chain cleavage enzyme. Early work by Ciba Geigy [24] led to the discovery of potent imidazole and triazole aromatase inhibitors. SAR around Ciba Geigy and Zeneca's compounds proved that fluoro or cyano substitution on the aryl ring gave optimal inhibition and selectivity. Shortly thereafter, Orion exemplified the use of long-chained diarylalkyltriazoles containing fluoro and cyano substituents, leading to the discovery of finrozole (**36**) [25], a representative of a new series of benzyl triazole inhibitors. For this long-chained diarylalkyltriazole chemotype, the optimal scaffold proved to be the α-ω-diarylalkyl, containing a three or four carbon bridge to the heterocycle (Table 17.1). Positioning the cyano group on the α-phenyl ring and the fluorine on the ω-phenyl ring (**33–36**) gave the best aromatase inhibition (Table 17.1). The most selective inhibitor, finrozole (**36**, Table 17.1), inhibited estrogen synthesis in both preclinical and human studies [26].

Recently, several efforts have focused on the development of new aromatase inhibitors. Reports by Simons [7], Hartmann [27], and Potter [8] describe triazoles containing new structural entities that exhibit aromatase inhibitory activity. Recent SAR of anastrozole- and letrozole-derived analogs described by Potter and coworkers exemplifies the use of a triazole inhibitor in which one of the cyano groups is replaced by a phenyl ring. The resulting biphenyl moiety is designed to mimic the steroidal A/C rings of the enzyme's natural ligand. The terminal phenyl ring is used as a scaffolding to investigate the effects of different hydrogen bond acceptors in varied positions relative to the triazole. The first group of triazoles described

Table 17.1 SAR of α-ω-diarylalkyl triazole inhibitors of human aromatase enzyme.

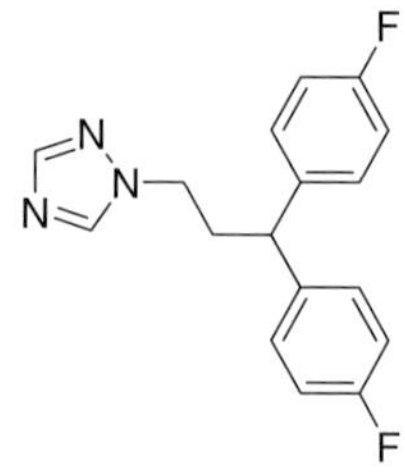

31

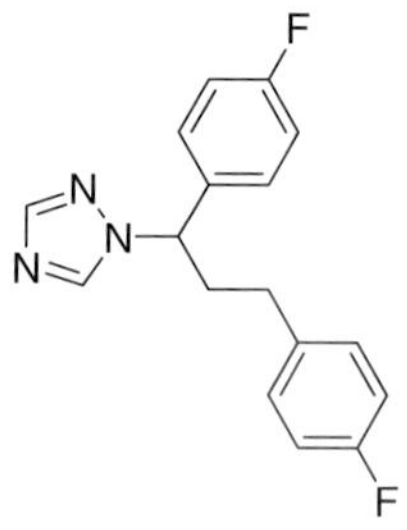

32

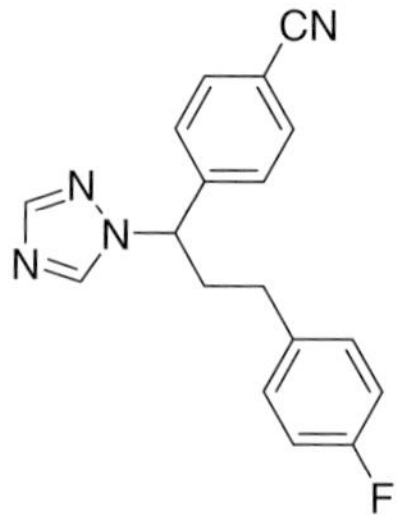

33

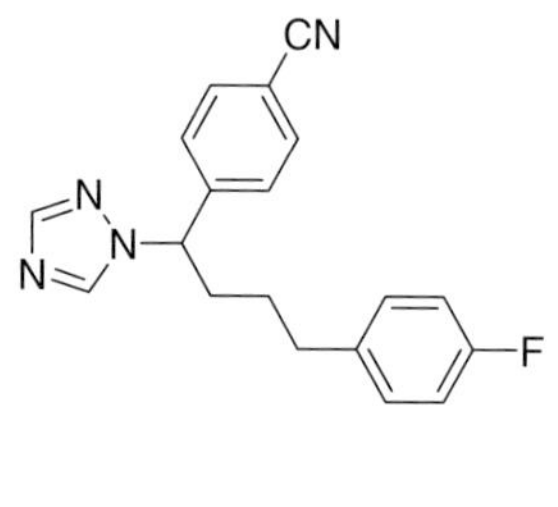

34

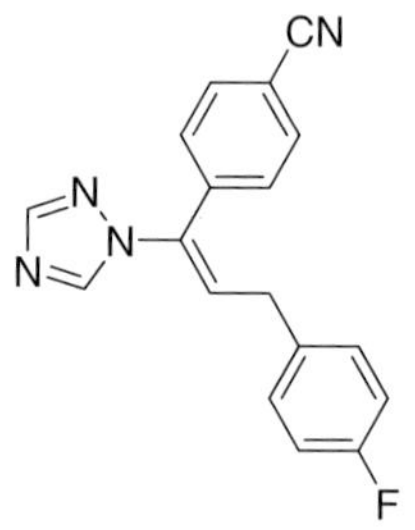

35

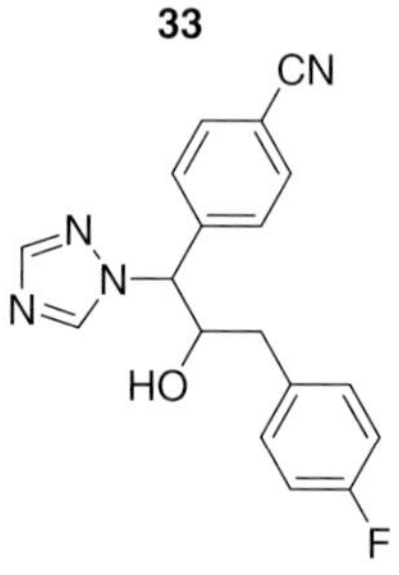

36

Compound	**31**	**32**	**33**	**34**	**35**	**36**
IC_{50} (μM)	36.0	4.0	0.19	0.12	0.05	0.22

is structurally related to anastrozole (**2a**), with a phenyl ring replacing one of the dimethylacetonitrile functional groups. This structural modification, which results in the loss of molecular symmetry, is well tolerated. Analogs substituted on the aryl ring with chloro (**38**) or cyano (**39**) retain inhibitory activity, and even unsubstituted **37** was equipotent with anastrozole (**2a**) (Table 17.2).

The next set of compounds explores triazoles (**40–48**) that contain a cyano group on the proximal ring positioned ortho to the biphenyl bridge (Table 17.3). Analogs bearing smaller functional groups (hydrogen, chloro, cyano, acetyl, and fluoro) at the 3- or 4-position of the distal aryl ring exhibited good potency in the *in vitro* assay ($IC_{50} < 5$ nM). Compound **45**, containing an acetyl group at the 3-position, proved to be the most potent of the analogs tested ($IC_{50} = 0.12$ nM). Compounds with bulkier substituents (napthyl (**47**), *tert*-butyl (**48**)) showed weaker potencies compared to those with smaller functional groups; however, all compounds showed similar or better potencies to that of anastrozole (**2a**).

Next, the effects of the nitrile position relative to the triazole moiety were studied (Table 17.4). In compound **49**, the nitrile group was repositioned to be ortho to the

Table 17.2 Inhibition of aromatase activity in JEG-3 cells for triazole inhibitors.

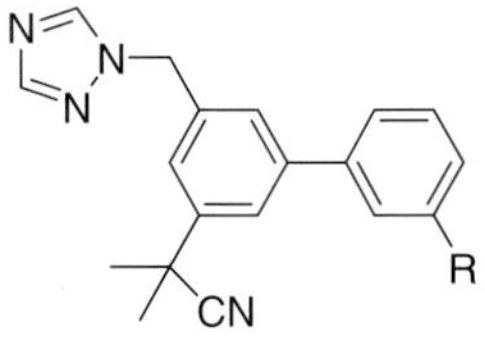

Compound	2a	37	38	39
R	–	H	Cl	CN
IC_{50} (nM)	1.5	0.5	2.2	1.5

Table 17.3 Inhibition of aromatase activity in JEG-3 cells for triazole inhibitors.

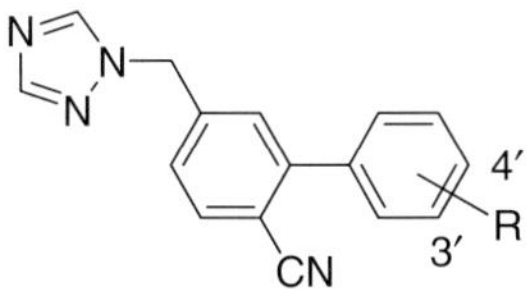

Compound	40	41	42	43	44	45	46	47	48
R	H	3′-Cl	4′-Cl	3′-CN	4′-CN	3′-$COCH_3$	4′-F	3′4′-Ph	4′-tBu
IC_{50} (nM)	0.2	0.22	0.45	0.2	0.33	0.12	0.2	2.2	4.0

Table 17.4 Inhibition of aromatase activity in JEG-3 cells for aromatase inhibitors with repositioned nitrile groups.

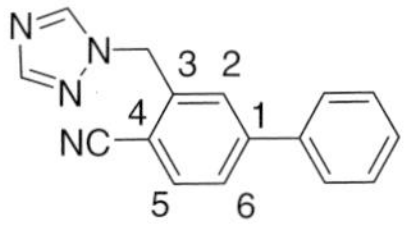

49

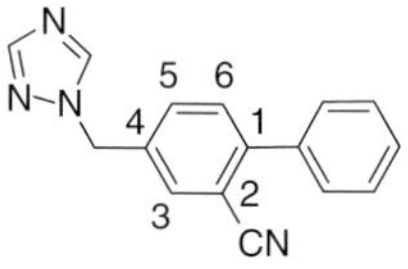

50

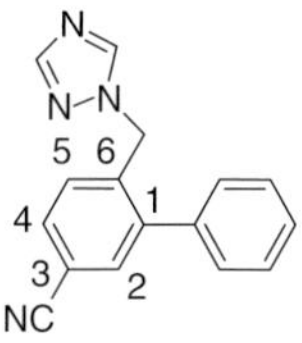

51

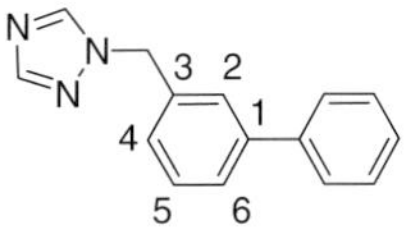

52

Compound	49	50	51	52
Inhibition (%)	68.8	98.6	99.0	70.0
IC_{50} (nM)	–	4.6	1.7	210

heterocycle as compared to the para orientation of compound **40**. This modification resulted in a significant loss of activity, suggesting that the nitrile in **49** is not capable of making an auxiliary interaction with neighboring amino acids in the aromatase active site. Altering the position of the nitrile group ortho to the pendant phenyl ring, as in **40**, but relocating the triazole to the 4-position of the aryl ring increased potency (**50**, Table 17.4). Positioning the nitrile meta to the pendant phenyl ring and adjusting the triazole to the 6-position of the aryl ring was also tolerated (**51**, Table 17.4). Lastly, arranging the triazole meta to the biphenyl ring junction in the absence of the nitrile (**52**) results in a loss of activity, confirming the importance of this group as potency driver. The SAR data set clearly shows that the spatial relationship between the triazole and nitrile group plays an important role in the compound's ability to inhibit the enzyme.

References

1. Santen, R.J., Samojlik, E., Lipton, A., Harvey, H., Ruby, E.B., Wells, S.A., and Kendall, J. (1977) *Cancer*, **39**, 2948–2958.
2. Brodie, A.M. (1994) *J. Steroid Biochem. Mol. Biol.*, **49**, 281–287.
3. Busch, B. (1906) *J. Prakt. Chem.*, **2**, 5433–5439.
4. Kaiser, D.W. and Peters, G.A. (1953) *J. Org. Chem.*, **18**, 196–202.
5. Ainsworth, C., Easton, N.R., Livezey, M., Morrison, D.E., and Gibson, W.R. (1962) *J. Med. Pharm. Chem.*, **51**, 383–389.
6. Furet, P., Batzl, C., Bhatnagar, A., Francotte, E., Rihs, G., and Lang, M. (1993) *J. Med. Chem.*, **36**, 1393–1400.
7. Saberi, M.R., Vinh, T.K., Yee, S.W., Griffiths, B.J., Evans, P.J., and Simons, C. (2006) *J. Med. Chem.*, **49**, 1016–1022.
8. Potter, B.V.L., Jackson, T., Woo, L.W.L., Trusselle, M.N., Purohit, A., and Reed, M.J. (2008) *ChemMedChem*, **3**, 603–618.
9. Edwards, P.N. (1988) EP 296749, June 14, 1988.
10. Edwards, N.P. (1988) US 4,935,437, June 10, 1988.
11. Pulla Reddy, M. (2005) WO 2005105762, May 4, 2005.
12. Gaitonde, A. (2005) WO 2006000836, June 22, 2005.
13. Patel, H.V. (2004) WO 2004076409, Feb. 5, 2004.
14. Benes, M. (2007) WO 2008034644, Sep. 9, 2007.
15. Brueggemeier, R.W., Hackett, J.C., and Diaz-Cruz, E.S. (2005) *Endocr. Rev.*, **26**, 331–345.
16. Simpson, E.R. (2003) *J. Steroid Biochem. Mol. Biol.*, **86**, 225–230.
17. *American Cancer Society 2009-2010 Facts and Figures*, American Cancer Society, Atlanta.
18. Johnston, S. and Stebbing, J. (2003) *Clin Evid.*, **10**, 1975–2002.
19. Noel, C.T., Reed, M.J., Jacobs, H.S., and James, V.H.T. (1981) *Steroid Biochem.*, **14**, 1101–1106.
20. Wood, P.M., Woo, L.W., Humphreys, A., Chander, S.K., Purohit, A., Reed, M.J., and Potter, B.V. (2005) *J. Steroid Biochem. Mol. Biol.*, **94**, 123–130.
21. Vinh, T.K., Ahmadi, M., Delgado, P.O., Perez, S.F., Walters, H.M., Smith, H.J., Nicholls, P.J., and Simons, C. (1999) *Bioorg. Med. Chem. Lett.*, **9**, 2105–2108.
22. Karkola, S., Holtje, H.D., and Wahala, K. (2007) *J. Steroid Biochem. Mol. Biol.*, **105**, 63–70.
23. Ghosh, D., Griswold, J., Erman, M., and Pangborn, W. (2009) *Nature*, **457**, 219–223.
24. Bhatnagar, A.S., Hausler, A., Schieweck, K., Lang, M., and Bowman, R. (1990) *J. Steroid Biochem. Mol. Biol.*, **37**, 1021–1027.

25. Karjalainen, A., Kalapudas, A., Sodervall, M., Pelkonen, O., and Lammintausta, R. (2000) *Eur. J. Pharm. Sci.*, **11**, 109–131.

26. Ahokoski, O., Irjala, K., Huupponen, R., Halonen, K., Salminen, E., and Scheinin, H. (1998) *Br. J. Clin. Pharmacol.*, **45**, 141–146.

27. Gobbi, S., Cavalli, A., Rampa, A., Belluti, F., Piazzi, L., Paluszcak, A., Hartmann, R.W., Recanatini, M., and Bisi, A. (2006) *J. Med. Chem.*, **49**, 4777–4780.

Part V
Inflammation and Gastrointestinal Diseases

Bioactive Heterocyclic Compound Classes: Pharmaceuticals,
First Edition. Edited by Jürgen Dinges and Clemens Lamberth.

18
Acetonide-Based Glucocorticoids for the Treatment of Asthma, Skin Inflammation, and Diseases of the Eye

Kevin P. Cusack, Vikram G. Kalthod, Rajarathnam E. Reddy, and Sanjay R. Chemburkar

18.1
Introduction

Acetonide is the generic name for compounds with a 2,2-dimethyl-[1,3]dioxolane ring system. The dioxolane ring system (**1**) (Figure 18.1) is a structural analog of tetrahydrofuran (THF) (**2**) and is one of the most commonly used protecting groups for 1,2-diols, catechols, and carbonyls [1]. Examples of the use of this protecting group strategy can be found as early as the 1930s when an improved procedure for the formation of acetals and ketals was reported [2]. 1,3-Dioxolane has also found utility as a masked formylating agent for aldehydes and ketones in the presence of SmI_2 (Scheme 18.1) [3].

In addition, the 1,3-dioxolane ring system is present in flavoring agents such as the apple-like flavor/fragrance ethyl 2-(2-methyl-1,3-dioxolan-2-yl)acetate (**3**), fragrances such as the woody fragrance 2,4-dimethyl-2-(5,6,7,8-tetrahydro-5,5,8,8-tetramethyl-2-naphthalenyl)-1,3-dioxolane (**4**), and preservatives such as the antislime agent for food contact paper 4-bromoacetoxymethyl-*m*-dioxolane (**5**) (Figure 18.2). Several classes of drugs including glucocorticoid (GC) compounds useful in asthma, dermatology, and ocular disease contain a 1,3-dioxolane (**1**) [4]. GCs are a steroidal class of molecules known to interact with the glucocorticoid receptor (GR) within the nuclear hormone receptor family to drive altered gene expression profiles [5]. The name glucocorticoid is derived from both the mechanism of action and structure. GCs are catabolic steroids known to be involved in glucose metabolism (gluco-) and are synthesized in the adrenal cortex (-cort-) under the control of the hypothalamic-pituitary-adrenal (HPA) axis [6]. GCs are also anti-inflammatory. In humans, the natural ligand for the GR is the steroid cortisol (**6**) (Figure 18.3), which is enzymatically converted to cortisone (**6a**) *in vivo* [7].

Several related synthetic GCs containing an acetonide ring formed across C16 and C17 of the steroidal D-ring are known. In total, there are 49 analogs with a

Bioactive Heterocyclic Compound Classes: Pharmaceuticals,
First Edition. Edited by Jürgen Dinges and Clemens Lamberth.

1 Dioxolane **2** Tetrahydrofuran

Figure 18.1 Basic structures and numbering system utilized for dioxolane (**1**) and tetrahydrofuran (**2**).

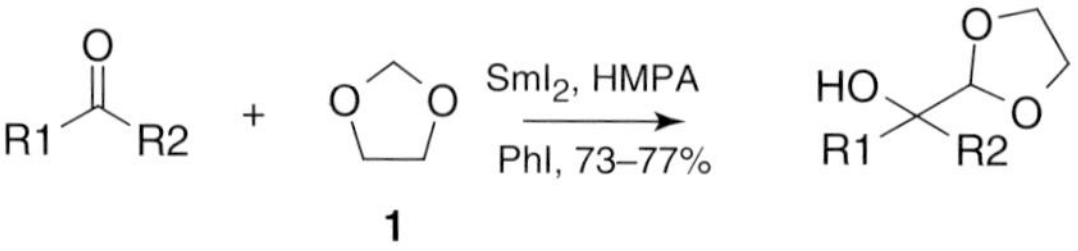

Scheme 18.1 Formylation of aldehydes and ketones with 1,3-dioxolane (**1**). HMPA, hexamethylphosphoramide.

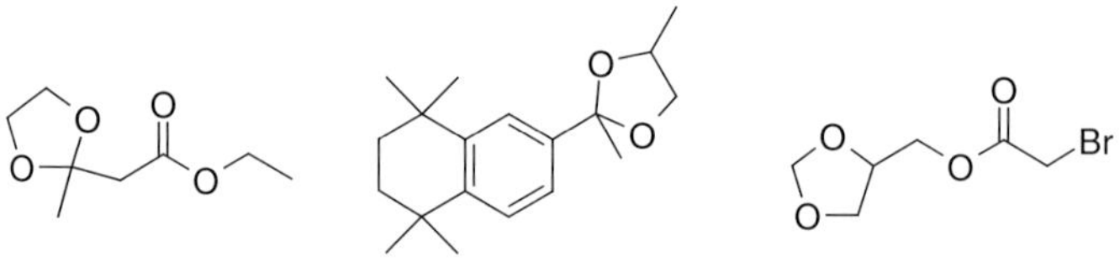

3 **4** **5**

Figure 18.2 1,3-Dioxolanes found in flavors **3**, fragrances **4**, and preservatives **5**.

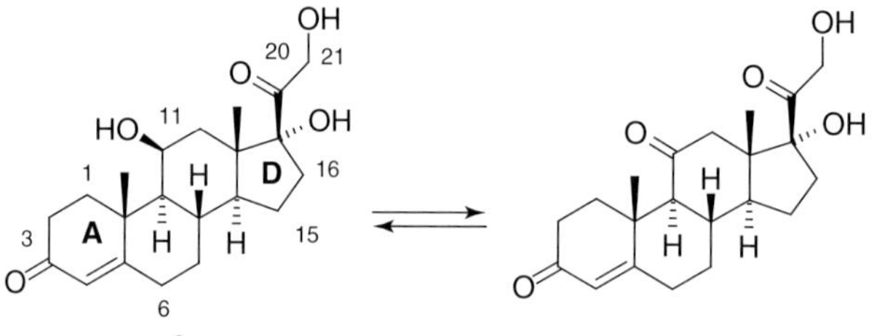

6 Cortisol **6a** Cortisone

Figure 18.3 Structure, ring nomenclature, and numbering of cortisol (**6**) and cortisone (**6a**).

steroidal backbone containing a D-ring-fused dioxolane in the World Drug Index (WDI). Relevant examples of drugs in this structural class are shown in Figure 18.4. The common feature is a 2-substituted-1,3-dioxolane ring. The substituent at the 2-position of the dioxolane ring is most frequently a geminal dimethyl but can also be a monoalkyl or spirocyclic alkyl group. The most widely prescribed members of this family are the corticosteroids budesonide (**7**) [8] and triamcinolone acetonide (**8**) [9] (Figure 18.4), which constitute the subset of analogs that are the focus of this chapter.

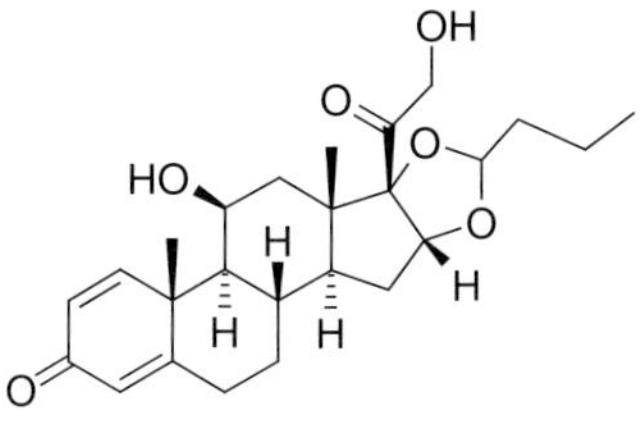

7
Budesonide

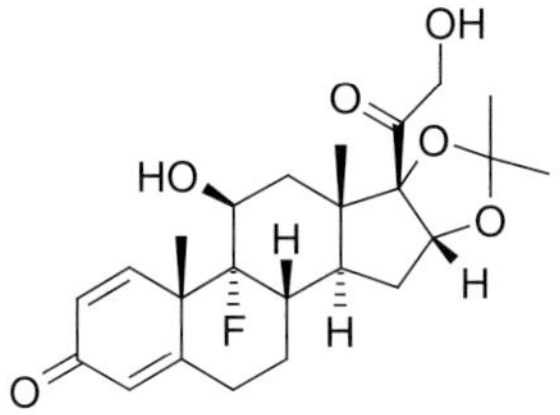

8
Triamcinolone acetonide

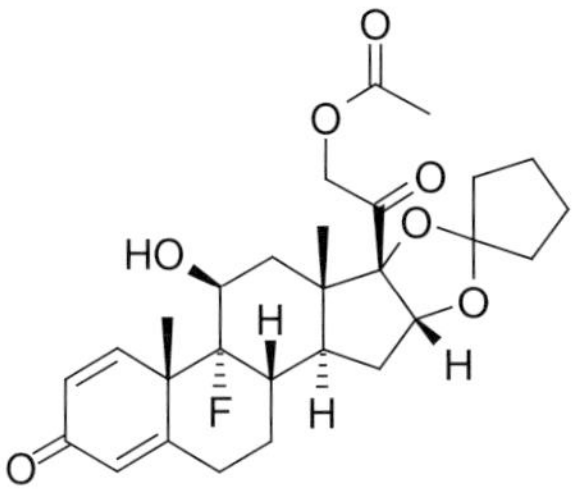

9
Amcinonide

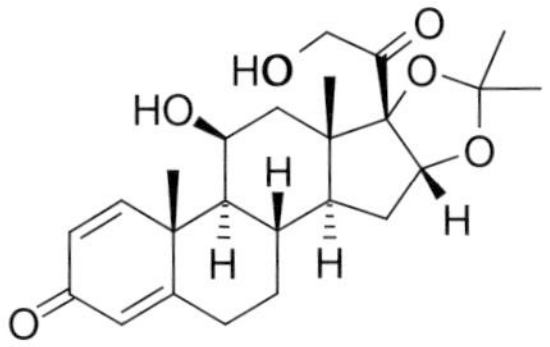

10
Desonide

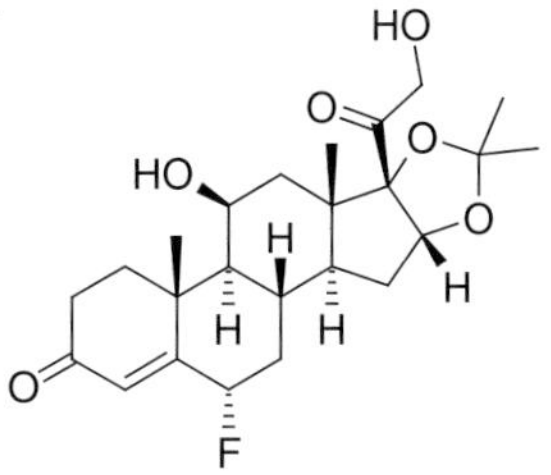

11
Fludroxycortide

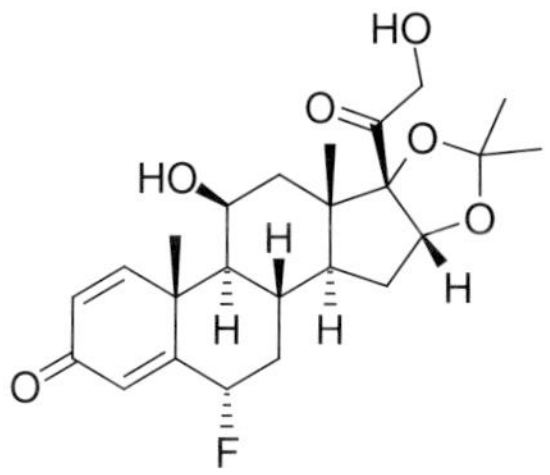

12
Flunisolide

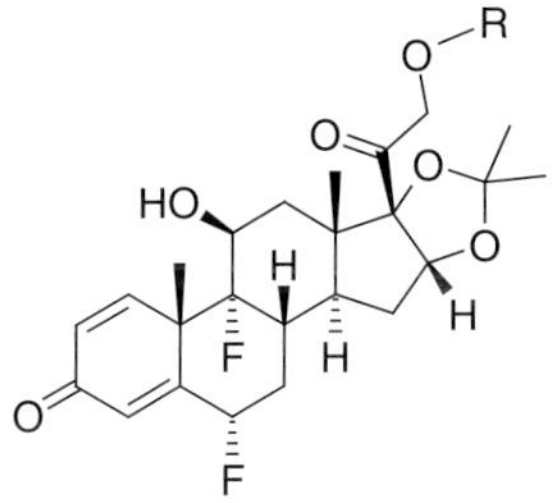

13a Flucinonide (R=Ac)
13b Flucinolone acetonide (R=H)

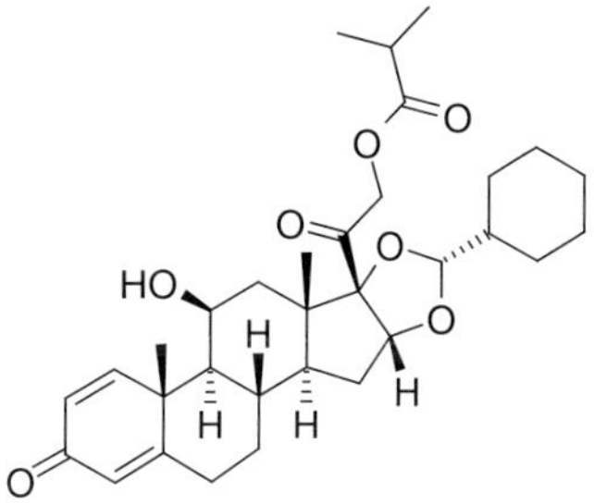

14
Ciclesonide

Figure 18.4 Examples of structures of dioxolane-containing steroids.

18.2 History

1,3-Dioxolane (**1**), a liquid, has traditionally found utility as a solvent or extractant. However, it also appears as a commonly used protecting group for alcohols and carbonyls. The 1,3-dioxolane ring is best described as a geminal diether or cyclic acetal and as such is unstable under aqueous acid conditions. In spite of this chemical instability, multiple steroidal therapeutics have been designed to contain a dioxolane ring and the dioxolane ring is most commonly fused to the steroidal D-ring. An examination of how researchers arrived at the concept of fusing a dioxolane ring to the steroid core requires an understanding of the development of steroid research and identification of the major breakthroughs that contributed to the modern GCs we know today. Three major areas of structure–activity relationships (SARs) are of particular importance. These include the importance of the 9α-F, the role of the C-16 hydroxyl, and the role of the C1–C2 double bond.

In the 1930s, the field of steroid research was still in its infancy. Research was limited in scope and based mainly in academic laboratories with little commercial interest. This paradigm changed following reports of the isolation of useful quantities of adrenal steroids by several groups in the United States and Europe including the research laboratories of Reichstein in Switzerland, Kendall at the Mayo Clinic, and Wintersteiner at Columbia [10]. Although the structures were not known at that time, the five compounds dubbed compounds A–E were later shown to include both cortisol (**6**) and cortisone (**6a**) (Figure 18.3). Commercial interest in steroid research rapidly began to build, and efforts in medicinal chemistry at the Squibb Research Institute and at Lederle, coupled with key contributions from Upjohn, Syntex, and Merck, eventually provided the foundation for some of the most successful steroid drugs in this class. Assembly of the first dedicated medicinal chemistry teams focusing on steroid research began in the early 1940s following the exodus of several key contributors from the war in Europe. Ironically, Percy Julian, one of the great American entrepreneurs who made significant contributions to this field, had to complete his education in Europe because of the Jim Crow policies prevalent in the United States at that time. Although no one could have anticipated where this journey would take them, each would make important contributions to this still relevant field of research. At Squibb, Erhard Fernholz, the coinventor of a process to convert stigmosterol (**18**) to progesterone, was recruited to lead this early effort. However, because of his untimely death in 1941 he did not live to see the fruits of his contributions, and Oskar Wintersteiner was selected to replace him [11]. In spite of a rather inconspicuous start due to competition from other projects at that time, the dedication of Wintersteiner and his assistant Milly Moore allowed a low level of periodic steroid research to continue at the Squibb Institute. The dedication and persistence of Wintersteiner and his group toward steroid research was critically important as the discovery of penicillin by Alexander Fleming in the early 1940s quickly dwarfed any interest in steroid research in the natural products group at Squibb. The steroid field was reinvigorated, however, in 1942 when the US government learned of a rumor from the Polish underground that a German scientist had unraveled the secret of the adrenal cortex and was able to produce an extract allowing Luftwaffe pilots to fly at 40 000 ft without detrimental effects. The US government now had a critical interest in the synthesis and testing of cortisol (**6**) and cortisone (**6a**) [11, 12]. The rumors were of course later proved to be untrue, but the ensuing efforts eventually provided sufficient quantities of these materials for human testing leading to the second important breakthrough in the GC field. Hensch and colleagues carried out the first human injection of compound E (cortisone) in 1948–1949 demonstrating potent anti-inflammatory activity in a clinical trial on patients with rheumatoid arthritis [13a]. Clinical material for this pivotal trial was supplied by a team of Merck scientists led by Sarett [13b,c]. It was a remarkable accomplishment involving nearly 40 chemical steps from desoxycholic acid. When Hensch, Kendall, and Reichstein were awarded the Nobel Prize in 1950 for their discovery of the active analogs of the adrenal gland, the race was officially on to commercialize steroids of this class.

Multiple medicinal chemistry groups were actively investigating potential synthetic processes to produce large amounts of cortisone. The synthesis of cortisone would prove elusive as robust synthetic methods were not yet available to produce the required 11-oxo analogs of a steroid core (Figure 18.3). Although the Squibb Institute was largely inactive in steroid research at that time, a series of fortunate events during the early 1940s left them well poised to respond to the challenge. Oskar Wintersteiner of the Squibb organic chemistry group sought out and invited a young researcher with an interest in corticosteroid research to join the Squibb Research Institute as head of the antibiotics and steroids department in 1944. The researcher was Josef Fried, a skilled scientist who had been passed over by other large pharmaceutical companies during the early 1940s.

Although Fried's early projects at Squibb did not center on corticosteroids, when the actions of cortisone in patients with rheumatoid arthritis were discovered Fried was serendipitously working in collaboration with Edward Freis on the isolation of steroidal natural products that contained 11-oxygenated products. In spite of their best efforts, these building blocks did not prove to be suitable starting materials for conversion to cortisone. Collaboration with Titus and Perlman of the fermentation group also did not immediately yield the desired oxidation method for the 11-position of the steroid core. Following several iterations of basic research involving the fermentation, isolation, and characterization of novel steroids from fermentation broths, the Squibb team had identified a method for the selective 16α-oxidation but could still not access the elusive 11-oxidation. These findings were reported in April 1952 as other groups began to report parallel successes [14]. A chemical process toward the 11α-hydroxylation was developed on a small scale by research teams at Syntex starting from diosgenin (**15**) [15], while simultaneously, the first scalable 11α-hydroxylation was demonstrated on progesterone core via utilization of *Rhizopus arrhizus* by Peterson and Murray from Upjohn. The fermentation route provided the first large quantities of commercially prepared cortisone for clinical trials [16]. Results of this research were actually reported by the teams from Upjohn and Squibb in the same journal and in the same month [17]. The Squibb research team expanded their efforts by utilizing fermentation broths of Reichstein's compound S with Perlman and Thoma in the fermentation group. It was known that compound S contained the desired side chain for cortisone (**6**) and cortisol (**6a**). Compound S was readily available from a startup company founded by Percy Julian in Mexico. The ever-inventive Julian had identified chemical routes to supply novel steroids from Mexican yam (*Dioscorea*). These efforts were quickly rewarded when it was shown that *Aspergillus* cultures could produce the desired 11α-epimer of cortisol, and the Squibb team reported these results in August 1952 [18].

Additional discoveries by Fried, Rosenkranz, and Djerassi and colleagues, such as the ability to produce 10-fold more potent cortisol analogs, again raised the bar in this field. This was particularly important as it demonstrated that synthetic

GCs could be prepared with potency values exceeding that of the natural ligand, thereby refuting claims by the Nobel laureate Kendall that this would never be possible [11, 15]. The recognition that 9α-fluorination was particularly relevant led Fried to file a patent application in 1954 on this important discovery. The application was challenged by other companies, and it was not until 1958 that the patent was granted without exception to the Squibb team. As it turns out, however, management at Squibb had lost interest in cortisol and its 9α-fluoro analog even after exciting efficacy was demonstrated for topical formulations in dermatology studies. This efficacy was not lost in Merck, and they quickly joined in the race with the production of a 50 kg batch of material. This action rekindled interest at Squibb, which feared losing out on the race to commercialize fludrocortisone (9α-F-cortisol). Once again, the low-level activities and passions of key scientists came to the forefront. As it turns out, Fried and his highly skilled assistant Emily Sabo had been supplying clinical material to Harvey Blank who had arranged for clinical studies to be conducted with researchers at the Oregon Medical School. It surely helped that Blank continued to support these studies at risk to all their careers. As a result of this foresight and dedication Squibb was able to rapidly assemble a package for FDA approval, thus beating out Merck.

With a leading position in steroid research, in particular, GCs, the Squibb group next set out to identify a selective GC. This was based on the realization that in spite of the remarkable activity of cortisone, long-term use was accompanied by severe side effects such as electrolyte disturbances [10]. This time around, however, it was international competition which involved empirical probing of nearly every position of the cortisol backbone to discover the necessary selectivity handles. A significant contribution came from publications out of the Bernstein laboratory at Lederle [19, 20] This group had discovered that addition of the 16α-hydroxyl group to not only 9α-fluorocortisol but also other analogs resulted in the elimination of the sodium-retaining activity, which was later shown to be related to mineralocorticoid receptor activity [21]. It was a bit of a coup for the Lederle group in that the role of the 16α-hydroxyl group went unnoticed or unclaimed by the research team at Squibb, who first demonstrated 16α-hydroxylation on the progesterone core. The gain in selectivity came at a price in potency, however, and resulted in an analog that was now only about 1.5 times as potent as cortisol (**6**). The next critical discovery contributing to the foundation of the acetonide-containing steroid drugs came about because of a failed fermentation, underscoring the success one can achieve with a combination of directed scientific research and an eye for serendipitous discovery [11]. In this case it was Fried and the group at Squibb who benefited from the observation of a competitor in the field. As described by Fried, a fermentation batch to produce 16α-hydroxy-9α-fluorocortisol showed greater salt retention when tested *in vivo*. Since this was easily explained by off target activity due to incomplete oxidation of the 16-position, Fried set out to block the properly hydroxylated material via acetonide formation in order to further examine the profile of the minor impurity. In this way, one can demonstrate the presence of nonhydroxylated impurity through masking of the 16α-hydroxy group, rendering the major product "inactive" and effectively increasing the concentration of the nonhydroxylated material when a high-enough dose is given to animals. This experiment was designed around anecdotal evidence from Bernstein of Lederle, who had reported the preparation of the acetonide analog at both the Gordon Conference on the Chemistry of Steroids and Natural Products in August 1957 [20] and the Laurentian Hormone Conference in September 1957 [22]. Bernstein, however, did not evaluate the biological activity of this acetonide analog as it lacked a 17α-hydroxyl group and according to available information or presumptions, should not be active in corticoid assays. To the surprise of the Squibb scientists the high dose actually resulted in the death of the animals in the experiment. It was later shown through a careful dose response study that the activity was due to the presence of the 16α, 17α-acetonide. Isolation and characterization of this material demonstrated that the acetonide actually drove a 40-fold improvement in corticoid activity and became the basis for today's highly prescribed budesonide (**7**) and triamcinolone acetonide (**8**). In addition to the increased activity and mineralocorticoid receptor selectivity, it was later demonstrated that the acetonide group provided increased skin permeability, making its topical use more effective [10]. The addition of the C1–C2 double bond that is present in both budesonide (**7**) and triamcinolone acetonide (**8**) was actually developed as far back as 1950 at Syntex [23] but like other discoveries in this field was not appreciated until a new team, in this case the research group of Hershberg at Squibb, recognized its value [24].

Interestingly, the development of the selective use of ketal-protecting groups for steroidal diketones was originally discovered in 1941 by Fernholz and Stavely at Squibb, although it was not further expanded upon [25, 26]. It was the contribution of scientists at Lederle and other laboratories that recognized the potential utility of this chemistry as it allowed selective functionalization at key positions of the steroidal core through migration of double bonds and concomitant oxidation reactions [27]. Through such manipulations the SARs of functionalized steroids could be derived and led to the discovery of triamcinolone at Lederle. Acknowledging the fact that triamcinolone acetonide (**8**) was developed following the discovery of the activity of the acetonide group by Fried and colleagues at Squibb, a comarketing agreement between Squibb and Lederle (then a division of American Cyanamid Company and the basis of the triamcinolone name) was devised. Triamcinolone acetonide (**8**) was marketed for topical use as Kenalog® by Squibb and Aristocort® by Lederle.

18.3 Synthesis

The active pharmaceutical ingredient (API) present in acetonide-based GC drugs (compounds **7–14**, Figure 18.4) has several common structural features around the steroidal backbone, which are mainly based on modification of rings B, C, and D and the type of the 1,3-dioxolane unit. Chemistry for the preparation of these APIs has evolved over a period of six decades, and a general theme has emerged

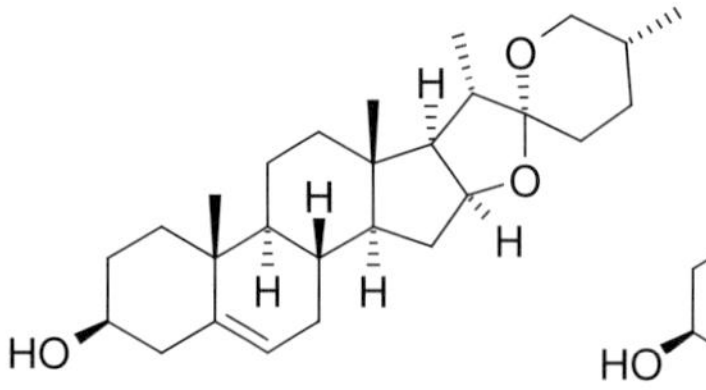

15
Diosgenin

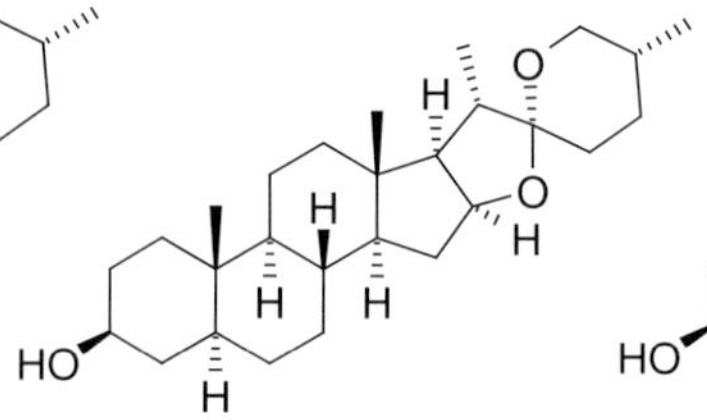

16
Tigogenin

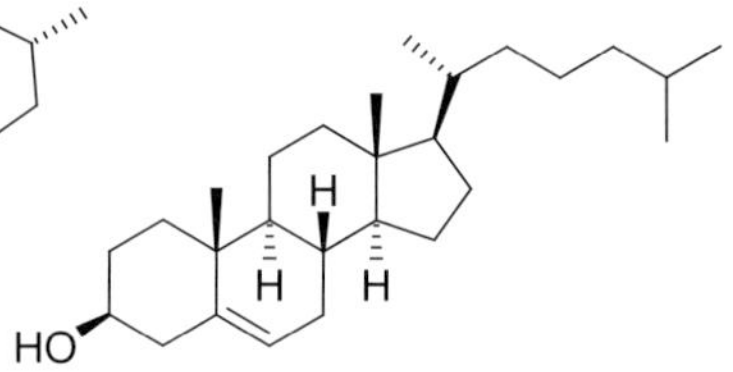

17
β-sitosterol

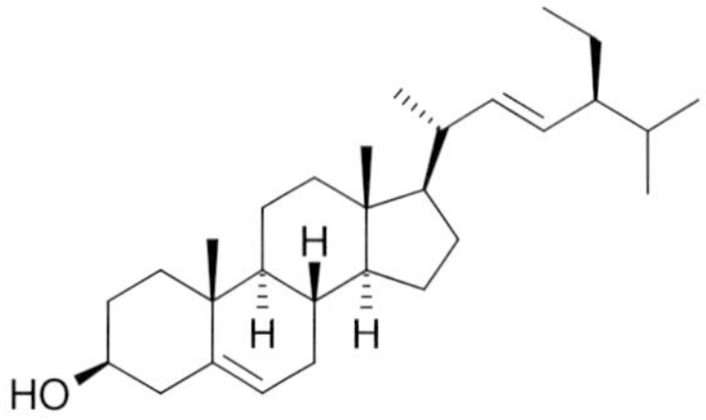

18
Stigmasterol

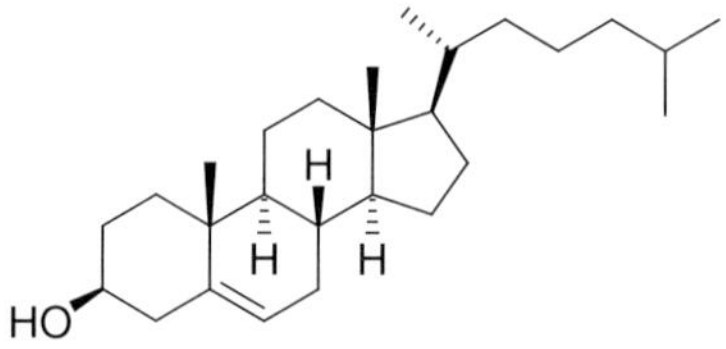

19
Cholesterol

Figure 18.5 Starting materials for production of steroidal intermediates and hormones.

through utilization of common intermediates followed by introduction of the 1,3-dioxolane group [11, 15, 16, 27]. In the golden era of steroidal chemistry during the mid-twentieth century, significant efforts were devoted to the identification of key starting materials (Figure 18.5), for example, diosgenin (**15**), tigogenin (**16**), β-sitosterol (**17**), stigmosterol (**18**), and cholesterol (**19**), and their conversion to key steroidal intermediates and hormones via novel technologies [11, 15, 16, 27, 28]. Although steroids were primarily produced in the 1940s from the extracts of animal source materials (e.g., tissue, plasma, or urine), the subsequent advancements in steroid chemistry became possible due to the remarkable discovery that common molds, such as those belonging to the genus *Rhizopus*, were capable of introducing oxygen at the 11-position of progesterone by a fermentation process [17, 29]. These cross-functional teams discovered ways to manipulate abundantly available raw materials from soybeans [30] and yam-based sources [15] for the production of key steroidal building blocks such as 16-dehydropregnenolone acetate (**20**, 16-DPA), 5-αH-16-DPA (**21**), androst-3-ene-3,17-dione (**22**) and its 1,2-dehydro-analog (**23**), as well as 9α-hydroxyandrost-4-ene-3,17-dione (**24**) and its 1,2-dehydro-analog **25**, all on commercial scale based on combinations of fermentation and chemical transformations (Figure 18.6) [11, 15, 16, 27]. Access to these building blocks enabled the wider research community to modify the structure of the steroidal backbone at will, leading to the efficient introduction of various functional groups, which facilitated the discovery of newer classes of steroidal drugs including the development of acetonide-based corticosteroids.

For example, several methodologies were developed [28, 31–34] toward the regeneration of the 17 side chain as well as for the introduction of the C17-hydroxyl

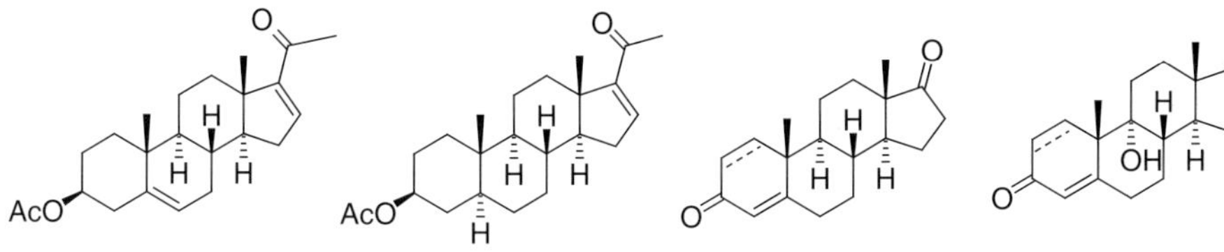

20
16-Dehydropregnenolone acetate (16-DPA)

21
5-αH-16-DPA

22
Androst-4-ene-3,17-dione
or **23**
1,2-dehydro analog

24
9α-Hydroxyandrost-4-ene-3,17-dione
or **25**
1,2-dehydro analog

Figure 18.6 Steroidal building blocks for synthesis of acetonide-based corticosteroids.

group in ring-D. One of the approaches, a silicon nucleophilic annealation process (SNAP, Scheme 18.2), involved dehydration of 9α-hydroxyandrost-4-ene-3,17-dione (**24**) using chlorosulfonic acid [16, 35] to introduce a double bond in ring-C followed by conversion to the corresponding cyanohydrin intermediate **26**. Silylation of the hydroxyl group in **26** using $ClSiMe_2CH_2Cl$ and a catalytic amount of dimethylaminopyridine (DMAP) in THF afforded the intermediate **27**, which was subsequently treated with lithium *N,N*-diisopropylamide (LDA) followed by acetic anhydride to afford **28** containing a C17 side chain with C21 and C17 hydroxyl groups. Alternatively, the silyl-protected intermediate **27** was treated with LDA followed by reduction with zinc in the presence of acid to afford key building block **29** containing the acyl side chain and hydroxyl group at the C17 position.

In another approach, the C17 side chain containing the C20 hydroxyl group as well as the C17 and C16 hydroxyl groups in ring-D was introduced starting from the fermentation-derived key building block, 9α-hydroxyandrost-4-ene-3,17-dione (**24**) [36, 37]. Accordingly, **24** was converted to the corresponding methyl ether (**30**) (Scheme 18.3), which was then treated with TosMIC reagent in the presence of a base followed by phosphorus oxychloride to afford **31**. Reaction of the olefin **31** with formaldehyde in the presence of a base followed by treatment with acetic acid produced **32** containing both a C17 side chain and a C21 hydroxyl group as its formate ester. The double bond in ring-D of intermediate **32** was oxidized with potassium permanganate to the corresponding 1,2-diol. The acetonide (**33**) was then obtained through acetalization of the C17,C16-diol with acetone in the presence of an acid catalyst. This intermediate is converted to triamcinolone acetonide (**8**) in six steps [37].

Triamcinolone acetonide (**8**) is also produced starting from **34** via introduction of a C9 fluoro group in ring-C and a 1,2-double bond in ring-A (Scheme 18.4). The starting material **34** was prepared by either fermentation or chemical transformations [19]. Compound **34** was oxidized with osmium tetraoxide to the corresponding C17 and C16 diols, followed by selective acetylation with acetic anhydride at the 16-position to form the diacetate **35**. Stereoselective fluorination at C9 position of the C-ring was achieved by a series of synthetic steps, which included selective bromination of the C9,C11-double bond with MeCONHBr, followed by the formation of the epoxide using potassium acetate. The epoxide **36** was then opened

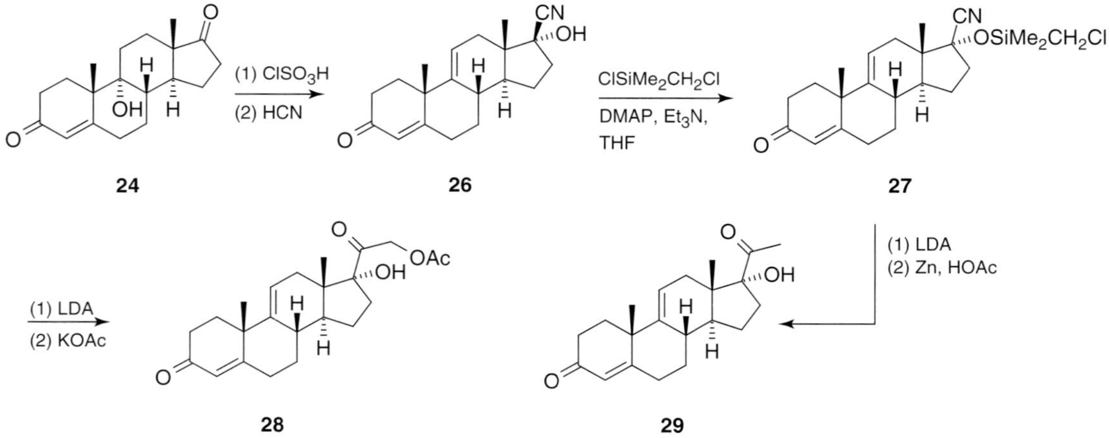

Scheme 18.2 Introduction of C17 side chain and C17 hydroxyl groups on steroids.

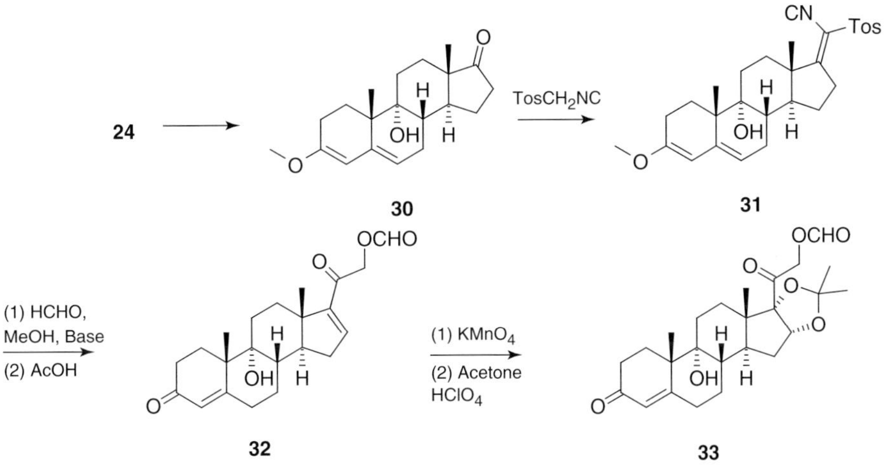

Scheme 18.3 Introduction of C17 side chain and C16 and C17 hydroxyl groups.

regioselectively with hydrofluoric acid to introduce the α-fluoride at the C9 position as well as the β-hydroxyl at the C11 position (**37**). Microbiological dehydrogenation using *Corynebacterium simplex* or *Nocardia corallina* produced the second double bond in the A-ring, from which triamcinolone acetonide (**8**) was formed by deacetylation at the C16 position, followed by reaction with acetone under acidic conditions to facilitate the conversion of the diol to the acetonide group.

Unlike budesonide (**7**), which is racemic at the acetonide C2 position, ciclesonide (**14**) is the enantiomerically pure (*R*)-epimer. The (*R*)-epimer has greater affinity for the GR than the corresponding (*S*)-epimer, and was therefore selected for clinical development [38, 39]. The synthesis of ciclesonide (**14**) is described in Scheme 18.5, which involves acylation of 16α-hydroxyprednisolone (**38**) with isobutyric anhydride in pyridine to provide the 16,17,21-triester **39**. This intermediate is treated with cyclohexane carbaldehyde and hydrochloric acid in dioxane to yield the cyclic ketal as a diastereomeric mixture. This diastereomeric mixture was then resolved by chromatography and subsequent fractional crystallization to yield ciclesonide (**14**) [38, 39].

Flunisolide (**12**) was synthesized (Scheme 18.6) from 6α-fluoro-9,11-epoxy-17,21-dihydroxypregn-1,4-diene-3,20-dione-17,21-diacetate (**40**). The fluoro group at the 6α-position was introduced using Selectfluor [40]. Reaction of the epoxide **40** with hydrobromic acid afforded the bromide **41** [41]. Debromination of **41** and elimination of the acetate on C17 produced the olefin **42**. The 1,3-dioxolane moiety was introduced through dihydroxylation of the C16,17-double bond (**43**) and treatment with acetone/$HClO_4$. Synthesis was completed via saponification of the acetate on C21 to provide flunisolide (**12**).

In addition to fermentation- and chemical-transformation-based technologies, recently the entire biosynthetic pathway to produce cortisol (**6**) has been engineered into yeast [42]. This approach may provide an opportunity to synthesize

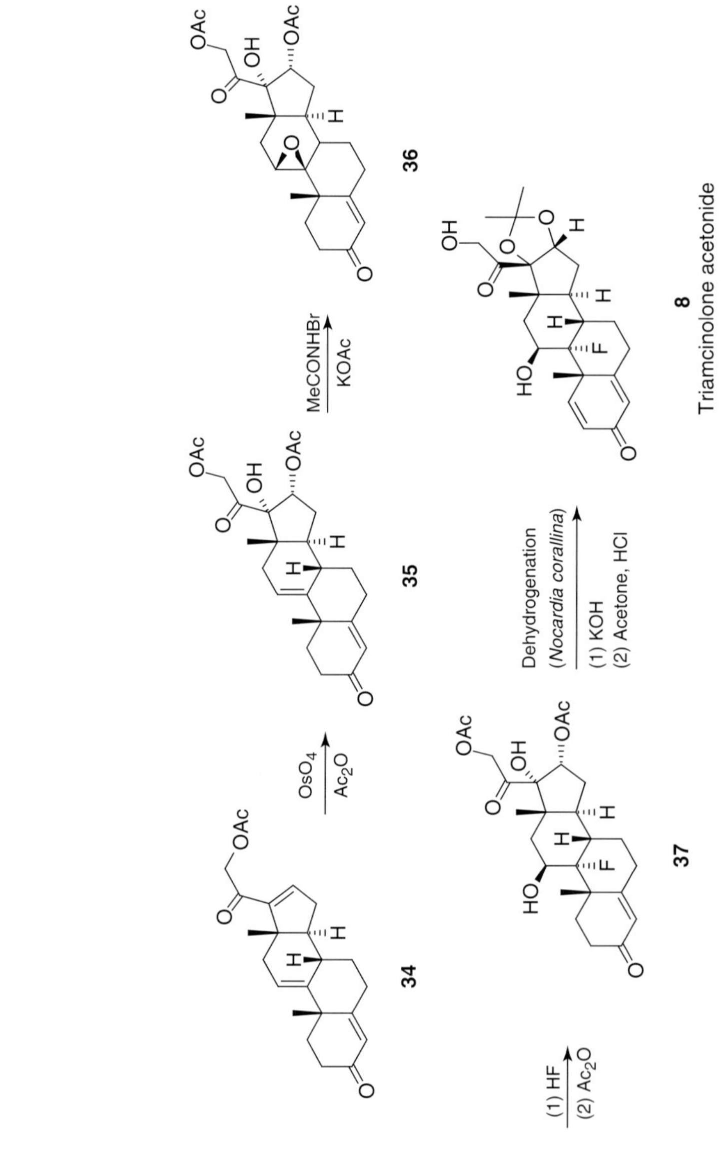

Scheme 18.4 Synthesis of triamcinolone acetonide (**8**).

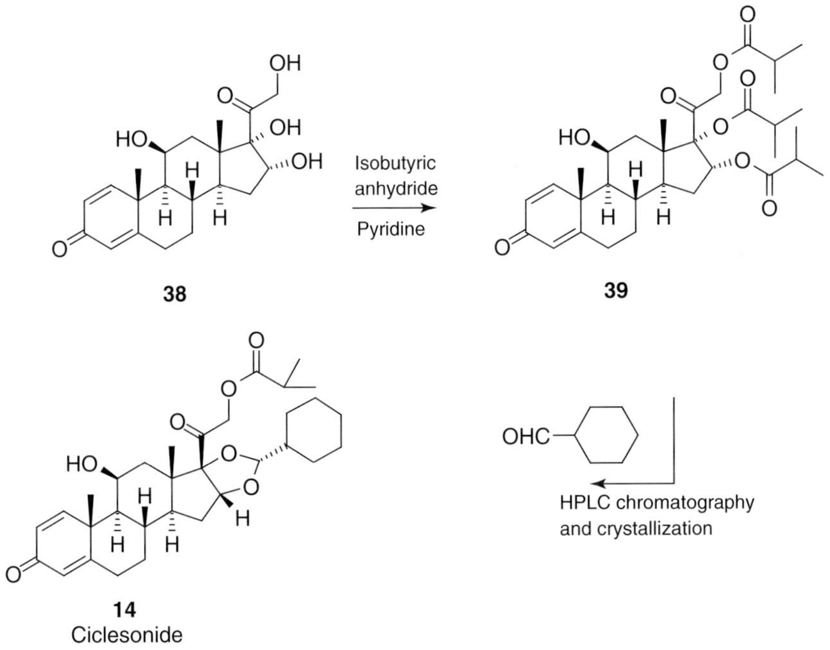

Scheme 18.5 Preparation of ciclesinode (**11**) from 16α-hydroxyprednisolone (**38**).

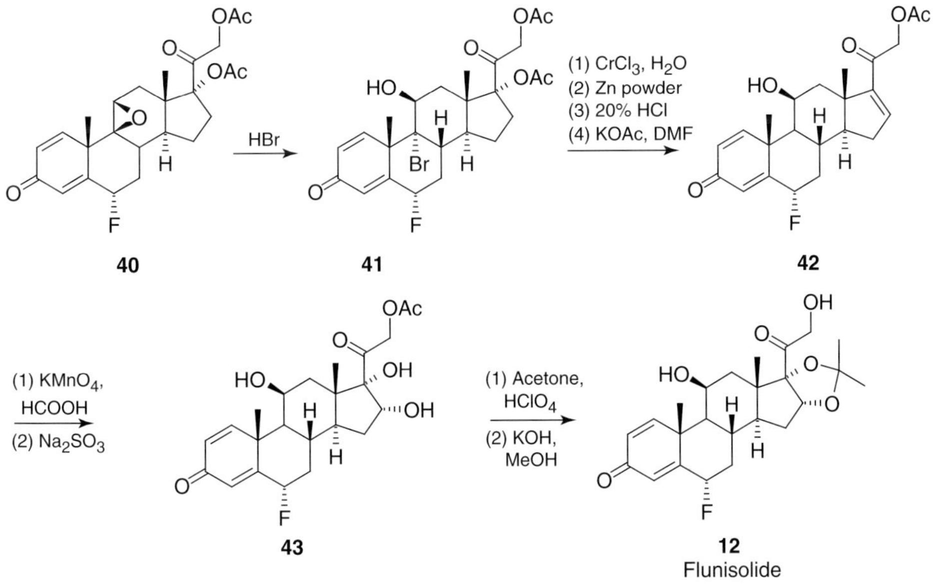

Scheme 18.6 Preparation of flunisolide (**12**).

corticosteroids without any limitations imposed by the natural sources of the starting materials. Advancements in molecular science may allow for economical and greener technologies for production of important intermediates that occur in the yam-based and soy-steroid-based approaches.

18.4 Mode of Action

Following the discovery of the anti-inflammatory properties of human cortisone in rheumatoid arthritis [13a] the GR has been cloned [43]. This and other advances have contributed to the development of our understanding of how GCs counter inflammation. Cortisol (**6**) is the natural ligand for GR in humans (corticosterone in some rodents). The binding of cortisol (**6**) (or a synthetic GC) to GR results in pleiotropic effects including, for example, metabolic as well as anti-inflammatory activities. The effectiveness of GCs has been demonstrated in the treatment of asthma, skin inflammation, and diseases of the eye. However, this broad activity gives rise to the undesirable side effect profile associated with prolonged administration of synthetic GCs. The side effects limiting both dose and duration of treatment for many patients include osteoporosis, diabetes, adrenal suppression, weight gain, hypertension, muscle wasting, cataracts, sleep disturbances, and central obesity [44].

Alternative splicing patterns and divergent translational start sites have resulted in the production of multiple isoforms of GR including GRα, GRβ, GRγ, GR-A, and GR-P [45]. Variations can occur at the N- and C-terminal domains of GR depending on the isotype, while the DNA binding domain (DBD) remains fairly constant. The most common GC binding isoform is GRα in contrast to GRβ, which cannot bind GCs. Although GRβ does not bind GCs, it can suppress the actions of GRα and has been implicated in GC resistance. Before induction by GC binding, GR is principally located in the cytoplasm in association with molecular chaperones such as heat shock proteins (Figure 18.7). Subsequent to GC binding, a conformational change in the receptor occurs resulting in the chaperones being shed and the nuclear localization signal being activated [46]. Once inside the nucleus, liganded GR operates as a transcription factor recruiting various cofactor proteins (coactivators and corepressors) that modify the transcription of its target genes. The ability of GR to interact with coactivators or corepressors is sensitive to the conformation of GR that is conferred by ligand binding. A GR agonist, such as cortisol (**6**), induces a GR conformation that shifts helix 12 so that it provides part of the binding surface for coactivators [46]. On the other hand, a GR antagonist such as RU 38486 destabilizes helix 12 packing and thus provides an alternative surface for corepressor binding [47]. The dynamic shifting of helix 12 between active and repressive positions in the presence of various GR ligands likely dictates the degree of transactivation on target genes. The separation of these two modes of GR activity may lead to a GC with fewer side effects [48].

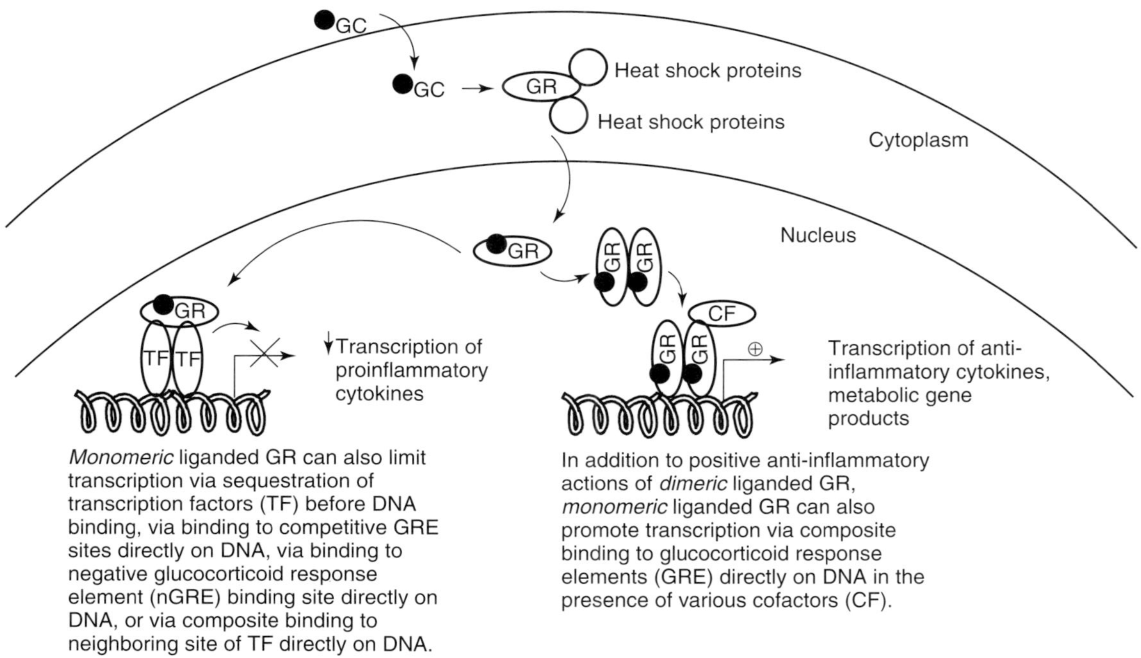

Figure 18.7 Model for glucocorticoid receptor modulation [45].

The temporal response to GCs can vary from minutes to days. The transcriptional mode of GC action occurs in hours to days, whereas there are more rapid GC-based cellular responses taking place within minutes, which are termed *nongenomic (nontranscriptional) effects*. In addition, GR can give rise to positive or negative transcriptional effects via a monomeric or dimeric form (Figure 18.7) resulting in a multifaceted control mechanism for gene regulation [45]. Adding to this complexity is the fact that although many cell types express GR, the number of receptors per cell varies, thus affecting overall steroid responsiveness [49].

18.4.1
Asthma

GC agonists are effective in the treatment of asthma. It is likely that there are multiple cellular targets for GCs, including inflammatory cells such as eosinophils, T lymphocytes, dendritic cells, and macrophages in addition to structural cells such as endothelial cells, epithelial cells, and airway smooth muscle cells (Figure 18.8). An examination of the expression pattern of GR in the lung reveals GR expression in multiple cell types with the highest density in endothelial and epithelial cells [49]. Airway epithelial cells are of particular importance as cellular targets for inhaled corticosteroids (ICSs) because of their localization and ability to express multiple inflammatory proteins that contribute to the complex inflammation of asthma [49]. In addition, action on structural cells contributes to increased leakiness and increased mucus secretion in an inflammatory setting.

Anti-inflammatory GC agonists such as triamcinolone acetonide (**8**) and flunisolide (**12**), originally developed for topical use, were found to be effective for

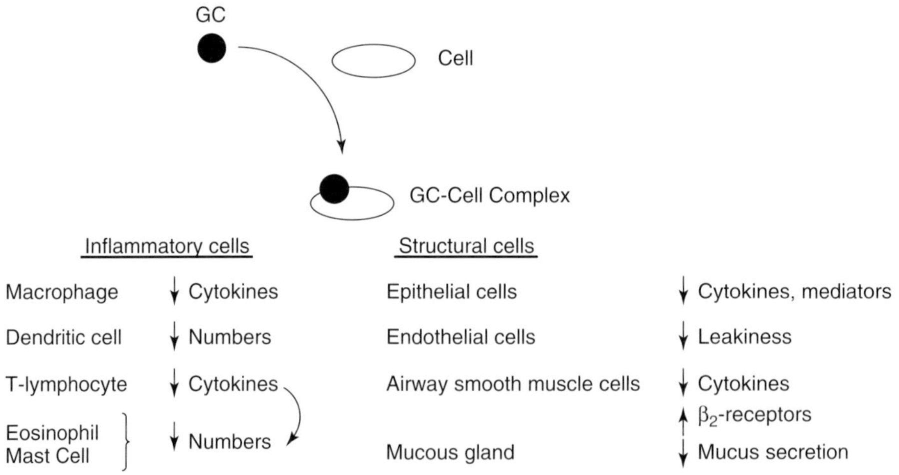

Figure 18.8 Glucocorticoid receptors are widely distributed in airways with changes noted upon GC binding [49].

the treatment of asthma and were subsequently developed and marketed as ICSs. Owing to their route of administration, ICS partition between the lungs and the oral cavity. The portion that reaches the lung deposits in the airways, where dissolution must occur before it can enter the lung tissue [50]. Agents such as budesonide (**7**), flunisolide (**12**), and triamcinolone acetonide (**8**) readily dissolve and are rapidly absorbed. In addition, fatty acid esters at C21 formed in parenchyma may provide increased retention for agents such as ciclesonide (**14**) and budesonide (**7**) [51]. More lipophilic agents such as fluticasone propionate, however, dissolve more slowly and are prone to clearance by airway ciliary action.

18.4.2
Skin Inflammation

Atopic dermatitis (AD), often referred to as *eczema*, is a chronic, inflammatory dermatologic condition. It commonly occurs in patients affected by asthma and/or allergic rhinitis and is associated with elevated serum IgE levels. Psoriasis is a general inflammatory skin condition, while seborrheic dermatitis is an inflammatory disorder affecting areas of the head and trunk. The standard of care for these inflammatory skin diseases is administration of topical glucocorticoids (TGs). The clinical effectiveness of GCs has been attributed to their anti-inflammatory, immunosuppressive, vasoconstrictive, and antiproliferative effects. Indeed, GRs are present in keratinocytes and fibroblasts where they are known to mediate production of various inflammatory mediators [52]. While highly effective as topical anti-inflammatory agents, chronic use of GCs results in unwanted side effects. These effects include localized skin thinning and a reduction in lipid content. For example, inhibition of IL-1α in keratinocytes is anti-inflammatory, while the same inhibition in fibroblasts has been noted to be antiproliferative and atrophogenic [52]. Vasoconstrictive effects of TGs may also contribute to their pharmacological effectiveness by diminishing local erythema. An important feature to consider is the ability of the applied TG to reach the target cells. TGs have to permeate the outer 10–20 μm of the epidermis, the stratum corneum (SC), which constitutes the major barrier to percutaneous absorption [52]. Therefore, the formulation is critical to the successful use of these agents.

18.4.3
Eye Disease

GCs may also be used for treating diseases of the eye such as uveitis [53, 54]. Uveitis has many potential causes, including infection with a virus, fungus, bacteria, or parasite; inflammatory disease affecting other parts of the body; or injury to the eye [53]. Similar to the treatment of lung diseases, treatment of diseases of the eye is often accompanied by multiple undesired side effects including cataracts, glaucoma, and infection, which limit their use. Uveitis was originally described as inflammatory processes of the middle layers of the eye but has been expanded to include intraocular inflammation at any site. There are also various clinical forms of ocular allergy that have an inflammatory component [54]. Environmental antigens are believed to initiate the inflammatory cascade. When these antigens reach the mucosal epithelium of the eye they trigger a cascade resulting in a Th1- and Th2-mediated immune responses. GC agonists are able to dampen these responses, resulting in diminished inflammation; however, the pleiotropic effects of GCs results in undesired activity as well. The challenge for current therapies is to identify more selective modulators of GR or to limit exposure to the target tissue bed while restricting systemic exposure in order to avoid off target effects. Related to this approach has been the design of soft steroids, which when applied topically partition between the local and systemic compartments. However, these agents are designed to rapidly metabolize to inactive forms on systemic exposure, thus limiting the ability to affect multiple tissues over time [55].

18.5
Structure–Activity Relationship

A limited number of steroidal GR agonist and antagonist ligands have been cocrystallized with the GR ligand binding domain, and high resolution crystallography data is available [46]. However, due to the plasticity of the receptor, the ability to conduct true structure-based drug design remains elusive and is not covered here. As SAR depends on the specific corticosteroid, the mode of delivery, and the target organ and species used for testing, a general SAR is presented to cover these applications with emphasis on activity in humans. Regardless, there is a high level of similarity between SAR for inhaled, topical, and ocular corticosteroids. The major differences arise around fine-tuning of the metabolism profile to limit systemic exposure of topically applied formulations.

Cortisol (**6**) that contains the required 11β hydroxyl was the first corticosteroid to be used widely in dermatology [12]. Cortisone (**6a**) is ineffective topically because of lack of conversion to the active form cortisol (**6**) in the skin. On the basis of the need to improve the topical activity of cortisol-like compounds cortisol (**6**) was structurally modified to achieve greater topical potency and improved safety. A combination of directed work and serendipitous discoveries led to a series of advances in this field, and the ability to track gains in potency and safety to specific structural changes revolutionized thinking around improving the benefit/safety profile of GC agonists. Fludrocortisone (9α-F cortisol) with a fluorine atom at the 9α position improved upon the activity of hydrocortisone, as did flucortolone with a 6α fluoro. However, both analogs displayed electrolyte imbalance related to activation of the mineralocorticoid receptor. Flumethasone with fluorine atoms at both the 6α and 9α positions was additive toward GR activity but again did not address the electrolyte imbalance. When it was serendipitously discovered that an acetonide group fused to the D-ring was beneficial for both activity and selectivity, this feature was rapidly incorporated into the most advanced analogs. For example, in flurandrenolone, an acetonide group was added at the C16,C17α positions. The effect of this group was more suitable properties for topical absorption as well as reduced electrolyte imbalance. Triamcinolone acetonide (**8**) further increased activity due to the addition of a double bond between C1 and C2 while retaining the suitable properties for topical absorption and reduction of the undesirable electrolyte imbalance due to the presence of the C16,C17-α acetonide group. Fluocinolone acetonide possessed exceptional topical activity resulting from a combination of all four features discussed above (6α- and 9α-F, double bond at C1 and C2, acetonide at C16, C17-α positions). It is worth noting that while C16, C17-α acetonide modification produced highly improved anti-inflammatory agents suitable for topical use, the addition of a 16α (or 16β) methyl produced highly effective agents suitable for oral use, in particular, dexamethasone invented by Merck. Further increases in lipophilicity due to addition of an acetate group at C21 led to the discovery of fluocinonide, which had improved topical activity and four times the potency of its precursor, fluocinolone acetonide [12]. The SAR for the 16,17-α acetals is summarized in Table 18.1. In general, increased functional activity results in an improved anti-inflammatory response. However, in these agents, side effects are not dissociated from anti-inflammatory activity and chronic use may still result in sufficient accumulation of drug to have increased side effect risk over time. Hence, an attempt was made to minimize side effects by lowering systemic exposure through more rapid metabolic clearance (the so-called soft steroids).

In addition to the role in altering potency, these same molecular changes also alter the metabolic profile. Addition of a 9α fluoro improves the metabolic stability of the 11β hydroxyl group, limiting its conversion to the inactive 11-keto metabolite [56]. Triamcinolone acetonide (**8**) as a result does not form a 11-keto metabolite but is 6β hydroxylated (inactive) and also converted to the 17-β carboxylate (inactive) [57]. 6α Fluorination increases functional potency but reduces half-life by permitting oxidative defluorination to form the inactive 6β hydroxyl metabolite [58]. 6α and 9α fluorination results in increased activity, but the half-life is reduced [59–61]. Budesonide (**7**) and ciclesonide (**14**), with hydrogen in 2-position of the acetals, are metabolized by opening of the acetals to the C16,C17-α dihydroxyl analogs (16α hydroxy prednisolone), which have reduced activity. In addition, budesonide (**7**) is also hydroxylated at the 6β position and on the propyl side chain of the acetal [62]. While budesonide (**7**) is racemic, ciclesonide (**14**) is the (*R*)-enantiomer. The (*R*)-enantiomer with an alpha alkyl substituent has higher functional potency compared to the (*S*)-enantiomer, but the half-life is lower for the (*R*)-enantiomer [59, 63]. Ciclesonide (**14**) forms hippuric acid as a metabolite, suggesting opening of the acetal. C16,C17-α acetals are progestinic compounds just like 17α esters [64, 65]. As circulating levels of these agents are low, progestinic activity may not be expressed as a side effect. Looking ahead, further advancements in our understanding of the SAR may be able to provide new agents with improved selectivity, true dissociation of the anti-inflammatory activity, and some of the troublesome side effects.

Table 18.1 SAR for corticosteroid acetonides.

	Position	Group	Effect on activity	Effect on metabolism
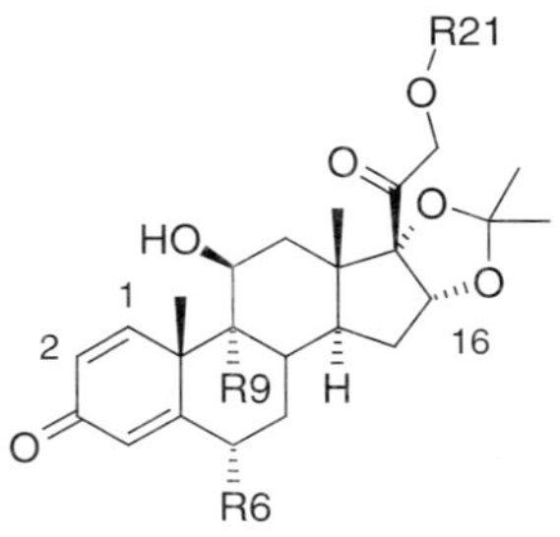	C1,C2	Double bond	Improved functional activity	Inhibited ability to reduce 11-ketone to 11β-OH
	R6	6α-Methyl 6α-Fluoro	Improved functional activity Improved functional activity	Increased half-life Decreased half-life
	R9	9α-Fluoro	Improved functional activity	Altered metabolic stability of 11β-OH
	C16,C17	16α,17α- Acetonide	Increased functional activity, improved selectivity versus MR but decreased selectivity toward PR when substituted	Hot spot for metabolism. (*R*)-Ciclesonide less stable than (*S*)-ciclesonide
	R21	$C(O)CH(CH_3)_2$	Cleaved off in target tissue	Increased tissue permeability

Abbreviations: MR, mineralocorticoid receptor; PR, progesterone receptor.

References

1. Green, T.W. and Wuts, P.G.M. (1991) *Protective Groups in Organic Synthesis*, vol. **5**, John Wiley & Sons, Inc., pp. 119–136.
2. Salmi, E.J. (1938) *Ber. Dtsch. Chem. Ges. Abt. B*, **71**, 1803–1808.
3. Matsukawa, M., Inanaga, J., and Yamaguchi, M. (1987) *Tetrahedron Lett.*, **28**, 5877–5878.
4. Roohk, D.J., Varady, K.A., Turner, S.M., Emson, C.L., Gelling, R.W., Shankaran, M., Lindwall, G., Shipp, L.E., Scanlan, T.S., Wang, J.C., and Hellerstein, M.K.J. (2010) *Pharm. Exp. Ther.*, **333**, 281–289.
5. Grad, I. and Picard, D. (2007) *Mol. Cell. Endocrinol.*, **275**, 2–12.
6. Cota, D. (2008) *J. Neuroendocrinol.*, **20**, 35–38.
7. Robinson-Rechavi, M., Garcia, H.E., and Laudet, V.J. (2003) *Cell Sci.*, **116**, 585–586.
8. Spencer, C.M. and McTavish, D. (1995) *Drugs*, **50**, 854–872.
9. Jeal, W. and Faulds, D. (1997) *Drugs*, **53**, 257–280.
10. Katz, M. and Gans, E.H. (2008) *J. Pharm. Sci.*, **97** (8), 2936–2947.
11. Fried, J. (1992) *Steroids*, **57**, 384–391.
12. Katz, M. and Gans, E.H. (2008) *J. Pharmaceutical Sci.*, **97**, 2936–2947.
13. (a) Hensch, P.S., Kendall, E.C., Slocumb, C.H., and Polley, H.F. (1949) *Proc. Staff Meetings Mayo Clinic*, **24**, 181–197; (b) Sarett, L.S. and Patchett, A.A. (2002) *National Academy of Sciences*, vol. **81**, National Academy of Sciences Press, Washington, DC; (c) Sarett, L.S., Patchett, A.A., and Steelman, S. (1963) in *Progress in Drug Research*, 7th edn, vol. **5** (ed. E. Jucker), Birkhauser Verlag, Basel, pp. 136–139.
14. Pearlman, D., Titus, E., and Fried, J. (1952) *J. Am. Chem. Soc.*, **74**, 2126.
15. Rosenkranz, G. (1992) *Steroids*, **57**, 409–418.
16. Hogg, J.A. (1992) *Steroids*, **57**, 593–616.
17. Peterson, D.H. and Murray, H.C. (1952) *J. Am. Chem. Soc.*, **74**, 1871–1872.
18. Fried, J., Thoma, R.W., Gerke, J.R., Herz, J.E., Donin, M.N., and Perlman, D. (1952) *J. Am. Chem. Soc.*, **74**, 3962–3963.
19. Bernstein, S., Lenhard, R.H., Allen, W.S., Heller, M., Littell, R., Stolar, S.M., Feldman, L.I., and Blank, R.H. (1959) *J. Am. Chem. Soc.*, **81**, 1689–1696.
20. Bernstein, S., Littell, R., Brown, J.J., and Ringler, I. (1959) *J. Am. Chem. Soc.*, **81**, 4573–4574.
21. Bernstein, S., Lenhard, R.H., Allen, W.S., Heller, M., Littell, R., Stolar, S.M., Feldman, L.I., and Blank, R.H. (1956) *J. Am. Chem. Soc.*, **78**, 5693–5694.
22. Fried, J., Borman, A., Kessler, W.B., Grabowich, P., and Sabo, E.F. (1958) *J. Am. Chem. Soc.*, **80**, 2338–2339.
23. Ringhold, H.J., Rosenkranz, G., and Sondheimer, F. (1956) *J. Org. Chem.*, **21**, 239–240.
24. Herzog, H.L., Nobile, A., Tolksdorf, S., Charney, W., Hershberg, E.B., and Perlman, P.L. (1955) *Science*, **121**, 176.
25. Fernholz, E. (1944) Cyclic ketals of 3-keto-17-oxy-cythrenes and method of preparing 3-keto-17-oxy-cythrenes. US Patent 2356154, Aug. 22, 1944.
26. Fernholz, E. (1945) Cyclic ketals of keto-cyclopentanoperhydrophenanthrenes and methods of preparing them. US Patent 2378918, June 26, 1945.
27. Bernstein, S. (1992) *Steroids*, **57**, 392–402.
28. Van Rheenan, V.H. and Sheppard, K.P. (1979) *J. Org. Chem.*, **44**, 1582–1584.
29. Peterson, D.H., Murray, H.C., Eppstein, S.H., Reinecke, L.M., Weintraub, A., Meister, P.D., and Leigh, H.M. (1952) *J. Am. Chem. Soc.*, **74**, 5933–5935.
30. Andor, A., Jekkel, A., Hopwood, D.A., Jeanplong, F., Ilkoy, E., Konya, A., Kurucz, I., and Ambrus, G. (2006) *Appl. Environ. Microbiol.*, **72**, 6554–6559.
31. Hessler, E.J. and Van Rheenan, V.H. (1980) US Patent 4216159.
32. Livingston, D.A., Petre, J.E., and Bergh, C.L. (1990) *J. Am. Chem. Soc.*, **112**, 6449–6450.
33. Kataoka, H., Watanabe, K., Miyazaki, K., Tahara, S., Ogu, K., Matsuoka, R., and Goto, K. (1990) *Chem. Lett.*, **19** (9), 1705–1708.
34. Van Leusen, D., Batist, J.N.M., Lei, J., Van Echten, E., Brouwer, A.C., and Van Leusen, A.M. (1994) *J. Org. Chem.*, **59**, 5650–5657.

35. Shephard, K.P. (1978) US Patent 4102907.
36. Van Leusen, D. and Van Leusen, A.M. (1984) *Tetrahedron Lett.*, **25**, 2581–2584.
37. Van Leusen, D. and Van Leusen, A.M. (1994) *J. Org. Chem.*, **59**, 7584–7538.
38. Mealy, N.E., Bayes, M., and Castaner, J. (2001) *Drugs Future*, **26**, 1133–1139.
39. Calatayud, J., Cond, J.R., and Luna, M. (1996) US Patent 5482934.
40. Chernyak, S., Zarbow, M., and Gutman, D. (2006) US patent 7098328 B2.
41. Li, Z. and Li, J. (2010) CN Patent 101759760 A.
42. Scczebara, F.M., Chandelier, C., Villeret, C., Masurel, A., Bourot, C.D., Blanchard, S., Groisillier, A., Testet, E., Costaglioli, P., Cauet, G., Degryse, E., Balbuena, D., Winter, J., Achstetter, T., Spagnoli, R., Pompon, D., and Dumas, B. (2003) *Nat. Biotechnol.*, **21**, 143–148.
43. Hollenberg, S.M., Weinberger, C., Ong, E.S., Cerelli, G., Oro, A., Lebo, R., Thompson, E.B., Rosenfeld, M.G., and Evans, R.M. (1985) *Nature*, **318**, 635–641.
44. Schacke, H., Docke, W.D., and Asadullah, K. (2002) *Pharmacol. Ther.*, **96**, 23–43.
45. Beck, I.M., Van den Berghe, W., Vermeulen, L., Yamamoto, K.R., Haegeman, G., and De Bosscher, K. (2009) *Endocr. Rev.*, **30**, 830–882.
46. Bledsoe, R.K., Montana, V.G., Stanley, T.B., Delves, C.J., Apolito, C.J., McKee, D.D., Consler, T.G., Parks, D.J., Stewart, E.L., Willson, T.M., Lambert, M.H., Moore, J.T., Pearce, K.H., and Xu, H.E. (2002) *Cell*, **110**, 93–105.
47. Kauppi, B., Jakob, C., Farnegardh, M., Yang, J., Ahola, H., Alarcon, M., Calles, K., Engstrom, O., Harlan, J., Muchmore, S., Ramqvist, A.K., Thorell, S., Ohman, L., Greer, J., Gustafsson, J.A., Carlstedt-Duke, J., and Carlquist, M. (2003) *J. Biol. Chem.*, **78**, 22748–22754.
48. Hu, X., Du, S., Tunca, C., Braden, T., Long, K.R., Lee, J., Webb, E.G., Dietz, J.D., Hummert, S., Rouw, S., Hegde, S.G., Webber, R.K., and Obukowicz, M.G. (2011) *Endocrinology*, **152** (8), 3123–3134 [Epub ahead of print].
49. Barnes, P.J. (2004) *Proc. Am. Thorac. Soc.*, **1**, 345–351.
50. Edsbacker, S. and Johansson, C.J. (2006) *Basic Clin. Pharmacol. Toxicol.*, **98**, 523–536.
51. Barnes, P.J., Pedersen, S., and Busse, W.W. (1998) *Am. J. Respir. Crit. Care Med.*, **157**, S1–53.
52. Wiedersberg, S., Leopold, C.S., and Guy, R.H. (2008) *Eur. J. Pharm. Biopharm.*, **68**, 453–466.
53. Hajj-Ali, R.A., Lowder, C., and Mandell, B.F. (2005) *Cleve. Clin. J. Med.*, **72**, 329–339.
54. Leonardi, A. (2005) *Expert Opin. Emerg. Drugs*, **10**, 505–520.
55. Cerasoli, F. (2006) *Chest*, **130**, 54S–64S.
56. Bush, I.E. and Mahesh, V.B. (1964) *Biochem. J.*, **93**, 236–255.
57. Argenti, D., Jensen, B.K., Hensel, R., Bordeaux, K., Schleimer, R., Bickel, C., and Heald, D. (2000) *J. Clin. Pharmacol.*, **40**, 770–780.
58. Teitelbaum, P.J., Chu, N.I., Cho, D., Tokes, L., Patterson, J.W., Wagner, P.J., and Chaplin, M.D. (1981) *J. Pharmacol. Exp. Ther.*, **218**, 16–22.
59. Andersson, P., Lihne, M., Thalen, A., and Ryrfeldt, A. (1987) *Xenobiotica*, **17**, 35–44.
60. Dahlberg, E., Thalen, A., Brattsand, R., Gustafsson, J.A., Johansson, U., Roempke, K., and Saartok, T. (1984) *Mol. Pharmacol.*, **25**, 70–78.
61. Brattsand, R., Thalen, A., Roempke, K., Kallstrom, L., and Gruvstad, E. (1982) *J. Steroid Biochem.*, **16**, 779–786.
62. Edsbacker, S., Andersson, P., Lindberg, C., Paulson, J., Ryrfeldt, A., and Thalen, A. (1987) *Drug Metab. Dispos.*, **15**, 403–411.
63. Stoeck, M., Riedel, R., Hochhaus, G., Hafner, D., Masso, J.M., Schmidt, B., Hatzelmann, A., Marx, D., and Bundschuh, D.S. (2004) *J. Pharmacol. Exp. Ther.*, **309**, 249–258.
64. Hughes, S.C., Shardlow, P.C., Hollis, F.J., Scott, R.J., Motivaras, D.S., Allen, A., and Rousell, V.M. (2008) *Drug Metab. Dispos.*, **36**, 2337–2344.
65. Child, K.J., English, A.F., Gilbert, H.G., Hewitt, A., and Woollett, E.A. (1968) *Arch. Dermatol.*, **97**, 407–410.

19
Benzimidazole-Based H^+/K^+-ATPase Inhibitors for the Treatment of Gastroesophageal Reflux Disease

Steve Swann

19.1
Introduction

2-(Pyridin-2-ylmethylsulfinyl)-1*H*-benzo[*d*]imidazoles, or more generally, sulfinylbenzimidazoles (**1**), are a class of molecules composed of two heterocycles, a pyridine and a benzimidazole, linked through a methylsulfinyl group (Figure 19.1).

This chemotype was originally discovered in the mid-1970s as a potent inhibitor of gastric acid secretion in the stomach [1]. Subsequently, it was discovered that molecules of this type bind irreversibly to the H^+/K^+-adenosine triphosphatase (H^+/K^+-ATPase) acid pump [2]. Since their initial discovery, proton pump inhibitors (PPIs) have become the largest selling medications in the history of the pharmaceutical industry (Prilosec® peak sales of $6 billion in 2000) and have improved the lives of millions of people suffering from gastric-acid-associated disorders. To date, there are six marketed PPIs including omeprazole [3] (**2**, Losec®, Prilosec®: AstraZeneca, 1994), pantoprazole [4] (**3**, Protonix®: Wyeth-Ayerst, 1994), lansoprazole [5, 6] (**4**, Prevacid®, Helicid®, Zoton®, Inhibitol®, Monolitum®: TAP, 1995), rabeprazole [7] (**5**, Pariet®, AcipHex®: Eisai/Janssen, 1999), esomeprazole [8] (**6**, Nexium®, Sompraz®, Zoleri®, Lucen®, Esopral®, Axagon®: AstraZeneca, 2001), and dexlansoprazole [9] (**7**, Kapidex®, Dexilant®: Takeda, 2009) (Figure 19.2).

This chapter describes the discovery of the sulfinylbenzimidazoles (**1**), the synthetic routes that have been developed for this chemotype, as well as the structure–activity relationships and key modifications to the core that have been explored in the past 25 years.

19.2
History

The first benzimidazoles [10, 11] were described by Hobrecker in 1872 [12], who identified the 2,5- and 2,6-regioisomers of dimethyl benzimidazole when reducing 2-nitro-4-methylacetanilide. The earliest report of a sulfinylbenzimidazole originated in 1962 from Gorczyca [13] at the Medical Academy in Krakow, Poland, who,

Bioactive Heterocyclic Compound Classes: Pharmaceuticals,
First Edition. Edited by Jürgen Dinges and Clemens Lamberth.

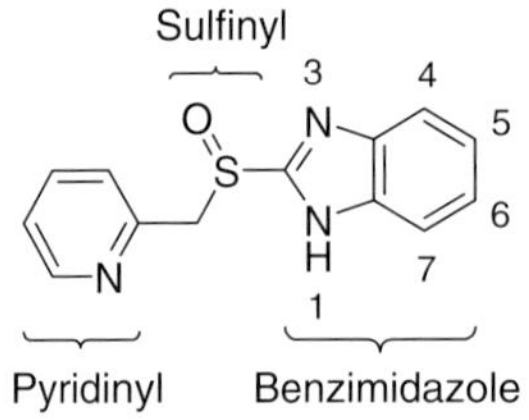

1

Figure 19.1 Basic structure and numbering for 2-(pyridin-2-ylmethylsulfinyl)-1*H*-benzo[*d*]imidazole (**1**).

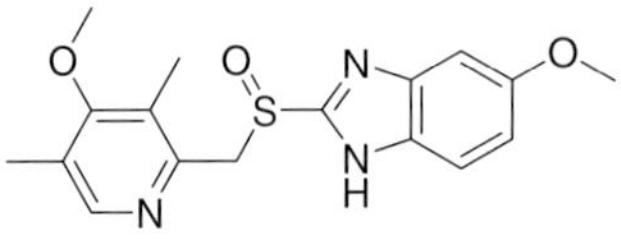

2
Omeprazole

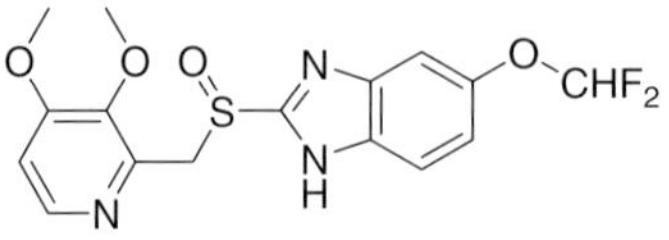

3
Pantoprazole

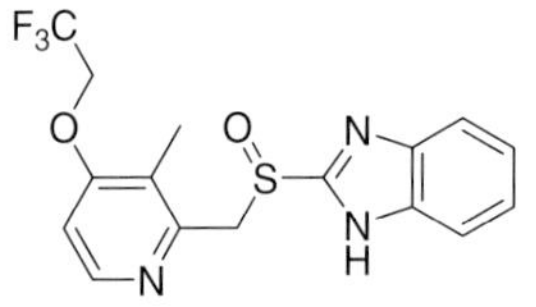

4
Lansoprazole

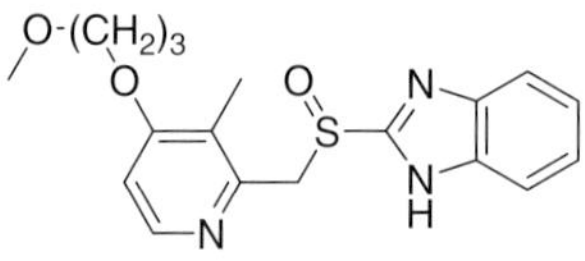

5
Rabeprazole

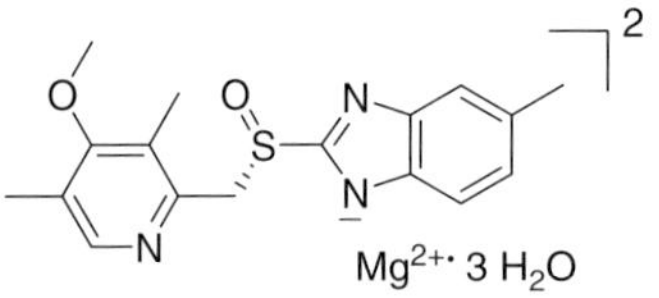

6
(*S*)-esomeprazole

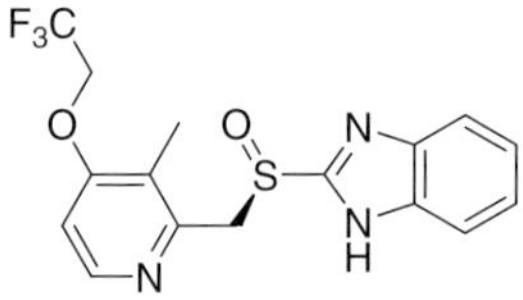

7
(*R*)-(+)dexlansoprazole

Figure 19.2 Structures of the marketed sulfinylbenzimidazoles: omeprazole (**2**) [3], pantoprazole (**3**) [4], lansoprazole (**4**) [5, 6], rabeprazole (**5**) [7], esomeprazole (**6**) [8], and dexlansoprazole (**7**) [9].

during her investigation of theophylline derivatives, introduced a side chain that contained this structural moiety. The discovery of pyridinylsulfinylbenzimidazoles (**1**) was first reported by a small research group (Hässle) within Astra Pharmaceuticals in the early 1970s, which was aiming to identify a drug to suppress gastric acid secretion for the treatment of patients with peptic ulcers [14]. Until then, the treatment options for peptic ulcers or gastritis were limited to the administration of antacids for temporary relief of the symptoms or a gastrectomy, which posed a potential for serious side effects. After a number of years of research, the Hässle group identified their first *in vivo* hit compound, H124/26 (**9**) (Figure 19.3) [1]. The 2-mercaptobenzimidazole was a cyclized derivative of the previously reported compound CMN 131 (**8**) [15], which was known to exhibit the desired antisecretory

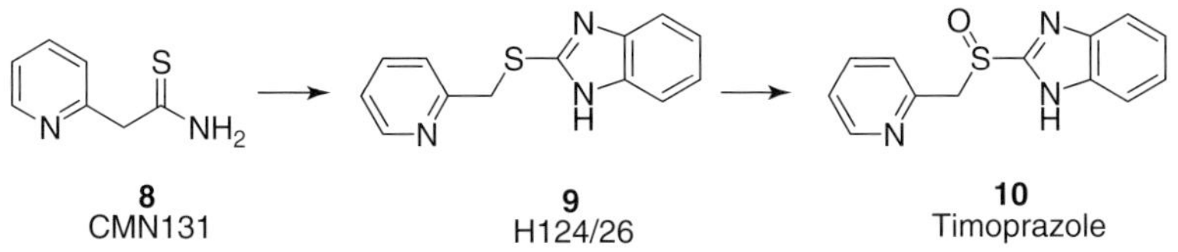

Figure 19.3 Evolution of timoprazole (**10**) from CMN131 (**8**).

profile but was acutely toxic. The scientists at Hässle discovered that cyclization of the thioamide into a 2-mercaptobenzimidazole eliminated the acute toxicity seen with CMN131 (**8**) while maintaining potent antisecretory effects.

Less than a year later, H 83/69 (**10**), the corresponding sulfoxide metabolite of **9**, was identified as having even more potent antisecretory activity than the parent compound [1]. This metabolite, generically termed *timoprazole* (**10**), represented the first ever reported sulfinylbenzimidazole with potent antisecretory activity *in vivo*. Unfortunately, it was later discovered that administration of this compound resulted in atrophy of the thymus as well as enlargement of the thyroid [14]. Additional optimization efforts around this core showed that modulating physical properties, specifically log *P*, could effectively differentiate the antisecretory activity from the thyroid and thymus toxicity [1]. These efforts yielded picoprazole (**11**), which exhibited potent antisecretory activity with no ancillary toxic effects (Figure 19.4). In addition, the scientists at Hässle found that modulation of the pK_a of the compound could enhance its accumulation in the acidic environment of the parietal cells in the stomach. Ultimately, in 1979, as a result of additional rounds of optimization, H168/68 (**2**) was synthesized, which was later generically named omeprazole (**2**) [16]. Omeprazole (**2**) contains a modified pyridine ring that is a full pK_a unit ($pKa \sim 3.9$) more basic than timoprazole (**10**) and a 5-methoxy group on the benzimidazole that enhances the stability of the molecule at neutral pH.

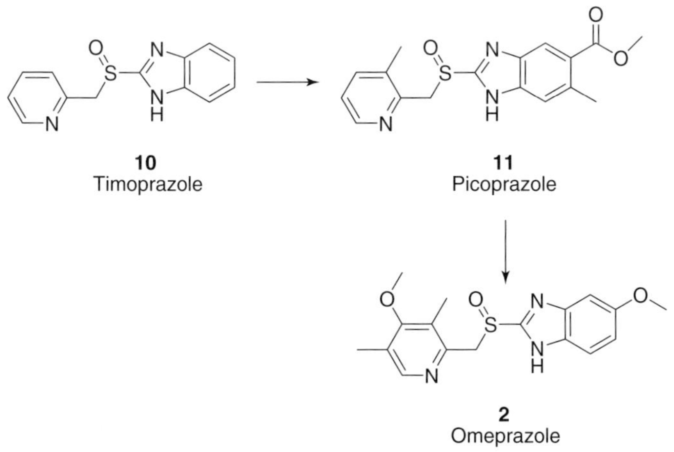

Figure 19.4 Structural evolution of omeprazole (**2**) from timoprazole (**10**).

The Investigational New Drug (IND) application for omeprazole (**2**) was filed in 1980, and the molecule was first launched in Europe as Losec (**2**) in 1988. In 1996, Losec (**2**) became the highest selling drug in the pharmaceutical industry grossing $6 billion at its peak sales.

The first ever blockbuster drug marketed for gastric-acid-associated disorders, Tagamet®, was introduced in Europe in 1976 and later approved by the FDA in 1979. Tagamet, discovered by Smith, Kline and French (SK&F), was a first-in-class compound, which inhibited the histamine 2 (H_2) receptor, a receptor in the gastric acid secretion pathway [17]. H_2 receptor antagonists represented the first major breakthrough in the short-term treatment of heartburn associated with acid indigestion. However, in comparison to the H_2 receptor antagonists, PPIs exhibit longer-lasting and more efficacious suppression of gastric acid *in vivo*. The clinical superiority of these compounds is the result of their direct inhibition of the gastric acid pump H^+/K^+-ATPase [16], which is the terminal step in the complex acid secretion pathway in parietal cells in the stomach.

19.3 Synthesis

A number of early patents were filed that described the synthesis of sulfinylbenzimidazoles as well as the associated *in vivo* antisecretory activity [18, 19]. The general synthetic strategy for this heterocycle involves the coupling of an appropriately substituted chloromethyl pyridine with a cognate 2-mercaptobenzimidazole, followed by oxidation of the sulfur to the sulfoxide. For the synthesis of omeprazole (**2**), the 2-chloromethyl pyridine precursor **15** is synthesized from 2,3,5-trimethylpyridine (**12**), which, in turn, is oxidized to the pyridine *N*-oxide and subsequently nitrated in the 4-position to form **13** (Scheme 19.1). The nitro group is then displaced by an *in situ* generated methoxide anion, and acetoxylation of the 2-methyl group using acetic anhydride generates the protected hydroxymethyl intermediate **14**. Ester hydrolysis and chlorination using thionyl chloride furnishes **15**.

The 2-mercaptobenzimidazole portion of the molecule is synthesized from 4-methoxybenzenediamine (**16**), which is cyclized using a xanthate ester to form **17** (Scheme 19.2). The 2-mercaptobenzimidazole and pyridine building blocks are then coupled in ethanol using NaOH, and the subsequent racemic oxidation of the resultant sulfide (**18**) is accomplished in the presence of *m*-chloroperoxybenzoic acid (*m*CPBA) to provide omeprazole (**2**).

A large number of patents were subsequently filed, which improved on several aspects of the original synthesis of omeprazole (**2**) and other sulfinylbenzimidazoles (**1**). For example, the oxidation of the sulfide to the sulfoxide in the final step of the synthesis was very problematic due to incomplete oxidation of the sulfur or overoxidation to the corresponding sulfone. Thus, a number of methods were reported that improved the yields of the oxidation step, which include performing the reaction at elevated pH [20] as well as incorporating solvents that aid in the

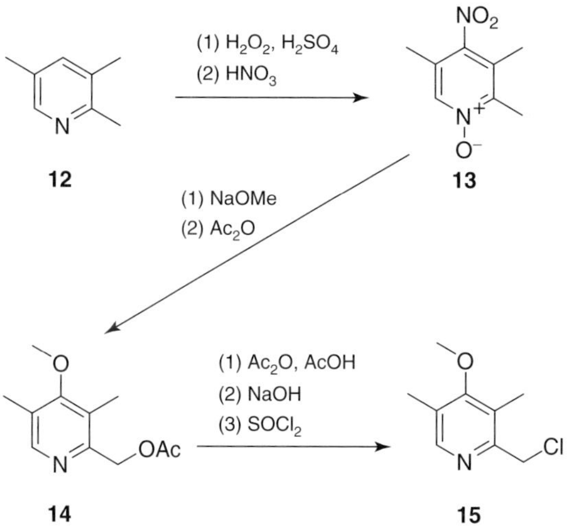

Scheme 19.1 Synthesis of the chloromethylpyridine precursor **15** of omeprazole (**2**).

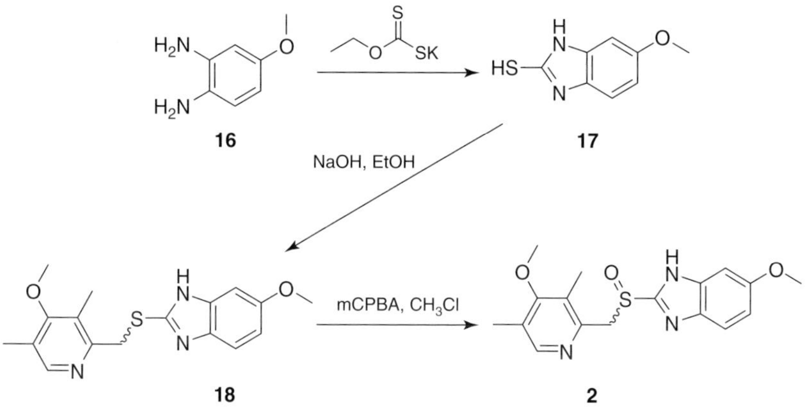

Scheme 19.2 Synthesis of the benzimidazole core **17** and final steps of the omeprazole (**2**) synthesis.

crystallization of the final product [21]. One unique approach for the oxidation of the sulfide was identified by Slemon *et al.* [22] who found that the α-amidothioether (**19**) is readily oxidized to the corresponding sulfoxide (**20**) with hydrogen peroxide in the presence of vanadium acetylacetonate with ~90% yield (Scheme 19.3). The sulfoxide (**20**) can then be hydrolyzed to the corresponding acid, which subsequently decarboxylates cleanly to form omeprazole (**2**).

A more recent approach reported by Bhalerao *et al.* [23] from Dr Reddy Laboratories described the synthesis of the omeprazole core via the coupling of the sulfonic ester (**22**) with the *in situ* generated Grignard reagent **21** (Scheme 19.4).

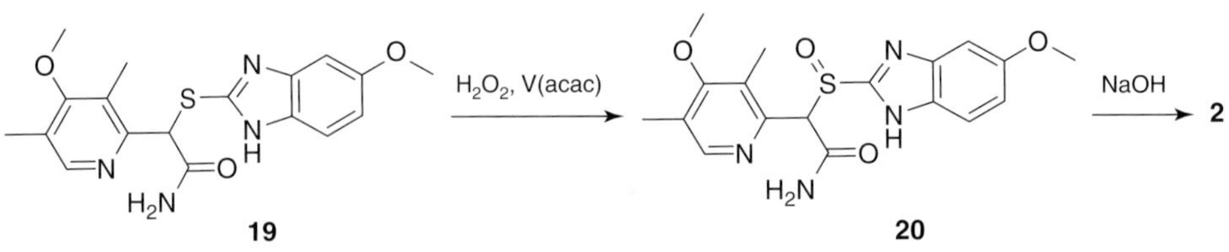

Scheme 19.3 Oxidation of the α-amidothioether (**19**) to the corresponding sulfone (**20**), followed by hydrolysis and decarboxylation to yield omeprazole (**2**).

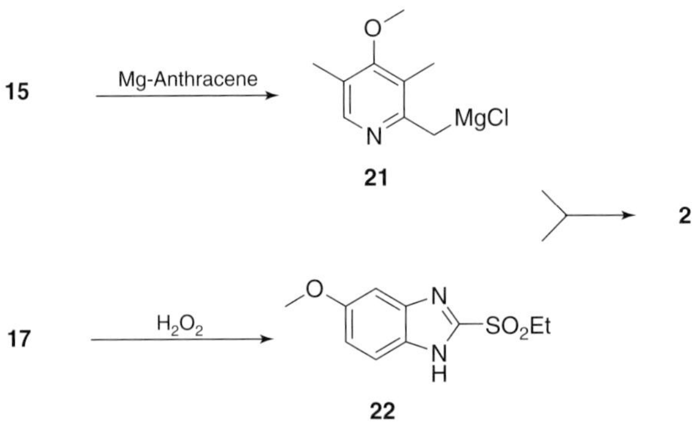

Scheme 19.4 Synthesis of omeprazole (**2**) by coupling the *in situ* generated Grignard reagent **21** and the cognate benzimidazole-2-sulfonic ester (**22**).

This unique approach simplified the synthesis by completely eliminating the problematic oxidation step.

In 2001, it was reported that the (*S*)-enantiomer of omeprazole, generically termed *esomeprazole* (**6**), exhibited a higher bioavailability and oral potency in humans, as compared to the corresponding (*R*)-enantiomer [24]. As a result, the synthesis and/or isolation of the enantiomerically pure forms of sulfinylbenzimidazoles (**1**) became of greater interest. The isolation of the individual enantiomers of omeprazole (**2**) was readily achieved on a small scale using simple HPLC [25]. However, on a large scale, the separations initially required the use of a chiral auxillary to facilitate the separation of the resulting diastereomeric mixture (Scheme 19.5) [26, 27]. This was accomplished by preparing the N-chloromethyl derivative **23** of omeprazole (**2**) and subsequent formation of the chiral ester (**24**) with (*R*)-(−)-mandelic acid, using phase-transfer catalysis conditions. Esomeprazole (**6**) was then isolated via reversed-phase HPLC of **24**, followed by hydrolysis of the mandelic ester.

More recently, methods for resolving the enantiomers of omeprazole (**2**) have been reported using various recrystallization conditions [28]. The first chiral synthesis of esomeprazole (**6**), which was marketed as Nexium®, was published by the process group at AstraZeneca in 2000 [29] and focused on an enantioselective oxidation of the sulfide in the terminal step of the synthesis (Scheme 19.6). The process started by using a previously reported synthesis described by Pitchen *et al.* [30], who used modified Sharpless oxidation conditions to cleanly oxidize prochiral

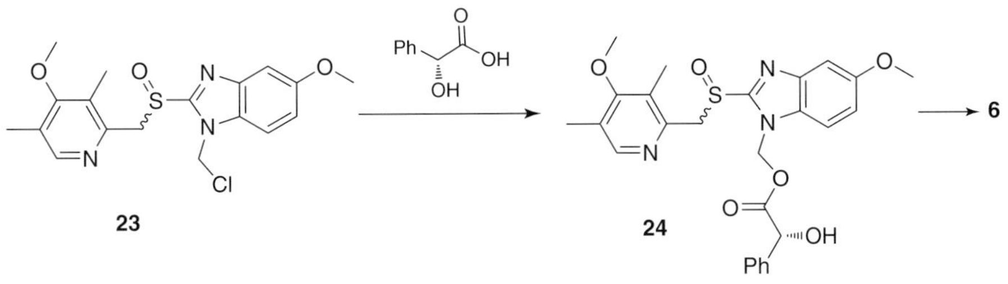

Scheme 19.5 Synthesis of esomeprazole (**6**) using *R*-(−)-mandelic acid as chiral auxiliary.

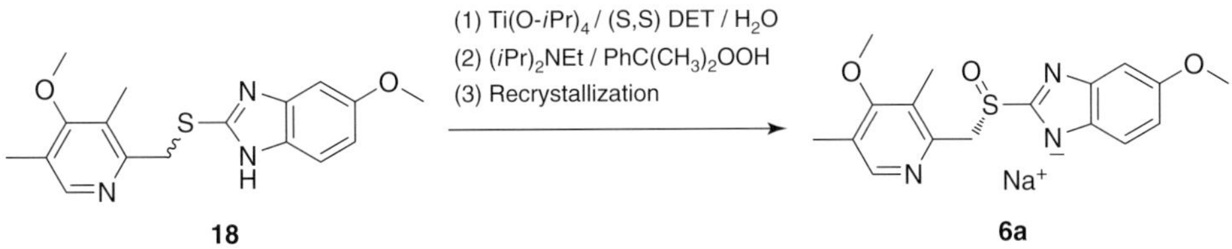

Scheme 19.6 Chiral oxidation used for the synthesis of (*S*)-esomeprazole (**6**).

sulfides into sulfoxides. The group at AstraZeneca eventually further modified these conditions by preforming the requisite titanium species in the presence of the sulfide (**18**) at high temperatures, as well as adding a basic tertiary amine. These modifications enabled the synthesis of the sodium salt of esomeprazole (**6**) in 94% enantiomeric excess.

Another method that was used for chiral sulfide oxidation in the omeprazole core was a bio-oxidation [31] using different microorganisms, namely *penicillium* species. This method gave enrichment levels of **6** between 56 and 99% *ee* depending on the organism and was also used for the corresponding chiral synthesis of pantoprazole (**3**) and lansoprazole (**4**). Unfortunately, this method is not suitable for larger-scale production processes.

19.4 Mode of Action

The gastric parietal cells contain the acid pump H^+/K^+-ATPase, which is located at the terminal step of the gastric acid secretion pathway [18]. The parietal excretion of gastric acid can be stimulated by neuronal (acetylcholine), paracrine (histamine), or endocrine signals (gastrin), all of which indirectly result in an increase of the pH in the stomach (Figure 19.5). The sulfinylbenzimidazoles (**1**) directly inhibit the gastric acid pump in the parietal cells, which results in a specific and long-lasting inhibition of acid secretion in the stomach.

After the discovery of the sulfinylbenzimidazoles (**1**) as potent PPIs, it became apparent that the nature of their inhibition was somehow linked to their instability

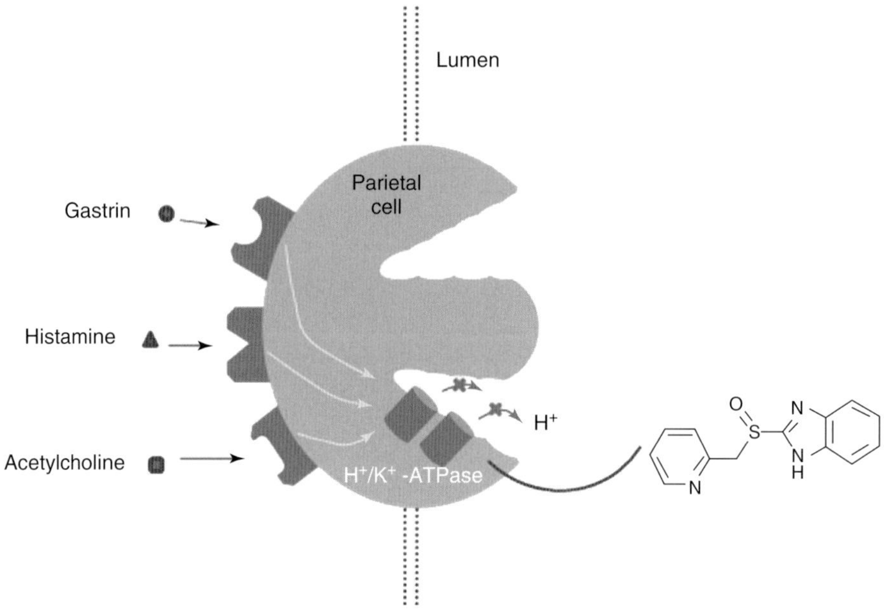

Figure 19.5 Regulation of gastric acid secretion in the parietal cells in the stomach.

in solution. It was later found that pretreatment of omeprazole (**2**) with acid was required for *in vitro* inhibition of H^+/K^+-ATPase. A number of subsequent studies around the acid decomposition of **2** and related sulfinylbenzimidazole analogs revealed that the principle inhibitory component is a reactive sulfenamide, which is formed on acid-catalyzed rearrangement of the sulfinyl core (Scheme 19.7) [32–35]. The sulfenamide (**27**) is a highly reactive intermediate that forms a disulfide adduct with reactive cysteines on the surface of H^+/K^+-ATPase, resulting in irreversible inhibition. The acid-catalyzed rearrangement of the omeprazole core commences with the reversible protonation of the benzimidazole ring. The rate-limiting step is a nucleophilic attack of the pyridine nitrogen on the C2 position of the benzimidazole. The resultant highly unstable dihydrobenzimidazole (**25**) rearomatizes by opening the five-membered heterocycle to give the sulfenic acid (**26**). The sulfenic acid is subsequently attacked by the benzimidazole nitrogen and on dehydration, forms the reactive sulfenamide species (**27**). This species can then form disulfide adducts with solvent-exposed cysteines on the surface of H^+/K^+-ATPase [36–39].

19.5 Structure–Activity Relationships

To date, all the marketed PPIs contain sulfinylbenzimidazole as their basic framework and only differ in the substituents on each of the heterocyclic rings. Because of the unique nature of the mode of action of these compounds, optimization of

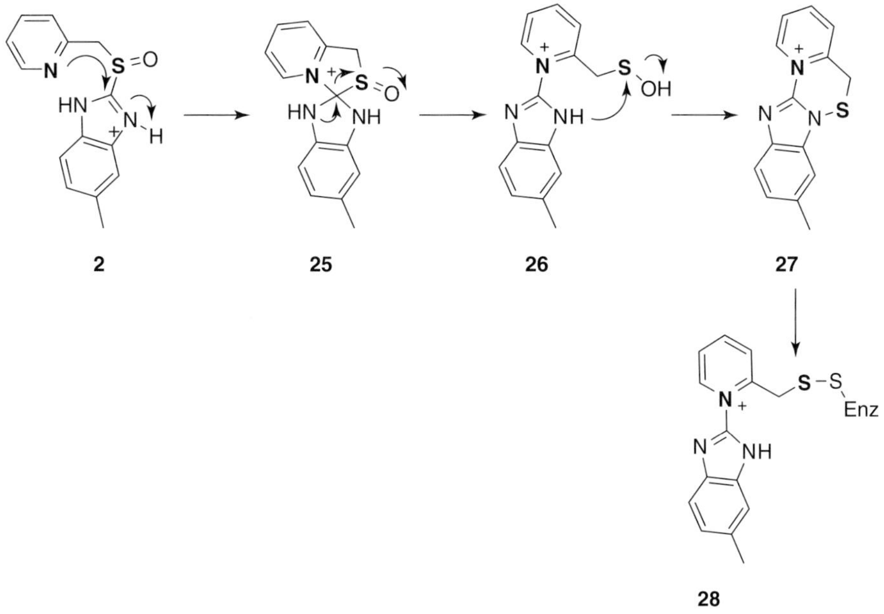

Scheme 19.7 Rearrangement of the sulfinylbenzimidazole core to the reactive sulfenamide (**27**) and its reaction with solvent-exposed cysteines on the surface of H^+/K^+-ATPase.

their *in vivo* antisecretory activity was largely empirical. Much of the early synthetic optimization work involved evaluation of different substituents on the sulfinylbenzimidazole core and the resultant effects on molecular stability, the propensity to rearrange to the active species, as well as physicochemical properties.

The original optimization of the sulfinylbenzimidazole core was undertaken where all three elements of the core, the two heterocycles and the linker, were systematically modified while monitoring antisecretory effects *in vivo* [40]. Electron-donating substituents on the pyridine ring, particularly at the 4-position, enhanced the antisecretory activity of the compounds through increased propensity to form the spirocyclic intermediate and the subsequent formation of the reactive sulfenamide (**27**) (Scheme 19.7). Conversely, substituents with less capacity to donate electrons resulted in a decrease in antisecretory activity. The significance of this structure–activity relationship is further reflected by the fact that all six of the marketed PPIs contain an alkoxy substituent at the 4-position of the pyridine ring. In addition, these groups increase the relative basicity, or pK_a, of the entire molecule, which in turn enhances their accumulation in the acidic environment of the parietal cells in the stomach. Conversely, because of the critical nature of forming the spirocyclic intermediate **25** in the activation pathway, bulky substituents at the 6-position of the pyridine ring resulted in a significant decrease in potency, due to the steric bulk around the nucleophilic nitrogen. As can be appreciated from Table 19.1 (column 1), replacement of the pyridine ring with a corresponding thiazole (**29**) or imidazole (**30**) as well as transposition of the nitrogen in the pyridine

Table 19.1 General structure–activity relationships around the mercaptobenzimidazole core.

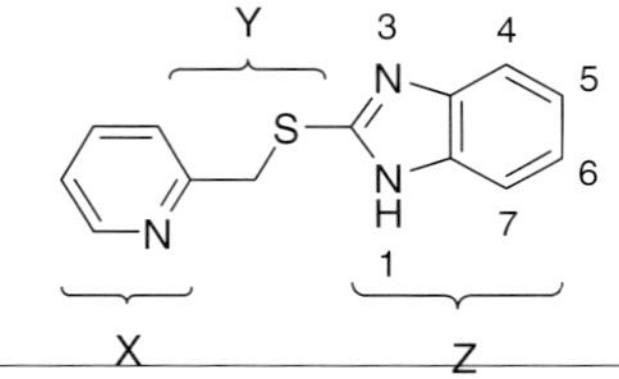

Compound	X	Relative activity	Compound	Y	Relative activity	Compound	Z	Relative activity
2	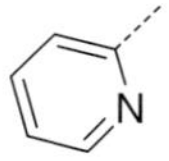	+ + +	2	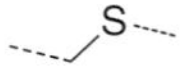	+ + +	2	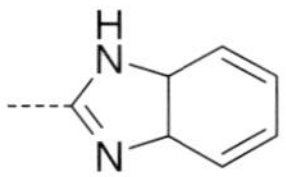	+ + +
29	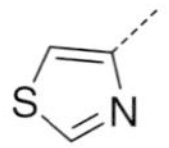	0	33		0	37	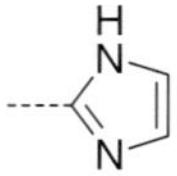	0
30	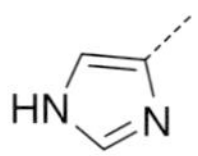	++	34	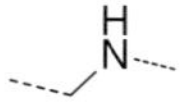	+	38	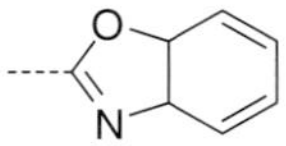	+
31	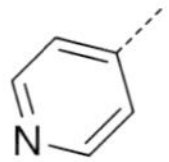	0	35	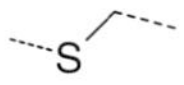	0	39	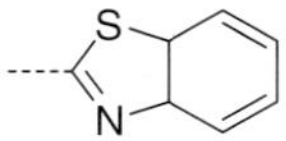	+
32	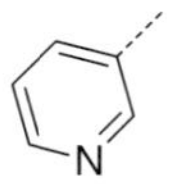	0	36	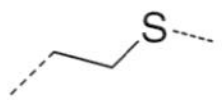	0	40	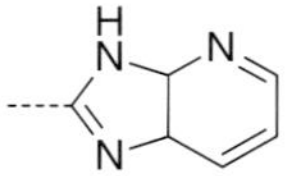	++

+ + +, highly active; 0, no observable activity.

ring to the corresponding 3- and 4-positions (**31** and **32**) resulted in significant decreases in activity.

Modifications to the sulfinyl group and the alkyl linker (compounds **33–36**, Table 19.1, column 4) also significantly reduced the activity of this class of compounds. Different oxidation states of the sulfur group, as well as transposition and heteroatom replacements in the linker (**33–35**), resulted in significant reduction of activity, a likely result of a decrease in the propensity for molecular rearrangement. Similarly, extension of the linker between the pyridine and the benzimidazole (**36**) resulted in acid-stable compounds with no antisecretory effects. Substitutions on the benzimidazole ring (data not shown) generally had modest impact on the potency of the sulfinylbenzimidazoles (**1**). Typically, electron-withdrawing groups caused a reduction in activity, whereas electron-donating groups increased activity as well as stability at neutral pH. By contrast, replacing the benzimidazole with other heterocycles such as imidazole (**37**), benzoxazole (**38**), benzothiazole (**39**), and even azabenzimidazole (**40**) consistently resulted in a decrease in activity (Table 19.1, column 7).

Since the initial discovery of omeprazole (**2**) in the 1970s and the subsequent introduction of other sulfinylbenzimidazole-based PPIs to the market, a large amount of work has been done to further explore modifications of the core of this class of molecules and their effects on antisecretory activity, as well as other clinically relevant phenotypes. Herling *et al.* [41] reported the synthesis of the thienoimidazole derivative **41** (Figure 19.6), which exhibited a comparable inhibitory profile to omeprazole (**2**) but was less stable in aqueous environment.

The analogous benzothiazoline (**42**), reported by Yoon *et al.* [42], was only sevenfold weaker than omeprazole (**2**) *in vitro*, despite the drastic modification of the core. Uchiyama *et al.* [43] reported the synthesis of TU-199 (**43**), an azabenzimidazole modification to the omeprazole core that they claim has more potent and longer-lasting antisecretory activity. Other modifications to the benzimidazole

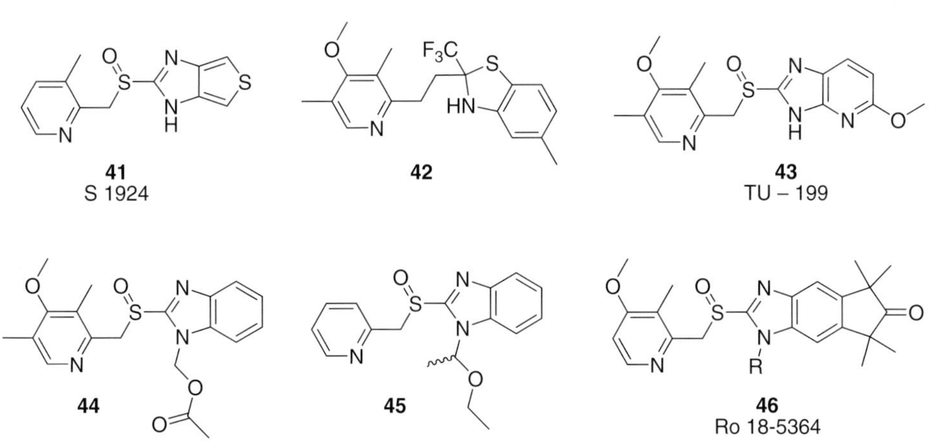

Figure 19.6 Structures of unique sulfinylbenzimidazole analogs that exhibit antisecretory activity.

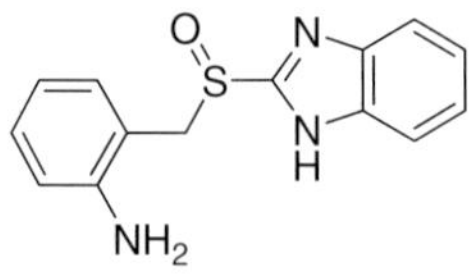

47
NC – 1300

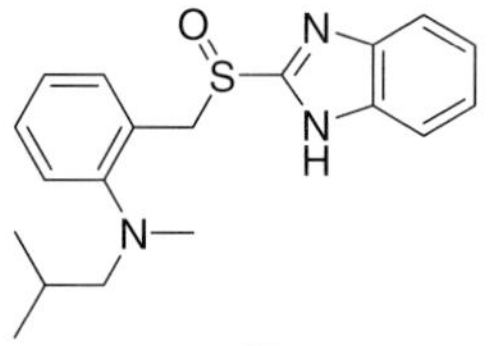

48
Leminoprazole

Figure 19.7 Structures of sulfinylbenzimidazoles NC-1300 (**47**) and leminoprazole (**48**), where the pyridine is replaced with a substituted phenyl ring.

include synthesis of the N-alkylated proforms **44** and **45** as well as ring fusion analogs such as Ro-18-5364 (**46**).

There are also a number of modifications to the pyridine ring of the sulfinylbenzimidazole core reported in the literature, many of which exhibit potent antisecretory activity. Modest modification such as replacement of the pyridine with a pyrimidine as well as synthesis of 2-amino benzyl analogs such as NC-1300 (**47**) [44] and leminoprazole (**48**) [45] (Figure 19.7) have been reported to show antisecretory activities similar to that of omeprazole (**2**).

When first discovered, the sulfinylbenzimidazoles (**1**) represented a unique class of compounds that exhibited efficacious inhibition of gastric acid excretion *in vivo*. Since then, PPIs have become the largest selling medications in the history of the pharmaceutical industry and have improved the lives of millions of people suffering from gastric-acid-associated disorders. The discovery of their novel mode of action served as a seminal advancement in the field of drug discovery, and their comprehensive use and effectiveness has revolutionized the status of care for gastric-acid-associated disorders.

References

1. Lindberg, P., Brändström, A., Wallmark, B., Mattsson, H., Rikner, L., and Hoffmann, K.J. (1990) *Med. Res. Rev.*, **10**, 1–54.
2. Fellenius, E., Berglindh, T., Sachs, G., Olbe, L., Elander, B., Sjostrand, S.E., and Wallmark, B. (1981) *Nature*, **290**, 159–161.
3. Brándström, A., Lindberg, P., and Junggren, U. (1985) *Scand. J. Gastroenterol.*, **20**, 15–22.
4. Kohl, B., Sturm, E., Senn-Bilfinger, J., Simon, W.A., Krueger, U., Schaefer, H., Rainer, G., Figala, V., and Klemm, K. (1992) *J. Med. Chem.*, **35**, 1049–1057.
5. Morini, G., Grandi, D., Arcari, M.L., and Bertaccini, G. (1995) *Gen. Pharmacol. Vasc. Syst.*, **26**, 1021–1025.
6. Tomiyama, Y., Morii, M., and Takeguchi, N. (1994) *Biochem. Pharm.*, **48**, 2049–2055.
7. (a) Morii, M., Takata, H., Fujisaki, H., and Takegucht, N. (1990) *Biochem. Pharm.*, **39**, 661–667; (b) Prakash, A. and Faulds, D. (1998) *Drugs*, **55**, 261–267.
8. Andersson, T., Röhss, K., Bredberg, E., and Hassan-Alin, M. (2001) *Aliment. Pharmacol. Ther.*, **15**, 1563–1569.
9. Wittbrodt, E.T., Baum, C., and Peura, D.A. (2009) *Clin. Exp. Gastroenetrol.*, **2**, 117–128.
10. Wright, J.B. (1951) *Chem. Rev.*, **48**, 397–541.
11. Wang, Z. (2010) *Phillips-Ladenburg Benzimidazole Synthesis*, John Wiley & Sons, Inc., pp. 2197–2199.

12. Hobrecker, F. (1862) *Ber. Dtsch. Chem. Ges.*, **5**, 920–924.
13. Gorczyca, M. (1962) *Diss. Pharm.*, **14**, 381–391.
14. Olbe, L., Carlsson, E., and Lindberg, P. (2003) *Nat. Rev. Drug Discov.*, **2**, 132–139.
15. Malen, C.E. and Danree, B.H. (1971) *J. Med. Chem.*, **14**, 244–246.
16. Olbe, L., Haglund, U., Leth, R., Lind, T., Cederberg, C., Ekenved, G., Elander, B., Fellenius, E., Lundborg, P., and Wallmark, B. (1982) *Gastroenterol*, **83**, 193–198.
17. (a) Brimblecombe, R.W., Duncan, W.A., Durant, G.J., Ganellin, C.R., Parsons, M.E., and Black, J.W. (1975) *Proc. Br. J. Pharmacol.*, **53**, 435–436; (b) Black J.W., Duncan, W.A., Durant C.J., Ganellin C.R., and Parsons E.M. (1972) *Nature*, **234**, 385–390.
18. Berntsson, P.B.M., Carlsson, S., Garberg, L.E., Junggren, U.F., Sjostrand, S.E., and VonWittken, S.G.W. (1977) (Aktiebolaget Astra, Hå̆ssle). US4045564, Aug. 30, 1977.
19. Junggren, U.K. and Sjostrand, S.E. (1981) (Hå̆ssle). US4255431,Mar. 10, 1981.
20. Brandstrom, A.E. (1991) (Aktiebolaget Astra). EP0533752, Dec. 12, 1991.
21. Milac, N.H. and Jereb, D. (2001) (LEK, Tovarna Farmacevtskih in Kemicnih Izdelkov. D.D.) US463651, July 31, 2001.
22. Slemon, C. and Macel, B. (1996) US5502195, Mar 26, 1996.
23. Bhalerao, D.S., Kondaiah, G.C.M., Dwivedi, N., Mylavarappu, R.K., Reddy, L.A., Roy, A., Nagaraju, G., Reddy, P.P., Bhattacharya, A., and Bandichhor, R. (2010) *Synth. Commun.*, **40**, 2983–2987.
24. (a) Andersson, T., Hassan-Alin, M., Hasselgren, G., Rohss, K., and Weidolf, L. (2001) *Clin. Pharmacokinet.*, **40**, 411–426; (b) Miner, P., Katz, P.O., Chen, Y., and Sostek, M. (2003) *Am. J. Gastroenterol.*, **98**, 2616–2620.
25. Erlandsson, P., Isaksson, R., Lorentzon, P., and Lindberg, P. (1990) *J. Chromatogr.*, **532**, 305–319.
26. Lindberg, P.L. and Von Unge, S. (1994) (Astra AB). WO9427988, Dec. 8, 1994.
27. Kohl, B. and Senn-Bilfinger, J. (1992) (BYK Gulden Lomberg Chemische Fabrik Gmbh.). DE 4035455, May 29, 1992.
28. Gerard, C., Guillaume, T., and Marie-Noelle, P. (2009) (Universite de Rouen). WO12201144, May 14, 2009.
29. Cotton, H., Elebring, T., Larsson, M., Li, L., Sörensen, H., and von Unge, S. (2000) *Tetrahedron: Asymmetry*, **11**, 3819–3825.
30. Pitchen, P., Dunach, E., Deshmukh, M.N., and Kagan, H.B. (1984) *J. Am. Chem. Soc.*, **106**, 8188–8193.
31. Holte, R.L.P., Reeve, C., and Taylor, S. (1996) (Aktiebolaget Astra) WO9617076, June 6, 1996.
32. Lindberg, P., Nordberg, P., Alminger, T., Brandstrom, A., and Wallmark, B. (1986) *J. Med. Chem.*, **29**, 1327–1329.
33. Brändström, A., Lindberg, P., Junggren, U., and Wallmark, B. (1986) *Scand. J. Gastroenterol.*, **21**, 54–56.
34. Lindberg, P., Brändström, A., and Wallmark, B. (1987) *Trends Pharmacol. Sci.*, **8**, 399–402.
35. Besancon, M., Simon, A., Sachs, G., and Shin, J.M. (1997) *J. Biol. Chem.*, **272**, 22438–22446.
36. Besancon, M., Shin, J.M., Mercier, F., Munson, K., Miller, M., Hersey, S., and Sachs, G. (1993) *Biochemistry*, **32**, 2345–2355.
37. Shin, J.M., Besancon, M., Prinz, C., Simon, A., and Sachs, G. (1994) *Aliment. Pharmacol. Ther.*, **8** (Suppl. 1), 11–23.
38. Shin, J.M., Besancon, M., Simon, A., and Sachs, G. (1993) *Biochim. Biophys. Acta.*, **1148**, 223–233.
39. Olbe, L. (1999) Proton Pump Inhibitors, in *Milestones in Drug Therapy* (eds M.J. Parnham and J. Bruinvels), Springer, Birkhauser, Basel, p. 257.
40. Brandstrom, A., Lindberg, P., and Junggren, U. (1985) *Scand. J. Gastroenterol. Suppl.*, **108**, 15–22.
41. Herling, A.W., Bickel, M., Lang, H.J., Weidmann, K., Rosner, M., Metzger, H., Rippel, R., Nimmesgern, H., and Scheunemann, K.H. (1988) *Pharmacology*, **36**, 289–297.

42. Yoon, S.H., Seo, S., Lee, Y., Hwang, S., and Kim, D.Y. (1998) *Bioorg. Med. Chem. Lett.*, **8**, 1909–1912.
43. Uchiyama, K., Wakatsuki, D., Kakinoki, B., Takeuchi, Y., Araki, T., and Morinaka, Y. (1999) *J. Pharm. Pharmacol.*, **51**, 457–464.
44. Okabe, S., Akimoto, Y., Yamasaki, S., and Nagai, H. (1988) *Dig. Dis. Sci.*, **33**, 1425–1434.
45. Takahashi, S., Nakamura, E., and Okabe, S. (1995) *J. Pharmacol. Exp. Ther.*, **275**, 1396–1401.

Part VI
Metabolic Diseases

Bioactive Heterocyclic Compound Classes: Pharmaceuticals,
First Edition. Edited by Jürgen Dinges and Clemens Lamberth.

20
Thiazolidinedione-Based Insulin Sensitizers:PPAR-γ Agonists for the Treatment of Type 2 Diabetes

Steven Richards

20.1
Introduction

Thiazolidinediones (**1**, TZDs, glitazones) are a drug class that provided a breakthrough for the management of type 2 diabetes beginning in the late 1990s [1]. The members of this drug class are named after the thiazolidine-2,4-dione heterocycle that comprises the head group of these compounds (Figure 20.1).

So far, three agents from this class have made it to market as antidiabetic agents: troglitazone (**2**, Rezulin,® Parke-Davis and Sankyo) [2], pioglitazone (**3**, Actos®: Takeda) [3], and rosiglitazone (**4**, Avandia®: Glaxo Smith Kline (GSK)) [4] (Figure 20.2).

20.2
History

TZD (**8**), also known as *α-μ-diketotetrahydrothiazole, 4-thiazolidinone,* and *Senfölessigsäure,* was first synthesized around the turn of the twentieth century [5, 6]. This heterocycle is structurally related to rhodanine (where the C2 carbonyl is a thiocarbonyl) and hydantoin (where the ring S is replaced by N). This similarity caused some confusion. The early syntheses of these compounds was well reviewed by F.C. Brown in 1961 [7]. The preparation of TZD (**8**) was described as shown in Scheme 20.1 [6]. Cyanothiocarbamate (**4**) was cyclized to TZD (**8**) in boiling concentrated HCl, presumably proceeding through the acyclic intermediate **7**. The authors confirmed the identity of the intermediate by heating ethylthiocarbamic acid (**5**) and chloroacetic acid (**6**). Intermediate **7** could be isolated under these conditions before spontaneously cyclodehydrating to form 2,4-TZD (**8**).

In 1914, Kučera reported the condensation of TZD (**7**) with piperonal to form the unsaturated nucleus **9** (Scheme 20.2) [8]. At that time, molecules such as these were often used as dyes.

Bioactive Heterocyclic Compound Classes: Pharmaceuticals,
First Edition. Edited by Jürgen Dinges and Clemens Lamberth.

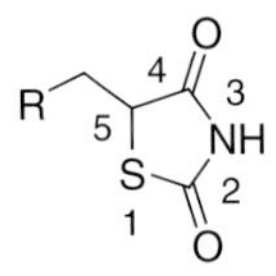

1

Figure 20.1 Basic structure and numbering convention for thiazolidinediones (1).

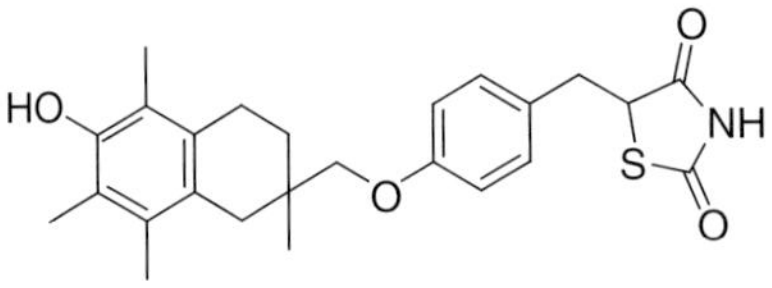

2
Troglitazone

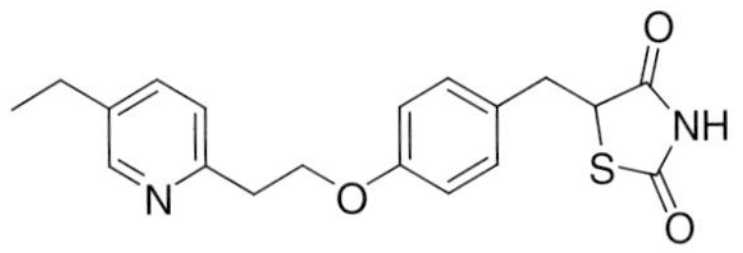

3
Pioglitazone

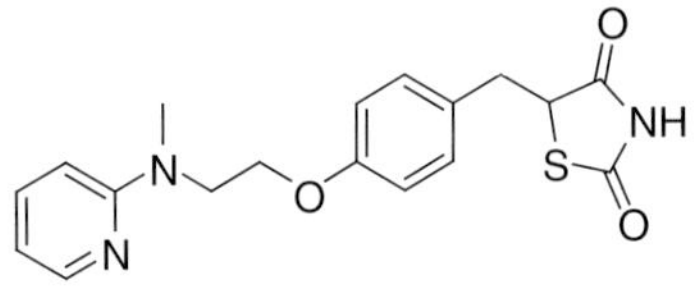

4
Rosiglitazone

Figure 20.2 Marketed thiazolidines for the treatment of diabetes.

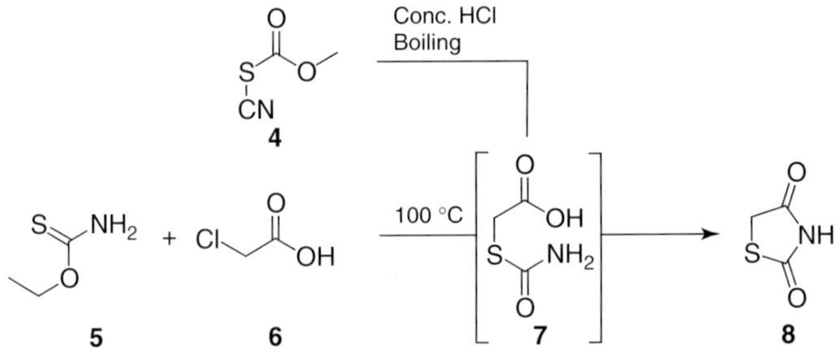

Scheme 20.1 Turn-of-the-century synthesis of thiazolidinedione (8).

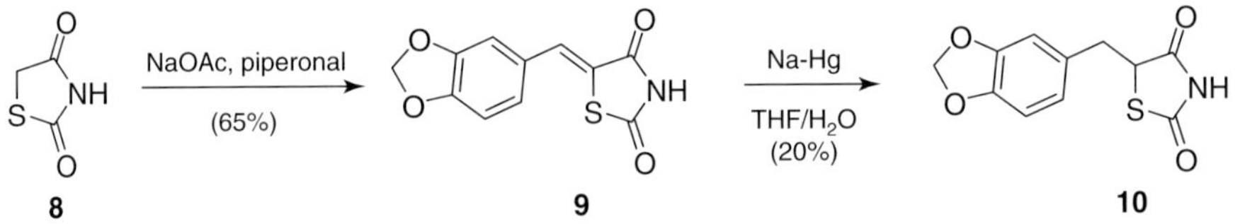

Scheme 20.2 Method to attach a side chain at 5-position of the thiazolidinedione (8).

Forty years later, the TZD scaffold was found to have antibacterial properties. For example, Brown et al. [9] demonstrated that a series of alkylidine TZDs, compounds **11–14** (Figure 20.3), were effective in arresting the growth of *Aspergillus niger*.

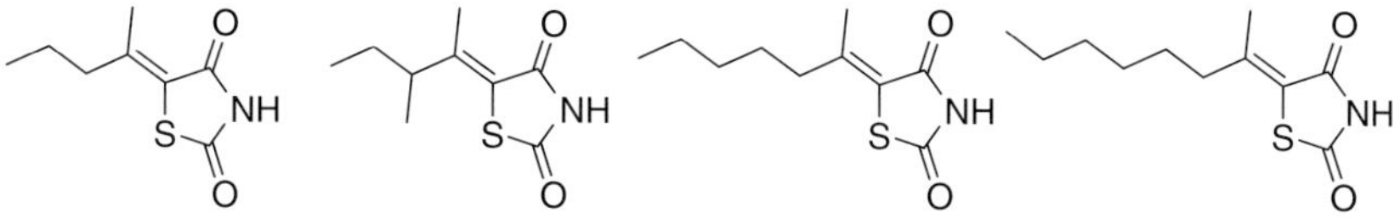

	11	12	13	14
%*A. niger* growth inhibition @ 50 ppm	54	63	71	24
% *A. niger* growth inhibition @ 250 ppm	100	100	100	49

Figure 20.3 Effect of alkyl chain length on *Aspergillus niger* growth inhibition.

The compounds were similar in potency to their corresponding rhodanines. Brown observed a trend toward decreased antibacterial activity with increasing length of the alkyl side chain (**13** vs. **14**). When investigated for activity against the malarial parasite *Plasmodium berghei* [10], the compounds had little to no activity [10]. With respect to the synthesis of **10**, which already represents the basic structure of the resent day antidiabetic agents, the authors noted that the sulfur containing heterocycle was not amenable to standard catalytic hydrogenation conditions. Instead, they opted to use sodium amalgam to obtain a modest amount of **10**. Despite some success, 5-substituted TZDs were not further investigated in the clinic for the treatment of bacterial or parasitic infections.

In 1980, researchers at Takeda first reported efforts to identify lipid-lowering agents with comparable activity to fibrates. They began by synthesizing a number of alkanoic acids with a pendant biaryl ether moiety. These compounds were evaluated subchronically in healthy Sprague Dawley (SD) rats and genetically obese, diabetic KKA^y mice. 3-[4-(4-chlorophenoxy)benzoyl] propionic acid (**15**) was found to lower plasma triglycerides in rats after four days of 0.1–0.2% dosing in chow (Figure 20.4) [11].

The effect was comparable to clofibrate, the first of several fibric acid derivatives to modulate lipid levels. Surprisingly, **15** also lowered blood glucose by more than 20%, compared to untreated controls, after four days of dosing in KKA^y mice. Noting "severe toxicity" the researchers abandoned **15** in favor of compounds with a better profile.

Replacement of the keto acid motif with the α-chloro ester provided **16**. This compound was about 30-fold more effective in lowering plasma triglycerides in SD rats than clofibrate [12]. Ester (**16**) also lowered blood glucose. Side effects were considered not severe at the efficacious dose. However, repeated higher doses were toxic to the liver and other organs. Modification of the biaryl ether moiety to the phenethyl ether (**17**) alleviated much of the toxicity without impacting efficacy. As such, **17** lowered plasma triglyceride and glucose levels but only caused a small increase in liver weight [13].

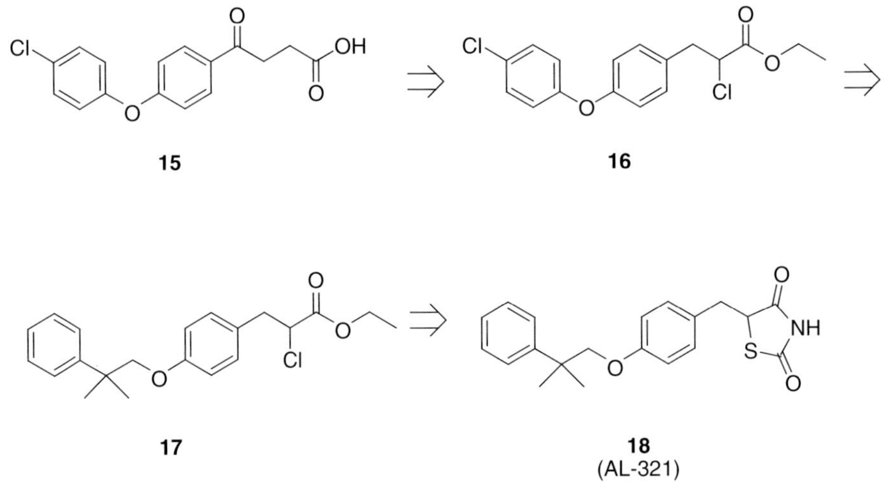

Figure 20.4 SAR leading to the discovery of the first thiazolidinedione, (**18**), with antidiabetic properties.

The first TZD to have antidiabetic properties ascribed to it was the compound AL-321 (**18**) [14]. The acidity of the heterocyclic ring was deemed important as *N*-methylation resulted in an inactive compound. The corresponding rhodanine and hydantoin analogs were both inactive. AL-321 (**18**), dosed at 0.1% (w/w) in chow, lowered plasma triglyceride and glucose levels in KKA^y mice by 50–69% relative to control.

Further efforts to improve the activity of AL-321 (**18**) focused on ether side chain modifications [15]. Removal of the gem-dimethyl substituents or modification of the aryl ring produced compounds with improved potencies. Unfortunately, these analogs were also more toxic, leading to increased liver weight and liver lipids after high or chronic dosing. Replacement of the gem-dimethyl phenethyl group with cycloalkyl groups was found to be beneficial. Ciglitazone (**19**, AD-3878, U-63287) was the result of this effort (Figure 20.5).

The compound maintained glucose-lowering efficacy comparable to AL-321 (**18**) and exhibited "extremely low toxicity." Ciglitazone (**19**) was effective in lowering nonfasting blood glucose in nine-week-old KKA^y mice after four days of dosing at 30–186 mg kg^{-1} [16]. In addition, this experiment revealed dose-dependent decreases in plasma insulin and triglycerides. These effects correlated with a decrease in insulin resistance. Compound **19** also exhibited robust activity in the leptin deficient, hyperphagic *ob/ob* mouse. Ciglitazone (**19**) lowered blood glucose in these animals after 72 h of dosing at 11 mg kg^{-1} day^{-1}in chow [17]. Pancreatic islet morphology was examined, and increased insulin content in ciglitazone (**19**)-treated animals was observed. In addition, an investigation of the fat pads from *ob/ob* mice showed increased insulin sensitivity through both increased number of insulin receptors and increased rate of insulin-stimulated glucose metabolism. Increased lipogenesis was also observed [18]. To confirm the

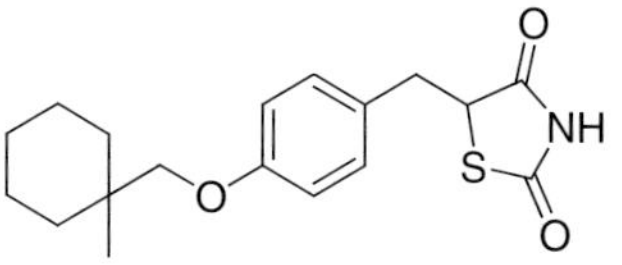

19
Ciglitazone

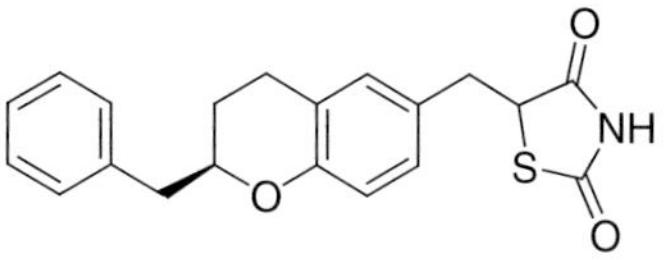

20
Englitazone

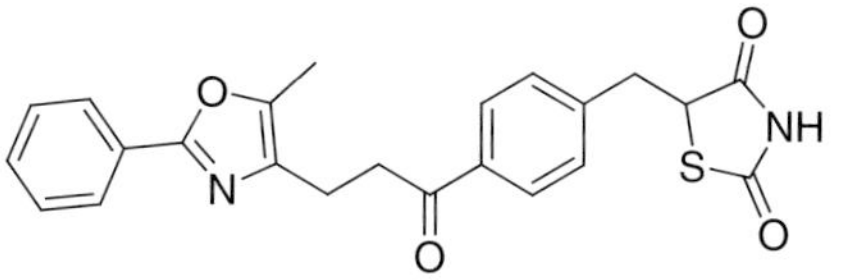

21
Darglitazone

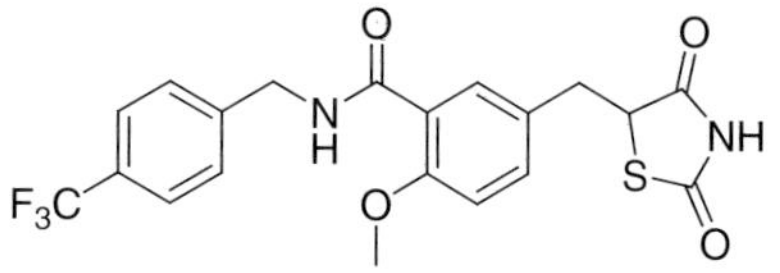

22
KRP-297

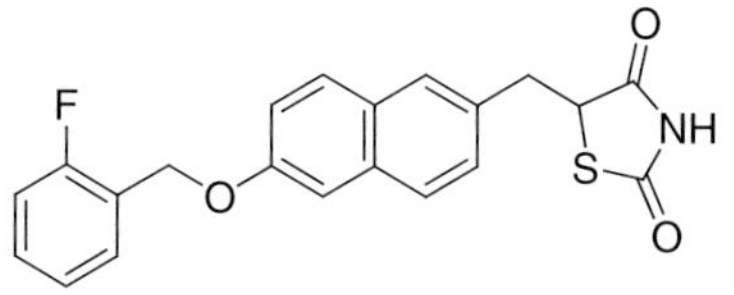

23
Netoglitazone

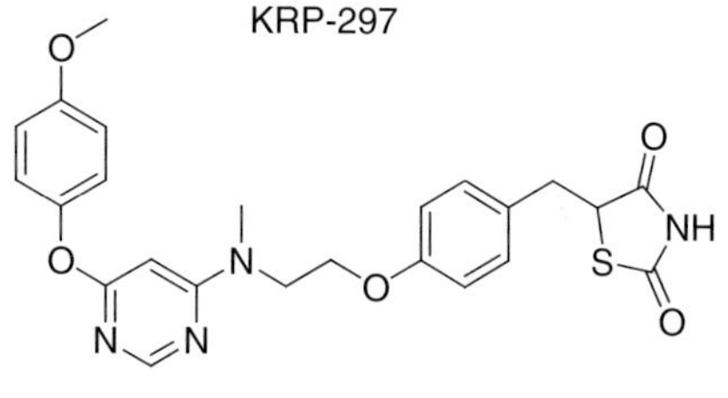

24
Lobeglitazone

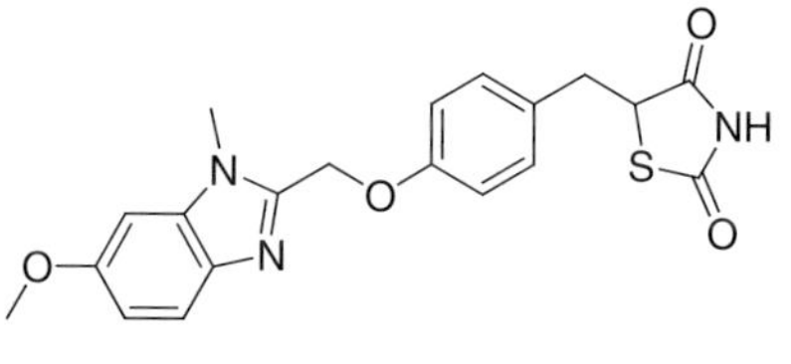

25
Rivoglitazone

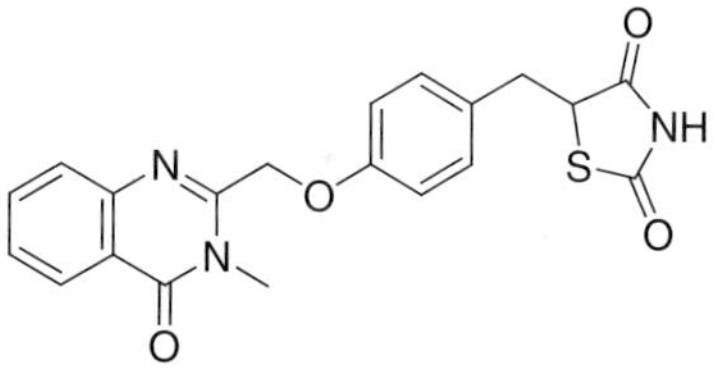

26
Balaglitazone

Figure 20.5 Additional well-characterized TZDs for the treatment of diabetes.

notion that the glucose-lowering effects of this new compound were driven by insulin-stimulated activity in fat, two additional animal models were examined. Normal rats were given a daily 300 mg kg^{-1} dose of ciglitazone (**19**) for nine days. Fasting blood glucose and plasma insulin levels were unchanged compared to control. Similarly, there was no difference in the rate of glucose disposal following an oral glucose challenge [16]. Streptozotocin-treated rats, animals that model the type 1, insulin-dependant, diabetic phenotype also did not benefit from ciglitazone (**19**) treatment [16]. Ciglitazone (**19**) ultimately did not make it to market because of an insufficient therapeutic window in patients [1]. However, as the first representative of a new class of antidiabetic agents, with a molecular target yet unknown, it was fundamentally important to the field.

Scientists at Sankyo, in a complementary effort to discover a treatment for the metabolic syndrome, targeted lipid peroxidation as well as lipid and glucose lowering. Their strategy for reducing lipid peroxidation involved building a hindered phenol, à la vitamin E, to act as a radical trap. They then planned to combine this phenol with the structural features of clofibrate and ciglitazone (**19**) to achieve plasma lipid and glucose lowering. To accomplish this, the team functionalized the antioxidant Trolox® to make a variety of phenol-containing ethers. The effectiveness of these compounds was initially measured chemically by comparing the inhibition of the peroxidation of ethyl linoleate with the activity of vitamin E. Biochemical testing of inhibition of lipid peroxidation was carried out in rat liver microsomes. Finally, the hypolipidemic and hypoglycemic effects were measured *in vivo* [2]. Troglitazone (**2**) was the culmination of these research efforts. It was potent in both chemical and *in vitro* lipid peroxidation assays. In KKA^y mice, troglitazone (**2**) lowered plasma glucose with an ED_{25} of 6 mg kg^{-1}. Troglitazone (**2**) was also active in *ob/ob* mice and Zucker fatty rats, lowering plasma glucose, triglycerides, and free fatty acids (FFAs). Increased insulin sensitivity was observed through improved glucose disposal during an oral glucose tolerance test and through decreased hepatic gluconeogenesis [19, 20].

Early clinical trial results in a group of 11 patients with type 2 diabetes were encouraging [21]. Eight of the subjects responded after 6–12 weeks of troglitazone (**2**) treatment (200 mg b.i.d.). They had decreased plasma glucose and insulin and an improved response to an oral glucose tolerance test. Larger trials followed in which many of the early trial results were recapitulated. Parker [22] provided an excellent review of these trial data in 2002. In addition to decreased blood glucose, reductions in glycated hemoglobin (HbA_{1C}) on the order of 1% were reported. While well tolerated, body weight gain was noted as a side effect. Troglitazone (**2**) was also evaluated as a combination therapy with standard of care (insulin, metformin, sulfonylureas, α-glucosidase inhibitors, and repaglinide). Of particular note are the combination trials with insulin and metformin [23, 24]. Both combinations provided synergistic benefit to patients. Troglitazone (**2**) was launched in March 1997 by Parke-Davis with the trade name Rezulin. Almost immediately after launch, concerns about the safety of the compound surfaced. A retrospective analysis of clinical trial data revealed increased liver toxicity with troglitazone (**2**) treatment [25]. This manifested in increased serum alanine aminotransferase (ALT) levels, which returned to normal on discontinuation of treatment. Severe hepatotoxicity forced the withdrawal of troglitazone (**2**) in Europe within weeks of its release. Two years later, troglitazone (**2**) was withdrawn from the US market [26].

Englitazone (**20**, CP 68722) arose from a Pfizer effort to make conformationally constrained ciglitazone (**19**) analogs [27]. Racemic englitazone (**20**) was potent in *in vitro* experiments measuring insulin sensitivity [28]. It was also efficacious in lowering glucose and insulin in the *ob/ob* mouse model after four days of dosing [29]. The *R* diastereomer proved to be significantly more potent *in vivo*, achieving equivalent efficacy to ciglitazone (**19**) at a lower, 10 mg kg^{-1}, dose. Englitazone (**20**) did not proceed through clinical trials because of adverse liver effects [1].

Nearly 10 years after the seminal discovery of ciglitazone (**19**), scientists from Takeda reported a follow-up to that effort. Pyridyl alkyloxy derivatives were initially investigated as part of the ciglitazone (**19**) discovery work. Most of the compounds reported were weaker than the alkyloxy derivatives [15]. Further investigation revealed the 2-pyridylethoxy moiety to be more potent *in vivo* than ciglitazone (**19**). As with the previous work, screening was done in diabetic KKA^y mice with the glucose and lipid-lowering ED_{25} of new compounds compared to ciglitazone (**19**). From this screening exercise pioglitazone (**3**) emerged. It had a sixfold lower ED_{25} for glucose and lipid lowering than ciglitazone (**19**). Pioglitazone (**3**) improved peripheral glucose utilization and decreased hepatic glucose output in Wistar fatty rats after a 3 mg kg^{-1} dose for seven days [30]. The pioglitazone (**3**) preclinical efficacy was predictive of efficacy in diabetic patients. One of the early clinical studies involved 20 patients treated for three months with 30 mg pioglitazone (**3**). After that period, they underwent a euglycemic hyperinsulinemic clamp experiment. The glucose disposal rate (R_d) in these patients was significantly improved. In addition, fasting plasma glucose and insulin were lowered. HbA_{1c} dropped almost 1%. Reductions in serum triglyceride and FFA levels were also achieved [31]. These results held up in expanded clinical trials. Additional clinical benefits included increased high-density lipoprotein (HDL) and the drug has been shown to be cardioprotective [32]. Pioglitazone (**3**) was launched in 1999 as Actos. In 2009, Actos generated $4.1 billion in sales [33]. Combination products include Actoplus Met® (pioglitazone (**3**) + metformin), Actoplus Met XR® (pioglitazone (**3**) + metformin extended release formulation), and Duetact® (pioglitazone (**3**) + glimepiride) [34]. In 2012, drug maker Ranbaxy Pharmaceuticals will launch a generic version of pioglitazone (**3**) [35].

Pfizer continued work in this area and developed a compound as a backup to englitazone (**20**). The first part of their research strategy involved replacement of the aryl ether moiety with a bioisosteric ketone. Next, they focused on replacing the distal phenyl group present in englitazone (**20**). The culmination of these research efforts was darglitazone (**21**, CP-86325). Darglitazone (**21**) was 100-fold more potent *in vivo* than ciglitazone (**19**) in lowering plasma glucose in ob/ob mice [36]. Efforts to determine a molecular target for darglitazone (**21**) were also reported [37, 38]. A dose of 25 mg of darglitazone (**21**) was given to 19 obese diabetic patients for 14 days. Significant reductions in plasma triglycerides and nonesterified fatty acids were observed. Insulin-dependant improvement in glycemia was also reported [39]. Despite this promising clinical study, development of darglitazone (**21**) was halted due to undisclosed reasons.

A team at Smith Kline Beecham observed that a hydroxylated metabolite of ciglitazone (**19**) was more potent *in vivo* than the parent compound. They reasoned that this increased activity was related to the increased hydrophilicity of the metabolite. This inspired a research effort focused on identifying a TZD with a lower $c\log P$. The discovery of rosiglitazone (**4**) was the result [40]. Compared to ciglitazone (**19**), the $c\log P$ of rosiglitazone (**4**) was 2 orders of magnitude lower. Rosiglitazone

(**4**) was potent in *ob/ob* mice, producing a statistically significant decrease in the glucose area under the curve at as low as 1–3 $mg\,kg^{-1}$ dose in chow. Comparable efficacy to ciglitazone (**19**) dosed at 1000 $mg\,kg^{-1}$ was achieved at a 10 $mg\,kg^{-1}$ dose [4]. Although rosiglitazone (**4**) was developed via an *in vivo* phenotypic screening approach, it's molecular target, peroxisome-proliferator-activated receptor-γ (PPAR-γ), was discovered [41]. The K_d for rosiglitazone (**4**) against a receptor construct was 43 nM. The determination that TZDs are PPAR-γ agonists would shape subsequent research efforts. Rosiglitazone (**4**) was advanced to clinical trials as a monotherapy and in combination with insulin, sulfonylureas, and metformin. The readers are referred to a review by Maliowski and Bolesta [42] for an excellent summary. In general, rosiglitazone (**4**) lowered HbA_{1C} from 0 to 1% and significantly reduced fasting plasma glucose. The most significant side effect reported was dose-dependant body weight gain. In 1999, GSK launched rosiglitazone (**4**) as Avandia. It is also available as Avandamet® (rosiglitazone (**4**) + metformin) and Avandaryl® (rosiglitazone (**4**) + glimepiride) [43]. In 2009, sales of Avandia (**4**) amounted to $772 million [33]. In 2007, a meta-analysis of 42 clinical trials with Avandia (**4**) revealed an increased risk of myocardial infarction and death from cardiovascular events [44]. This resulted in the FDA adding a black box warning to the Avandia (**4**) label. In 2010, European regulators recommended the withdrawal of Avandia (**4**) from the market [45]. In the same year, FDA restricted the use of the drug to patients who "cannot control their diabetes on other medications"[46].

The Japanese pharmaceutical company, Kyorin Pharmaceuticals, developed a dual-active TZD ligand, KRP-297 (**22**) [47]. The PPAR-γ agonist activity of KRP-297 (**22**) manifested itself in *db/db* mice in the form of reduced plasma glucose, insulin, triglycerides, and nonesterified fatty acids. In addition, pancreatic insulin content decreased and β-cell morphology improved after treatment [48]. In 2000, Merck in-licensed KRP-297 (**22**, MK-0767) [49]. The compound has progressed through phase III clinical trials but has not yet entered the market.

In 1998, Mitsubishi-Tokyo Pharmaceuticals reported their discovery of netoglitazone (**23**, MCC-555). Researchers revealed **23** to be more potent than rosiglitazone (**4**) and pioglitazone (**3**) in lowering plasma glucose in KKA^y mice. This is despite the fact that netoglitazone (**23**) has lower affinity for PPAR-γ than rosiglitazone (**4**) [50]. As such, this putative partial agonist was able to lower hepatic glucose output but not whole body glucose in Zucker fatty rats [51]. The clinical status of this compound has not been published.

Lobeglitazone (**24**, CKD-501) was discovered by Chong Kun Dang Pharmaceuticals. It is an analog of rosiglitazone (**4**), in which the pyridine ring has been elaborated to a pyrimidine aryl ether. Lobeglitazone (**24**) was more potent than rosiglitazone (**4**) and pioglitazone (**3**) in KKA^y mice [52]. Lobeglitazone (**24**) has completed phase II clinical trials.

Another exquisitely potent TZD recently discovered is rivoglitazone (**25**, CI-1037, CS-011, R-119702) by Daiichi Sankyo. After 14 days of dosing in Zucker fatty rats, rivoglitazone (**25**) lowered blood glucose with an ED_{50} 178-fold lower than pioglitazone (**3**). Rivoglitazone (**25**) also dose dependently increased plasma adiponectin levels in *db/db* mice [53]. A phase II trial also showed increased plasma adiponectin after 14 days of dosing at 10 mg [54]. Rivoglitazone (**25**) was tested in a 26-week phase IIb study versus pioglitazone (**3**). The drug did indeed lower fasting plasma glucose and HbA_{1C} at 1–3 mg doses. The 3 mg dose was comparable to a 45 mg dose of pioglitazone (**3**). However, the 2 and 3 mg doses were associated with significant edema and weight gain [55]. A larger trial examined 0.5–1.5 mg doses of rivoglitazone (**25**) compared with a 30 mg dose of pioglitazone (**3**) for 12 weeks. Reductions in HbA_{1C} were modest (0.11–0.22%) but the incidences of edema were markedly reduced [56].

Balaglitazone (**26**, DRF-2593), a product of Dr. Reddy's Laboratories, Nordic Bioscience, and Novo Nordisk, is a partial activator of PPAR-γ. The compound was more potent than rosiglitazone (**4**) in rodents, producing a fivefold lower ED_{50} in *db/db* mice for glucose lowering than rosiglitazone (**4**) [57]. The researchers also investigated the potential for water retention and bone loss and found **26** to have no effect on either parameter [58]. Balaglitazone (**26**) has completed phase III clinical trials.

Mitoglitazone (no structure publically available), also known as *MSDC-0160*, is under development for the treatment of diabetes. Metabolic Solutions Development Company claims that this TZD is PPAR-γ sparing, acting instead through an unnamed mitochondrial target. As such, mitoglitazone is an insulin sensitizer without the negative effects of PPAR-γ activation (edema and weight gain). The compound has completed phase IIa clinical trials [59].

20.3 Synthesis

The TZD moiety has several unique features that influence its reactivity and physicochemical properties. The aqueous p*K*a of substructure **1** ranges from 6 to 8 [60]. This characteristic allows for drug formulation as a sodium salt, as in the cases of englitazone (**20**) and netoglitazone (**23**). The second acidic site on the TZD nucleus is the proton at C5. In cases in which enantiopure TZD drugs are dosed, racemization occurs *in vivo* [61]. One proposed mechanism to explain the racemization invokes a reversible S oxidation followed by tautomerization of the corresponding enol sulfoxide [62]. This assertion was supported by quantum chemical calculations [63]. Thus the preparations of TZDs are racemic at C5.

Many glitazones were made via one of the two routes outlined in Scheme 20.3. In the first route, the appropriately functionalized halopropionate (**27**) was cyclized with thiourea under acidic conditions to form thiazolidine imine (**28**). The imine was then hydrolyzed under acidic conditions to form the desired glitazone.

The second route involved condensation of an aldehyde (**29**) with TZD (**8**), as was reported near the turn of the twentieth century [8]. Typically, amine bases such as pyrrolidine were used to affect this aldol-type condensation. The resulting five to six unsaturated TZDs (**30**) are formed with exclusive *Z* regiochemistry. In the case of the unsaturated pioglitazone precursor, photolysis provided the *E* isomer in a 2 : 3 ratio [3]. The reduction of the 5,6-double bond via hydrogenation

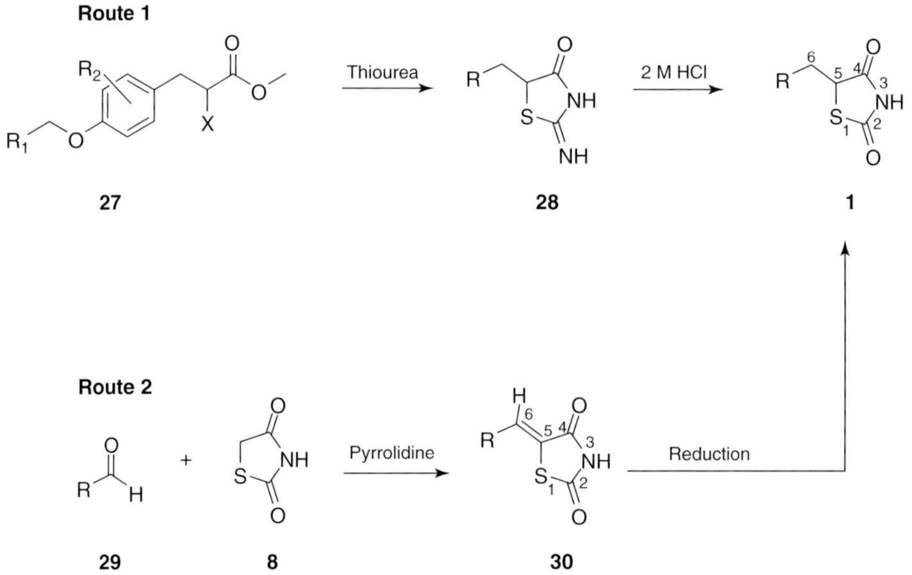

Scheme 20.3 General synthesis routes for the preparation of TZDs.

has been reported [64]. However, due to the risk of reducing the thiazolidine ring itself and the hazards of scaling up hydrogenation reactions, alternative methods were required. Researchers at GSK used $LiBH_4$ and pyridine in refluxing tetrahydrofuran (THF) [64]. Since the pyridine is required for the reduction, the authors suggest that a tetrakis-(dihydrohydropyridyl)boronate intermediate forms *in situ* and is critical for the reaction to proceed. The scale-up route to lobeglitazone (**24**) featured a transfer hydrogenation using Hantzsch ester (diethyl 1,4-dihydro-2,6-dimethyl-3,5-pyridinedicarboxylate) and silica gel as a catalyst [65].

The halopropionates (**27**) used in the synthesis of ciglitazone (**19**), troglitazone (**2**), and KRP-297 (**22**) were prepared as shown in Scheme 20.4 [66]. Appropriately functionalized nitrophenyl ethers were reduced using catalytic hydrogenation. The resulting anilines were subjected to Meerwein arylation, a procedure that involves diazotization, subsequent coupling of the diazonium with methylacrylate, and concomitant halogenation [2].

The (*R*)-benzyldihydropyranyl substituent in englitazone (**20**) required the aldehyde coupling partner to be made via a chiral synthesis. The first-generation

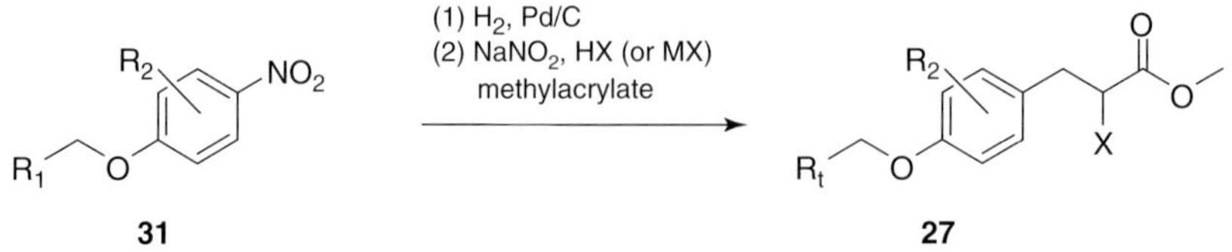

Scheme 20.4 General synthesis of halopropionates (**28**).

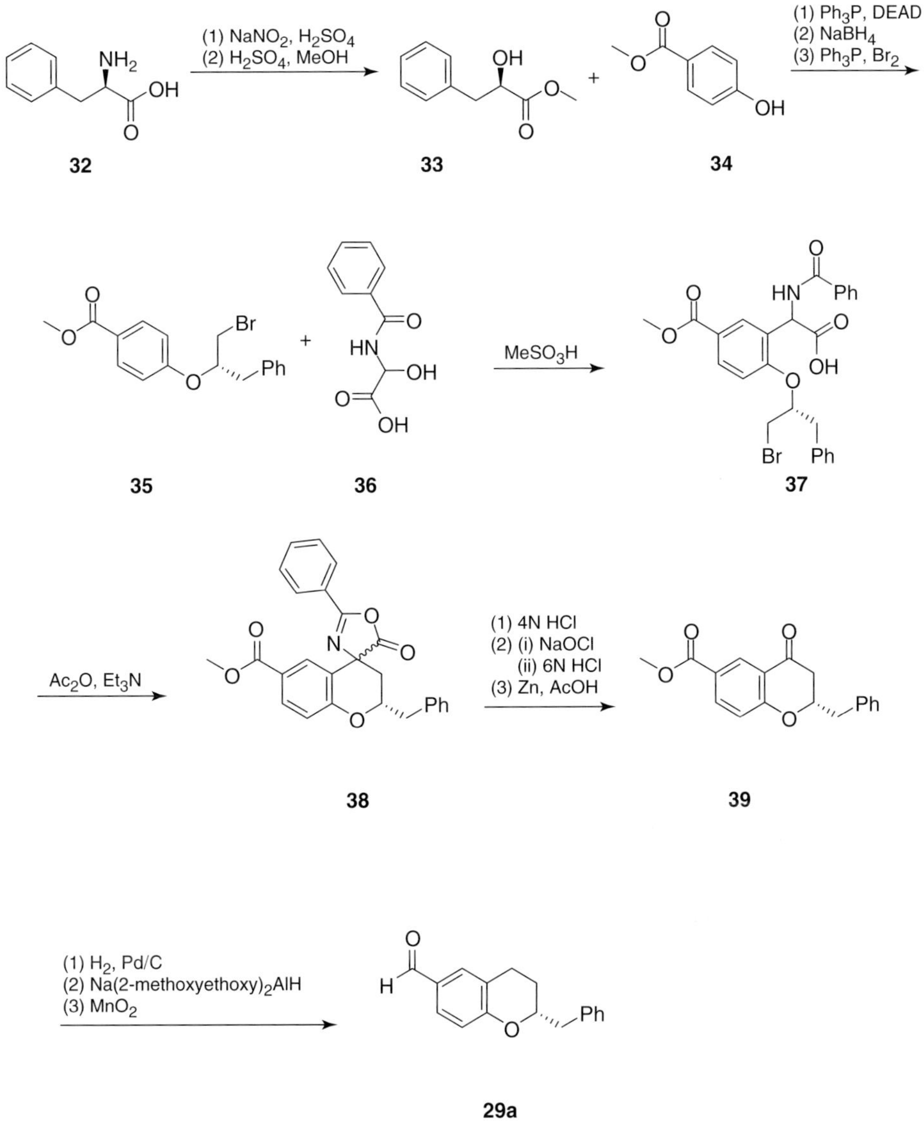

Scheme 20.5 First-generation process route to prepare the englitazone (**20**) building block **29a**.

process route is shown in Scheme 20.5 [67]. Phenylalanine (**32**) was diazotized and hydrolyzed to provide, after esterification, hydroxyphenylpropionate (**33**). The sequence proceeded with retention of configuration. Mitsuobu inversion with 4-hydroxymethylbenzoate (**34**) followed. The alkyl ester was selectively reduced with $NaBH_4$ in 10% aqueous THF and taken on crude to the bromination step. Bromide (**35**) was alkylated with α-hydroxyhippuric acid phenyl amide (**36**). Acid

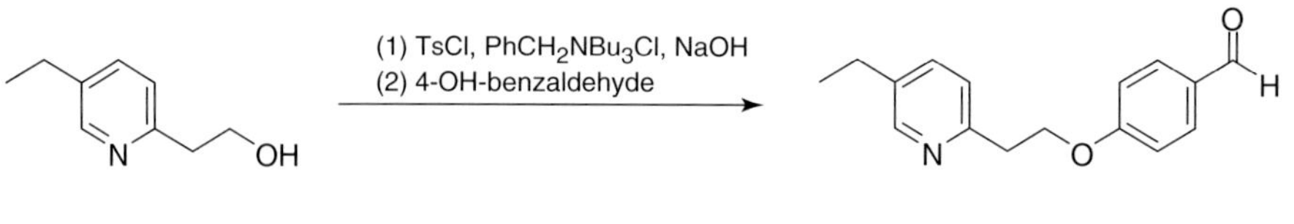

40 29b

Scheme 20.6 Preparation of the pioglitazone aldehyde (**29b**).

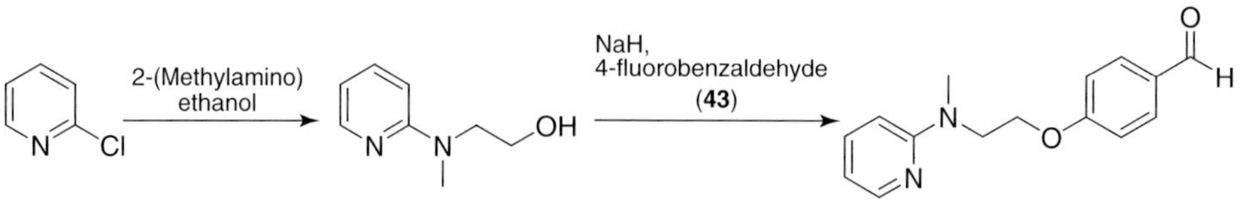

41 42 29c

Scheme 20.7 Rosiglitazone aldehyde (**29c**) preparation.

(**37**) was then subjected to a cyclization/spirocyclization cascade, mediated by acetic anhydride to form **38**. The azalactone was decomposed to an enamide using 4 M HCl. Oxidative cleavage of that enamide provided ketone (**39**). Spurious chlorinated by-products, formed during the oxidation step, were reduced using Zn in AcOH to improve the overall yield of **39**. The ketone was subjected to catalytic hydrogenolysis, and the methyl ester was converted in two steps to aldehyde (**29a**). The authors reported the preparation of 570 g of **29a** using this route.

The pioglitazone aldehyde (**29b**) synthesis is shown in Scheme 20.6. Tosylation of hydroxyethylpyridine (**40**) was accomplished under phase-transfer conditions. In the same pot, the tosylate was then trapped with 4-hydroxybenzaldehyde to furnish **29b** [64].

The rosiglitazone aldehyde coupling partner (**29c**) was prepared in a similar manner (Scheme 20.7). 2-Chloropyridine (**41**) was aminated with 2-(methylamino)ethanol. The hydroxyl pyridine (**42**) was treated with NaH and the resulting alkoxide trapped with 4-fluorobenzaldehyde (**43**) [40].

Netoglitazone (**23**) was prepared differently. Instead of condensing an aldehyde with TZD (**8**), the coupling was carried out using a benzyl iodide. The synthesis of iodide (**46**) is shown in Scheme 20.8.

6-Hydroxy-2-naphthaldehyde (**44**) was coupled with 2-fluorobenzyl alcohol (**45**) under Mitsunobu conditions. The aldehyde was then reduced to the alcohol and converted to iodide (**46**) [68].

The process synthesis of the aldehyde coupling partner for lobeglitazone (**24**) (Scheme 20.9) was very similar to the medicinal chemistry route, with some key modifications [66]. The first improvement was the formation of hydroxypyrimidine (**49**) via a two-step, one-pot procedure. Thus 3,5-dichloropyrimidine (**47**) was monoalkylated with 4-methoxyphenol (**48**) using KF in warm DMF. To this intermediate was added 2-(methylamino)ethanol with continued heating to furnish

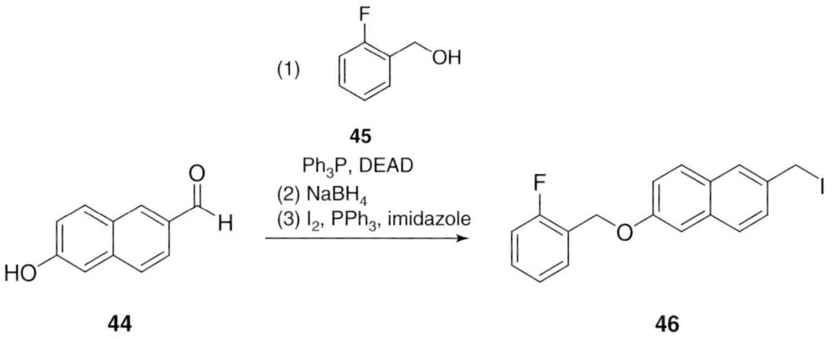

Scheme 20.8 Netoglitazone iodide (**46**) preparation.

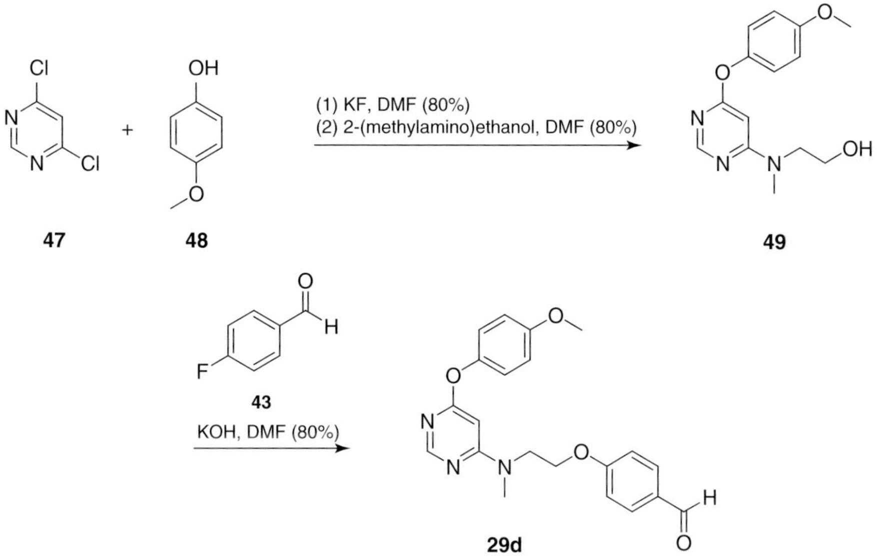

Scheme 20.9 Preparation of lobeglitazone aldehyde (**29d**).

49. Alcohol (**49**) was coupled with 4-fluorobenzaldehyde (**43**) using KOH. Both coupling steps were accomplished without using NaH on scale. Using this route, 1.82 kg of **29d** was prepared.

Since darglitazone (**21**) contains an aryl ketone instead of an ether, the preparation of the aldehyde coupling partner **29e** differed from what was shown above. The route is shown in Scheme 20.10. In this case, the β-ketoester (**51**) was deprotonated with NaH and alkylated with chloromethyloxazole (**50**). Aqueous HCl treatment then served to hydrolyze and decarboxylate the aliphatic ester as well as to unmask the aldehyde [36].

The reported synthesis of rivoglitazone (**25**) did not rely on the late-stage introduction of the TZD moiety. Instead, the TZD head group was carried through much of the synthesis. Highlights of the route are shown in Scheme 20.11.

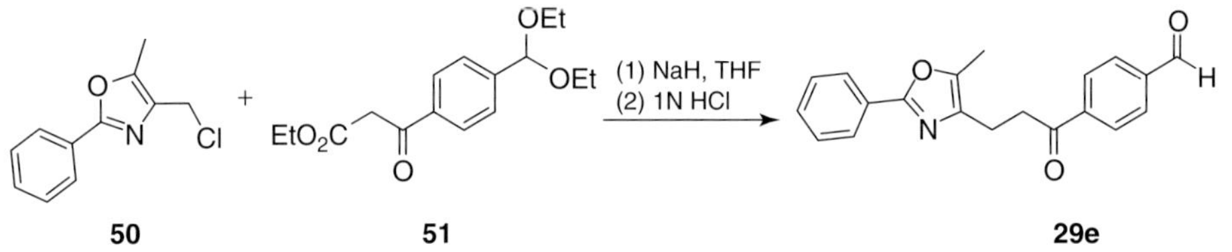

Scheme 20.10 Darglitazone aldehyde (29e) preparation.

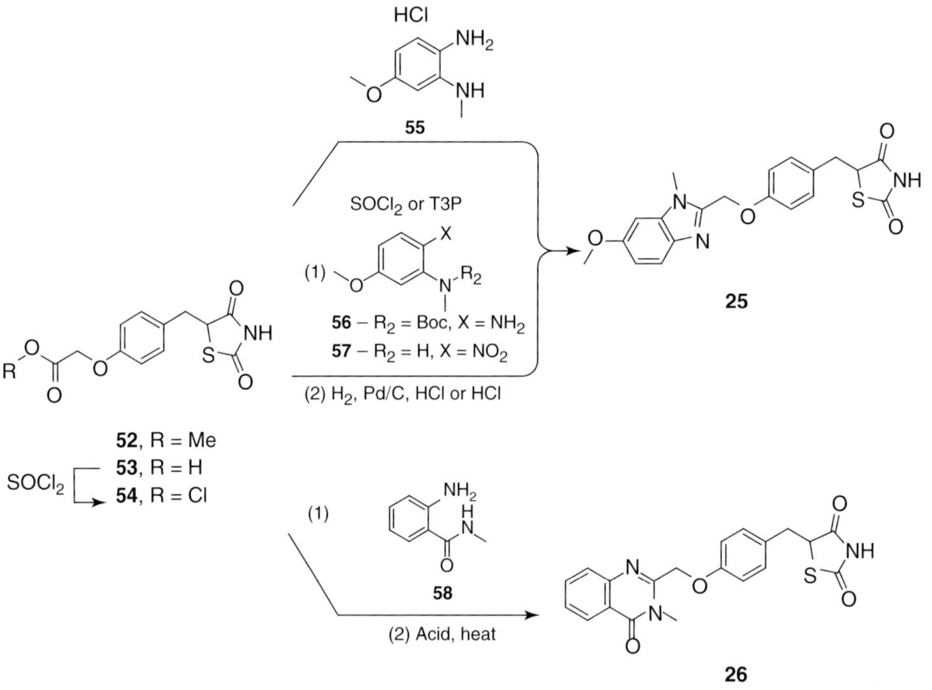

Scheme 20.11 Rivoglitazone (25) and balaglitazone (26) syntheses.

The benzimidazole moiety was assembled in a couple of different ways. In one route, the methyl ester (**52**) was condensed with aniline (**55**) under acidic conditions to yield rivoglitazone (**25**). A more stepwise approach involved coupling the acid first to aniline (**56**) or (**57**). Coupling to **57** occurred through the intermediate acid chloride (**54**). Coupling with **56** was accomplished using propane phosphonic acid anhydride (T3P®). Deprotection of the Boc group or hydrogenation of the nitro group led to cyclodehydration to form **25** [69].

Similarly, balaglitazone (**26**) was prepared via acid (**53**) (Scheme 20.11). Formation of the acid chloride (**54**) and coupling with anthranilic amide (**58**) was followed by cyclodehydration to form **26** [70].

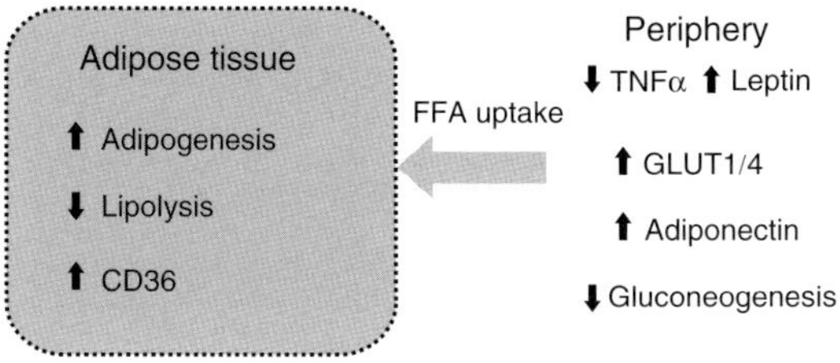

Figure 20.6 Diabetes-relevant effects of PPAR-γ activation.

20.4 Mode of Action

PPAR-γ is a nuclear receptor that natively forms a heterodimer with the retinoid X receptor (RXR). Three isoforms have been identified in humans, termed *PPAR-γ1*, *PPAR-γ2*, and *PPAR-γ3*. The tissue distribution of these isoforms differs. All three isoforms are present in adipose tissue. PPAR-γ1 and PPAR-γ3 are also distributed more broadly [71]. Activation of this PPAR-γ/RXR complex by small-molecule ligands leads to conformational changes that disengage corepressors (such as NCoR and SMRT) and enhance binding of coactivators (such as p300/CBP, PGC − 1α, and SRC1) [71]. Once the activated complex reaches the PPAR response element of a target gene, transcription begins [72].

PPAR-γ activation fundamentally promotes adipocyte generation from fibroblasts. The adipocytes generated through this process are smaller and more insulin sensitive than normal adipocytes [73]. In addition, PPAR-γ activates a host of other gene transcription events, many of which remain unknown [71]. A summary of the consequences of PPAR-γ activation relevant to the treatment of diabetes is shown in Figure 20.6.

In adipose tissue, increased adipogenesis drives decreased lipolysis, that is, generation of FFAs. This is coupled with an increase in the expression of fatty acid transporter CD36. CD36 takes up FFA into adipocytes, lowering FFA concentrations in other tissues [71]. The decrease in tissue FFA concentrations improves skeletal muscle glucose utilization (Randle cycle) and decreases gluconeogenesis [71]. In addition, the proinflammatory cytokine TNFα, which has been linked to insulin resistance, is reduced. Levels of the antiorexigenic hormone leptin are increased [71]. Finally, glucose transporter expression, both constitutively active GLUT1 and inducible GLUT4, increases [71].

While TZDs activate PPAR-γ, it is not completely clear if they are antidiabetic agents solely because of that activity. There are differences in the clinical benefits obtained by using different TZDs. For example, rosiglitazone (**4**) and pioglitazone (**3**) have differential effects on HDL, low-density lipoprotein (LDL), and cholesterol levels [71]. In addition, the broad spectrum of gene expression that PPAR-γ regulates potentially narrows the therapeutic window for TZDs. In addition to mechanism-related edema and weight gain, glitazone treatment has been linked to osteoporosis and increased fracture risk [74, 75]. The tendency for PPAR-γ agonists to elicit a myriad of target-related effects, coupled with the potential for poorly understood off-target effects has likely prompted some researchers to advocate for the removal of TZDs (**1**) from screening collections [76].

20.5 Structure–Activity Relationship

Most antidiabetic glitazone (**1**) structure–activity relationship (SAR) development was done via phenotypic screening in rodents. Ciglitazone (**19**) was often used as the reference standard. While comparative PPAR-γ binding and transactivation assay data have been published, those data have only influenced recent programs [71]. Specific improvements in *in vivo* potency were covered in the Section 20.2 of this chapter. This section serves to highlight the general SAR trends.

SAR development efforts were focused on three major parts of the glitazone scaffold (Figure 20.7) [61]. Since PPAR-γ has an affinity for fatty acids, it is not surprising that glitazones (**1**) have a similar morphology, namely, an acidic head group (**A**) attached to a hydrophobic linker and tail region (**B**, **C**). Very little modification of the head group was tolerated. When AL-321 (**18**) was methylated at N3, eliminating the acidic functionality, all hypoglycemic and triglyceride activity was lost [14]. The rhodanine analog of AL-321 (**18**) was also significantly less potent. 5,6-Unsaturated analogs, like substructure **30**, had variable impacts on *in vivo* potency. For example, the 5,6-unsaturated analog of rosiglitazone (**4**) was 10-fold less potent in mice [4]. However, the 5,6-unsaturated pioglitazone (**3**) was equipotent to its parent compound [61]. Modification of the methylene linker between the TZD head group and the aryl group was deleterious to activity. Several examples of inactive analogs with larger substituents in the 5–6 linker region are presented in the ciglitazone (**19**) discovery work [15]. The **B** ring phenyl substituent was optimal. Replacement with a pyridyl ring resulted in analogs with reduced activity [15]. Thiophene analogs of darglitazone (**21**) were very active but less potent than darglitazone (**21**) itself [36]. Modification of the X linker group between the phenyl group and the lipophilic tail was tolerated, as evidenced by darglitazone (**21**) and netoglitazone (**23**). However, O to S replacement resulted in significantly lower activity [3, 36]. The lipophilic tail, R, was most tolerant to substitution. Aromatic heterocycles tended to provide more active compounds than aliphatic or aromatic carbocycles.

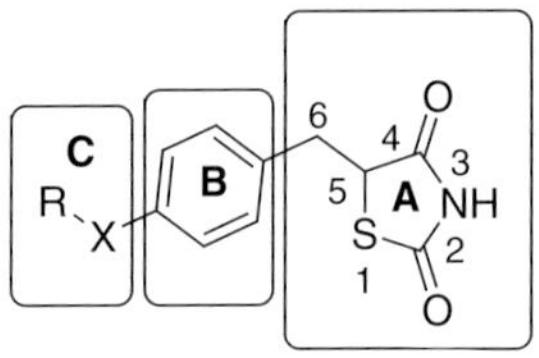

Figure 20.7 Regions of the TZD scaffold that underwent SAR optimization.

References

1. Day, C. (1999) *Diabetic Med.*, **16**, 179–192.
2. Yoshioka, T., Fujita, T., Kanai, T., Aizawa, Y., Kurumada, T., Hasegawa, K., and Horikoshi, H. (1989) *J. Med. Chem.*, **32**, 421–428.
3. Momose, Y., Meguro, K., Ikeda, H., Hatanaka, C., Oi, S., and Sohda, T. (1991) *Chem. Pharm. Bull.*, **39**, 1440–1445.
4. Cantello, B.C.C., Cawthorne, M.A., Cottam, G.P., Duff, P.T., Haigh, D., Hindley, R.M., Lister, C.A., Smith, S.A., and Thurlby, P.L. (1994) *J. Med. Chem.*, **37**, 3977–3985.
5. Volhard J. (1874) *J. Prakt. Chem.*, **9**, 6–30.
6. Wheeler, H. and Barnes, B. (1900) *Am. Chem. J.*, **24**, 60–82.
7. Brown, F.C. (1961) *Chem. Rev.* **61**, 463–521.
8. Kucera, F.. (1914) *Montash. Chem.* **35**, 137–156.
9. Brown, F.C., Bradsher, C.K., and Chilton, S.W. (1956) *J. Org. Chem.*, **21**, 1269–1271.
10. Burton, W.H., Budde, W.L., and Cheng, C.C. (1970) *J. Med. Chem.*, **13**, 1009–1012.
11. Kawamatsu, Y., Saraie, E., Imamiya, K., Nishikawa, K., and Hamuro, Y. (1980) *Arzneim.-Forsch.*, **30**, 454–459.
12. Kawmatsu, Y., Asakawa, H., Saraie, T., Imamiya, E., Nishikawa, K., and Hamuro, Y. (1980) *Arzneim.-Forsch.*, **30**, 585–589.
13. Kawamatsu, Y., Asakawa, T., Saraie, K., Mizuno, E., Imamiya, E., Nishikawa, K., and Hamuro, Y. (1980) *Arzneim.-Forsch.*, **30**, 751–758.
14. Sohda, T., Mizuno, K., Tawada, H., Sugiyama, Y., Fujita, T., and Kawamatsu, Y. (1982) *Chem. Pharm. Bull.*, **30**, 3563–3573.
15. Sohda, T., Mizuno, K., Imamiya, E., Sugiyama, Y., Fujita, T., and Kawamatsu, Y. (1982) *Chem. Pharm. Bull.*, **30**, 3580–3600.
16. Fujita, T., Sugiyama, Y., Taketomi, S., Sohda, T., Kawamatsu, Y., Iwatsuka, H., and Suzuoki, Z. (1983) *Diabetes*, **32**, 804–810.
17. Chang, A.Y., Wyse, B.M., Gilchrist, B.J., Peterson, T., and Diani, A.R. (1983) *Diabetes*, **32**, 830–838.
18. Chang, A.Y., Wyse, B.M., and Gilchrist, B.J. (1983) *Diabetes*, **32**, 839–845.
19. Fujiwara, T., Yoshioka, S., Yoshioka, T., Ushiyama, I., and Horikoshi, H. (1988) *Diabetes*, **37**, 1549–1557.
20. Fujiwara, T., Okuno, A., Yoshioka, S., and Horikoshi, H. (1995) *Metabolism*, **44**, 486–490.
21. Suter, S., Nolan, J., Wallace, P., Gumbiner, B., and Olefsky, J. (1992) *Diabetes Care*, **15**, 193–203.
22. Parker, J. (2002) *Adv. Drug Delivery Rev.*, **54**, 1173–1197.
23. Schwartz, S., Raskin, P., Fonseca, V., and Graveline, J.F. (1998) *New Engl. J. Med.*, **338**, 861–866.
24. Inzucchi, S.E., Maggs, D.G., Spollett, G.R., Page, S.L., Rife, F.S., Walton, V., and Shulman, G.I. (1998) *New Engl. J. Med.*, **338**, 867–872.
25. Watkins, P.B. and Whitcomb, R.W. (1998) *New Engl. J. Med.*, **338**, 916–917.
26. Scheen, A.J. (2001) *Diabetes Metab.*, **27**, 305–313.
27. Clark, D.A., Goldstein, S.W., Volkmann, R.A., Eggler, J.F., Holland, G.F., Hulin, B., Stevenson, R.W., Kreutter, D.K., Gibbs, E.M., Krupp, M.N., Merrigan, P., Kelbaugh, P.L., Andrews, E.G., Tickner, D.L., Suleske, R.T., Lamphere, C.H., Rajeckas, F.J., Kappeler, W.H., McDermott, R.E., Hutson, N.J., and Johnson, M.R. (1991) *J. Med. Chem.*, **34**, 319–325.
28. Kreutter, D.K., Andrews, K.M., Gibbs, E.M., Hutson, N.J., and Stevenson, R.W. (1990) *Diabetes*, **39**, 1414–1419.
29. Stevenson, R.W., Hutson, N.J., Krupp, M.N., Volkmann, R.A., Holland, G.F., Eggler, J.F., Clark, D.A., McPherson, K., Hall, K.L., Danbury, B.H., Gibbs, E.M., and Kreutter, D.K. (1990) *Diabetes*, **39**, 1218–1227.
30. Sugiyama, Y., Shimura, Y., and Ikeda, H. (1990) *Arzneim.-Forsch.*, **40**, 436–439.
31. Yamasaki, Y., Kawamori, R., Wasada, T., Sato, A., Omori, Y., Eguchi, H., Tominaga, M., Sasaki, H., Ikeda, M.,

Kubota, M., Ishida, Y., Hozumi, T., Baba, S., Uehara, M., Shichiri, M., and Kaneko, T. (1997) *Tohoku J. Exp. Med.*, **183**, 173–183.
32. Forst, T., Hanefeld, M., and Pfützner, A. (2011) *Expert Opin. Pharmacother.*, **12**, 1571–1584.
33. Njardarson Group (2011) *http://cbc.arizona.edu/njardarson/group/top-pharmaceuticals-poster* (accessed June 2011).
34. Actos (2011) *http://www.actos.com/actos/whatisit.aspx* (accessed June 2011).
35. Rambaxy (2011) *http://www.ranbaxyusa.com/newsdisp150310.aspx* (accessed June 2011).
36. Hulin, B., Clark, D.A., Goldstein, S.W., McDermott, R.E., Dambek, P.J., Kappeler, W.H., Lamphere, C.H., Lewis, D.M., and Rizzi, J.P. (1992) *J. Med. Chem.*, **35**, 1853–1864.
37. Hulin, B., Lau, C.L., and Gibbs, E.M. (1993) *Bioorg. Med. Chem. Lett.*, **3**, 703–706.
38. Zhang, B., Szalkowski, D., Diaz, E., Hayes, N., Smith, R., and Berger, J. (1994) *J. Biol. Chem.*, **269**, 25735–25741.
39. Chaiken, R.L., Eckert-Norton, M., Pasmantier, R., Booden, G., Ryan, I., Gelfand, R.A., and Lebovitz, H.E. (1995) *Diabetologia*, **38**, 1307–1312.
40. Cantello, B.C.C., Cawthorne, M.A., Haigh, D., Hindley, R.M., Smith, S.A., and Thurlby, P.L. (1994) *Bioorg. Med. Chem. Lett.*, **4**, 1181–1184.
41. Lehmann, J.M., Moore, L.B., Smith-Oliver, T.A., Wilkison, W.O., Wilson, T.M., and Kliewer, S.A. (1995) *J. Biol. Chem.*, **270**, 12953–12956.
42. Malinowski, J.M. and Bolesta, S. (2000) *Clin. Therapeut.*, **22**, 1151–1168.
43. Avandia (2011) *http://www.avandia.com/* (accessed June 2011).
44. Nissen, S.E. and Wolski, K. (2007) *New Engl. J. Med.*, **356**, 2457–2471.
45. European Medicines Agency (2011) *http://www.ema.europa.eu/ema/index.jsp?curl=pages/medicines/human/public_health_alerts/2010/09/human_pha_detail_000020.jsp&murl=menus/medicines/medicines.jsp&mid=* (accessed June 2011).
46. FDA (2011) *http://www.fda.gov/Drugs/DrugSafety/PostmarketDrugSafetyInformationforPatientsandProviders/ucm226956.htm* (accessed June 2011).
47. Murakami, K., Tobe, K., Ide, T., Mochizuki, T., Oshashi, M., Akanuma, Y., Yazaki, Y., and Kadowaki, T. (1998) *Diabetes*, **47**, 1841–1847.
48. Yajima, K., Hirose, H., Fujita, H., Seto, Y., Fujita, H., Ukeda, K., Miyashita, K., Kawai, T., Yamamoto, Y., Ogawa, T., Yamada, T., and Saruta, T. (2003) *Am. J. Physiol. Endocrinol. Metab.*, **284**, E966–E971.
49. Merck Annual Report 2000 (2011) *http://www.anrpt2000.com/innovation1.htm* (accessed June 2011).
50. Reginato, M.J., Bailey, S.T., Krakow, S.L., Minami, C., Ishii, S., Tanaka, H., Mitchell, A., and Lazar, M.A. (1998) *J. Biol. Chem.*, **273**, 32679–32684.
51. Upton, R., Widdowson, P.S., Ishii, S., Tanaka, H., and Williams, G. (1998) *Br. J. Pharmacol.*, **125**, 1708–1714.
52. Lee, H.W., Kim, B.Y., Ahn, J.B., Kang, S.K., Lee, J.H., Shin, J.S., Ahn, S.K., Lee, S.J., and Yoon, S.S. (2005) *Eur. J. Med. Chem.*, **40**, 862–874.
53. Kanda, S., Nakashima, R., Takahasi, K., Tanaka, J., Ogawa, J., Ogata, T., Yachi, M., Araki, K., and Ohsumi, J. (2009) *J. Pharmacol Sci*, **111**, 155–166.
54. Schimke, K. and Davis, T. (2007) *Curr. Opin. Invest. Drugs*, **8**, 338–344.
55. Truitt, K.E., Goldberg, R.B., Rosenstock, J., Chou, H.S., Merante, D., Triscari, J., and Wang, A.C. (2010) *Curr. Med. Res. Opin.*, **26**, 1321–1331.
56. Kong, A.P.S., Yamasaki, A., Ozaki, R., Saito, H., Asami, T., Ohwada, S., Ko, G.T.C., Wong, C.K., Leung, G.T.C., Lee, K.F., Yeung, C.Y., and Chan, J.C.N. (2011) *Diabetes Obes. Metab.*, accepted article, doi: 10.1111/j.1463-1326.2011.01411x
57. Larsen, P.J., Lykkegaard, K., Larsen, L.K., Fleckner, J., Sauerberg, P., Wasserman, K., and Wulff, E.M. (2008) *Eur. J. Pharmacol.*, **596**, 173–179.
58. Henriksen, K., Byrjalsen, I., Nielsen, R.H., Madsen, A.N., Larsen, L.K., Christiansen, C., Beck-Nielsen, H., and Karsdal, M.A. (2009) *Eur. J. Pharmacol.*, **616**, 340–345.

59. Metabolic Solutions Development Company (2011) *http://www.msdrx.com/pipeline.aspx* (accessed June 2011).
60. Mohler, D.L., Shen, G., and Dotse, A.K. (2000) *Bioorg. Med. Chem. Lett.*, **10**, 2239–2242.
61. Hulin, B., McCarthy, P.A., and Gibbs, E.M. (1996) *Curr. Pharm. Des.*, **2**, 85–102.
62. Hulin, B., Newton, L.S., Lewis, D.M., Genereux, P.E., Gibbs, E.M., and Clark, D.A. (1996) *J. Med. Chem.*, **39**, 2897–3907.
63. Bharatam, P.V. and Khanna, S. (2004) *J. Phys. Chem.*, **108**, 3784–3788.
64. Giles, R.G., Lewis, N.J., Quick, J.K., Sasse, M.J., Urquhart, M.M.J., and Youssef, L. (2000) *Tetrahedron*, **56**, 4531–4537.
65. Lee, H.W., Ahn, J.B., Kang, S.K., Ahn, S.K., and Ha, D. (2007) *Org. Process Res. Dev.*, **11**, 190–199.
66. Nomura, M., Kinoshita, S., Satoh, H., Maeda, T., Murakami, K., Tsunoda, M., Miyachi, H., and Awano, K. (1999) *Bioorg. Med. Chem. Lett.*, **9**, 533–538.
67. Urban, F.J. and Moore, B. (1992) *J. Heterocycl. Chem.*, **29**, 431–438.
68. Sorbera, L.A. (2002) *Drugs Future*, **27**, 132–139.
69. Davies, S., Castañer, R., and Bolós, J. (2008) *Drugs Future*, **33**, 926–931.
70. Lohray, V.B., Lohray, B.B., Paraselli, R.B., Gurram, R.M., Ramanujam, R., Chakrabarti, R., and Pakala, S.K.S. (1997) (Dr. Reddy's Research Foundation). WO 9741907, June 30, 1997.
71. Henke, B.R. (2004) in *Progress in Medicinal Chemistry*, vol. **42** (eds F.D. King, G. Lawton, and A.W. Oxford), Elsevier B.V., Amsterdam, pp. 1–53.
72. Doshi, L.S., Brahma, M.K., Bahirat, U.A., Dixit, A.V., and Nemmani, K.V.S. (2010) *Expert Opin. Invest. Drugs*, **19**, 489–512.
73. Spiegelman, B.M. (1998) *Diabetes*, **47**, 507–514.
74. Lecka-Czernik, B. (2010) *Curr. Osteoporos. Rep.*, **8**, 178–184.
75. Grey, A. (2009) *Diabetes Obes. Metab.*, **11**, 275–284.
76. Baell, J.B. and Holloway, G.A. (2010) *J. Med. Chem.*, **53**, 2719–2740.

Index

Bioactive Heterocyclic Compound Classes: Pharmaceuticals,
First Edition. Edited by Jürgen Dinges and Clemens Lamberth.

V

Z